AF389143

Smart Urban Water Networks

Smart Urban Water Networks

Editors

Armando Di Nardo
Dominic L. Boccelli
Manuel Herrera
Enrico Creac
Andrea Cominola
Riccardo Taormina
Robert Sitzenfrei

MDPI • Basel • Beijing • Wuhan • Barcelona • Belgrade • Manchester • Tokyo • Cluj • Tianjin

Editors
Armando Di Nardo
Università degli Studi della
Campania "Luigi Vanvitelli"
Italy

Dominic L. Boccelli
University of Arizona
USA

Manuel Herrera
University of Cambridge
UK

Enrico Creac
University of Pavia
Italy

Andrea Cominola
Technische Universität Berlin
and Einstein Center Digital
Future
Germany

Riccardo Taormina
Delft University of Technology
(TU Delft)
The Netherlands

Robert Sitzenfrei
University of Innsbruck
Austria

Editorial Office
MDPI
St. Alban-Anlage 66
4052 Basel, Switzerland

This is a reprint of articles from the Special Issue published online in the open access journal *Water* (ISSN 2073-4441) (available at: https://www.mdpi.com/journal/water/special_issues/smart_water_system).

For citation purposes, cite each article independently as indicated on the article page online and as indicated below:

LastName, A.A.; LastName, B.B.; LastName, C.C. Article Title. *Journal Name* **Year**, *Volume Number*, Page Range.

ISBN 978-3-0365-0978-5 (Hbk)
ISBN 978-3-0365-0979-2 (PDF)

Contents

About the Editors

Armando Di Nardo (professor), Ph.D. and Associate Professor with a National Academic Qualification as a Full Professor at the Department of Engineering of University of Campania "Luigi Vanvitelli" and Associate Researcher at Institute of Complex Systems of Italian National Research Council. Engineer, expert in smart water networks, optimizations of water resources management and remediation, decision support system, hydroinformatic and hydraulic infrastructures. Coordinator of an European Action Group of EIP (European Innovation Partnership) on Water, titled CTRL+SWAN (Cloud Technologies & Real Time Monitoring + Smart Water Network). Cofounder of the academic spinoffs Environmental Technologies srl. and MedHydro srl. Author of 180 papers in national and international journals and conferences, with over 90 SCI papers, 3 books and 15 book chapters. The major research fields include water resources management, water network partitioning, groundwater remediation carried out with innovative approaches based on heuristic optimization techniques, complex network theory, spectral theory, etc. Creator and developer of the registered software SWANP (Smart Water Network Partitioning and Protection). Cofounder of three academic spinoff companies: Med.Hydro, ARTEMA and Environmental Technologies. Company advisor in the fundraising and development of research projects. Social innovator, cofounder of the foundation non-profit IISE and of the startup incubator HUB spa, experimenting with coworking, crowdsourcing, crowdfunding, startup acceleration in South of Italy. Jury member in startup and company competitions. Copywriter, blogger and writer for blogs and newspapers.

Dominic L. Boccelli is a Professor and Department Head of the Civil and Architectural Engineering and Mechanics at the University of Arizona. Prior to joining the University of Arizona, Dominic was an Associate Professor at the University of Cincinnati, where he was the founder and director of the Interdisciplinary Water Cluster. His research is broadly focused in the area of urban water infrastructure. More specifically, his research has emphasized drinking water distribution systems focused on real-time modeling, contamination warning systems, and water quality issues associated primarily with disinfectant dynamics and by-product formation. Dr. Boccelli's research has been funded by the National Science Foundation, the Water Research Foundation, the US Environmental Protection Agency, and the Binational Science Foundation (a partnership between the US and Israel). He and his co-author received the Best Research-Oriented Paper in the Journal of Water Resources Planning and Management (2016), and he received an Endevaour Executive Fellowship (2015) in support of his sabbatical at the University of Adelaide. He currently serves on the editorial boards of the Journal of Hydroinformatics, and the *Journal of the American Water Works Association*.

Manuel Herrera (Ph.D.) is a Research Associate in distributed intelligent systems at the Institute for Manufacturing at the University of Cambridge. Currently, he is the recipient of the Frank Hansford-Miller Fellowship 2021, awarded by the Statistical Society of Australia (WA Branch). Manuel's research focuses on predictive analytics and complex (adaptive) networks for smart and resilient critical infrastructure. His interdisciplinary profile has proven to be successful in terms of the number and quality of publications; having a high academic impact. Manuel's latest work deals with the AI-enabled management and maintenance of the UK national infrastructure. This is the case of his participation in granted projects on telecommunication systems, urban water, and 5G ports.

Manuel is a fellow of the Royal Statistical Society and a member of associations such as the Institute of Electrical and Electronics Engineers and the International Water Association.

Enrico Creac Ph.D., Associate Professor, obtained his Ph.D. in Hydraulic Engineering in 2006 and has researched topics pertinent to water and environmental systems for over fifteen years. His career began at the Universities of Catania and then Ferrara, Italy. From May 2014 to June 2015, he took up a Research Fellow post at the University of Exeter and became an Assistant Professor at the University of Pavia in September 2015. He has been an Associate Professor at the University of Pavia since September 2018. He has been an Honorary Senior Research Fellow at the University of Exeter and Adjunct Senior Lecturer at the University of Adelaide. In September 2018, he obtained the Full Professor Qualification in the Hydraulics, Hydrology and Hydraulic Infrastructure Sector. He has been a lecturer in hydraulic infrastructures at both undergraduate and postgraduate levels and has published more than 100 Scopus indexed papers, most of which are in international ISI indexed journals. He is the Associate Editor of the Journal of Water Resources Planning and Management—ASCE and participated in/coordinated with various national and international research projects. Research interests: design and management of water distribution systems; sewer and irrigation systems; numerical modeling of shallow waters and sediment transport; demand analysis; protection of water distribution systems from contamination.

Andrea Cominola (Ph.D.) is an Assistant Professor of Smart Water Networks at the Einstein Center Digital Future and Technische Universit'at Berlin. In early 2017, he received his Ph.D. in Information Technology at Politecnico di Milano (Environmental Intelligence group, formerly Natural Resources Management group) for his dissertation on "Modelling residential water consumers' behavior. From smart metered data to demand management". He is a Principal Investigator or Investigator of several national and international projects on urban water and energy demand modelling and management, smart meters, leakage and anomaly detection, behavioral modelling, data mining, and machine learning for water and coupled human-environment systems analysis. He is the author of 15 scientific publications in peer-reviewed international journals, 2 book chapters, 47 publications/presentations in international conferences, and a reviewer for several international journals. Since 2019, he has served as an Associate Editor for the Journal of Water Resources Planning and Management (American Society of Civil Engineers) and, since 2020, of the ICE's Water Management journal. He has also been a Guest Editor of Special Issues in different journals. Among other recognitions, in 2020, he was awarded the Early Career Research Excellence (ECRE) biennial award by the International Environmental Modelling and Software Society (iEMSs).

Riccardo Taormina Taormina (Ph.D.) joined TU Delft's Department of Water Management in July 2019 as an Assistant Professor in Urban Water Infrastructure (tenure track). He has a BSc and an MSc in Environmental Engineering from Politecnico di Torino, and a Ph.D. in Civil Engineering from The Hong Kong Polytechnic University. Prior to joining TU Delft, Riccardo worked as a Postdoc at the iTrust Research Center of Singapore University of Technology and Design. His research activity involves the use of data, computational modelling, and machine learning/AI for solving complex water problems. He is also an expert in the cyber-physical security of water infrastructure, which was the main topic of his postdoctoral experience. Riccardo codirects AidroLab, TU Delft's AI lab in sustainable water management. He is the coauthor of over 20 peer-reviewed publications in international journals. Since 2020, he has served as an Associate Editor of AQUA—*Water Infrastructure, Ecosystems and Society,* and since 2021, he has also been an Associate Editor of the ASCE *Journal of Water Resources Planning and Management (JWRPM).* He is also the recipient of the 2019 Best Research Paper Award from JWRPM.

Robert Sitzenfrei is full professor of urban water management at the University of Innsbruck. He received his Ph.D. in Civil Engineering from the Faculty of Engineering Sciences in 2010. As a researcher at the Department of Infrastructure Engineering in Innsbruck, he led and co-led basic and applied research projects in all areas of urban water management, publishing more than 170 articles in peer-reviewed journals, books and scientific conferences. He continuously serves as a scientific committee member of different international conferences and is/was a guest editor of Special Issues in different journals. In 2015, he became an Assistant Professor and subsequent Associate Professor at the University of Innsbruck. In 2018, he was appointed Full Professor of Urban Water Management at the University of Innsbruck. His research interests are in integrated and interdisciplinary approaches, resilience, urban water networks, complex network analysis, transition modelling, smart (water) cities, water and energy, sustainability, etc.

Editorial

Smart Urban Water Networks: Solutions, Trends and Challenges

Armando Di Nardo [1,2,*], Dominic L. Boccelli [3], Manuel Herrera [4], Enrico Creaco [5], Andrea Cominola [6,7], Robert Sitzenfrei [8] and Riccardo Taormina [9]

[1] Department of Engineering, Università degli studi della Campania Luigi Vanvitelli, Via Roma 29, 81031 Aversa, Italy

[2] Institute of Complex Networks (Italian National Research Council), Via dei Taurini, 19, 00185 Roma, Italy

[3] Office of Research and Development, National Homeland Security, Research Center, U.S. Environmental Protection Agency, MS 163, 26 W. Martin Luther King Dr., Cincinnati, OH 45268, USA; dboccelli@email.arizona.edu

[4] Institute for Manufacturing, Department of Engineering, University of Cambridge, 17 Charles Babbage Road, Cambridge CB3 0FS, UK; amh226@cam.ac.uk

[5] Department of Engineering and Architecture, Università degli Studi di Pavia, Via Ferrata 3, 27100 Pavia, Italy; creaco@unipv.it

[6] Chair of Smart Water Networks, Technische Universität Berlin, Straße des 17. Juni 135, 10623 Berlin, Germany; andrea.cominola@tu-berlin.de

[7] Einstein Center Digital Future, Wilhelmstraße 67, 10117 Berlin, Germany

[8] Unit of Environmental Engineering, Department of Infrastructure Engineering, University of Innsbruck, Technikerstraße 13, 6020 Innsbruck, Austria; sitzenfrei@uibk.ac.at

[9] Department of Water Management, Delft University of Technology, Stevinweg 1, 2628 CN Delft, The Netherlands; r.taormina@tudelft.nl

* Correspondence: armando.dinardo@unicampania.it

Citation: Di Nardo, A.; Boccelli, D.L.; Herrera, M.; Creaco, E.; Cominola, A.; Sitzenfrei, R.; Taormina, R. Smart Urban Water Networks: Solutions, Trends and Challenges. *Water* **2021**, *13*, 501. https://doi.org/ 10.3390/w13040501

Academic Editor: Enedir Ghisi

Received: 8 February 2021
Accepted: 13 February 2021
Published: 15 February 2021

Publisher's Note: MDPI stays neutral with regard to jurisdictional claims in published maps and institutional affiliations.

1. Introduction

This Editorial presents the paper collection of the Special Issue (SI) on Smart Urban Water Networks. The number and topics of the papers in the SI confirms the growing interest of operators and researchers for the new paradigm of Smart Networks as part of the more general Smart City. The SI showed that digital information and communication technology (ICT), with the implementation of smart meters and other digital devices, can significantly improve the modelling and the management of urban water networks, contributing to a radical transformation of the traditional paradigm of water utilities. The paper collection in this SI includes different crucial topics such as reliability, resilience, and performance of water networks, innovative demand management, and the novel challenge of real time control and operation, along with their implications for cyber-security. The SI collected fourteen papers that provide a wide perspective about solutions, trends, and challenges in the contest of smart urban water networks. Some solutions have already been implemented in pilot sites (i.e., for water network partitioning, cyber-security, and water demand disaggregation and forecasting) while further investigations are required for other methods, e.g., the data-driven approaches for real time control. In all cases, a *new deal* between academia, industry, and governments must be embraced to start the new era of smart urban water systems.

The deployment of digital information and communication technologies (ICTs) in different aspects of urban life has contributed to generating the notion of the Smart City [1], recently recognized in the scientific and technical international community as a city where the use of ICT allows making "the critical infrastructure components and services—which include city administration, education, healthcare, public safety, real estate, transportation, and utilities—more intelligent, interconnected, and efficient" [2]. The implementation of new monitoring and control sensor technologies and the availability of high computational power changed the traditional approach to studying, designing, and managing water systems and enabled the development of new data-driven approaches fed by big data. The availability of low-cost devices, controlled by remote systems, is pushing the operators of urban water systems to fill the technological gap with other network utilities (i.e., electricity, gas, Internet, etc.), as reported in [3]. This transformation, triggered by ICTs, has also generated the new concept of the smart water network (SWAN) as a key subsystem of the Smart City.

In the more general framework of Industry 4.0, the recent development of Internet of Things (IoT) technologies applied to smart grids opens further novel opportunities in the management of water network systems and beyond. At the current state, it is possible to imagine novel solutions based on digital innovations to study, analyze, assess, and improve traditional approaches for leakage reduction, pressure management, optimal maintenance, water quality protection from accidental and intentional contamination, network calibration, water use identification, water demand modelling and management, water network partitioning, adaptive and dynamic control, as well as the new challenges raised in the digital era (e.g., cyber-security). This leads to a transformation of the traditional operational criteria and contributes to an increase in the resilience of urban water systems. In addition, by analyzing the cross-links between the urban water infrastructure and other systems (i.e., power grids, urban drainage, smart homes, etc.) it is possible to account for multi-sectoral interconnections in planning and management decisions for more resilient Smart Cities.

The objective of this Special Issue is to gather contributions advancing scientific and technical methodologies, technologies, and best practices that advance smart urban water networks by leveraging the increasingly available computational power in simulation, IoT systems, and smart meter devices. Through this open access journal, a wide community of researchers, operators, and water utilities can have access to a collection of recent cutting-edge contributions showing how some key operational challenges of water networks can be improved by coupling ICT technologies, physically-based mathematical procedures and data-driven techniques (i.e., identification, optimization, complex network theory, etc.). In other terms, this will foster the digitization of urban water networks towards the concept of smart cities and societies.

The papers in this Special Issue provided heterogeneous contributions to the topics proposed by the Editors in the call, showing a large variety of implemented and potential solutions, current trends, and challenges that remain open for future research. In the following sections, the paper collection is presented, highlighting the main proposed novelties.

2. Special Issue Paper Collection

The keywords suggested by the Editors of this Special Issues tried to identify some potential fields of applications that are being transformed by digital innovation in smart urban water networks. In this editorial, it is worth reporting some of them to attest the effort of synthesis and offer to researchers and operators a possible map of innovation in current cutting-edge research on urban smart networks, with topics including: optimal network design and management, novel modeling approaches, application of IoT, adaptive automatic control of urban water network, machine learning and big data for water utilities management, characterization and modeling of water demands at different spatial and temporal scales, divide-and-conquer techniques for water network partitioning, innovative metrics for resilience computation, actions to protect water distribution network from accidental and intentional contamination, novel approaches for water safety plans, data-driven water demand modeling, non-intrusive load monitoring, water and energy nexus, end use disaggregation of water consumption, water demand user profiling, behavioral modelling and water-energy demand management, innovative decision support systems, hydroinformatic applications, innovative intermittent uses in drought periods, pump and turbine implementations, disaggregated pricing and tariff policing, and cyber-security applications.

Many of these concepts were addressed and discussed in the papers collected in this Special Issue, which was mainly dedicated to water distribution networks but, as will be shown below, also hosted some contributions on water drainage systems, highlighting how the digital transition is affecting all subsystems involved in the urban water cycle.

Specifically, the Special Issue on Water Journal collected 12 papers, which in this editorial and, consistently, in the on-line paper collection, are categorized in four main topics: (1) Reliability, Resilience, and Performance, (2) Smart Urban Water Demand Management, (3) Smart Real Time Control and Operation, and (4) Cyber-Security in Water Systems.

Two further contributions bring the total number of submitted papers to 14: a timely review on the Smart Water Systems, inserted as the overview of the Special Issue; and a review on the state-of-the-art literature on cyber-security in the water sector, which systematically presents the existing works in this fast growing field and identifies outstanding issues.

2.1. Overview on Smart Water Systems

The first paper of the Special Issue offered the opportunity to rethink the framework of smart water systems. The design and construction of such smart water systems are still not standardized enough for massive applications, and there is a lack of consensus on the overall transformative framework.

Some authors identified from their comprehensive literature review on smart water techniques the lack of a general architecture and a systematic framework to successfully guide real-world deployment of smart water systems [4]. To fill this gap, they suggested a novel approach consisting of five layers: (i) instrument layer, (ii) property layer, (iii) function layer, (iv) benefit layer, and (v) application layer, including two newly-defined metrics, i.e., smartness and cyber wellness. Therewith, the aim of the authors was to stimulate the implementation of smart water systems in practice as a joint work of academia, industry and government.

2.2. Reliability, Resilience and Performance

Some papers of the Special Issue deal with the subtopic of reliability of water networks. In these papers, some specific water network management issues, including network partitioning, and protection from contamination and other critical events are addressed, and a comparative analysis of some reliability indices was also provided.

Another interesting problem faced in the Special Issue is the optimization of fault examination in water distribution networks. It is essential to automatically detect faults (e.g., leaks, blockages) in water distribution systems to avoid or reduce the loss of resources, non-revenue water, and operational costs. In [5] was proposed an inverse transient-based optimization approach to identify such faults. They tested their approach with models of two hypothetical water distribution systems and found that their algorithm is proven reliable and efficient in detecting faults. In the paper [6], the authors reviewed the state-of-the-art literature on water networks partitioning in district metered areas (DMAs) and provided a comprehensive overview of existing methods and approaches. They classified these methods in two steps: clustering algorithms (dividing the network) and dividing procedures (identifying the optimal positions of gate valves and flow meters). Six of the most widely adopted clustering algorithms were presented and discussed in-depth, and future research gaps were identified (e.g., considering devices, such as pumps, operations under abnormal conditions).

Furthermore, [7] presented a strategy for reducing the impacts of contamination events in water distribution systems. The authors developed a hybrid strategy which is based on water network partitioning and the installation of sensors. By testing the framework on a real water distribution system, they showed how to reduce the impact of any kind of critical events.

Finally, in the literature, there are numerous reliability indices to evaluate the performance of water distribution systems. However, the choice of which one to use is often challenging, as they rely on different assumptions and some of them are correlated. In this regard, a very useful comparative analysis of reliability indices and hydraulic measures was carried out by [8], who investigated nine different reliability indices and six different hydraulic measures with 17 hypothetical networks with various topological features under different supply scenarios. They found that selecting the indices according to the defined goals is essential and, accordingly, give guidance on how to choose the right indices for different water network configurations.

2.3. Smart Urban Water Demand Management

The Special Issue hosted some papers on the topic of water demand management, which is showing an increasing interest in the technical and scientific community.

A comprehensive review of urban water consumption datasets at multiple spatial and temporal scales was proposed by [9]. The recent technological developments and increasing number of pilot studies in smart water metering is resulting in an increasing availability of high-resolution metering datasets for research applications. Motivated by the need for tracking the type and accessibility of the existing water consumption datasets in the rapidly evolving field of smart metering, the authors reviewed and collected available dataset sources and classified them according to spatial and temporal scale, and dataset accessibility. In the work [9], the authors found that the existing datasets are very heterogeneous in terms of temporal and spatial scales, and they can serve different purposes depending on the scale of interest, data resolution, and related analytics, including, for instance, water demand forecast, end use disaggregation, behavioral modeling. After assembling the catalogue of existing smart meter datasets and characterizing them with the above mentioned criteria, the authors formulated a series of recommendations to support future research efforts and encourage the open access publication of smart water meter data.

A spatial aggregation effect on water demand peak factor was also in the Special Issue. The single water consumption is a random and highly volatile process. However, when aggregating a large number of consumers, temporal, but also spatial trends and patterns, can be observed. In the work [10] the peak factor for the water demand consumption as a function of spatial data aggregation on the basis of the statistical analysis of data of 1000 households was investigated. They found an empirical relation for estimating the peak factors. Furthermore, they proposed a procedure to analyze smart meter data regarding the occurring water demand peak factors and give guidance for network operators on how to process their data for design and operation.

In another contribution based on a least square support vector machine [11], the authors established a forecasting chaotic time series for short-term water demand with a forecasting horizon of one day and a time step length of 15 min. To improve the quality of the forecast, they transformed the time series of differences between the forecasted and measured data to a chaotic time series and implemented an error correction module to improve the accuracy. By testing this hybrid model on three real-world supply areas, they showed an improvement of the obtained forecasting solutions regarding mean absolute percentage error.

Another interesting paper on the smart water grid for micro-trading rainwater was proposed in the special issue by [12]. While there might be a local urban water shortage, local excess water might be available in supply areas. For non-potable water, some authors proposed to establish a smart water grid which allows to trade rainwater on a local level [12]. For doing that, they envisioned a distribution network connecting residential rainwater tanks that would enable to buy and sell rainwater on a local level (e.g., for irrigation purposes), and which would be monitored and controlled via numerous smart water sensors. In a hydraulic feasibility study, they analyzed these micro-trading and showed that water and energy savings are feasible across different climates.

2.4. Smart Real Time Control and Operation

Some papers inserted in the Special Issue regarded the innovative topic of implementation of real time control and operation of smart devices in water networks. This aspect represents one of the main operational challenges for water utilities to definitively shift towards the paradigm of a smart water network.

A first contribution of this section of the Special Issue dealt with the optimal placement of pressure sensors using fuzzy logic. Indeed, smart pressure sensors can be used to detect leakage in water distribution systems. However, it is challenging to find a suitable location for such sensors to gain the maximum benefit, while considering budget and other constraints. In [13] the optimal placement of pressure sensors in water distribution systems was investigated by considering the nodal sensitivity to leakage, data uncertainties and node entropy in order to cover a maximum area by a sensor. The authors successfully showed the application of their approach to a benchmark system and also to a real-world case study.

In the work of [14], the authors presented an interesting application on real-time pressure control by analyzing different stochastic consumptions. As known, pressure management in water distribution systems is important to supply water in sufficient quantity and quality in a cost efficient and reliable way. If there is an excess pressure in a water distribution system, pressure control valves can be used, but the challenge is to cope with many different water consumption states. The researchers performed a numerical investigation of flow-dependent pressure controllers from the literature and assessed their performance based on a stochastic demand model to mimic realistic conditions. They found that different controller schemes perform quite similar. Therefore, they suggested using the scheme with a simple structure without performing any forecast of future demand.

2.5. Cyber-Security in Water Systems

One paper regarded the very interesting topic of the state of the art of cyber-security in water systems [15]. It is clear that also in water systems the evolution from isolated bespoke systems to those that use general-purpose computing hosts, IoT sensors, edge computing, wireless networks, artificial intelligence, and IoT devices will increase significantly the risk of cyber-attacks. The authors highlighted the importance of protecting water infrastructure from malicious entities that can conduct industrial espionage and sabotage against these systems. The review of [15] focused on the aspects of the system vulnerability, of the actual measures, and the perspective to improve the cyber-security of water systems. The authors found that the majority of cyber-security studies were carried out on drinking water systems, others on drinking water treatment systems, and only a few on non-drinking water systems (i.e., canal automation systems used for irrigation and wastewater systems). However, while the impacts of cyber-physical attacks are increasingly discussed in the literature, only few studies address the problem of how to efficiently protect micro components in smart water systems. Therefore, it was concluded that further works should specifically focus on making smart water systems reliable and safe. To successfully enable smart water systems in practice, future research should focus on efficiently protecting micro components by including cyber-physical components in the resilience assessment of urban water systems.

Finally, the last two papers hosted in the Special Issue were not fully aligned to the topic of water distribution networks, but they are very interesting in the more general paradigm of smart networks and big data collection with innovative smart sensors.

The first paper proposed the usefulness of hydrological time-series water depth clustering that can be extended to other smart measures. Specifically, clustering of recorded information is a meaningful statistical method to gain knowledge out of a multitude of real-time measured data. For urban drainage systems, where an increasing number of sensors are installed, this information might also be of great interest for the detection and forecasting of flooding events. The researchers [16] investigated how data-driven unsupervised machine learning algorithms can be used to group hydraulic-hydrological data of measurements in storm water drainage systems. By investigating different clustering and performance evaluation methods, suggestions are given about what kind of method should be applied according to the type of detection events (e.g., short-duration or long duration). This can be implemented as a flood early warning system.

Although not aligned to the topic of water distribution network, the last paper on IoT for wastewater treatment plants also provided useful suggestions to the technical and scientific community about the application of wireless sensor networks that can be a promising approach for different fields of urban water management. In [17] was presented a low-cost IoT system for water quality monitoring for wastewater treatment plants at a close-to-market stage. With a novel ion chromatography detection method, they integrated and tested a nitrate and nitrite analyzer under real conditions. The results of comparing laboratory and low-cost IoT systems revealed the reliability of the proposed device.

3. Discussion

The interest of researchers for the Special Issue was high with 14 published papers (4 review papers and 10 research papers).

The review papers showed that some topics, such as innovative procedures for water network partitioning [18], smart meters and tools for water demand measuring [19], end use disaggregation and forecasting [9], and applications for cyber-security [15] are already available for water utilities. However, as appropriately reported in [4], and confirmed by [15], more coordination between academia, industry, and government is required to guide real world deployment of smart urban water systems. In order to meet the demands of industry and government and successfully turn this new paradigm into practice, the researchers [4] showed that it is necessary to obtain a consensus from conceptual, technical, and practical perspectives. However, also for more consolidated innovations (like softwares, best practices, and procedures) no comprehensive consensus exists. Accordingly, the five-layer framework proposed by the authors aims to simplify the implementation of smart water technologies in novel solutions and case studies, and, for the first time, to better characterize the peculiar features of smart water systems.

Besides presenting new approaches and solutions to smart water networks, the works presented in this SI also highlight the open challenges that should be prioritized in future research.

First, the achievement of a shared definition of resilience of water systems and a shared formulation performance indices for the management represent a key priority to further advance the concept and standardization of smart water networks. In fact, with the help of smart meters and the analysis of big data it will be possible to define novel metrics and consequently improve calibration phases and maintenance plans and better face water crisis periods through water demand management. With reference to the latter point, this Special Issue highlights that the technologies and the methodologies proposed are mature to start pilot sites on a large scale. It is worth highlighting that in [4] was identified that widely applied concepts of resilience of urban water infrastructure are lacking smart components and that there is a need for novel concepts for smart water systems as these are even more complex than traditional systems. By defining two conceptual metrics (smartness and cyber wellness), a first step in this direction was taken, but comprehensive further research is required to successfully tackle these short-comings in current smart water applications.

Further, the more advanced challenges delineated in the SI are the topic regarding real-time control and operation of water systems, also with the possibility to activate dynamic changes in the network operation using smart devices controlled in real-time (e.g., regulation valves, on-off valves).

The availability of a large amount of data collected by smart sensors in IoT framework brings up valuable information and knowledge from the system and speeds up the spreading of data-driven applications in water industries. Some solutions offer new visions when well calibrated hydraulic models are difficult to obtain. In these cases, it could be possible by analyzing the learning system behavior only using data collected from hydraulic, maintenance, and economic information (i.e., length of pipes, diameter, type of material, age, flow, costs, etc.) and the know-how of the operators recorded in maintenance journals (i.e., date and time, type and causes of disservice) without any physical modelling.

This aspect is very interesting and data-driven approaches also represent a new challenge for the future of smart water systems.

4. Conclusions

This Special Issue shows that the multi-faceted paradigm of smart urban water networks can be declined in different ways and applications. While the digital transformation of water networks still presents several open challenges, many solutions can be considered ready to be implemented by water utilities and operators. However, the technological transfer from research laboratories to the water market is still slow for many reasons, mainly due to the delay of the standardization processes and a common regulatory framework.

Overall, the papers collected in this SI offer to the technical and scientific community a wide overview of the solutions and possibilities offered by the implementation of smart meters, IoT, innovative modelling, and simulation approaches fostered in the last years by the availability of high computational power and new digital technologies. The digital transition of water networks towards smart systems is an ongoing and incremental process. Yet, radical changes have been already observed in the last years and more advances leveraging the state of the art, including the contributions presented in this SI, can be expected if a *new deal* between academia, industry, and governments will be embraced to reap and materialize all the benefits of the digital transformation.

Author Contributions: Writing—original draft preparation, A.D.N. and R.S.; review and editing, all authors. All authors have read and agreed to the published version of the manuscript.

Funding: This research received no direct external funding.

Institutional Review Board Statement: The study did not require ethical approval.

Informed Consent Statement: No required.

Data Availability Statement: Not applicable.

Conflicts of Interest: The authors declare no conflict of interest.

References

1. Chourabi, H.; Nam, T.; Walker, S.; Gil-Garcia, J.R.; Mellouli, S.; Nahon, K.; Pardo, T.; Scholl, H.J. Understanding Smart Cities: An Integrative Framework. In Proceedings of the 45th Hawaii International Conference on System Sciences (HICCS 2012), Maui, HI, USA, 4–7 January 2012; pp. 2289–2297.
2. Washburn, D.; Sindhu, U.; Balaouras, S.; Dines, R.A.; Hayes, N.M.; Nelson, L.E. *Helping CIOs Understand "Smart City" Initiatives: Defining the Smart City, Its Drivers, and the Role of the CIO*; Forrester Research, Inc.: Cambridge, MA, USA, 2010.
3. Stewart, R.A.; Nguyen, K.; Beal, C.; Zhang, H.; Sahin, O.; Bertone, E.; Vieira, A.S.; Castelletti, A.; Cominola, A.; Giuliani, M.; et al. Integrated intelligent water-energy metering systems and informatics: Visioning a digital multi-utility service provider. *Environ. Model. Softw.* **2018**, *105*, 94–117. [CrossRef]
4. Li, J.; Yang, X.; Sitzenfrei, R. Rethinking the Framework of Smart Water System: A Review. *Water* **2020**, *12*, 412. [CrossRef]
5. Lin, C.-C.; Yeh, H.-D. An Inverse Transient-Based Optimization Approach to Fault Examination in Water Distribution Networks. *Water* **2019**, *11*, 1154. [CrossRef]
6. Khoa Bui, X.; Marlim, M.S.; Kang, D. Water Network Partitioning into District Metered Areas: A State-Of-The-Art Review. *Water* **2020**, *12*, 1002. [CrossRef]
7. Ciaponi, C.; Creaco, E.; Di Nardo, A.; Di Natale, M.; Giudicianni, C.; Musmarra, D.; Santonastaso, G.F. Reducing Impacts of Contamination in Water Distribution Networks: A Combined Strategy Based on Network Partitioning and Installation of Water Quality Sensors. *Water* **2019**, *11*, 1315. [CrossRef]
8. Jeong, G.; Kang, D. Comparative Analysis of Reliability Indices and Hydraulic Measures for Water Distribution Network Performance Evaluation. *Water* **2020**, *12*, 2399. [CrossRef]
9. Di Mauro, A.; Cominola, A.; Castelletti, A.; Di Nardo, A. Urban Water Consumption at Multiple Spatial and Temporal Scales. A Review of Existing Datasets. *Water* **2021**, *13*, 36. [CrossRef]
10. Del Giudice, G.; Di Cristo, C.; Padulano, R. Spatial Aggregation Effect on Water Demand Peak Factor. *Water* **2020**, *12*, 2019. [CrossRef]
11. Wu, S.; Han, H.; Hou, B.; Diao, K. Hybrid Model for Short-Term Water Demand Forecasting Based on Error Correction Using Chaotic Time Series. *Water* **2020**, *12*, 1683. [CrossRef]
12. Ramsey, E.; Pesantez, J.; Fasaee, M.A.K.; DiCarlo, M.; Monroe, J.; Berglund, E.Z. A Smart Water Grid for Micro-Trading Rainwater: Hydraulic Feasibility Analysis. *Water* **2020**, *12*, 3075. [CrossRef]

13. Francés-Chust, J.; Brentan, B.M.; Carpitella, S.; Izquierdo, J.; Montalvo, I. Optimal Placement of Pressure Sensors Using Fuzzy DEMATEL-Based Sensor Influence. *Water* **2020**, *12*, 493. [CrossRef]
14. Page, P.R.; Creaco, E. Comparison of Flow-Dependent Controllers for Remote Real-Time Pressure Control in a Water Distribution System with Stochastic Consumption. *Water* **2019**, *11*, 422. [CrossRef]
15. Tuptuk, N.; Hazell, P.; Watson, J.; Hailes, S. A Systematic Review of the State of Cyber-Security in Water Systems. *Water* **2021**, *13*, 81. [CrossRef]
16. Li, J.; Hassan, D.; Brewer, S.; Sitzenfrei, R. Is Clustering Time-Series Water Depth Useful? An Exploratory Study for Flooding Detection in Urban Drainage Systems. *Water* **2020**, *12*, 2433. [CrossRef]
17. Martínez, R.; Vela, N.; el Aatik, A.; Murray, E.; Roche, P.; Navarro, J.M. On the Use of an IoT Integrated System for Water Quality Monitoring and Management in Wastewater Treatment Plants. *Water* **2020**, *12*, 1096. [CrossRef]
18. Di Nardo, A.; Di Natale, M.; Di Mauro, A.; Martínez Díaz, E.; Blázquez Garcia, J.A.; Santonastaso, G.F.; Tuccinardi, F.P. An advanced software to design automatically permanent partitioning of a water distribution network. *Urban. Water J.* **2020**, *17*, 259–265. [CrossRef]
19. Walker, D.; Creaco, E.; Vamvakeridou-Lyroudia, L.; Farmani, R.; Kapelan, Z.; Savić, D. Forecasting domestic water consumption from smart meter readings using statistical methods and artificial neural networks. *Procedia Eng.* **2015**, *119*, 1419–1428. [CrossRef]

Article

Cyber-Attack Detection in Water Distribution Systems Based on Blind Sources Separation Technique

Bruno Brentan [1,2], **Pedro Rezende** [3], **Daniel Barros** [1], **Gustavo Meirelles** [1,2], **Edevar Luvizotto Jr.** [2,3] **and Joaquín Izquierdo** [2,*]

[1] Hydraulic Engineering and Water Resources Department, School of Engineering, Federal University of Minas Gerais, Belo Horizonte 31270-901, Brazil; brentan@ehr.ufmg.br (B.B.); danielbezerrab@gmail.com (D.B.); gustavo.meirelles@ehr.ufmg.br (G.M.)

[2] Fluing-Institute for Multidisciplinary Mathematics, Universitat Politècnica de València, 46022 Valencia, Spain; edevar@fec.unicamp.br

[3] Laboratory of Computational Hydraulics, School of Civil Engineering, Architecture and Urban Design, University of Campinas, Campinas 13083-889, Brazil; pedronr27@gmail.com

* Correspondence: jizquier@upv.es

Abstract: Service quality and efficiency of urban systems have been dramatically boosted by various high technologies for real-time monitoring and remote control, and have also gained privileged space in water distribution. Monitored hydraulic and quality parameters are crucial data for developing planning, operation and security analyses in water networks, which makes them increasingly reliable. However, devices for monitoring and remote control also increase the possibilities for failure and cyber-attacks in the systems, which can severely impair the system operation and, in extreme cases, collapse the service. This paper proposes an automatic two-step methodology for cyber-attack detection in water distribution systems. The first step is based on signal-processing theory, and applies a fast Independent Component Analysis (fastICA) algorithm to hydraulic time series (e.g., pressure, flow, and tank level), which separates them into independent components. These components are then processed by a statistical control algorithm for automatic detection of abrupt changes, from which attacks may be disclosed. The methodology is applied to the case study provided by the Battle of Attack Detection Algorithms (BATADAL) and the results are compared with seven other approaches, showing excellent results, which makes this methodology a reliable early-warning cyber-attack detection approach.

Keywords: water distribution systems; cyber-attack detection; blind sources separation; FastICA

Citation: Brentan, B.; Rezende, R.; Barros, D.; Meirelles, G.; Luvizotto, E., Jr.; Izquierdo, J. Cyber-Attack Detection in Water Distribution Systems Based on Blind Sources Separation Technique. *Water* **2021**, *13*, 795. https://doi.org/10.3390/w13060795

Academic Editors: Marco Franchini and Francesco De Paola

Received: 30 December 2020
Accepted: 11 March 2021
Published: 14 March 2021

Publisher's Note: MDPI stays neutral with regard to jurisdictional claims in published maps and institutional affiliations.

1. Introduction

In recent decades, urban areas all over the world have not stopped growing and becoming increasingly dense. Consequently, virtually all urban services are in dire need to become more efficient and accessible to all citizens. Water distribution systems (WDSs), which are obviously among the main urban components, have undergone many changes. In this paper, we focus on the connection between WDS physical and cyber layers, thus turning WDSs into cyber-physical systems. The physical layer of a WDS (pipes, valves, pumps, reservoirs, etc.) can be remotely controlled and monitored by the cyber layer, which allows the implementation of predictive control, and early-warning systems in case of anomalies. As a result, the efficiency of urban water systems is improved.

Cyber-physical systems may considerably improve the operation of water companies, but they will also increase the possibilities for system failure. This is chiefly because cyber layers can include gates that may be easily violated during various kinds of attacks (e.g., information access for damaging the entire water distribution process) [1]. Attackers can access programmable logic controllers and change pump and valve schedules, operational points, and/or corrupt data in SCADA systems. This could threaten the creation and expansion of smart cities that depend on the reliability of cyber systems [2].

9

Scenarios of cyber-physical attacks in water systems have already become a reality. According to the United States Department of Homeland Security, in 2015, 25 cyber-attacks were disclosed in various water systems [3]. In Israel, three attacks happened between 2019 and 2020. The first attack in 2019 managed to change the free chlorine level and, consequently, harmed the water quality of the system. In 2020 the attacks changed pumping operational points, bringing high pressure to the system and associated increasing leakage [4]. In their Systematic Review of the State of Cyber-Security in Water Systems, Tuptuk et al. [5] compile a set of cyber-physical attacks occurred between 2000 and 2020 that have been made public. Most of them were remotely performed and even a recent one used cryptocurrency mining for the attack. The examples of cyber-attacks in the USA and Israel show that despite a system may be highly protected, attackers manage to find their ways to enter the system and eventually produce chaos. Consequently, even virtually fully secure SCADA systems need additional mechanisms to try to close any access gate to the system and minimize the impact of any security breach.

With the aim of improving the reliability of cyber-physical systems, special attention has been given by researchers to the topic, as shown by the promotion of dedicated events. One milestone on cyber-physical system analysis applied to water systems was the International Workshop on Cyber-Physical Systems for Smart Water Networks, in 2015 [6]. The works in that conference mainly focused on data acquisition via SCADA system and the security of the system. Nevertheless, no cyber-physical failure detection methodologies were proposed. However, recently, the detection of malicious attacks in WDSs has become a problem highly faced by researchers and managers, and has been the subject of recommendations from various protection agencies (e.g., Environmental Protection Agency—EPA, from USA). The main objective of this kind of developments is the reduction of the system vulnerability, thus narrowing the potential damage to the physical layer.

Considering the importance of the problem, the Battle of the Attack Detection Algorithms (BATADAL) [7] was organized in a special session of the World Environmental and Water Resources Congress, in Sacramento, California on 21–25 May 2017. The challenge was proposed for comparing possible approaches in detection attacks. Several solutions, concisely described in the next section, were presented.

According to the above-mentioned systematic literature review [5], the vast majority of works in cyber-attack detection, including the ones presented in BATADAL, are based on machine learning, developing classifiers or auto-encoder algorithms. However, the authors of [5] pinpoint the need for targeting at other fields of study for building increased confidence on the algorithms. An alternative, exploited in other research fields, is the use of signal-detection models. These kinds of models handle a mixture of true signal and noisy data. When applied for cyber-attack detection, the main objective of a signal-detection model is to separate attack from normal data, which helps detect abnormal situations accurately and efficiently. One example of signal-processing data applied to detect anomalies in cyber-physical systems is the application of Independent Component Analysis (ICA) [8]. This algorithm separates original signals into components or sources by suitably demixing them. The demixing and consequent separation of signals can help highlight anomalies, thus easing their identification.

Moreover, for automatic identification, the application of statistical control processes such as cumulative sum (CUSUM) and abrupt change point detection (ACPD) have shown to be very useful tools.

Considering the substantial number of applications of ICA for anomaly detection problems in various research fields, and the simultaneous lack of applications in water distribution, this paper proposes a two-stage algorithm for cyber-attack detection in water distribution systems. In the first stage, hydraulic time series acquired by a SCADA system are processed by the ICA algorithm. The resulted signals, so-called sources, are highly affected by cyber-attacks, as shown in the results. This feature is used for automatic detection in the second stage, using an ACPD algorithm. The methodology is applied to the BATADAL case study, and the results are compared, under the same framework, including case study, objectives and metrics, with other approaches presented in the Battle. All seven attacks hidden on the test data sets used in the event are detected by this methodology, thus resulting in a reliable early-warning cyber-attack detection algorithm. Regarding the limitations of this approach, we must mention that some attack scenarios have been detected too late, which is a limitation, otherwise, typical of any detection evaluation methodology. However, overall, the methodology can be considered a novel non-machine-learning-based approach in the field of cyber-attack detection in WDSs.

2. Related Work

The recent literature presents several data analysis and computational modelling techniques aimed at developing early-warning systems for cyber-attack detection in water systems. For example, in [9] a classification algorithm is developed using Support Vector Machines for identifying cyber-attacks in water systems. The authors propose a simple one-class classification approach based on a truncated Mahalanobis distance. The algorithm is tested on a real dataset from a water distribution system in France. Hidden Markov chains are used in [10] for analyzing and detecting anomalies in the SCADA system of a water supply system. Normal behavior was first modelled and then modified with generated abnormal data to simulate potential attack detection. Not only water distribution systems, but also water treatment plants have been used for investigating cyber-attacks. Attacks in Programmable Logic Controllers (PLC) are designed by [11] for better comprehension of the impacts in the produced water.

In BATADAL, seven solutions, coming from research groups from all over the world, were presented, which were ranked based on time-detection and classification accuracy of the events. As our approach in this paper is directly competing with those seven solutions, to make it clear its novelty, we concisely describe the methodologies used in the other solutions. Those contributions together with several papers derived from the event, which we also mention later, can be considered a state-of-the-art literature on the subject, which can be enlarged with [5].

A two-stage method based on feature vector extraction and classification was proposed in [12]: vector extraction was applied to multidimensional hydraulic data, and safety classification was performed by random forests, the machine-learning algorithm developed by [13]. In [14] recurrent neural networks (RNNs) were used for hydraulic state estimation of network district metered areas and, based on the RNN output, a statistical control process was applied for detecting abrupt changes in the residual time series.

The authors in [15] use first operational variables to check whether physical and/or operating rules have been violated, and the generated set of flagged events feeds a deep learning method based on a convolutional variational auto-encoder to calculate the probability for measured data being anomalous.

In [16] also two detection methods were proposed: one evaluates consistency of the SCADA data and verifies the relation between actuator rules (e.g., pump/valve operation) and the measured data; then, the second method uses principal component analysis (PCA) for separating the hydraulic time series into normal and abnormal data.

A three-stage detection method was presented in [17]: the first step detects outliers in the data, focusing on single sensor analysis; the second stage employs a multilayer perceptron to detect SCADA data nonconformity to normal operation; and the third stage finds anomalies affecting multiple sensors.

Another three-module method was presented in [18]: the first module evaluates the consistency of the data against the set of control rules; the second applies statistical analysis to identify anomalous behaviors; then, the anomalies are confirmed by the third module, which finds correlations between hydraulic variables.

Finally, a model-based approach using EPANET for hydraulic simulations was developed in [19]; analyses of the residual time series between simulated and measured data from SCADA system detected the anomalies, and a multilevel classification algorithm was implemented to classify the residual time series into normal and abnormal events.

BATADAL opened a fruitful discussion among various research groups around the world. Following the cyber-attack detection paradigm, new approaches have been presented in the literature after that Battle. For example, work [1] points to multisite detection approaches based on simultaneous analysis for an efficient warning system. In this work the authors present a joint data-model-based approach for cyber-attack detection: the model of the water network is used for inference from the observational data. Exploring the capacity of machine-learning techniques, in [20] a model for detecting anomalies in a water system controlled by SCADA using various machine-learning techniques is presented. The model classifies events including physical failures and cyber-attacks. As another example, research [21] has tested a set of machine-learning algorithms, highlighting the performance of extreme learning machine for classifying normal and abnormal data from multisite sensors.

Despite many devoted efforts to detect cyber-attacks on WDSs in recent years, the primary focus, as observed in the literature, has been mainly on machine learning and optimization techniques. The techniques of signal-processing for cyber-physical attack detection is still not well explored in the literature, especially in water distribution.

Work [22] investigates the application of Independent Component Analysis (ICA) for stealthy false data attack detection without prior knowledge of any power grid topology. The separated signal by ICA is used for detecting virtually unobservable attacks. The authors in [23] apply ICA for obtaining the fundamental traffic components and, in a second stage, the components are classified by machine-learning-inferred decision trees. Still on ICA applications, work [24] develops an algorithm to characterize hidden structures in fused residuals. Suppression of possible noisy content in residuals—to decrease the likelihood of false alarms—is achieved by performing the residual analysis solely on the dominant parts of a so-called demixing matrix.

In the water resources field, ICA has been applied to drought analysis, exploring hydrological data [25]. Also, in [26] the application of ICA to assess and estimate leakage in water distribution networks is proposed. The algorithm is tested on data acquired in a leakage experimental platform. Water demand is forecasted using a principal component model, and ICA is applied for developing climate predictors in [27].

Once demixed by ICA, source signals can be treated for automatically detecting anomalies, and this inspired us to apply ICA and then ACPD to the automatic detection of cyber-physical attacks. In this line, still within urban hydraulics, but with a different purpose, automatic identification of pipe bursts has been developed using statistic control processes applied to hydraulic parameters (e.g., pressure nodal pressure and flow in pipes) [28] or jointly to water demand forecasting [29]. Also, to improve the capacity of burst and leakage detection, work [30] proposes ACPD applied to filtered signals of consumption data.

After the Introduction, the structure of the paper is the following. The Materials and methods are presented in the next section. Then a section is devoted to the case study, and includes the obtained results and a discussion. The paper closes with the Conclusions section.

3. Materials and Methods

The methodology for cyber-attack detection proposed in this paper is based on two separate techniques. The first one comes from the signal-processing field and applies a Blind Source Separation (BSS) algorithm, which makes use of Independent Component Analysis. This technique produces the segregation of the original measured signals, affected by the attacks, into independent components. These components can be detected using a statistical control method, which corresponds with the other technique in this work: an abrupt change point detection algorithm is applied to the separate signals to accurately detect the start and the end times of the attacks, which helps characterize the attacks. Let us first concisely describe these techniques.

3.1. Independent Component Analysis-ICA

ICA is a methodology for multivariate signal-processing based on the statistical independence property. ICA techniques seek to uncover the independent source signals from a set of observations that are composed of linear mixtures of the underlying sources. The sources are the data projected onto some new axes that must be discovered. Accordingly, this process is known as blind source separation, a category of algorithms that try to decompose mixed signals into their original sources. A classical example of separation of a mixed signal is the cocktail party in which a band is playing [31]. Invited people to the cocktail are not listening each instrument of the band separately, but the combination of all the instruments, voices and noises of the environment. Is it possible to separate each sound's source captured by the microphones? To answer the question, BSS algorithms are proposed that try to isolate each source.

Let us consider N time series each consisting of M samples (measured points). The aim is to find a transformation of these time series into a new representation in which independent components are identified and separated.

Formally, we represent the N measured time series

$$\mathbf{X}_i = (x_{i1}, x_{i2}, \cdots, x_{iM})^T, i = 1, \cdots, N \tag{1}$$

compactly by a matrix $\mathbf{X}$ whose rows are the transposed time series

$$\mathbf{X} = \begin{pmatrix} x_{11} & x_{12} & \cdots & x_{1M} \\ \vdots & \vdots & \ddots & \vdots \\ x_{N1} & a_{N2} & \cdots & x_{NM} \end{pmatrix}. \tag{2}$$

This $N \times M$ matrix is supposed to be a linear combination of the original signals, which can also be represented by another $N \times M$ matrix $\mathbf{S}$ with similar structure to $\mathbf{X}$, i.e., the rows of $\mathbf{S}$ are the transposed of the original time series $\mathbf{S}_i = (s_{i1}, s_{i2}, \cdots, s_{iM})^T$. The linear combination may be expressed by

$$\mathbf{X} = A\mathbf{S}, \tag{3}$$

where A, so-called mixing matrix, is the matrix representing the linear transformation. Keeping the analogy of the cocktail party, $\mathbf{X}$ corresponds to the sounds listened by the guests and $\mathbf{S}$ to the original sounds. The main objective of ICA is to determine the mixing matrix A and the original sources $\mathbf{S}$. This task is formulated as an inverse and dual problem. First, a demixing matrix W must be found and then, based on this matrix, the source vector is calculated by

$$\mathbf{S} = W\mathbf{X}. \tag{4}$$

Since the problem is highly underdetermined, the direct calculation of W or A is not possible. An estimate $\mathbf{Y} \approx \hat{\mathbf{S}}$ of the sources is made instead by calculating a demixing matrix W, which acts on X such that

$$\mathbf{Y} = W\mathbf{X} = \hat{S}. \tag{5}$$

and $W \approx A^{-1}$.

To perform this approximation, the process in the ICA algorithm uses some factorization of the observed data (mainly singular value decomposition), and high order statistics (such as the fourth moment, kurtosis) to measure signal-noise separation. From a statistical point of view, the separated signals must be independent, and the independent components must have a non-Gaussian distribution [32]. Based on this non-Gaussian nature, to calculate W, most ICA methods estimate the inverse of A, allowing the calculation of the source vector. The trick behind this process is to find that A^{-1} that maximizes the non-Gaussian nature of the independent components. Usually, this process is done based on maximum-likelihood estimation, maximization of the output entropy or minimization of mutual information in the output [33].

In this paper, the non-Gaussian nature is measured based on the the concept of negentropy, as presented and discussed by [32] in the algorithm called fastICA. The idea behind negentropy comes from the Information Theory. Gaussian-distributed data has entropy H equal to zero, while non-Gaussian-distributed data has non-negative entropy. Negentropy J is calculated as:

$$J(\mathbf{x}) = H(\mathbf{x_{gauss}}) - H(\mathbf{x})),$$ (6)

where $\mathbf{x_{gauss}}$ is a Gaussian random variable with the same covariance as $\mathbf{x}$.

The fastICA algorithm is based on a fixed-point scheme for finding $W \approx A^{-1}$ through maximization of the negentropy. In addition, based on that matrix, it is possible to approximately rebuild the source vector as written in (5).

3.2. Abrupt Change Point Detection-ACPD

After sources separation by fastICA, it is expected that one of the sources will be affected by the cyber-attack. For detecting this change, an algorithm of abrupt change point detection (ACPD) is applied. ACPD is performed by evaluating one or more statistical parameters of the time series, so-called control variables.

For a formal definition, following the ACPD algorithm proposed by [34], let us first identify, among the separate signals provided by fastICA, that one that best represents the kind of signal we are interested in. In our case, we must identify that series mainly representing non-periodic behavior. Let $\mathbf{Y}^{(1)} = (y_{11}, y_{12}, ..., y_{1M})^T$, one of the signals obtained by (5), be our series of interest, where M is the size of the time series. The algorithm tries to identify the various, say m, change points in this time series, which are positioned at indexes $\tau_1, ..., \tau_m$. Each position τ_i corresponds to an integer value between 1 and $M - 1$ and splits the time series into intervals $[\tau_i, \tau_{i+1}]$.

A common approach to estimating $\boldsymbol{\tau} = (\tau_1, ..., \tau_m)$ is by minimizing the objective function:

$$\sum_{i=1}^{m+1} f(\tau_i, \tau_{i+1}) + \beta p(m),$$ (7)

where $f(\tau_i, \tau_{i+1})$ is a cost function related to the time series in the interval $[\tau_i, \tau_{i+1}]$. Several cost functions have been proposed in the literature, such as log-likelihood [35], quadratic loss or cumulative sums [36]. Moreover, $\beta p(m)$ is a penalty function to avoid overfitting. The most common choice, according to [34], is a linear variation $p = \beta m$. This constraint allows the method to estimate a vector $\boldsymbol{\tau}$ corresponding to a trade-off between the minimization of the cost function (found by a large-size $\boldsymbol{\tau}$) and the minimization of the penalty function (found by a small-size $\boldsymbol{\tau}$) [37].

The entire process can be summarized as follows:

- A point is chosen and the time series is divided into two intervals.
- For each interval, a control variable (mean, standard deviation, root-mean-square, etc.) is computed.
- For each point within the interval, deviations of control variables are computed.
- The deviations are summed for all the intervals to calculate the total residual error, and the objective function (7) is evaluated.
- Vary the division point to minimize the total residual error.

The result of this process is exactly the set of components of $\boldsymbol{\tau}$. For this work, each component of the source's signal $\mathbf{Y}$ found by fastICA obtained by (5) is evaluated by the ACPD algorithm, and the vector $\boldsymbol{\tau}$ corresponds to the start and the end times of an attack.

3.3. Automatic Detection of Cyber-Attacks in WDSs

Following the formalization given for fastICA and ACPD algorithms, this section presents the application of both methods for disclosing cyber-attacks in WDSs. First, based on the available data set, the input time series for fastICA are selected. Hydraulic measurements (e.g., pressure, flow and tank level) are considered in this work as input data, which are combined to get the best input arrangement. After a trial-and-error process, we have identified that decomposing the signal into two components will be enough to suitably identify the effects of the attacks. Indeed, the results presented for the case study confirm this assumption. From the software development point of view, the data is processed in Python language and makes use of the package SKLEARN.

The non-periodic component of the demixed signal is then used as the input for the ACPD algorithm. This second process is responsible for automatically identifying the start and end time of the anomalies, thus allowing the disclosure of the attack. The output of this process is the exact interval of time where the water network was subjected to an attack. With this outcome, it is possible to apply the performance evaluation metrics considered in BATADAL, and then, to compare the ability of the proposed algorithm with other approaches. In this stage, the demixed data is processed in the MATLAB programming environment, and makes use of several tools in the toolbox of Signal-processing. For a better understanding, Figure 1 presents the flowchart of the complete methodology.

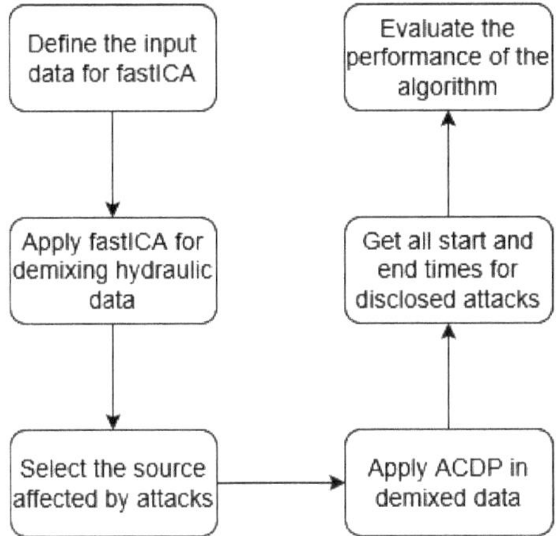

Figure 1. Flowchart of the complete methodology for disclosing cyber-attacks applying fastICA and ACPD algorithms.

3.4. Performance Evaluation

In addition to the BATADAL data sets, the performance evaluation also follows the criteria and metrics presented in [7], namely time-to-detection (TTD) and single classification rate (SCR).

TTD is the time required by the algorithm to find an attack and can be calculated as:

$$TTD = t_0 - t_d,\tag{8}$$

where t_0 is the time when an attack is detected, and t_d is the time when the attack really started. When an attack is detected, TTD varies in the interval $[0, \Delta t]$, where Δt is the total attack duration. For calculating the total TTD under several attack scenarios, work [7] presents a score for the specific attack detection calculated by (9):

$$S_{TTD} = 1 - \frac{1}{n_a} \sum_i^{n_a} \frac{TTD_i}{\Delta t_i},\tag{9}$$

where n_a is the number of attack scenarios.

An ideal algorithm for cyber-attack detection must be able not only to quickly disclose the attacks, but also to not produce false positive warnings. For evaluating the accuracy of the algorithm, the true positive rate, TPR (10), and the true negative rate TNR (11), are calculated based on a confusion matrix. Both rates are combined for calculating the SCR (12):

$$TPR = \frac{TP}{TP + FN}, \tag{10}$$

$$TNR = \frac{TN}{TN + FP}, \tag{11}$$

$$SCR = \frac{TPR + TNR}{2}, \tag{12}$$

where TP and TN are the numbers of true positive and true negative time stamps, respectively. FP and FN are the numbers of false positive and false negative time stamps.

Criteria (9) and (12) are considered by [7] and the final score S is calculated as a weighted sum of S_{TTD} and SCR (13)

$$S = \gamma S_{TTD} + (1 - \gamma)SCR, \tag{13}$$

the real number γ being used to build a suitable convex combination. For equally weighted criteria $\gamma = 0.5$.

4. Case Study

The methodology presented in this paper is applied to the case study posed in BATADAL [7], which uses the water network D-town (Figure 2) and considers potential attacks to pump stations and pressure and tank level sensors, as indicated in the figure. The network is composed of 429 pipes, 388 junction nodes, 7 tanks, 1 reservoir, 11 pumps and 5 valves.

Three data sets are provided by BATADAL generated via epanetCPA [38], a MATLAB toolbox for cyber-attack design and hydraulic simulation. Please note that due to obvious security reasons, studies of cyber-physical attacks are usually conducted using simulated data that reproduce real-world conditions [5]. In the case of BATADAL, hourly pressure, flow, tank level and control device status are provided in the data sets. The first data set corresponds to one year of data without cyber-attacks. The second data set is based on a set of 492 h. This data set unfolds an entire, well-labeled cyber-attack, and other six cyber-attacks partially or completely hidden. Finally, the third data set has 7 new attacks distributed along 407 h of data.

The application of the methodology starts by selecting the combination of data to be used as input for fastICA from the available data. Since the water network is naturally divided into small district metered areas according to its topology, eight combinations of data are used as input for the ICA algorithm. These combinations consider the hydraulic connections of the system and are summarized in Table 1.

Table 1. Description of control and measuring devices for fastICA application

Combination	Measured Element	Type of Data
A	J300, J289	Pressure
B	J307, J302	Pressure
C	V2, T2, J422	Flow, Tank Level and Pressure
D	T1, PU1, PU2, PU3	Tank Level and Flow
E	J256, T3, PU4, PU5	Pressure, Tank Level and Flow
F	J415, T4, PU6, PU7	Pressure, Tank Level and Flow
G	J306, T5, PU8, PU9	Pressure, Tank Level and Flow
H	J317, T6, T7, PU10, PU11	Pressure, Tank Level and Flow

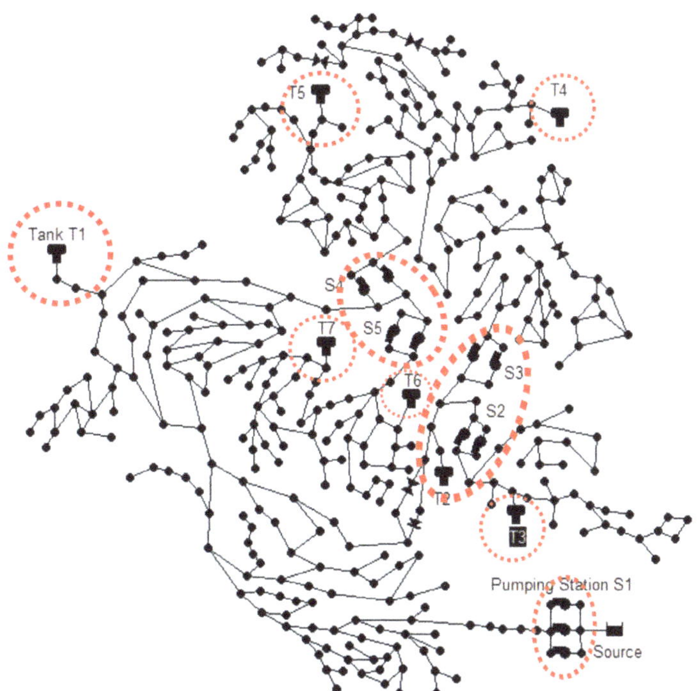

Figure 2. D-town water network topology highlighting potential attack locations.

Using the combinations presented in Table 1, the algorithm fastICA is applied, which separates each combination into 2 (approximate) sources. To illustrate the signal separation, Figure 3a presents the original data for combination B, and Figure 3b presents the separated signals, split into two sources. In the separated sources (Figure 3b), an abnormal trend of the time series is discovered in the test data set.

This behavior is repeated for other combinations. One source has a periodic trend, as a typical behavior of a WDS, while the second source is similar to a random noise. This second one is, usually, highly affected by the attacks and is considered by the detection algorithm to identify abrupt changes.

For automatic detection of the changes in the separated signals, ACDP is applied. The algorithm evaluates the second source, highly affected by the attacks, and allows a more accurate detection of the anomalies. Applying ACDP to the sources obtained from all combinations (Table 1), the start and end time indexes of the attacks are obtained.

The entire process may be summarized as follows. First, a combination of hydraulic time series is selected and is processed by fastICA (Figure 4a); this algorithm splits the time series and produces two sources that are processed by ACDP (Figure 4b). Finally, ACDP is launched to locate the time interval when the attack occurred (Figure 4c), allowing the water company to start actions for mitigating the impacts of the attack. Figure 4c shows in detail the attack corresponding to combination F. It is possible to observe the delay in detecting the attack (interval between the first black and the green lines). As described in [7], this attack is related to changes of tank T4 signal. Even though these changes are not easily identified in the original data, as shown in Figure 4a, after fastICA processing, source signal 1 clearly reveals the change in data, allowing ACDP to disclose the attack.

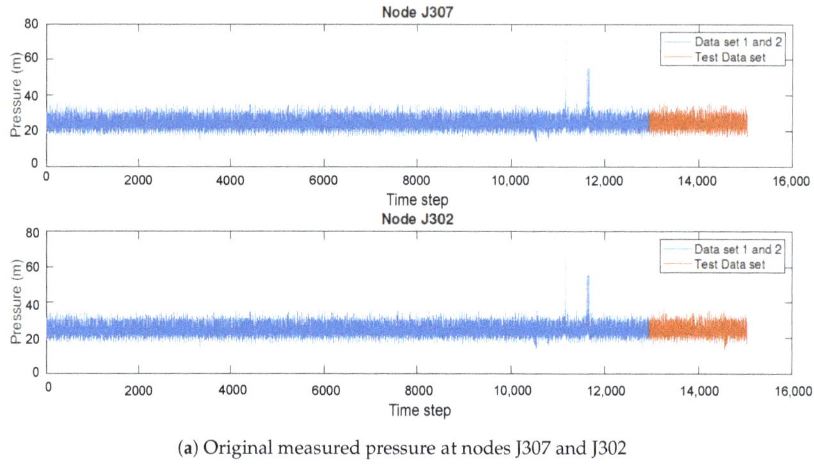

(**a**) Original measured pressure at nodes J307 and J302

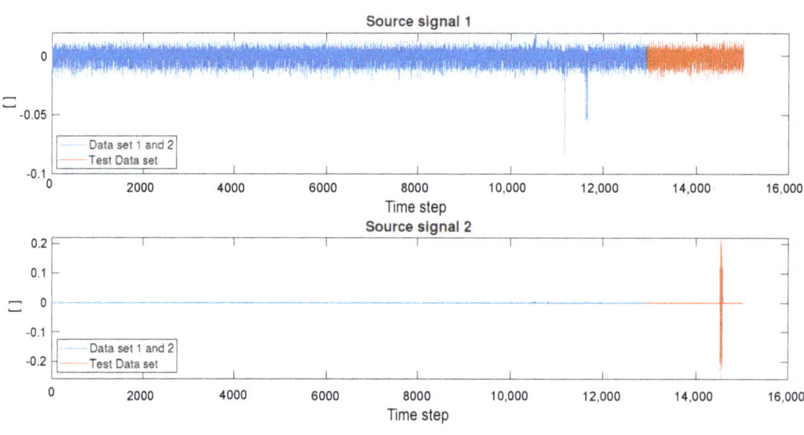

(**b**) Separated signals from J307 and J302

Figure 3. Comparison between mixed and separated pressure signal—combination B.

Still for illustrating the joint capability of fastICA and ACDP, Figure 5a shows original measured data of pumps PU8 and PU9, node J306 and tank level T5. The joint process by fastICA and ACDP applied to the corresponding test data set reveals that no attacks are found in the sources. This fact corroborates the accuracy of the algorithm, mainly in terms of false positives minimization, since according to [7], there were no attacks occurring in the test data set.

The ACDP applied to all sources and combinations for the test data set resulted in the identification of 7 cyber-attacks, i.e., all the attacks were disclosed by the proposed methodology. Figure 6 presents the confusion matrix with the numbers of TP, TN, FP and FN.

Based on the confusion matrix, it is possible to calculate $TPR = 0.966$ and $TNR = 0.980$, resulting in a $SCP = 0.973$. Compared to the seven teams that presented solutions for BATADAL, the value of SCP is the second higher, the first team having obtained $SCP = 0.975$, virtually identical. Comparing the TPR, the methodology of the present work gets the highest scores, showing its efficiency to find abnormal scenarios.

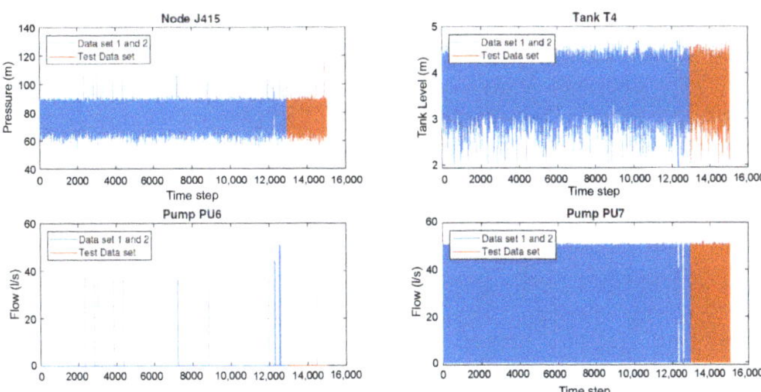

(**a**) Original measured pressure at node J415, tank level at Tank T4 and flow at pumps PU6 and PU7

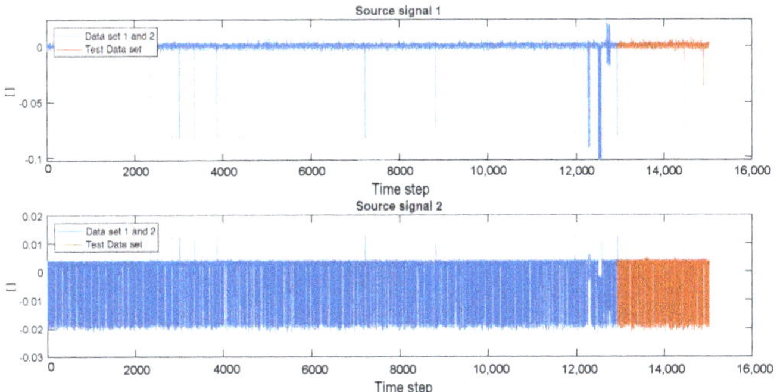

(**b**) Separated signals from node J415, tank level at Tank T4 and flow at pumps PU6 and PU7

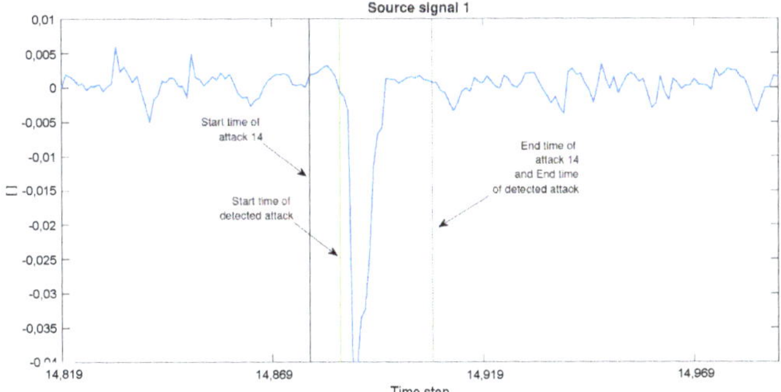

(**c**) Detail of ACPD algorithm applied to test data set using signal one of fastICA applied to node J415, tank level at Tank T4 and flow at pumps PU6 and PU7

Figure 4. Complete data processing, illustrating fastICA and ACPD applied to Combination F.

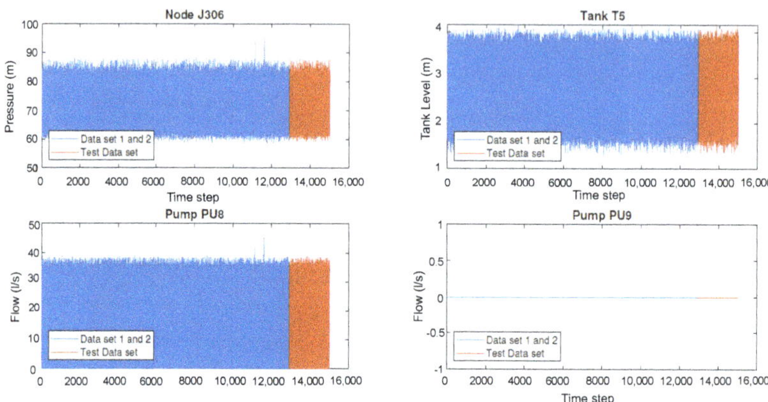

(**a**) Original measured pressure at node J306, tank level at Tank T5 and flow at pumps PU8 and PU9

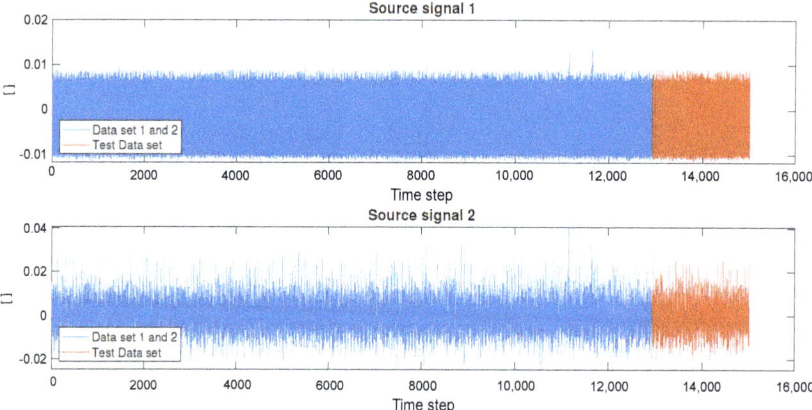

(**b**) Separated signals from node J306, tank level at Tank T5 and flow at pumps PU8 and PU9 processed by ACDP

Figure 5. Original and processed data for combination G.

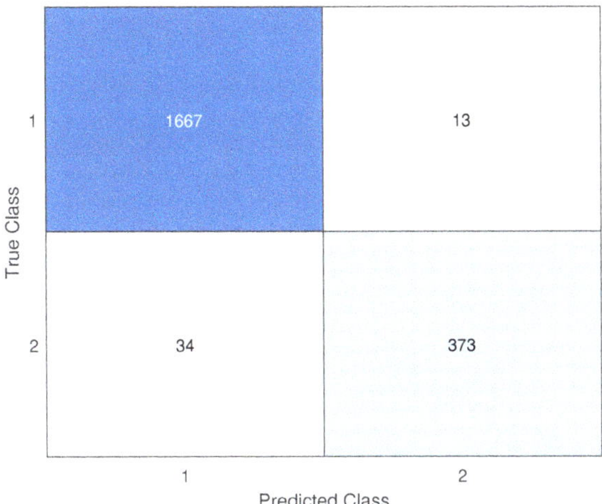

Figure 6. Confusion matrix for the test data set presenting the number of true positives and negatives on the main diagonal and the false negatives and false positives on the counterdiagonal

The results in terms of TTD, are summarized in Table 2. Four out of the seven attacks are detected immediately or in a maximum of 1 h later. The rest is detected in a maximum of 10 h later, as shown in the table. Based on these values, the score for the other metric proposed in BATADAL, namely S_{TTD}, is calculated, resulting in 0.913. Compared to the other teams, this value is the lowest and shows that despite the accuracy of the methodology, for some abnormal scenarios, early warnings cannot be suitably obtained. Based on both metrics SCR and S_{TTD} the final score is calculated, resulting in 0.973. This final score is the second highest, when compared with the seven teams that presented solutions in BATADAL.

Table 2. Summarized results for the test data sets presenting start and end time date for each attack

Attack Label	Start Date	Start Time	End Date	End Time
8	16 January 2017	10	19 January 2017	4
9	30 January 2017	8	2 February 2017	2
10	9 February 2017	3	10 February 2017	9
11	12 February 2017	11	13 February 2017	17
12	24 February 2017	9	28 February 2017	3
13	10 March 2017	13	13 March 2017	16
14	26 March 2017	3	27 March 2017	1

5. Conclusions

The security of water distribution systems has become increasingly complex due to the rapid rise of telemetry and remote controls. The growing number of reported cyber-attacks in WDSs has also created an important need for new, fast and efficient methodologies for early-warning systems that help guarantee WDS security.

Most efforts devoted to detecting cyber-attacks in WDSs have primary focused on machine-learning and optimization techniques. Statistical analysis of measured data can provide valuable results for quick detection of anomalies. However, as attested in [5], studies from other fields are necessary to build confidence in the models. In this paper, we focus on signal-processing. Among the signal-processing techniques based on statistical analysis, fastICA is explored in this work. FastICA has shown to be a powerful tool for hydraulic data analysis, mainly under abnormal conditions. The signal separation follows a trend, where one signal is more related to a typical periodical oscillation of the system, and the second one is more related to a random process. The latter is highly affected by abnormal conditions and, consequently, it is a possible input for detection algorithms. The application of fastICA to hydraulic time series (e.g., tank level) allowed to clearly highlight the attacks against the studied water system. These attacks cannot be easily disclosed in the original time series; however, this task becomes easier after processing the data by a BSS algorithm.

Change point detection algorithms are useful for automatic statistical changes in time series, and can be used for early-warning systems. In this work, the ACPD algorithm is applied to the separate signals resulted from fastICA for automatically defining changes in data, which are seen to correspond to cyber-attacks. The methodology applied to the BATADAL case study resulted in the detection of the seven attacks with high accuracy and few false positives. We claim that the methodology can be perfectly applied to any real system, as long as the water utility can measure at least one of the hydraulic parameters, namely flow, pressure and tank level.

Nevertheless, some attack scenarios have been detected too late, which is a limitation, otherwise typical of most risk evaluation methodologies. Special attention to this kind of attacks should be paid, requiring more investigation for developing ultimate conclusions about the global efficiency of the methodology. Future works, more than ratifying the efficiency of detection algorithms, should go deeper into the cyber-physical problem, investigating the causes of the attacks, optimally placing grids of dedicated sensors, and timely responding to prevent the occurrence of damage. Optimal sensor placement is still an only recently and partially formed subject. Accordingly, efforts should be devoted to expanding and enriching this field by producing novel and efficient methodologies to help fully develop this field of research.

Author Contributions: Conceptualization, B.B. and E.L.J.; methodology, P.R., B.B. and D.B.; software, B.B. and P.R.; validation, G.M. and D.B.; formal analysis, J.I. and E.L.J.; writing—original draft preparation, B.B. and P.R.; writing—review and editing, G.M. and J.I.; supervision, J.I. and E.L.J. All authors have read and agreed to the published version of the manuscript.

Funding: This research received no external funding.

Institutional Review Board Statement: Not applicable.

Informed Consent Statement: Not applicable.

Data Availability Statement: Not applicable.

Conflicts of Interest: The authors declare no conflict of interest.

References

1. Taormina, R.; Galelli, S. Deep-learning approach to the detection and localization of cyber-physical attacks on water distribution systems. *J. Water Resour. Plan. Manag.* **2018**, *144*, 04018065. [CrossRef]
2. Adepu, S.; Palleti, V.R.; Mishra, G.; Mathur, A. Investigation of cyber attacks on a water distribution system. In *International Conference on Applied Cryptography and Network Security*; Springer: Berlin/Heidelberg, Germany, 2020; pp. 274–291.
3. Clark, R.M.; Panguluri, S.; Nelson, T.D.; Wyman, R.P. Protecting drinking water utilities from cyberthreats. *J. Am. Water Work. Assoc.* **2017**, *109*; 50–58. [CrossRef]
4. Water Infrastructure: When States and Cyber Attacks Rear Their Ugly Heads, Howpublished. Available online: https://www.stormshield.com/news/water-infrastructure-when-states-and-cyber-attacks-rear-their-ugly-heads (accessed on 25 January 2021).
5. Tuptuk, N.; Hazell, P.; Watson, J.; Hailes, S. A Systematic Review of the State of Cyber-Security in Water Systems. *Water* **2021**, *13*, 81. [CrossRef]
6. *CySWater'15: Proceedings of the 1st ACM International Workshop on Cyber-Physical Systems for Smart Water Networks*; Association for Computing Machinery: New York, NY, USA, 2015.
7. Taormina, R.; Galelli, S.; Tippenhauer, N.O.; Salomons, E.; Ostfeld, A.; Eliades, D.G.; Aghashahi, M.; Sundararajan, R.; Pourahmadi, M.; Banks, M.K.; et al. Battle of the attack detection algorithms: Disclosing cyber attacks on water distribution networks. *J. Water Resour. Plan. Manag.* **2018**, *144*, 04018048. [CrossRef]
8. Comon, P. Independent Component Analysis. In *Internat. Signal Processing Workshop on High-Order Statistics, Chamrousse, France, 10–12 July 1991*; Lacoume, J.L., Ed.; Higher-Order Statistics, Elsevier: Amsterdam, The Netherlands, 1992; pp. 29–38.

9.   Nader, P.; Honeine, P.; Beauseroy, P. Detection of cyberattacks in a water distribution system using machine learning techniques. In Proceedings of the 2016 Sixth International Conference on Digital Information Processing and Communications (ICDIPC), Beirut, Lebanon, 21–23 April 2016; pp. 25–30.

10. Zohrevand, Z.; Glasser, U.; Shahir, H.Y.; Tayebi, M.A.; Costanzo, R. Hidden Markov based anomaly detection for water supply systems. In Proceedings of the 2016 IEEE International Conference on Big Data (Big Data), Washington, DC, USA, 5–8 December 2016; pp. 1551–1560.

11. Adepu, S.; Mathur, A. An investigation into the response of a water treatment system to cyber attacks. In Proceedings of the 2016 IEEE 17th International Symposium on High Assurance Systems Engineering (HASE), Orlando, FL, USA, 7–9 January 2016; pp. 141–148.

12. Aghashahi, M.; Sundararajan, R.; Pourahmadi, M.; Banks, M.K. Water Distribution Systems Analysis Symposium–Battle of the Attack Detection Algorithms (BATADAL). In *World Environmental and Water Resources Congress 2017*; American Society of Civil Engineers: Sacramento, CA, USA, 2017; pp. 101–108.

13. Breiman, L. Random forests. *Mach. Learn.* **2001**, *45*, 5–32. [CrossRef]

14. Brentan, B.M.; Campbell, E.; Lima, G.; Manzi, D.; Ayala-Cabrera, D.; Herrera, M.; Montalvo, I.; Izquierdo, J.; Luvizotto, E., Jr. On-line cyber attack detection in water networks through state forecasting and control by pattern recognition. In *World Environmental and Water Resources Congress 2017*; American Society of Civil Engineers: Sacramento, CA, USA, 2017; pp. 583–592.

15. Chandy, S.E.; Rasekh, A.; Barker, Z.A.; Shafiee, M.E. Cyberattack detection using deep generative models with variational inference. *J. Water Resour. Plan. Manag.* **2019**, *145*, 04018093. [CrossRef]

16. Giacomoni, M.; Gatsis, N.; Taha, A. Identification of cyber attacks on water distribution systems by unveiling low-dimensionality in the sensory data. In *World Environmental and Water Resources Congress 2017*; American Society of Civil Engineers: Sacramento, CA, USA, 2017; pp. 660–675.

17. Abokifa, A.A.; Haddad, K.; Lo, C.; Biswas, P. Real-time identification of cyber-physical attacks on water distribution systems via machine learning–based anomaly detection techniques. *J. Water Resour. Plan. Manag.* **2019**, *145*, 04018089. [CrossRef]

18. Pasha, M.F.K.; Kc, B.; Somasundaram, S.L. An approach to detect the cyber-physical attack on water distribution system. In *World Environmental and Water Resources Congress 2017*; American Society of Civil Engineers: Sacramento, CA, USA, 2017; pp. 703–711.

19. Housh, M.; Ohar, Z. Model-based approach for cyber-physical attack detection in water distribution systems. *Water Res.* **2018**, *139*, 132–143. [CrossRef]

20. Hindy, H.; Brosset, D.; Bayne, E.; Seeam, A.; Bellekens, X. Improving SIEM for critical SCADA water infrastructures using machine learning. In *Computer Security*; Springer: Berlin/Heidelberg, Germany, 2018; pp. 3–19.

21. Choi, Y.H.; Sadollah, A.; Kim, J.H. Improvement of Cyber-Attack Detection Accuracy from Urban Water Systems Using Extreme Learning Machine. *Appl. Sci.* **2020**, *10*, 8179. [CrossRef]

22. Esmalifalak, M.; Nguyen, H.; Zheng, R.; Han, Z. Stealth false data injection using independent component analysis in smart grid. In Proceedings of the 2011 IEEE International Conference on Smart Grid Communications (SmartGridComm), Brussels, Belgium, 17–20 October 2011; pp. 244–248.

23. Palmieri, F.; Fiore, U.; Castiglione, A. A distributed approach to network anomaly detection based on independent component analysis. *Concurr. Comput. Pract. Exp.* **2014**, *26*, 1113–1129. [CrossRef]

24. Lughofer, E.; Zavoianu, A.C.; Pollak, R.; Pratama, M.; Meyer-Heye, P.; Zörrer, H.; Eitzinger, C.; Radauer, T. On-line anomaly detection with advanced independent component analysis of multi-variate residual signals from causal relation networks. *Inf. Sci.* **2020**, *537*, 425–451. [CrossRef]

25. Ndehedehe, C.E.; Agutu, N.O.; Okwuashi, O.; Ferreira, V.G. Spatio-temporal variability of droughts and terrestrial water storage over Lake Chad Basin using independent component analysis. *J. Hydrol.* **2016**, *540*, 106–128. [CrossRef]

26. Gao, J.; Qi, S.; Wu, W.; Li, D.; Ruan, T.; Chen, L.; Shi, T.; Zheng, C.; Zhuang, Y. Study on leakage rate in water distribution network using fast independent component analysis. *Procedia Eng.* **2014**, *89*, 934–941. [CrossRef]

27. Moradkhani, H.; Meier, M. Long-lead water supply forecast using large-scale climate predictors and independent component analysis. *J. Hydrol. Eng.* **2010**, *15*, 744–762. [CrossRef]

28. Jung, D.; Kang, D.; Liu, J.; Lansey, K. Improving resilience of water distribution system through burst detection. In *World Environmental and Water Resources Congress 2013: Showcasing the Future*; Elsevier: Amsterdam, The Netherlands, 2013; pp. 768–776.

29. Bakker, M.; Jung, D.; Vreeburg, J.; Van de Roer, M.; Lansey, K.; Rietveld, L. Detecting pipe bursts using Heuristic and CUSUM methods. *Procedia Eng.* **2014**, *70*, 85–92. [CrossRef]

30. Christodoulou, S.E.; Kourti, E.; Agathokleous, A. Waterloss detection in water distribution networks using wavelet change-point detection. *Water Resour. Manag.* **2017**, *31*, 979–994. [CrossRef]

31. Comon, P. Contrasts, independent component analysis, and blind deconvolution. *Int. J. Adapt. Control. Signal Process.* **2004**, *18*, 225–243. [CrossRef]

32. Hyvärinen, A.; Oja, E. Independent component analysis: Algorithms and applications. *Neural Netw.* **2000**, *13*, 411–430. [CrossRef]

33. Ziehe, A. Blind source separation based on joint diagonalization of matrices with applications in biomedical signal processing. Ph.D. Thesis, Universitat Potsdam, Potsdam, Germany, April 2005.

34. Killick, R.; Fearnhead, P.; Eckley, I.A. Optimal detection of changepoints with a linear computational cost. *J. Am. Stat. Assoc.* **2012**, *107*, 1590–1598. [CrossRef]

35. Horváth, L. The maximum likelihood method for testing changes in the parameters of normal observations. *Ann. Stat.* **1993**, *35*, 671–680. [CrossRef]
36. Inclan, C.; Tiao, G.C. Use of cumulative sums of squares for retrospective detection of changes of variance. *J. Am. Stat. Assoc.* **1994**, *89*, 913–923.
37. Lavielle, M. Using penalized contrasts for the change-point problem. *Signal Process.* **2005**, *85*, 1501–1510. [CrossRef]
38. Taormina, R.; Galelli, S.; Tippenhauer, N.O.; Salomons, E.; Ostfeld, A. Characterizing cyber-physical attacks on water distribution systems. *J. Water Resour. Plan. Manag.* **2017**, *143*, 04017009. [CrossRef]

Article

A Smart Water Grid for Micro-Trading Rainwater: Hydraulic Feasibility Analysis

Elizabeth Ramsey*, Jorge Pesantez, Mohammad Ali Khaksar Fasaee, Morgan DiCarlo, Jacob Monroe and Emily Zechman Berglund

Civil, Construction, and Environmental Engineering, North Carolina State University, Raleigh, NC 27695, USA; jpesant@ncsu.edu (J.P.); mkhaksa@ncsu.edu (M.A.K.F.); mdicarl@ncsu.edu (M.D.); jgmonroe@ncsu.edu (J.M.); emily_berglund@ncsu.edu (E.Z.B.)

* Correspondence: evramsey@ncsu.edu

Received: 30 September 2020; Accepted: 29 October 2020; Published: 2 November 2020

Abstract: Water availability is increasingly stressed in cities across the world due to population growth, which increases demands, and climate change, which can decrease supply. Novel water markets and water supply paradigms are emerging to address water shortages in the urban environment. This research develops a new peer-to-peer non-potable water market that allows households to capture, use, sell, and buy rainwater within a network of water users. A peer-to-peer non-potable water market, as envisioned in this research, would be enabled by existing and emerging technologies. A dual reticulation system, which circulates non-potable water, serves as the backbone for the water trading network by receiving water from residential rainwater tanks and distributing water to households for irrigation purposes. Prosumer households produce rainwater by using cisterns to collect and store rainwater and household pumps to inject rainwater into the network at sufficiently high pressures. The smart water grid would be enabled through an array of information and communication technologies that provide capabilities for automated and real-time metering of water flow, control of infrastructure, and trading between households. The goal of this manuscript is to explore and test the hydraulic feasibility of a micro-trading system through an agent-based modeling approach. Prosumer households are represented as agents that store rainwater and pump rainwater into the network; consumer households are represented as agents that withdraw water from the network for irrigation demands. An all-pipe hydraulic model is constructed and loosely coupled with the agent-based model to simulate network hydraulics. A set of scenarios are analyzed to explore how micro-trading performs based on the level of irrigation demands that could realistically be met through decentralized trading; pressure and energy requirements at prosumer households; pressure and water quality in the pipe network.

Keywords: rainwater harvesting; water trading; dual reticulation; decentralized water supply; water distribution system; agent-based modeling; urban water management; smart city

---

1. Introduction

Urban water utility systems around the world are increasingly pressured by limited water resources, growing urban demand, and impacts from climate change. The United Nations projects that, by 2025, 1.8 billion people will be living in regions with absolute water scarcity, and two-thirds of the world population could be living under water-stressed conditions [1]. Supply-side strategies for urban water management are limited in water-scarce regimes, because they require large investments to construct new infrastructure and develop new resources. Demand-side strategies, on the other hand, extend existing resources by reducing demands through conservation campaigns, pricing strategies, and restrictions. Demand-side strategies that rely on continued demand reduction, however, are ultimately limited by hardened demands that cannot be reduced further. As water scarcity in urban centers increases with population growth and climate change, new technologies, advanced management strategies, and diverse water sources must emerge to create new efficiencies in water supply and use. Innovative programs can utilize new technologies and data that have emerged as part of smart cities initiatives [2]. For example, smart meters provide capabilities to collect sub-hourly water flow and consumption data in real-time [3], and automated control systems operate infrastructure components remotely and efficiently [4–6].

This manuscript explores a novel management strategy for improving water efficiency in urban areas by supplementing non-potable purposes of the total demand using alternative water sources. A smart water grid is presented here as a water network that is shared by multiple diverse users, who can either produce or consume water. The concept presented in this manuscript builds on an existing dual reticulation system, which pumps non-potable reclaimed water back to a community via a second parallel pipe network and reduces demands for high quality treated water, as compared with a conventional urban water cycle (Figure 1a,b). We propose that the existing non-potable water network can be used as a smart water grid to facilitate micro-trading, where households can exchange water within a peer-to-peer network. Households generate water through rainwater harvesting, putting rainwater "back on the grid" by pumping water into the non-potable water infrastructure system, and purchase water from neighbors by withdrawing water from the pipe network. Smart technologies, such as smart meters, blockchain, smart contracts, and automated infrastructure, would provide the necessary capabilities to allow real-time trading within a smart water grid. Within a smart city paradigm, a rainwater micro-trading program re-envisions the urban water cycle by allowing households to act as prosumers, who produce and sell water within their community (Figure 1c). By allowing households to trade rainwater, a new efficiency is introduced in the water cycle that offsets requirements to treat and pump reclaimed water from a centralized facility. This offset can create energy savings and save reclaimed water for use in other non-potable applications.

In the energy sector, micro-trading has been demonstrated as a viable market for decentralized resource production, in which households can generate energy through solar photovoltaic cells, store energy in batteries, and sell and transmit excess energy to neighbors through existing power distribution infrastructure [7,8]. Micro-trading water has a number of nuanced constraints that may limit its adoption by utilities and community members. For example, new infrastructure at households is needed to enable storage, sensing, treatment, and trading water, and household participation may vary based on climate and economics. The research presented in this manuscript focuses on the performance of centralized infrastructure and takes a simulation-based approach to evaluate the effects on water savings, energy savings, nodal pressure, and water quality. Reclaimed water networks are designed to maintain pressures and flows, and performance of a smart water grid may decrease due to new flows associated with produced water. The ability of prosumers to contribute water to a non-potable water network is facilitated by household-level pump systems, and the introduction of these decentralized flows into a pipe network affects flows and pressures in the network. High pressures in the network can limit the contribution of water from households that must overcome pressure heads through small pumps, and low pressures may emerge in times of low production and high demands. This research also evaluates savings in energy and water associated with a smart water grid. Micro-trading can reduce demands for produced reclaimed water, resulting in water savings. The energy required to run household pumps across a network, however, must be compared with the energy requirements of treating and pumping water from a centralized facility.

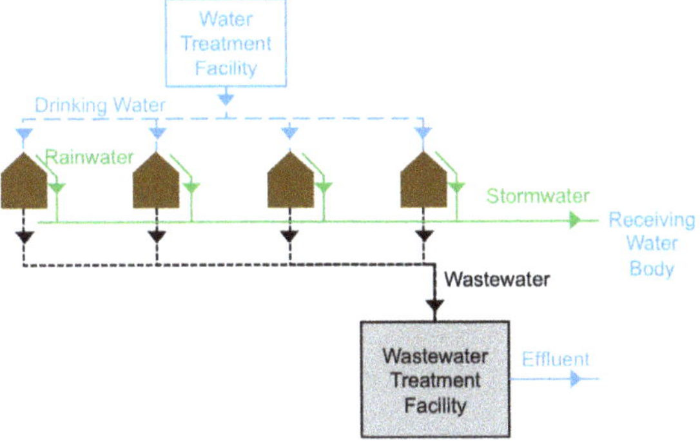

(**a**) Conventional urban water infrastructure.

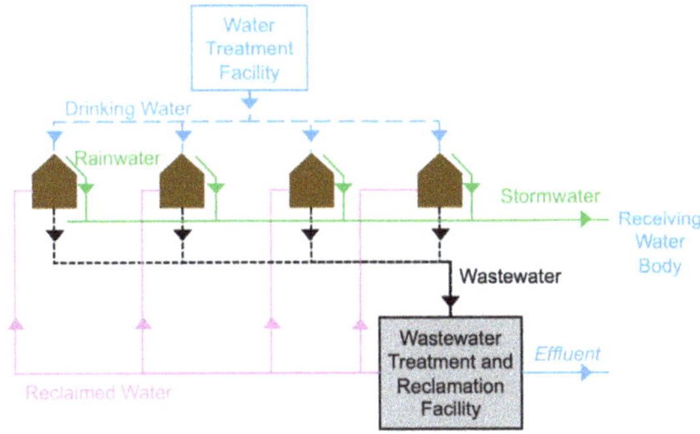

(**b**) Dual reticulation system.

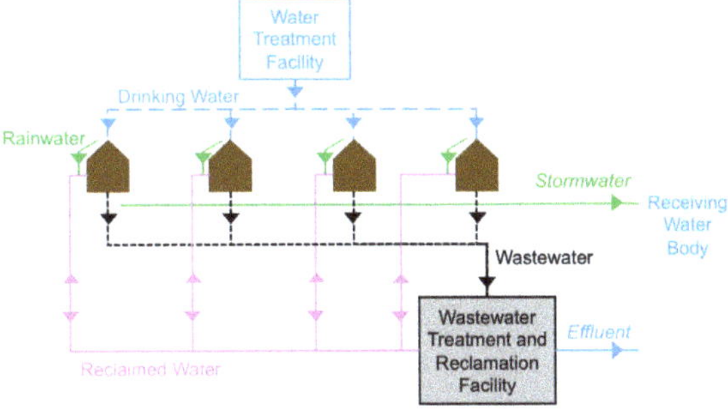

(**c**) Smart water grid.

Figure 1. (**a**) Conventional urban drinking water, wastewater, and stormwater systems. (**b**) A dual reticulation system closes the loop in the urban water cycle by treating wastewater and providing it for non-potable household uses. Effluent flows from the wastewater facility are reduced. (**c**) A smart water grid increases water efficiency in the urban water cycle by allowing households to contribute rainwater to the dual reticulation system. Effluent and stormwater flows are reduced.

The goal of this manuscript is to explore the feasibility of a smart water grid based on the performance of the centralized infrastructure and energy demands. This research develops a simulation framework that couples agent-based modeling and hydraulic models to test energy consumption, water consumption, and nodal water pressure in a smart water grid. Agent-based models simulate the individual behaviors and interactions of a population of agents to explore emergent system-level dynamics [9,10]. Agent-based modeling has been applied in water resources management to simulate a population of water consumers as agents with the purpose of exploring the emergence of system-level performance due to micro-level interactions [11]. A few studies couple agent-based models with hydraulic simulation to explore how changes in household demands affect system-level performance, such as pressure and flow directions [12–14]. Further, agent-based modeling has been applied to simulate markets for trading natural resources [11,15–17], and recent research applied agent-based modeling to simulate how households trade solar generated energy in a peer-to-peer energy market [8]. In the formulation developed in this research, consumer households are simulated as agents that exert irrigation demands, and prosumer households are simulated as agents that store rainwater, and pump rainwater into a pipe network. A hydraulic model is developed of a reclaimed water network that serves a small community of non-potable water consumers and prosumers. Output from the agent-based model specifies flows into and out of the water network at each node, and these outputs are used as negative and positive demand patterns for the hydraulic model. The modeling framework is applied for an illustrative case study that was developed based on realistic infrastructure data. Water consumption, nodal pressure, and energy consumption are evaluated for the network of water users for scenarios with and without micro-trading. System performance is tested for 126 scenarios across different climates and concentrations of prosumers to explore how precipitation and participation affect feasibility. The results demonstrate that a smart water grid is feasible, generating energy and water savings that vary in magnitude based on local climate and the level of community participation.

2. Background

2.1. Dual Reticulation Networks

Only a fraction of water used for urban purposes needs to be potable quality, and reclaimed water can serve as an alternative water source for non-potable applications [18]. Reclaimed water is wastewater that has been treated to levels lower than potable water quality and can be used for nonpotable applications, including washing, cooling, gardening, toilet flushing, and lawn irrigation [19]. Reclaimed water can be provided to a large group of consumers through a dual reticulation system, which includes two parallel pipe networks: the primary water network distributes potable water, and the secondary network conveys reclaimed water [20]. A centralized utility typically manages water reclamation programs to ensure that treatment standards are met and to distribute reclaimed water. Dual reticulation programs have been implemented in cities in the United States, Japan, and Australia [21]. Reclaimed water products can help conserve high-quality water produced by utilities for essential purposes.

Dual reticulation systems impose high capital costs, but costs can be offset by a reduction in demands that are exerted on aquifers and surface water sources, leading to improved ecosystem health and drought resilience [22]. While it is difficult to quantify the externalities in a cost-benefit analysis of non-potable water systems [23,24], dual reticulation systems can offset the use of potable water, creating savings in utility energy costs for water treatment [25]. Energy savings can be sufficient to offset the capital costs required for building a dual pipe infrastructure [26]. Dual reticulation systems may also create benefits by reducing the need for infrastructure investment for the main potable system [27] and systematically encouraging conservation by adding new value to water. The cost-benefit analysis of dual reticulation systems, however, can vary widely based on the characteristics of a location, such as infrastructure design, topography, energy sources, quality of source water, and existing infrastructure [28]. The distance between users and a water reclamation plant and the amount of uphill pumping can limit the feasibility of dual systems. A study of four U.S. cities that recycle water found that there were significant economic barriers to implementing dual-reticulation systems. Challenges were cited, including diminishing returns, due to the lack of additional large consumers of nonpotable water near the treatment plant; commitments to return treated effluent for instream flows; more efficient options for selling recycled water for cooling and industrial processes; lack of clear and convergent regulations around water reuse programs [22].

2.2. Rainwater Harvesting Systems

Rainwater harvesting systems can be installed at the household level to capture roof runoff, providing an alternative source to meet non-potable water demands [29]. Harvesting rainwater is a millennia-old agricultural practice, with increasing implementation in modern cities with large population demands, including in Adelaide and Addis Ababa [29,30]. Rainwater harvesting is used widely because it provides easy collection with low cost, treatment, and maintenance requirements [31]. Captured rainwater can be applied on-site or on a larger scale for community purposes, and communal rainwater tanks may be economically feasible [32]. Rainwater harvesting may be feasible for individual users, subject to the specific water demand and roof area [31,33], but its economic feasibility may be limited for some households because it does not provide a continuous supply of water and needs to be supplemented with other sources [34]. As an alternative source, however, rainwater can provide significant volumes of water, and it is estimated that up to 80% of rainfall could be harvested from urban rooftops in the U.S. [35]. For example, a study of California water indicates that recycling irrigation runoff water (priced at ($0.43–1.21 per 1000 gallons) was a cost-effective alternative to using the region's municipal water ($2.39–2.91 per 1000 gallons) [36]. Rainwater harvesting has other benefits, and it can reduce stormwater infrastructure costs by reducing peak flows. Liang et al. [37] showed that implementing smart rainwater harvesting systems reduces peak system flows by 35% to 85%.

There are practical limitations and costs that prevent the transition to the wide-scale use of rainwater infrastructure. Rainwater harvesting is shown to have long payback periods before benefits outweigh costs, with economic returns that are very sensitive to local policy, water quality concerns, and government rebates [38,39]. Grants from local initiatives and environmental agencies can reduce capital costs. For example, rebates of $0.50 per gallon of installed tank capacity were used to incentivize rainwater harvesting in Barbados [38]. The costs and benefits indicate that green infrastructure solutions have market value and should be strongly considered, though the economics of purchasing tanks and pumps, as required in the smart water grid, may need economic incentives to encourage wide-spread adoption. The system that is proposed in this research creates a new efficiency in the use of rainwater by providing the means to share rainwater within a community and creating a reliable source of non-potable flows by integrating rainwater harvesting within a reclaimed water reticulation system.

2.3. Micro-Trading in Water Markets and Smart Technologies

Economists have argued that scarce water can be allocated more efficiently through water markets, rather than through centralized control [40]. Large-scale water rights markets have been operating for decades among utilities and agricultural users [41], and new markets that trade conserved water are emerging as a strategy for demand-side management to create value and new incentives around conservation activities [42,43]. Decentralization of water services is seen as an approach to support a sustainable future for urban water management [44,45], where water can be supplied or treated at small-scale plants, rather than at centralized locations. Efficiencies may be gained through decentralized systems because water or wastewater does not need to be transported over long distances, and resources can be re-used to meet demands on-site. A few studies explore decentralized markets for water supply that allow for trading among households. Haddad [46] proposed a cap-and-trade water program among residential end-users, where a cap is used to grant each customer with use-rights to available water. Water conservationists could sell or rent unused use-rights, and customers would be heavily penalized for using water in excess of their use-rights. Water customers would call a toll free number to execute transfers through a specially-trained broker. Haddad's micro-trading system was criticized based on the complexities of making initial allocation of use-rights to users; expected sizeable transaction costs; lack of household expertise and willingness to engage in a market; impact of a water-use market on economic development in the area [47,48]. More recently, a decentralized water supply system was developed in Western Australia, in which consumers contribute recycled water, including stormwater and greywater, to a groundwater resource through garden bores [49]. By contributing recycled water, consumers become prosumers and gain credits in their water use accounts. The program is enabled through smart meters, which record water consumption and contributions at 10-minute intervals, and the shallow aquifer provides a pathway in the urban water cycle among households and utility.

The viability of micro-trading has been enabled by the emergence of blockchain technology in water markets. Blockchain technology is an information and communications technology (ICT) capable of addressing some of the challenges in implementing peer-to-peer markets [50]. A blockchain is a distributed ledger that provides a platform for digital transactions without a trusted third-party organization [51,52]. Data structures are both immutable and cryptographically verifiable, promising security, accuracy, authentication and traceability of transactions [53]. Blockchain can also reduce transaction costs associated with third-party brokers, though some fees may be necessary to maintain a critical centralized infrastructure [8]. Smart contracts can be used with blockchain, where smart contracts work as simple scripts encoded on the blockchain that contain predefined directions for automating workflows on recorded data and finalizing the settlement of financial transactions between buyers and sellers [54]. Smart contracts automate micro-trading to allow for rapid reconciliation between consumers and prosumers and reduce the time and associated cost of trading, which may increase participation. As an emerging technology, blockchain has applications in water resources management, supporting data sharing among utilities with assurances of confidentiality and commercial sensitivity; linking flowmeter sensor data with water resources mapping, billing, and operations; facilitating trading and tracking of water credits among large-scale users; allowing households to buy water in a market of competitive water providers; serving as a stable currency to enhance security of water supply [55–57]. Blockchain provides a ledger system that can support peer-to-peer markets for micro-trading water, and a few examples exist to date. Melbourne, Australia, has proposed a rainwater micro-trading program that would be under-girded by blockchain technologies [55]. The program would assign apartments with a quota of free rainwater from a communal tank, and excess rainwater would be conveyed to a large water recycling plant to supply treated water for non-drinking uses [58,59]. In another example, a proof-of-concept model was developed to simulate blockchain-enabled trading of virtual water among homes, where water could be sold by low-consuming households to households that want to exceed a daily limit on water consumption [60]. Similar to Haddad's cap-and-trade system [46], customers would trade water rights. Whereas Haddad's program would allow customers to trade water on a monthly or seasonal basis, the market proposed by Alcarria et al. [60] relies on smart meters, blockchain, and smart contracts, and the functionality of these technologies would allow customers to make daily decisions about trading water.

The smart water grid proposed in this research would rely on smart connected technologies, similar to the systems described above. Smart water meters are needed to record the exact flow rate and time of consumed and produced water. Water that is available during times of peak demands is inherently of higher value than water that is available at times of low demand. Prosumers can invest in large tanks to store water and release it during periods of high consumption, and precise meters are needed to record high resolution of trades. Automated infrastructure components, such as digitally operated pumps and valves, are needed to update flows into the network from prosumers when a trade is negotiated. Blockchain and smart contract technologies are needed to support micro-trading. Unlike other micro-trading systems described above, however, the smart water grid relies on a centralized pipe network to convey traded rainwater among households, and the focus of this research in on a feasibility analysis to explore how the hydraulics of the pipe network would be affected by decentralized buying and selling of rainwater. The feasibility of a smart water grid will also be affected by the availability and functionality of smart technologies, including blockchain and smart contracts, and the effects of benefits and costs of infrastructure, new technologies, and water and energy savings on market efficiency. While market efficiency and smart technologies are not included in the modeling framework that is described in this research, they should be explored in future research to further test the feasibility of the smart water grid.

2.4. Agent-Based Modeling for Water Infrastructure

Agent-based modeling simulates the behaviors and micro-interactions of a population of autonomous and heterogenous agents to model and study system-level phenomena [9,10]. Agent-based modeling has been applied to simulate a range of water resources planning problems by representing water users, stakeholders, and decision-makers as agents to capture decisions and behaviors around water use, water supply, wastewater services, and stormwater runoff [11]. Agent-based models have been applied to represent a population of residential water users that adapt their water consumption based on economics, climate, policies, and social influence [61–71]. These models simulate household decisions to use water and reduce consumption by adopting water-efficient technologies and restricting water use. Some frameworks couple agent-based modeling with the water supply system to capture feedback between the availability of water resources and decisions to conserve water [64–67,70]. Other agent-based models couple a population of agents with the hydraulic simulation of a water distribution system to evaluate how network flows are impacted by changing demands. Models capture water use changes during a water supply contamination event, based on exposure to the contaminant, communication from public officials, and social influence of peers [12,13,72–77]. Another set of studies uses agent-based modeling coupled with hydraulic simulation to evaluate how flows in a reclaimed water network and a potable water network change as customers adopt or resist water reuse programs [14,27,78]. Agent-based modeling has also been applied to model trading within natural resource markets, where agents use cost information to seek trades, negotiate, and adapt their preferences for trading permits with other agents. A few modeling studies couple an agent-based model with a water quality simulation model to assess water quality impacts of permit-trading strategies on river and estuary systems [11,15]. Other agent-based models capture the decisions of polluters to bid and sell permits in an emissions market, and these models are applied to assess the effect of trading on air quality [16,17]. More recently, an agent-based modeling was applied to simulate peer-to-peer markets by modeling households as agents that buy and sell energy in a residential smart energy grid [8]. In the research presented in this manuscript, agent-based modeling is loosely coupled with hydraulic simulation modeling to assess network performance metrics that are affected by agent behaviors to trade water. A simple market is simulated, where consumer agents buy rainwater when it is needed for irrigation, and prosumer agents meet demands when they have rainwater that is stored. The price of rainwater is not considered in this simulation, because the focus of the model is on the hydraulic feasibility of the network when trades are made. Additional research is needed to develop cost information and simulate how households make decisions to participate, bid, and execute micro-trades in a rainwater market.

3. Agent-Based Modeling Framework

The modeling framework presented in this research loosely couples an agent-based model with a hydraulic simulation model (Figure 2). The agent-based model represents households as either consumers, which purchase water and withdraw it from the system, or prosumers, which pump collected rainwater into the network when a buyer has committed to purchase the water. Prosumer water input and consumer demands are used to modify an input file for the hydraulic simulation, which calculates water flows and pressures in the pipe network based on production and consumption of water at households. The agent-based model is described following the Overview Design Details (ODD) protocol [79]. The ODD protocol provides a clear and succinct approach to describe agent-based models by describing purpose, entities, state variables, and scales as part of the Overview; process overview and design concepts as part of the Design; input, initialization, implementation, and submodels as part of the Details.

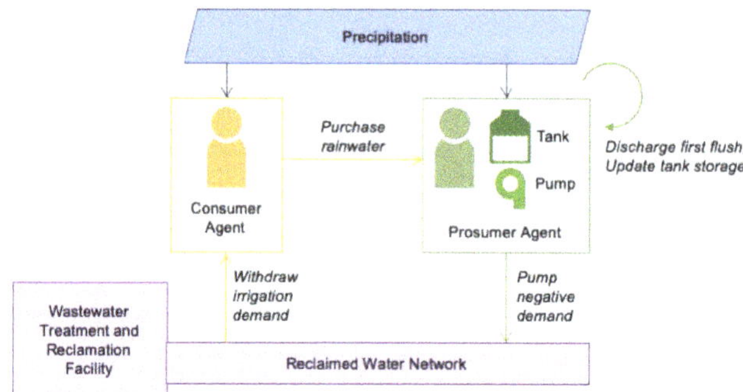

Figure 2. Agent-based modeling framework couples consumer and prosumer agents with a reclaimed water network. Image credit: Water Tank by Carlos Ochoa from the Noun Project.

3.1. Overview

3.1.1. Purpose

The purpose of the agent-based model is to simulate rainwater trading among consumer and prosumer agents facilitated through a reclaimed water network and to evaluate how water resources, energy consumption, and hydraulic performance of the network are affected by micro-trading.

3.1.2. Entities, State Variables, and Scales

Agents represent individual prosumer and consumer households. Each consumer agent is assigned a lawn area to calculate irrigation demand and is assigned a time of day for exerting demand. Prosumer agents are each assigned a rainwater tank volume capacity, a catchment area, and a small pump with a given exit pipe diameter, length, and roughness coefficient to add harvested rainwater to the hydraulic network. Parameters are used as input to the model (Table 1). State variables are updated dynamically (Table 2). For each prosumer, tank storage is updated due to precipitation, flushing requirements, and water released into the network. The model operates on an hourly time scale.

Three system-level state variables are used to represent the depth of hourly precipitation (P_t), total precipitation depth over the preceding 24-h period (BP_{24}), and the time step at which rain begins (TR).

Table 1. Parameters for consumer and prosumer agents.

Agent	Parameter	Description	Setting for Case Study
Consumer	TI_c	Time of day for irrigation demand	Section 3.3 and Figure 3 [80]
Consumer	DI	Daily irrigation demand	Equation (8)
Consumer	f	Irrigation factor	1.0
Consumer	k	Crop factor	0.7
Consumer	ET	Evapotranspiration	281.25 mm/month
Consumer	ρ	Household density	721 housing units/km^2
Consumer	U	Ratio of unpaved land	0.9
Consumer	L	Irrigable area of lawn	494.9 m^2 (Equation (10))
Consumer and Prosumer	A	Roof area	46.5 m^2 [81]
Prosumer	F	Required first flush	1.62 L/m^2
Prosumer	V	Rainwater harvesting tank capacity	5392 L [82]

Table 2. State variables for consumer and prosumer agents.

Agent	State Variable	Description	Calculation
Consumer	$D_{t,c}$	Hourly demand	Step 3
Consumer	$CQ_{t,c}$	Flows received from centralized system	Step 4
Consumer	$WA_{t,c}$	Water age at node in the network	Step 6
Consumer and Prosumer	$TW_{t,c,g}$	Traded rainwater	Step 4
Prosumer	$S_{t,g}$	Rainwater storage	Step 2
Prosumer	$VF_{t,g}$	Flushed volume	Step 2
Prosumer	$Q_{t,g}$	Flow from household pump	Step 4
Prosumer	$h_{t,g}$	Pressure at node in the network	Step 6

3.1.3. Process Overview and Scheduling

The following steps are executed at each hourly time step, t, of the simulation. The execution time step is labeled t_E. The agent-based model requires hourly precipitation (P_t) as input.

Step 1. Update system-level state variables. Based on the value of P_t, the values for BP_{24} and TR are updated. The precipitation over the preceding 24-h period (BP_{24}) is a binary variable that takes a value of one if there is a precipitation greater than zero in 24-h period before the execution time:

$$BP_{24} = \begin{cases} 1, & \text{if } \sum_{t=t_E-24}^{t_E-1} P_t > 0 \\ 0, & \text{otherwise} \end{cases} \tag{1}$$

where t is the time step and t_E is the current execution time step. Rain time, TR, is the first time step when the precipitation is greater than zero if the precipitation over the preceding 24-h period is zero.

$$TR = t_E, \text{ if } P_{t_E} > 0 \ \& \ BP_{24} = 0 \tag{2}$$

Step 2. Prosumer agents update rainwater storage values. Each prosumer agent g calculates rainwater storage volume, $S_{t,g}$, at time t based on the runoff from the roof catchment and the volume of water flushed for the first flush diversion:

$$S_{t,g} = min(V, S_{t-1,g} + P_t \times A - VF_{t,g}) \tag{3}$$

where A is the roof area; V is the capacity of the rainwater tank; $VF_{t,g}$ is the volume of water flushed from the rainwater tank at time step t for agent g. Prosumers can accumulate a maximum volume of water equivalent to the capacity of the rainwater tank (V); any excess volume is released as runoff. If no precipitation falls in the previous 24 hours before a distinct rain event begins, the prosumer agent is required to discard a first flush volume. The volume of rainwater that should be flushed ($VF_{t,g}$) ensures that a prosumer agent flushes a volume equal to $F \times A$ after the rain event begins, where F is the required first flush rate, and A is the roof area. The agent can flush the total volume ($F \times A$) over multiple time steps, if needed.

$$VF_{t,g} = \begin{cases} min(S_{t,g}, F \times A - \sum_{t=TR}^{t_E} VF_{t,g}), & \text{if } \sum_{t=TR}^{t_E-1} VF_{t,g} < F \times A \\ 0, & \text{otherwise} \end{cases} \tag{4}$$

Step 3. Consumer agents exert irrigation demands. If no precipitation fell in the previous 24 h ($BP_{24} = 0$), each consumer agent c exerts a daily irrigation demand (DI) at time step TI_c. The hourly demand exerted by each consumer agent c is assigned using Equation (5):

$$D_{t,c} = \begin{cases} DI, & \text{if } t = TI_c \ \& \ BP_{24} = 0 \\ 0, & \text{otherwise} \end{cases} \tag{5}$$

The value of DI is calculated to initialize the model, as described in Section 3.4.1.

Step 4. Prosumer and consumer agents trade rainwater. Each consumer agent with non-zero demand at time step t is randomly paired with a prosumer agent with $S_{t,g} > 0$. A consumer agent c receives traded water ($TW_{t,c,g}$) from prosumer agent g up to its demand, $D_{t,c}$. If the consumer agent has a non-zero volume of unmet demand, it is randomly matched with other prosumer agents until the total volume of traded water it receives is equal to $D_{t,c}$ or until no prosumers have stored rainwater. For time steps when prosumers cannot meet consumer demands, consumer demands are met using water that was reclaimed through the centralized treatment plant. The flow ($Q_{t,g}$) that a prosumer pumps into the network at each time step is the sum of traded water ($TW_{t,c,g}$) that is purchased by consumer agents.

$$Q_{t,g} = \sum_{c=1}^{C_{t,g}} TW_{t,c,g} \tag{6}$$

where $C_{t,g}$ is the number of consumers that prosumer g supplies at time step t. The total volume of water purchased by consumer agent c is supplemented by flows from the centralized system ($CQ_{t,c}$) at time step t to meet its demand:

$$\sum_{g=1}^{G_{t,c}} TW_{t,c,g} + CQ_{t,c} = D_{t,c} \tag{7}$$

where $G_{t,c}$ is the number of prosumers that sell water to consumer agent c at time step t.

Step 5. Increase time step. In this step, $t_E = t_E + 1$. The agent-based model is executed to simulate trades for a total of T time steps to simulate a one-month period. If the simulation time is reached (e.g., $t_E = T$) go to Step 6. Otherwise, go to Step 1.

Step 6. Execute hydraulic simulation model. The dataset of negative demands ($Q_{t,g}$ for all prosumers) and positive demands ($D_{t,c}$ for all consumers) are used as input for the hydraulic simulation model. Section 3.4.2 details the method for running the hydraulic simulation model.

Step 7. Calculate hydraulic effects and energy consumption for the infrastructure system. Methods for calculating energy consumption and water age are described in Sections 3.4.3 and 3.4.4, respectively.

3.2. Design Concepts

3.2.1. Decision-Making

Agents use simple heuristics to make decisions. Prosumer agents are uniformly sampled to meet consumer water demands until consumer water demands are met or no prosumers have rainwater remaining in their tanks. Consumer agents do not use information about network location, amount of available water, or cost associated with purchasing water to select a prosumer agent for trading. Prosumers are simple reactive agents and release water when matched with consumer agents.

3.2.2. Stochasticity

There is little stochasticity in consumer and prosumer behaviors. Households in the network are randomly assigned as consumer and prosumer agents, and consumer agents select among prosumers with uniform probability to buy water.

3.2.3. Sensing

Consumer agents know the volume of water that is stored by each prosumer agent, and both consumer and prosumer agents have exact information about precipitation depths.

3.2.4. Interaction

Consumer and prosumer agents exchange water directly. Trades are not constrained by spatial location, and any prosumer agent can trade with any consumer agent. Consumer agents do not interact with other consumer agents, and prosumer agents do not interact with other prosumer agents.

3.3. Details: Initialization, Input, and Implementation

The agent-based model is initialized with 2016 households, using a pre-specified ratio of prosumers to consumers. Parameter values are specified in Table 1. Values for the first flush volume as reported in the literature (e.g., Gikas and Tsihrintzis [83]), and rainwater harvesting regulations [84,85], vary in the range of 0.11–1.02 L/m^2. The value used in the prosumer model represents a conservative estimate of first flush. All tanks are empty at the beginning of the simulation, and the number of hours since the previous flush is set to 24, which forces prosumer agents to flush tanks before beginning trades. Each prosumer agent is assigned a household pump model, with a pump efficiency of 75%. Consumer households are each assigned a time of day to irrigate their lawns (TI_c), derived from the diurnal irrigation pattern for dual-reticulated systems reported by Willis et al. [80]. For each discrete value of TI_c, Figure 3 specifies the number of consumer agents that are randomly selected from the pool of consumer agents without replacement and assigned the selected value for TI_c. For example, if the consumer agent pool consists of 100 agents, then $TI_c = 19$ for 21 randomly selected agents. The agent-based model requires hourly precipitation data as input, which are needed to calculate demands and trades at hourly intervals.

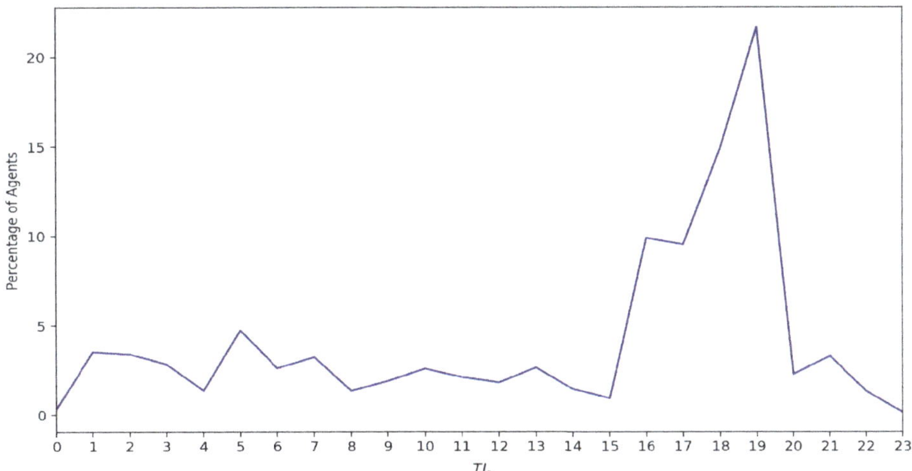

Figure 3. Consumer agents are assigned a value for TI_c using the distribution of values shown here.

The agent-based model is implemented in Multi-Agent Simulator Of Neighborhoods (MASON) [86], a Java-based discrete-event multi-agent simulation library. The code is published by Ramsey [87]. The output from MASON was used to create input for the hydraulic simulation submodel, which is described below.

3.4. Details: Submodels

3.4.1. Consumer Daily Irrigation Demand Submodel

The consumer daily demand volume, DI, is calculated using the outdoor water demand model [88].

$$DI = \frac{f \times L \times ((k \times ET) - r)}{days} \tag{8}$$

where f is an irrigation factor indicating frequency of watering; L is the irrigable lawn area (m^2); k is a crop coefficient; ET is evapotranspiration (mm/month); r is effective rainfall (mm/month); $days$ is the number of days per month. Effective rainfall represents the precipitation that penetrates the soil and thereby reduces the water demand of plants. It is calculated as a function of total measured monthly rainfall P_{month} (mm/month) [88], as:

$$r = \begin{cases} P_{month} & \text{if } P_{month} < 25 \text{ mm} \\ 0.504 \times P_{month} + 12.4 & \text{if } 25 \leq P_{month} \leq 152 \text{ mm} \\ 89.0 & \text{if } P_{month} > 152 \text{ mm} \end{cases} \tag{9}$$

The monthly demand value is converted and reported as a daily demand. The irrigation factor (f) is set at 1.0, because it is assumed all households that opt to connect to the system are frequent irrigators. The crop coefficient (k) is set as 0.7 to represent lawn. The value of L (irrigable lawn area) is calculated as

$$L = (\frac{1}{\rho} - A) \times U \tag{10}$$

where the household density is ρ (unit per m^2), roof area is represented as A (m^2), and the ratio of unpaved land is U (dimensionless).

3.4.2. Hydraulic Simulation Submodel

The pipe network is simulated using EPANET, which is a software application that calculates the movement and fate of drinking water constituents within water distribution systems [89]. Each household (consumer or prosumer agent) is represented in the network using three nodes: one node represents the street-level metered connection to the non-potable water network; a second node represents the irrigation demand node; a third node represents the negative demand node that allows a household to contribute rainwater to the network. The negative demand node represents an onsite rainwater harvesting tank and a pump that is used to put rainwater back into the network. The dataset of negative demands, or positive flows into the network $(-Q_{p,t})$, are placed at negative demand nodes, corresponding to each prosumer and time step. The dataset of positive demands $(D_{c,t})$ are placed at irrigation demand nodes, corresponding to each consumer and time step. Demands are used to modify the EPANET input file, and the hydraulic model is run for a one-month period to calculate network flows and pressure values.

3.4.3. Energy Consumption Submodel

Energy requirements for the water infrastructure network are based on three energy components: energy consumed by prosumers to pump rainwater into the network ($E_{prosumers}$), energy used to pump water from the centralized treatment plant (E_{system}), and energy required to treat wastewater (E_{treat}). The total energy required by the system (E_{total}) is the sum of the three components. Energy is reported in kilowatt hours (kWh).

The energy consumed by the prosumers to pump water into the network is calculated as:

$$E_{prosumers} = \sum_{g=1}^{G} \sum_{t=1}^{T} \gamma \times Q_{t,g} \times h_{t,g} \times \Delta_t \tag{11}$$

where γ is the specific weight of water (kN/m^3); $Q_{t,g}$ is the flow rate of pumped water from prosumer g, as defined above (m^3/s); $h_{t,g}$ is the pressure head at the negative demand node for prosumer g at time step t, which is the head required by the pump (m); G is the number of prosumers in the system; T is the number of simulated time steps; Δ_t is the time step, or one hour in this application.

The energy consumed to pump water from the centralized system is calculated as:

$$E_{system} = \sum_{t=1}^{T} \gamma \times Q_{S,t} \times H_t \times \Delta_t \tag{12}$$

where $Q_{S,t}$ is the flow rate of water pumped from the reservoir to the system (m^3/s); H_t corresponds to the head (m) gained by the pump at time step t.

The volume of reclaimed water that is offset by rainwater contributions can result in energy savings through a reduction in the volume of water that must be treated. Treating water to high standards is energy intensive, and supplementing the reclaimed water network with rainwater decreases the volume of wastewater that water treatment facilities need to process for household consumption. The energy required to treat wastewater can vary based on influent water quality, facility hydraulics, and treatment processes employed, and a value of 0.343 kWh/m^3 is adopted in this study [25]. The energy required to treat wastewater is calculated as:

$$E_{treat} = e_{treat} \times V_O \tag{13}$$

where V_O is the total volume of treated water that is pumped into the system over the simulated time; e_{treat} is the unit energy required to treat wastewater (0.343 kWh/m^3).

3.4.4. Water Age Submodel

The water age of the system is a surrogate metric for water quality [90]. Water age is calculated using water quality calculations in EPANET, which are executed at small time steps to reduce error. The weighted water age is calculated using Equation (14) at consumer nodes across the network.

$$WA_S = \frac{\sum_{c=1}^{C} \sum_{t=1}^{T_{WA}} b_{c,t} \times D_{c,t} \times (WA_{c,t} - WA_{lim})}{\sum_{c=1}^{C} \sum_{t=1}^{T_{WA}} D_{c,t}} \tag{14}$$

where WA_S is the weighted average water age above the limit for the system (hours). The acceptable limit for water age (WA_{lim}) is 48 h [90]. $WA_{c,t}$ represents the water age at consumer node c and time step t, reported in hours. The binary variable $b_{c,t}$ represents if the water age at node c and time step t exceeds the limit, where $k_{c,t} = 1$ if the water age is greater than the limit, and $k_{c,t} = 0$ otherwise. The time step for calculating water quality is 15 min, and the total number of time steps (T_{WA}) is 2880 for simulation of a 30-day month.

4. Virtual Network: Wolfpack City

"Wolfpack City" was developed as a virtual non-potable water distribution network (Figure 4) with realistic hydraulic design parameters and is used in this research to simulate a micro-trading program. It is assumed that each household receives potable water to meet high-quality end uses via a separate potable water system that is not modeled in this framework; simulations for Wolfpack City are specifically for non-potable water supply and demand. Wolfpack City represents a population of 2016 households, which exert irrigation demands based on rainfall and evapotranspiration. One demand is exerted as a constant flow (0.0078 m^3/s or 125 GPM) to represent an industrial demand, such as a cooling process.

4.1. Non-Potable Network System

Household elevations in Wolfpack City range from 282 to 312 m (Figure 4). The source represented by a reservoir with a head of 297 m is a reclaimed water treatment plant and pumps water to the network using a set of pumps and a tank. The pump station includes a main pump and eight additional parallel pumps that are controlled by the water level of the tank. The main pump delivers up to 0.0227 m^3/s and a gained head of around 56 m. The parallel pumps operate to meet the intermittent demand exerted by consumers and deliver up to 0.050 m^3/s each. The pump efficiency is simulated as 75%. The tank is initialized at full capacity.

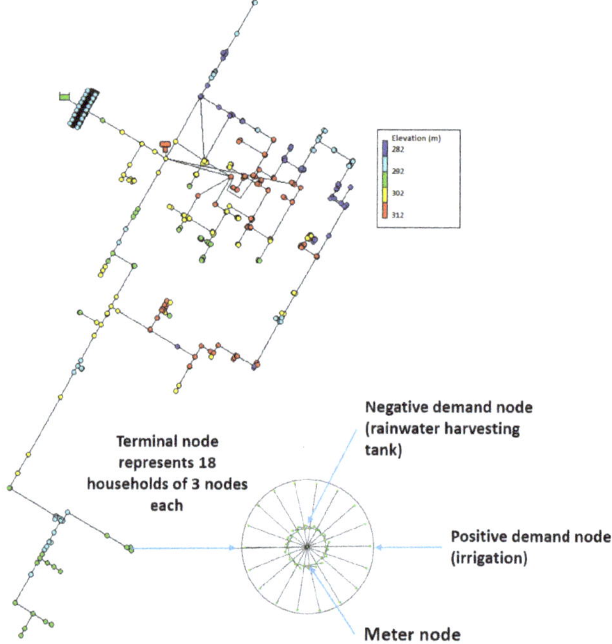

Figure 4. Wolfpack City water model. Each terminal node represents 18 households, which are represented using three nodes each: a meter node, positive demand node, and negative demand node.

4.2. Climate Data

Local climate data are needed to initialize rainwater harvesting tank storage and irrigation demand values for Wolfpack City. We selected the location of Wolfpack City based on the maximum potential rainwater yield at various locations across the U.S. using Equation (15) [91]:

$$Y = C \times LA \times RT \times P_{ann} \tag{15}$$

where Y is maximum potential rainwater harvesting yield; C is a runoff coefficient, assigned a value of 0.75 [91]; LA is land area; RT is the percentage of land cover which is rooftop; P_{ann} is average annual precipitation. The percentage rooftop is calculated based on land area, housing density, and average roof size [81,92]. We estimated the potential rainwater harvesting yield at 10 locations that are spread across regions of the U.S. using publicly available land cover data [92] and thirty year precipitation averages [93]. Cities that were included in the analysis are Baltimore, Maryland; Branson, Missouri; Dallas, Texas; Denver, Colorado; Fargo, North Dakota; Phoenix, Arizona; Raleigh, North Carolina; San Diego, California; Seattle, Washington. Cities that report the highest value for Y are Seattle (24.1

million m^3), Dallas (13.5 million m^3), and Phoenix, Arizona (12.2 million m^3).The values for each city are shown in Appendix A (Table A1).

Seattle, Washington, was selected as a climate region for Wolfpack City, and observations of precipitation and evapotranspiration in April 2020 are used to create climate scenarios. The data used in this study were recorded at USGS Station 12113346 for Springbook Creek at Orillia, WA [92]. Over April 2020, evapotranspiration was recorded as 281.25 mm/month. Seattle's household density (ρ) is 721 housing units/km^2, and other parameters needed for the agent-based model, such as roof size and ratio of unpaved land to total land area, are determined using national averages (Table 1) [81,92]. Using Equation (10), consumer irrigable lawn area (L_I) is calculated as 494.9 m^2.

4.3. Modeling Scenarios

A set of scenarios are developed to explore the performance of the smart water grid for variations in model characteristics. This analysis explores changes in the number of households that join the market as prosumers to assess how many prosumers are needed to meet demands exerted by consumers. The participation of households as prosumers varies from 0% to 100% of the total number of households by increments of 5%. Households are randomly assigned as prosumer and consumer agents to meet the scenario definition. This analysis also explores changes in precipitation, as precipitation is expected to affect the amount of rainwater traded. Higher depths of rainfall and more frequent precipitation events lead to greater volumes of water stored, but reduce the amount of water needed for irrigation, as consumers do not exert demands immediately following precipitation events. Precipitation scenarios are explored where the depth of rainfall at each hourly time step is multiplied by a factor of 0.5, 1.0, 2.0, 3.0, 4.0, and 5.0. The total number of time steps with precipitation greater than zero is held constant to maintain consistency across simulations; for example, if 5.0 mm is recorded during the first hour of Seattle's April 2020 precipitation data, then a total of 15.0 mm falls during that time step in a scenario with a precipitation factor of $3P$. Evapotranspiration (ET) values are held constant throughout all scenarios. According to the Blaney–Criddle equation, ET is correlated with temperature, wind, and daylight hours, which we assume as constant across all precipitation scenarios [94]. A total of 126 scenarios (six precipitation settings and 20 prosumer settings) are generated, and scenarios are labeled as the percentage of households that participate as prosumers in the market and the factor used to adjust rainfall depths. Each scenario was simulated over the one-month horizon using MASON, which required approximately 40 s to run using a 3.1 GHz Dual-Core Intel Core i5. Output from the agent-based model was used to create the input file depicting the demands of that scenario for EPANET, which required approximately 6.7 min when run using a 2.9 GHz 4-Core Intel Xeon W-2102.

5. Results

The results presented below demonstrate the dynamics of water storage, water consumption, energy consumption, and hydraulic performance for an example scenario. Subsequently, the performance of the ABM and the network across all scenarios are reported and explored.

5.1. Scenario 20% − 1P

Scenario 20% − 1P simulates rainfall using Seattle's April 2020 precipitation data (1P), and 20% of agents (403 of 2016 total agents) are initialized as prosumers. The time series plots of precipitation, irrigation demands and volume of traded rainwater pumped into the network are shown in Figure 5. The simulated horizon includes six distinct rainfall events of varying volumes, followed by a reduction in immediate irrigation demands. The reduction in irrigation demands represents that consumers do not exert irrigation demands within 24 h of a rainfall event. Traded rainwater injections into the reclaimed water network spike after this 24 h period. The highest peaks in irrigation demand each day correspond with the irrigation patterns shown in Figure 3, peaking at hour 19 (or 7:00 p.m.) each day. A total of 525 m^3 of harvested rainwater are pumped into the system during the simulation, which is the equivalent of the daily water demand exerted by 141 consumer households (8.7% of consumers). The volume of rainwater that is traded is a small fraction (less than 1%) of the total volume of water consumed, as shown in Table 3. The flows of water produced at the treatment plant and consumed at nodes for Scenario $20\% - 1P$ are simulated using EPANET (Figure 6), demonstrating that the network satisfies exerted demands without significant excess production; the difference in water produced and water consumed is 7% of the water consumed (Table 3). For this scenario, six of the nine pumps are turned on to meet demands. The centralized system maintains a minimum production volume when consumer demands are zero to meet the constant demand of 0.0079 m^3/s (or 20,477 m^3 over the one-month period).

The energy consumed for system-level pumping, reclaimed water treatment, and prosumer pumping for Scenario $20\% - 1P$ are calculated and compared with the energy required for Scenario $0\% - 1P$ (no prosumers) in Table 3 to allow for the examination of energy savings between scenarios with and without rainwater trading. The energy consumed by prosumers for household-level pumping increases by 41 kWh when 20% of households function as prosumers, compared with Scenario $0\% - 1P$. Residential demands are 20% lower in Scenario $20\% - 1P$, due to the number of consumers that switch to prosumers. The increase in energy for household pumping corresponds to a decrease in energy consumed by system-level pumping and treatment. There is a decrease of 2409 kWh in energy consumed by system-level pumps for Scenario $20\% - 1P$, and a total reduction in energy consumption of 11% when compared with Scenario $0\% - 1P$. Unit energy consumed is calculated as the total energy required per unit volume of water produced by both centralized and decentralized processes (Table 3). Unit energy provides an assessment of the energy efficiency of the system. The unit energy for Scenarios $20\% - 1P$ and $0\% - 1P$ are the same value (0.37 kWh/m^3), and the the addition of 403 prosumers to the network does not impact energy efficiency of meeting demands.

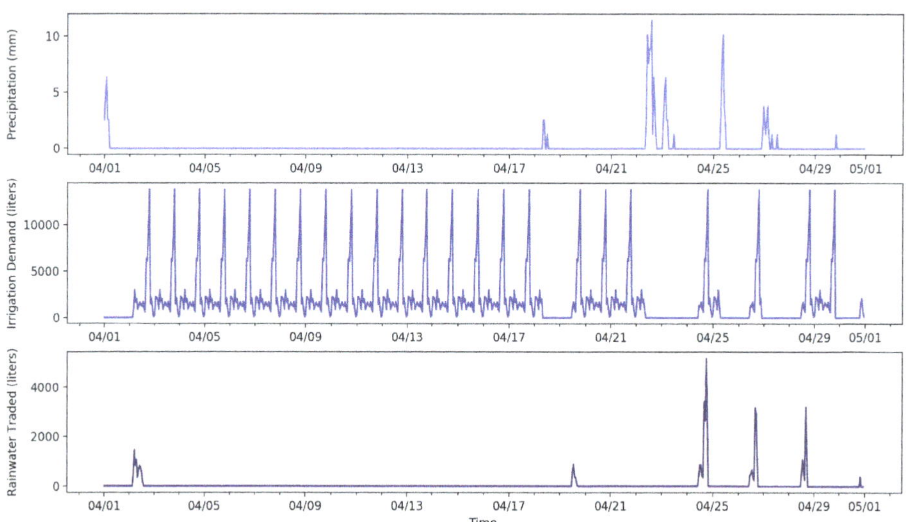

Figure 5. Precipitation, irrigation demand volume, and traded rainwater volume for Scenario $20\% - 1P$.

Pressure is affected by reductions in consumer demands and the injection of water at terminal nodes through household-level pumping. The maximum pressure in the network occurs when no demands are exerted and is the same value for Scenarios $0\% - 1P$ and $20\% - 1P$. As shown in Table 3, the minimum pressure when 20% of households are prosumers is slightly increased, compared to Scenario $0\% - 1P$. The distribution of pressures at a time step with low pressures is shown in Figure 7. The time step shown is 9 p.m. on a day that is preceded by 24 h without rainfall, and pressures remain at or near the minimum (7 m). In the central area of the city, elevations cover a 30 m range, and the pressures are around 20 m at nodes that are located at high elevations. The southern section of the network is at a lower elevation, and pressures remain in the range of 25–35 m during this period of relatively low pressure.

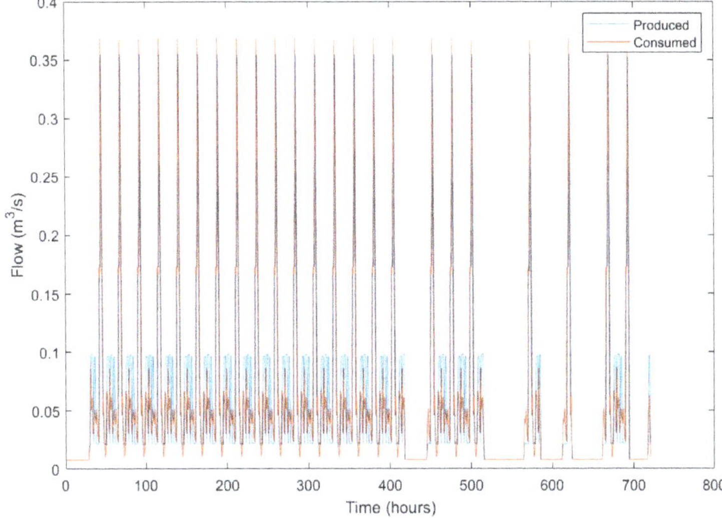

Figure 6. EPANET output reports flow of water produced by the treatment facility and flow of water consumed at nodes for Scenario $20\% - 1P$.

Table 3. Metrics reported for Scenarios $0\% - 1P$ and $20\% - 1P$.

	Scenario $0\% - 1P$	Scenario $20\% - 1P$
Volume of water consumed (m^3)	187,679	154,269
Volume of water produced (m^3)	183,895	164,015
Volume of traded rainwater (m^3)	0.00	525.06
Energy consumed by prosumer pumping ($E_{prosumers}$) (kWh)	0.00	40.97
Energy consumed by system-level pumping (E_{system}) (kWh)	20,836	18,427
Energy consumed by treatment (E_{treat}) (kWh)	47,307	42,193
Total energy consumed (E_{total}) (kWh)	68,143	60,661
Unit energy consumption (kWh/m^3)	0.37	0.37
Water age (WA_S) (h)	20.15	19.23
Minimum pressure (m)	3.00	6.30
Maximum pressure (m)	70.71	70.71

Water age is calculated for Scenarios $20\% - 1P$ and $0\% - 1P$ using Equation (14), calculated over all irrigation nodes, and is reported to explore water quality. It is expected that the water age of Scenario $20\% - 1P$ would be higher than the water age of Scenario $0\% - 1P$, because the injection of water at households may increase the residence time of water in the system, and there are fewer agents consuming water in the network. The water age of Scenario $20\% - 1P$ is marginally less than the water age of Scenario $0\% - 1P$ (a difference of less than 1 h). This difference may be due to the process of calculation. The water age is calculated only at consumer nodes, and there are fewer households acting as consumers in Scenario $20\% - 1P$, leading to a marginal reduction in water age. In addition, the age of water entering the system due to household-level pumping is initialized at zero hours, which does not account for the time that the water resides in the household-level rainwater harvesting tank.

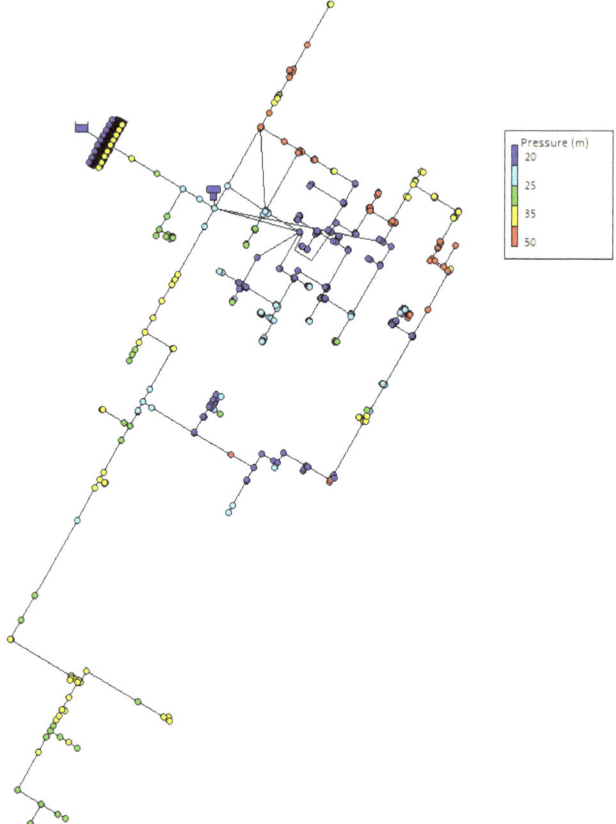

Figure 7. Lowest nodal pressures for Scenario 20% − 1*P* at 9:00 p.m. on a day with no rainfall in the previous 24 h.

5.2. Performance Analysis across All Scenarios

The total volume of traded rainwater, total number of trades, and percentage of irrigation demand met for each of the 126 scenarios (six levels of precipitation and 21 levels of prosumers) are shown in Figure 8. For all scenarios where no agents are prosumers (0%) or all households are prosumers (100%), no trades occur because of the homogeneity of agents. For all other scenarios, an increase in rainfall volume corresponds to an increase in volumes of trades (Figure 8a) and number of trades (Figure 8b). During higher precipitation volume scenarios, prosumer agents can harvest higher volumes of rainwater, which allows some consumer agents to satisfy demands through trading. For higher precipitation, the peaks in volume and number of trades correspond with lower percentages of prosumer agents. During scenarios with precipitation volume 1*P*, for example, the maximum volume of traded rainwater corresponds with a prosumer ratio of 80%, compared to 50% for precipitation volume 4*P*. This is because for lower rainfall depths, a higher number of prosumers are needed to participate in the market to offset the demand exerted by a consumer. The highest number of trades occur for scenarios with precipitation volume 1*P*. Consumer agents buy rainwater from multiple prosumers at each time step to meet demands, leading to a high number of trades (Figure 8b). At low precipitation depth, 0.5*P*, the number of trades is relatively low, because prosumer agents are unable to harvest higher volumes of rainwater, resulting in reduced ability to trade. The volume of rainwater that is traded decreases for higher numbers of prosumer agents because fewer agents are consumers to exert demands. The system does not meet total irrigation demands for any of the 126 scenarios (Figure 8c), however, because irrigation demands of consumers are substantially higher than harvested rainwater volumes. Across all precipitation volumes, the highest percentage of demand that is met through trading occurs when 95% of agents are prosumers.

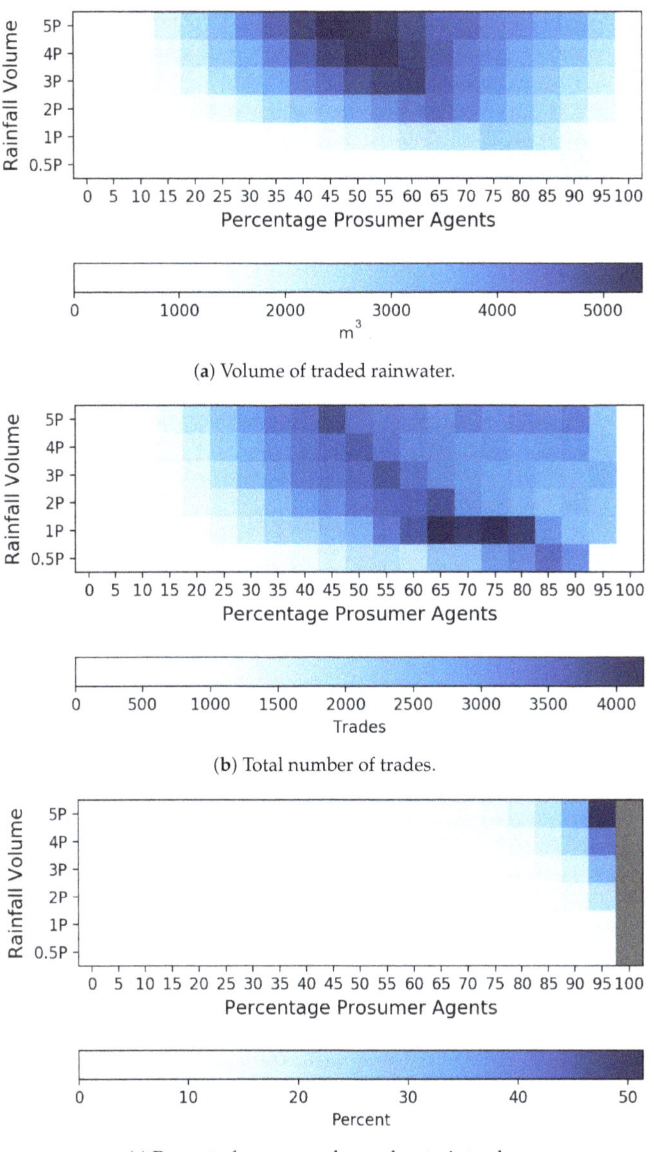

(**a**) Volume of traded rainwater.

(**b**) Total number of trades.

(**c**) Percent of consumer demand met via trades.

Figure 8. Total volume of traded rainwater (m³), total number of trades, and total percentage of irrigation demand satisfied by traded rainwater across 126 scenarios of varying rainfall depth and percentage of prosumer agents.

EPANET simulations were used to evaluate energy consumption, pressure, and water age. The amount of energy consumed by household-level pumping (Figure 9a) follows the same pattern as the volume of traded rainwater (Figure 8a) with the highest consumption of energy at high rainfall depths and around 50% prosumers. Scenarios with lower percentages of prosumer agents and low rainfall volumes require more system-level energy consumption (Figure 9b). System-level energy is orders of magnitude greater than energy consumed by household-level pumping, and there is a large reduction in system-level energy requirements for higher numbers of prosumers. This trend emerges because prosumer agents do not irrigate, which reduces the energy requirements of pumping from decentralized locations. Energy consumed for treatment is high for this system, approximately twice the energy required for pumping. Treating wastewater to non-potable standards is an energy-intensive activity, and these numbers reflect that cost. In this research, we assume that the volume of wastewater is treated to match the unmet demands in the systems, and the energy cost of treating excess wastewater that is not needed for reuse is not included in this framework. However, wastewater that is released to the environment or used for other recycling purposes would also need to be treated, and a holistic assessment of the interconnections between water, wastewater, and reclaimed water systems may use an alternative approach to holistically account for energy costs. Based on the sum of energy requirements for the smart water grid system (Figure 9d), a higher number of prosumers leads to higher savings in energy.

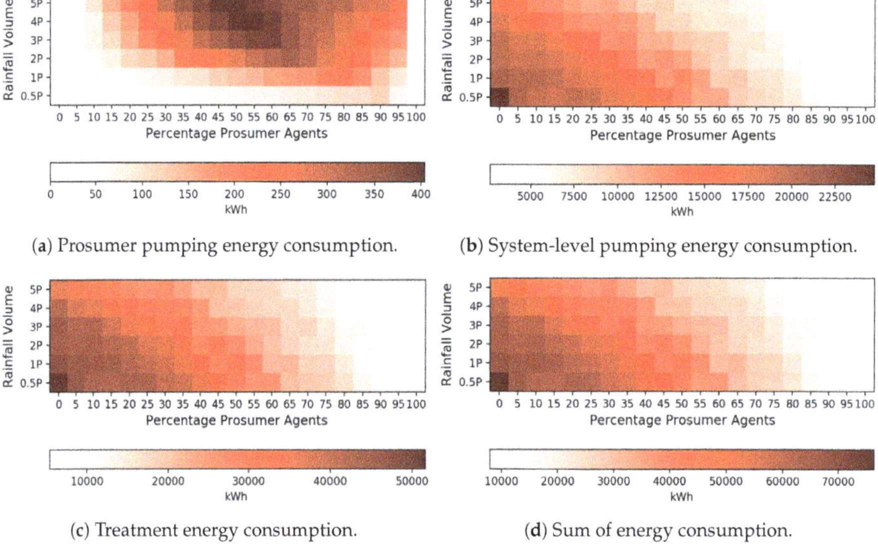

Figure 9. Energy consumption for (**a**) prosumers for pumping water at households, (**b**) system-level pumping, (**c**) treatment of water at the treatment plant, and (**d**) the sum of energy consumption for prosumers, system-level pumping, and water treatment.

The system's unit energy is calculated as the energy consumed by three processes (household-level pumping, system-level pumping, and treatment) per unit volume of water produced by the centralized system and prosumer pumping (Figure 10). Unit energy is used to represent the efficiency of the system in meeting demands. For scenarios of high percentages of prosumers and high precipitation depths, the water contributed by prosumers reduces the amount of water required by the centralized system, leading to higher energy efficiency. For example, unit energy generally increases across 1*P* scenarios, which corresponds with the reduction in treatment and pumping energy shown in

Figure 9b,c. Household-level pumps inject water in the network at peak demand times, which offsets the need to use the additional pumps that operate in parallel to the main pump. As a result, the number of pumps that operate to provide water from the centralized system decreases with increasing numbers of prosumers (nine pumps are needed for 0–5% prosumers; seven pumps: 10–15%; six pumps: 20–25%; five pumps: 30–40%; four pumps: 45–60%; three pumps: 65–80%; two pumps: 85–95%; one pump: 100%). Further, pumping water from terminal nodes in the network requires less energy than pumping from the centralized treatment plant, as there is less head loss to overcome when the water is pumped from near-by terminal nodes.

Unit energy does not increase monotonically across the $1P$ scenarios, however, because of the infrastructure complexities of water production. For some $1P$ scenarios, water production exceeds demands on the centralized system by up to 18%, while in others, water production drops to 91% of the demand exerted on the centralized system (that is, total demand minus the demands that are met by household-level pumping). In cases where the water demanded exceeds the water produced by the centralized system, the water storage tanks meet the remaining demand because they are initialized at full capacity. At 80% prosumers, the system reaches a minimum unit energy, and, subsequently, the energy required per unit volume increases with increasing percentages of prosumers. This trend mirrors the change in volume of traded rainwater across scenarios of increasing numbers of prosumers, shown in Figure 8a, which reaches a peak at 80% prosumers for the $1P$ scenario. When consumers comprise less than 20% of agents, they demand less water than prosumers produce when tanks are full. Because low volumes of water are traded and the bulk of demand is at the constant demand, which is met by the centralized system, the system requires higher unit energy. Higher precipitation depths ($2P$, $3P$, $4P$, and $5P$ scenarios) lead to improved energy efficiency because prosumers can produce more water to offset demands. Increasing the volume of water that can be provided by prosumers increases benefits to both water and energy savings through the smart water grid.

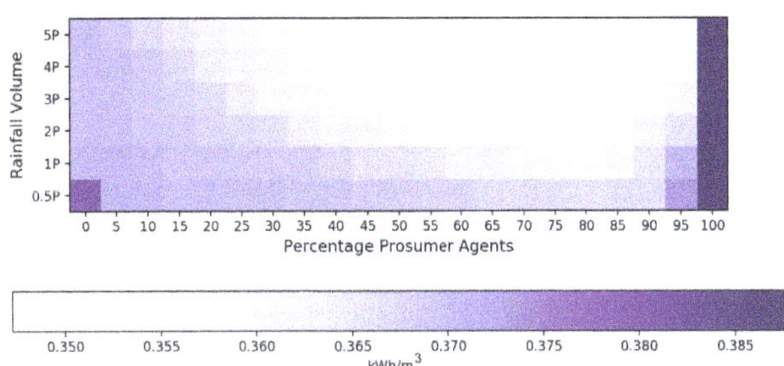

Figure 10. Energy consumed per unit volume of water produced. Energy is calculated as the sum of all three energy expenditures: household-level pumping, centralized pumping, and treatment. Produced water is calculated as the sum of water produced by the centralized system and the water produced by prosumer agents.

The minimum and maximum pressure across all consumer irrigation nodes and over all time steps are reported for each scenario. The minimum pressure across the scenarios varies between 5.0 and 30 m of head (Figure 11), while the maximum pressure is approximately the same at around 50 m for all scenarios. The minimum pressure is low at high numbers of consumers (low percentage of prosumers). Scenarios where the minimum pressure values fall below 7 m (approximately 10 psi) may be considered as infeasible because the pressure is not high enough to meet irrigation purposes. For $1P$ rainfall scenarios, scenarios are infeasible until the percentage of prosumer agents reaches around 30%. When the number of prosumers is lower than 30%, consumers require a large volume of water from the centralized system, and pressures in the central part of the system drop to values less than 7 m. With higher precipitation depths, the amount of water that is provided by prosumer agents increases, resulting in relatively higher minimum pressure values. For higher precipitation scenarios, prosumers contribute high volumes of water, leading to increasing pressure. For example, for $5P$ rainfall scenarios, approximately 15% of agents need to be prosumers to ensure a feasible system.

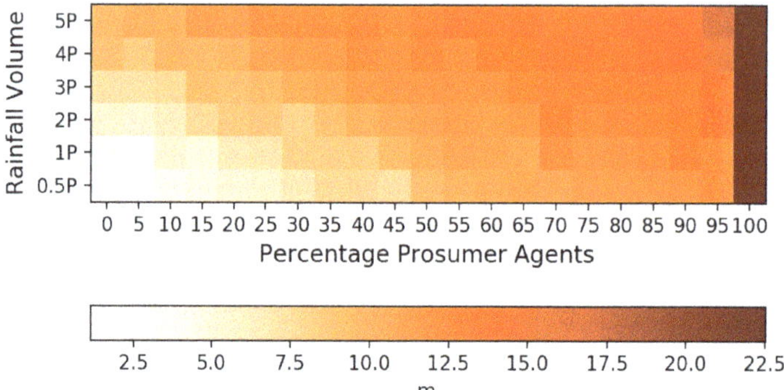

Figure 11. Minimum head (m) across consumer nodes for all scenarios.

As the number of prosumers increases, there are benefits in energy savings and meeting pressure requirements, as shown above. There is, however, an expected drop in water quality based on the residence time of water in the network. As the system has intermittent consumption for irrigation purposes, water age increases in scenarios with high precipitation values and high percentages of prosumers (Figure 12). Water age values represent the average number of hours exceeding the water age requirement of 48 h, and this number is between 20 and 30 h when the percentage of prosumers is less than 80%. The effect on water age does not grow significantly until the percentage of prosumers reaches 80%, and water age increases dramatically with additional prosumers. Tradeoffs among water age, energy consumption, and pressure may govern how a water micro-trading market should be designed. Results show for $1P$ scenarios, that water saving is at a maximum at 80%; unit energy is minimum at 80%; pressure requirements constrain the network to function only if 30% or more households join as prosumers; water age requirements may constrain the percentage of prosumers to less than 80%.

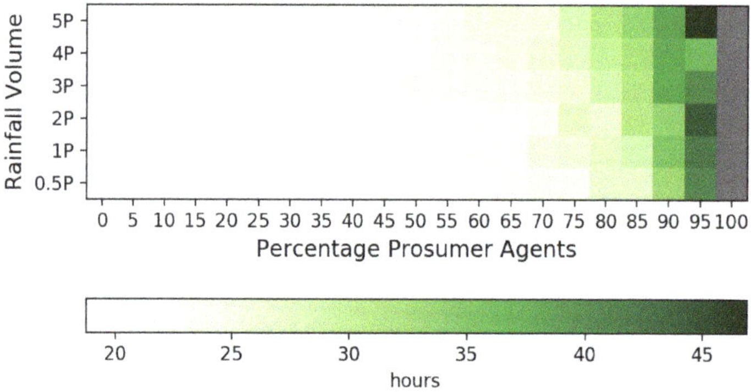

Figure 12. Water age calculated for all scenarios using Equation (14).

6. Discussion

This study proposes and tests the hydraulic feasibility of a smart water grid for micro-trading rainwater through a peer-to-peer non-potable water market that allows residential households to capture, use, sell, and buy rainwater within a network of water users. In this research, we explore the impact of the depth of rainfall and the distribution of consumers and prosumers on the performance metrics that are used to evaluate the hydraulic feasibility of the smart water grid. In related research, an agent-based modeling framework was developed to explore how the numbers of prosumers and consumers affect performance of a peer-to-peer household-level energy trading market, and results demonstrated that the presence of too many prosumers in the market led to market inefficiencies [8]. In the smart water grid, however, the volume of water that is required by consumers for irrigation is much higher than the volume of water that is produced by prosumers. Prosumers could not completely satisfy consumer demands for any of the simulated scenarios, and production from the centralized system was required to meet demands.

The volume of water that is produced by prosumers drives the performance of the smart water grid, with respect to both water and energy savings. The total volume of traded water increases and the energy required to pump demands decreases for scenarios with higher numbers of prosumers and higher rainfall depths, leading to a more energy efficient system. The energy requirements at households to pump water from rainwater tanks is lower than the energy required at the system level to pump the same volume of water from the water treatment plant. Pumping water from the central treatment plant requires the operation of extra parallel pumps, which are not needed when prosumers contribute water to the system. In addition, water that is pumped into the network at terminal nodes by prosumers does not need to overcome headlosses or elevation losses when consumer nodes are located nearby. Energy savings are also associated with treating smaller volumes of wastewater to nonpotable standards. In this model, the efficiencies of household pumps and systems were assumed to be equal (75%). Household pumps, however, may more realistically have lower efficiency than large pumps, which would change the analysis of energy consumption. The model developed in this research does not account for energy requirements of onsite treatment that could be required at prosumer households to treat rainwater. It is expected, however, that treatment of rainwater to meet nonpotable standards would require much less energy than treatment of wastewater. Pressure and water quality constraints are also explored in this research, as they are affected by increased trading and show tradeoffs based on the number of prosumers. For the simulations conducted in this research, consumer and prosumer agents are assigned randomly at nodes across the network, and the results are specific to one random realization. Further research can explore how clustering of consumer or prosumer agents at nodes in the network could affect pressure, energy, and water age through multiple realizations of initializing consumer and prosumer agents.

In this research, rainfall and evapotranspiration the data that are used to simulate demands are from the Seattle, WA, USA area, and we explore how higher precipitation can lead to a more efficient market. The modeling framework presented in this manuscript uses the theoretical outdoor water demand model, which may overestimate the amount of water required by households. Other climates may lead to differences in rainwater exchanges, and new methods for estimating irrigation may be needed to more accurately represent household behaviors.

6.1. Smart Technologies

The system that is conceptualized in this research would be possible through smart technologies that can record and account for water flows and peer-to-peer transactions in real-time. Smart water flow meters are needed to sense and record water contributed to and withdrawn from the water network at each household on sub-hourly time steps. AMI that includes smart water flow sensors can allow urban water managers to accurately and continuously account for non-potable water use and pumping [95]. Blockchain technologies can be applied to create a ledger to record transactions between peers, and smart contracts can be built on top of a distributed ledger to facilitate the settlement of water trades. Automated valves and pumps would be needed to automatically execute trades by releasing water at the household into the network. The integration of these technologies creates a smart water grid that enables new water and energy savings through decentralized water sources. These technologies were not modeled as functioning entities in this framework, and future research can explore how these technologies would be adopted and would function in a smart water grid.

Introducing this array of new technologies can create new vulnerabilities to failure that may occur due to internet disruption, power outages, and malfunctioning controllers. Fail-safe protocols and technological solutions are needed to account for loss of water or trades that are not fulfilled.

6.2. Semi-Centralized Infrastructure

The smart water grid is a semi-centralized system and relies on both decentralized and centralized infrastructure to provide diverse sources of non-potable water. A pre-existing dual reticulation system is required as part of the envisioned system to circulate non-potable water that is generated at lower water quality for end uses such as irrigation and flushing toilets. Constructing a secondary pipe network within an existing water supply system is typically cost-prohibitive, and implementation of the smart water grid may be better incorporated into new systems, such as the network that is conceptualized for Fisherman's Bend near Melbourne, Australia [59]. A smart water grid for micro-trading rainwater could also be implemented using a shared aquifer [49] or water trucks to provide conveyance of traded water. Micro-trading rainwater can also use a water rights structure, where households buy the rights to use rainwater from a community source. Rainwater harvesting provides a sustainable source of water by recycling runoff; however, rainwater is an unreliable source, and it is likely that rainwater would be unavailable at times when irrigation water is needed most, such as during droughts. By taking a semi-decentralized approach that incorporates rainwater tanks into a reclaimed water network, diverse sources are utilized to meet potable demands. Diverse portfolios of water sources can lead to reliable water supply systems [96], and the smart water grid would meet demands during low rainfall by circulating reclaimed water and reusing rainwater when it is available. Other benefits of decentralized water management associated with a smart water grid may include offsetting household energy costs [97] and reducing stormwater flows [37].

6.3. Peer-to-Peer Markets and Cost-Benefit Analysis

Criticisms of Haddad's early depiction of micro-trading [46] argued that households may not have the expertise, interest, or time to trade water [47]. Haddad argued that households regularly make complex decisions around finances and could readily bid on water prices [48]. A smart water grid would involve transaction costs and costs associated with rainwater tanks and smart meters, creating considerable economic barriers. Research around a peer-to-peer energy trading market found that households were engaged in bidding on energy, but they became disengaged due to the structure of transaction costs that created market inefficiency [8]. The buy-in for consumers to invest in rainwater cisterns and pumps could contribute to challenges in implementing a smart grid, as researchers have demonstrated that the payback period of rainwater harvesting systems alone can be 20–30 years [31,98]. Water resources are projected to become increasingly scarce [1], however, and a study conducted in India suggests that people, particularly in drought-prone areas, may be willing to invest in rainwater harvesting systems or other creative and environmentally-friendly water alternatives [99]. As described above, rainwater harvesting alone provides an intermittent source of water, and the cost-benefit analysis of purchasing a rainwater harvesting tank and pump to join a smart grid would be altered, because participants are granted access to continuous water supply. It is expected that micro-trading could function as an efficient market in a water-scarce urban environment. Further analysis is needed to explore the cost and benefits associated with the economics of the infrastructure and participation in the water market. The agent-based model can be extended in further research to capture economic decisions of households to join the market and to buy and sell water. New modeling mechanisms can be included in the framework to capture the interplay among demands, climate, trading adaptations, and infrastructure performance.

6.4. Water Quality

Water reuse programs have historically been challenged in garnering public support, due to the "yuck factor", or perception that treated wastewater is dirty or unsafe [100,101]. Water quality of water withdrawn through the smart water grid should be managed to mitigate public health risks and enhance positive perceptions of water quality. The quality of rainwater is generally accepted as high enough for irrigation and toilet flushing, especially when first flushing is used to remove contaminants that are washed off of roofs [38,83]. The quality of harvested water is expected to degrade, however, as the water moves through a pipe network and is stored in a tank. We include a first flush diversion that would improve the quality of water entering the network, but further research is needed to better represent the quality of water that is pumped from rainwater harvesting tanks and the fate and transport of contaminants in the network. Filtration and treatment systems can be installed at the point-of-entry, and research is needed to explore how to enhance household expertise in the operation and maintenance of treatment technology. For example, new research explores how real time control of the operation of biofilters can improve microbial removal from stormwater [102].

7. Conclusions

Smart water sensors can be integrated with water distribution infrastructure, distributed ledger technology, smart contracts, and automated control to support novel decentralized water markets to improve water savings in urban environments. This research demonstrates how a semi-decentralized water supply system can create water and energy savings by deploying existing smart city technologies and decentralized infrastructure within a centralized reclaimed water distribution system. In this research, we report the feasibility of a reclaimed water system that is augmented by prosumers, who pump harvested rainwater into the network at decentralized nodes. We construct an all-pipe hydraulic model for a hypothetical community to simulate demands exerted for non-potable uses at households and to evaluate hydraulics in the dual reticulation network. An agent-based modeling approach is developed to simulate household behaviors, including storing, pumping, trading, and withdrawing non-potable water. The agent-based model is loosely coupled with hydraulic modeling, and negative and positive demands of prosumers and consumers are used to modify input to the pipe network model. The simulation framework is applied using climate parameters for a location in the northwest U.S., which was selected based on the potential yield expected from rainwater harvesting, and a theoretical outdoor water demand model is used to simulate irrigation demands at households. Multiple scenarios are explored to demonstrate the feasibility of peer-to-peer non-potable water trading, and tradeoffs among the volume of traded water, energy savings, satisfaction of pressure constraints, and water quality are explored. The smart water grid is a complex system, and energy outcomes emerge based on reductions in the volume of water required by households, the volume of water provided by prosumers, and the dynamics of centralized water distribution infrastructure. The water and energy efficiency of scenarios depends on the reduction in the volume of water that is provided by the centralized system through traded rainwater. Higher volumes of water produced by prosumers increases both water and energy savings. Analysis demonstrates that there is a lower bound on the number of households that should participate as prosumers to meet pressure requirements, and an upper bound on prosumers to protect water quality. This research develops a novel water management system designed to further the use of decentralized infrastructure and smart city technologies in improving the sustainability of the built environment.

Author Contributions: Conceptualization: E.Z.B.; methodology: E.R., J.P., M.A.K.F., J.M.; formal analysis: E.R., E.Z.B., J.P.; data curation: M.D., E.R., J.P., M.A.K.F.; writing—original draft preparation: E.R., E.Z.B., M.D., M.A.K.F., J.P.; writing—review and editing: E.Z.B., E.R.; resources: E.Z.B., J.M.; supervision: E.Z.B. All authors have read and agreed to the published version of the manuscript.

Funding: This work was supported in part by North Carolina State University through a 2019 University of Adelaid—NC State Starter Grant.

Conflicts of Interest: The authors declare no conflicts of interest.

Appendix A

Table A1. Maximum potentialrainwater harvesting yield values for 10 U.S. cities.

City	State	LA (km^2)	RT (%)	P_{ann} (cm)	Y (million m^3)
Baltimore	Maryland	209.6	19.7%	29.1	12.0
Branson	Missouri	53.4	2.2%	28.3	0.3
Dallas	Texas	881.9	8.2%	18.7	13.5
Denver	Colorado	396.3	10.1%	29.4	11.7
Fargo	North Dakota	126.4	5.5%	13.7	1.0
Phoenix	Arizona	1338.2	6.1%	14.8	12.2
Raleigh	North Carolina	370.1	6.6%	29.4	7.2
San Diego	California	842.2	8.5%	8.8	6.3
Seattle	Washington	217.4	19.8%	56.0	24.1
Tulsa	Oklahoma	509.6	5.1%	27.7	7.2

References

1. United Nations. *UN-Water Annual Report 2015*; Technical Report; United Nations: Geneva, Switzerland, 2015.
2. Berglund, E.Z.; Monroe, J.G.; Ahmed, I.; Noghabaei, M.; Do, J.; Pesantez, J.E.; Fasaee, M.A.K.; Bardaka, E.; Han, K.; Proestos, G.T.; et al. Smart Infrastructure: A Vision for the Role of the Civil Engineering Profession in Smart Cities. *J. Infrastruct. Syst.* **2020**, *26*, 03120001. [CrossRef]
3. Giurco, D.P.; White, S.B.; Stewart, R.A. Smart Metering and Water End-Use Data: Conservation Benefits and Privacy Risks. *Water* **2018**, *2*, 461–467. [CrossRef]
4. Palleti, V.R.; Kurian, V.; Narasimhan, S.; Rengaswamy, R. Actuator network design to mitigate contamination effects in water distribution networks. *Comput. Chem. Eng.* **2018**, *108*, 194–205. [CrossRef]
5. Page, P.R.; Zulu, S.; Mothetha, M.L. Remote real-time pressure control via a variable speed pump in a specific water distribution system. *J. Water Supply Res. Technol* **2019**, *68*, 20–28. [CrossRef]
6. Munir, M.; Bajwa, I.; Cheema, S. An intelligent and secure smart watering system using fuzzy logic and blockchain. *Comput. Electr. Eng.* **2019**, *77*, 109–119. [CrossRef]
7. Hansen, P.; Morrison, G.; Zaman, A.; Liu, X. Smart technology needs smarter management: Disentangling the dynamics of digitalism in the governance of shared solar energy in Australia. *Energy Res. Soc. Sci.* **2020**, *60*, 1–13. [CrossRef]
8. Monroe, J.; Hansen, P.; Sorell, M.; Berglund, E.Z. Agent-Based Model of a Blockchain Enabled Peer-to-Peer Energy Market: Application for a Neighborhood Trial in Perth, Australia. *Smart Cities* **2020**, *3*, 1072–1099. [CrossRef]
9. Holland, J.H. *Hidden Order: How Adaptation Builds Complexity*; Basic Books: New York City, New York, USA, 1995.
10. Holland, J.H. *Emergence: From Chaos to Order*; OUP: Oxford, UK, 2000.
11. Berglund, E.Z. Using agent-based modeling for water resources planning and management. *J. Water Resour. Plan. Manag.* **2015**, *141*, 04015025. [CrossRef]
12. Zechman, E.M. Agent-based modeling to simulate contamination events and evaluate threat management strategies in water distribution systems. *Risk Anal.* **2011**, *31*, 758–772. [CrossRef] [PubMed]
13. Shafiee, M.E.; Zechman, E.M. An agent-based modeling framework for sociotechnical simulation of water distribution contamination events. *J. Hydroinf.* **2013**, *15*, 862. [CrossRef]
14. Kandiah, V.; Berglund, E.; Binder, A. Cellular Automata Modeling Framework for Urban Water Reuse Planning and Management. *J. Water Resour. Plan. Manag.* **2016**, *142*, 04016054. [CrossRef]
15. Zhang, Y.; Wu, Y.; Yu, H.; Dong, Z.; Zhang, B. Trade-offs in designing water pollution trading policy with multiple objectives: A case study in the Tai Lake basin, China. *Environ. Sci. Policy* **2013**, *33*, 295–307. [CrossRef]
16. Huang, H.; Ma, H. An agent-based model for an air emissions cap and trade program; A case study in Taiwan. *J. Environ. Manag.* **2016**, *183*, 613–621. [CrossRef]
17. Peng, Z.S.; Zhang, Y.L.; Shi, G.M.; Chen, X.H. Cost and effectiveness of emissions trading considering exchange rates based on an agent-based model analysis. *J. Clean. Prod.* **2019**, *219*, 75–85. [CrossRef]

18. Okun, D.A. Distributing reclaimed water through dual systems. *J. Am. Water Work. Assoc.* **1997**, *89*, 52–64. [CrossRef]

19. U.S. Environmental Protection Agency. *2012 Guidelines for Water Reuse*; Technical Report EPA/600/R-12/618; Office of Wastewater Management, Office of Water: Washington, DC, USA, 2012.

20. Grigg, N.; Rogers, P.; Edmiston, S. *Dual Water Systems: Characterization and Performance for Distribution of Reclaimed Water*; Technical Report 4333; Water Research Foundation: Denver, CO, USA, 2013.

21. Hambly, A.C.; Henderson, R.K.; Baker, A.; Stuetz, R.; Khan, S. Cross-connection detection in Australian dual reticulation systems by monitoring inherent fluorescent organic matter. *Environ. Technol. Rev.* **2012**, *1*, 67–80. [CrossRef]

22. Hess, D.; Collins, B. Recycling water in U.S. cities: Understanding preferences for aquifer recharging and dual-reticulation systems. *Water Policy* **2019**, *21*, 1207–1223. [CrossRef]

23. Grant, S.; Saphores, J.; Feldman, D.; Hamilton, A.; Fletcher, T.; Cook, P.; Stewardson, M.; Sanders, B.; Levin, L.; Ambrose, R.; et al. Taking the 'waste' out of 'wastewater' for human water security and ecosystem sustainability. *Science* **2012**, *337*, 681–686. [CrossRef]

24. Molinos-Senante, M.; Hernandez-Sancho, F.; Sala-Garrido, R. Economic feasibility study for new technological alternatives in wastewater treatment processes: A review. *Water Sci. Technol.* **2012**, *65*, 898–906. [CrossRef] [PubMed]

25. Barker, Z.; Stillwell, A.; Berglund, E. Scenario Analysis of Energy and Water Trade-Offs in the Expansion of a Dual Water System. *J. Water Resour. Plan. Manag.* **2016**, *142*, 05016012. [CrossRef]

26. Stillwell, A.; Twomey, K.; Osborne, R.; Greene, D.; Pedersen, D.; Webber, M. An integrated energy, carbon, water, and economic analysis of reclaimed water use in urban settings: A case study of Austin, Texas. *J. Water Reuse Desalination* **2011**, *1*, 208–223. [CrossRef]

27. Kandiah, V.; Berglund, E.Z.; Binder, A.R. An agent-based modeling approach to project adoption of water reuse and evaluate expansion plans within a sociotechnical water infrastructure system. *Sustain. Cities Soc.* **2019**, *46*, 101412. [CrossRef]

28. Kavvada, O.; Horvath, A.; Stokes-Draut, J.; Hendrickson, T.; Eisenstein, W.; Nelson, K. Assessing location and scale of urban nonpotable water reuse systems for life-cycle energy consumption and greenhouse gas emissions. *Environ. Sci. Technol.* **2016**, *50*, 13184–13194. [CrossRef]

29. Bawden, T. Water sensitive urban design technical manual for the greater Adelaide region. *Aust. Plan.* **2009**, *46.4*, 8–9. [CrossRef]

30. Herslund, L.; Backhaus, A.; Fryd, O.; Jørgensen, G.; Jensen, M.; Limbumba, T.; Yeshitela, K. Conditions and opportunities for green infrastructure—Aiming for green, water-resilient cities in Addis Ababa and Dar es Salaam. *Landsc. Urban Plan.* **2018**, *180*, 319–327. [CrossRef]

31. Imteaz, M.; Paudel, U.; Ahsan, A.; Santos, C. Climatic and spatial variability of potential rainwtaer savings for a large coastal city. *Resour. Conserv. Recycl.* **2015**, *105*, 143–147. [CrossRef]

32. Cook, S.; Sharma, A.; Chong, M. Performance analysis of a communal residential rainwater system: A case study in Brisbane, Australia. *J. Water Resour. Manag.* **2013**, *27*, 4865–4876. [CrossRef]

33. Berwanger, H.; Ghisi, E. Investment feasibility analysis of rainwater harvesting in the city of Itapiranga, Brazil. *Int. J. Sustain. Hum. Dev.* **2014**, *2*, 104–114.

34. Jung, K.; Lee, T.; Choi, B.; Hong, S. Rainwater harvesting system for continuous water supply to the regions with high seasonal rainfall variations. *Water Resour. Manag.* **2015**, *29*, 961–972. [CrossRef]

35. Lee, J.; Bae, K.H.; Younos, T. Conceptual framework for decentralized green water-infrastructure systems. *Water Environ. J.* **2018**, *32*, 112–117. [CrossRef]

36. Pitton, B.; Hall, C.; Haver, D. A cost analysis for using recycled irrigation runoff water in container nursery production: A Southern California nursery case study. *Irrig. Sci.* **2018**, *36*, 217–226. [CrossRef]

37. Liang, R.; Di Matteo, M.; Maier, H.R.; Thyer, M.A. Real-Time, Smart Rainwater Storage Systems: Potential Solution to Mitigate Urban Flooding. *Water* **2019**, *11*, 23. [CrossRef]

38. Campisano, A.; Butler, D.; Ward, S.; Burns, M.; Friedler, E.; Debusk, K.; Han, M. Urban rainwater harvesting systems: Research, implementation and future perspectives. *Water Resour.* **2017**, *115*, 195–209. [CrossRef]

39. Imteaz, A.; Moniruzzaman, M. Spatial variability of reasonable government rebates for rainwater tank T installations: A case study for Sydney. *Resour. Conserv. Recycl.* **2018**, *133*, 112–119. [CrossRef]

40. Grafton, R.; Libecap, G.; McGlenno, S.; Landry, C.; O'Brien, B. An integrated assessment of water markets: A cross-country comparison. *Rev. Environ. Econ. Policy* **2011**, *5*, 219–239. [CrossRef]

41. Garrick, D.; Whitten, S.; Coggan, A. Understanding the evolution and performance of water markets and allocation policy: A transaction costs analysis framework. *Ecol. Econ.* **2013**, *88*, 195–205. [CrossRef]

42. Rinaudo, J.; Calatrava, J.; Byans, M.D. Tradable water saving certificates to improve urban water use efficiency: An ex-ante evaluation in a French case study. *Aust. J. Agric. Resour. Econ.* **2016**, *60*, 422–441. [CrossRef]

43. Gonzales, P.; Ajami, N.; Sun, Y. Coordinating water conservation efforts through tradable credits: A proof of concept for drought response in the San Francisco Bay area. *Water Resour. Res.* **2017**, *53*, 7662–7677. [CrossRef]

44. Leigh, N.; Lee, H. Sustainable and Resilient Urban Water Systems: The Role of Decentralization and Planning. *Sustainability* **2019**, *11*, 918. [CrossRef]

45. Sharma, A.K.; Burn, S.; Gardner, T.; Gregory, A. Role of decentralized systems in the transition of urban water systems. *Water Sci. Technol.* **2010**, *10*, 577–583.

46. Haddad, B. Economic Incentives for Water Conservation on the Monterey Peninsula: The Market Proposal. *J. Am. Water Resour. Assoc.* **2000**, *36*, 1–15. [CrossRef]

47. Agthe, D.E.; Billings, R.B. Discussion "Economic Incentives for Water Conservation on the Monterey Peninsula: The Market Proposal" by Brent M. Haddad. *J. Am. Water Resour. Assoc.* **2000**, *36*, 931–932. [CrossRef]

48. Haddad, B. Reply to Discussion by Donald E. Agthe and R. Bruce Billings "Economic Incentives for Water Conservation on the Monterey Peninsula: The Market Proposal". *J. Am. Water Resour. Assoc.* **2000**, *36*, 933–934. [CrossRef]

49. Fornarelli, R.; Anda, M.; Dallas, S.; Schmack, M.; Dawood, F.; Byrne, J.; Morrison, G.; Fox-Reynolds, K. Enabling residential hybrid water systems through a water credit–debit system. *Water Supply* **2019**, *19*, 2131–2139. [CrossRef]

50. Mengelkamp, E.; Notheisen, B.; Beer, C.; Dauer, S.; Daueri, D.; Weinhardt, C. A blockchain-based smart grid: Towards sustainable local energy markets. *Comput. Sci.* **2018**, *33*, 207–214. [CrossRef]

51. Nakamoto, S. Bitcoin: A Peer-to-Peer Electronic Cash System. 2008. Available online: www.bitcoin.org (accessed on 31 October 2020).

52. Xu, X.; Weber, I.; Staples, M.; Zhu, L.; Bosch, J.; Bass, L.; Pautasso, C.; Rimba, P. A taxonomy of blockchain-based systems for architecture design. In Proceedings of the IEEE International Conference on Software Architecture (ICSA), Gothenburg, Sweden, 3–7 April 2017; pp. 243–252.

53. Andoni, M.; Robu, V.; Flynn, D.; Abram, S.; Geach, D.; Jenkins, D.; McCallum, P.; Peacock, A. Blockchain technology in the energy sector: A systematic review of challenges and opportunities. *Renew. Sustain. Energy Rev.* **2019**, *100*, 143–174. [CrossRef]

54. Christidis, K.; Devetsikiotis, M. Blockchains and Smart Contracts for the Internet of Things. *IEE* **2016**, *4*, 2292–2303. [CrossRef]

55. Aquatech. Blockchain's Potential to Disrupt water Supply Remains Divisive. 2018. Available online: https://www.aquatechtrade.com/news/utilities/blockchains-potential-to-disrupt-water-supply-remains-divisive/ (accessed on 31 October 2020).

56. Sobrinho, R.; Garcia, J.; Maia, A.; Romeiro, A. Blockchain technology and complex flow systems as opportunities for water governance innovation. *Rev. Bras. Inov.* **2019**, *18*, 157–176.

57. Pee, S.J.; Nang, J.H.; Jang, J.W. A Simple Blockchain-based Peer-to-Peer Water Trading System Leveraging Smart Contracts. In Proceedings of the International Conference Internet Computing and Internet of Things, Las Vegas, NV, USA, 30 July–2 August 2018; pp. 63–68.

58. Carey, A. Future Melbourne Residents Could Trade in Rainwater to Cut down on Bills. 2017. Available online: https://www.theage.com.au/national/victoria/future-melbourne-residents-could-trade-in-rainwater-to-cut-down-on-bills-20170905-gyb5mj.html (accessed on 31 October 2020).

59. South East Water. Fishermans Bend: A Water Sensitive Community. 2020. Available online: https://southeastwater.com.au/CurrentProjects/Projects/Pages/Fishermans-Bend.aspx (accessed on 31 October 2020).

60. Alcarria, R.; Bordel, B.; Robles, T.; Martín, D.; Ángel Manso-Callejo, M. A Blockchain-Based Authorization System for Trustworthy Resource Monitoring and Trading in Smart Communities. *Sensors* **2018**, *18*, 3561. [CrossRef]

61. Athanasiadis, I.N.; Mentes, A.K.; Mitkas, P.A.; Mylopoulos, Y.A. A hybrid agent-based model for estimating residential water demand. *Simulation* **2005**, *81*, 175–187. [CrossRef]

62. Schwarz, N.; Ernst, A. Agent-based modeling of the diffusion of environmental innovations—An empirical approach. *Technol. Forecast. Soc. Chang.* **2009**, *76*, 497–511. [CrossRef]

63. Galán, J.M.; López-Paredes, A.; Del Olmo, R. An agent-based model for domestic water management in Valladolid metropolitan area. *Water Resour. Res.* **2009**, *45*, W05401. [CrossRef]

64. Giacomoni, M.; Kanta, L.; Zechman, E. Complex adaptive systems approach to simulate the sustainability of water resources and urbanization. *J. Water Resour. Plan. Manag.* **2013**, *139*, 554–564. [CrossRef]

65. Kanta, L.; Zechman, E. Complex Adaptive Systems Framework to Assess Supply-Side and Demand-Side Management for Urban Water Resources. *J. Water Resour. Plan. Manag.* **2014**, *140*, 75–85. [CrossRef]

66. Giacomoni, M.; Berglund, E. A Complex Adaptive Simulation Framework for Evaluating Adaptive Demand Management for Urban Water Resources Sustainability. *J. Water Resour. Plan. Manag.* **2015**, *141*, 04015024. [CrossRef]

67. Kanta, L.; Berglund, E.Z. Exploring tradeoffs in demand-side and supply-side management of urban water resources using agent-based modeling and evolutionary computation. *Systems* **2015**, *3*, 287–308. [CrossRef]

68. Koutiva, I.; Makropoulos, C. Modelling domestic water demand: An agent-based approach. *Environ. Model. Softw.* **2016**, *79*, 35–54. [CrossRef]

69. Koutiva, I.; Makropoulos, C. Exploring the effects of domestic water management measures to water conservation attitudes using agent based modelling. *Water Sci. Technol. Water Supply* **2017**, *17*, 552–560. [CrossRef]

70. Mashhadi Ali, A.; Shafiee, M.E.; Berglund, E.Z. Agent-based modeling to simulate the dynamics of urban water supply: Climate, population growth, and water shortages. *Sustain. Cities Soc.* **2017**, *28*, 420–434. [CrossRef]

71. Darbandsari, P.; Kerachian, R.; Malakpour-Estalaki, S. An Agent-based behavioral simulation model for residential water demand management: The case-study of Tehran, Iran. *Simul. Model. Pract. Theory* **2017**, *78*, 51–72. [CrossRef]

72. Zechman, E.M. Integrating evolution strategies and genetic algorithms with agent-based modeling for flushing a contaminated water distribution system. *J. Hydroinf.* **2013**, *15*, 798. [CrossRef]

73. Shafiee, M.E.; Berglund, E.Z. Agent-based modeling and evolutionary computation for disseminating public advisories about hazardous material emergencies. *Comput. Environ. Urban Syst.* **2016**, *57*, 12–25. [CrossRef]

74. Shafiee, M.E.; Berglund, E.Z. Complex Adaptive Systems Framework to Simulate the Performance of Hydrant Flushing Rules and Broadcasts during a Water Distribution System Contamination Event. *J. Water Resour. Plan. Manag.* **2017**, *143*, 04017001. [CrossRef]

75. Shafiee, M.E.; Berglund, E.Z.; Lindell, M.K. An Agent-based Modeling Framework for Assessing the Public Health Protection of Water Advisories. *Water Resour. Manag.* **2018**, *32*, 2033–2059. [CrossRef]

76. Monroe, J.; Ramsey, E.; Berglund, E. Allocating countermeasures to defend water distribution systems against terrorist attack. *Reliab. Eng. Syst. Saf.* **2018**, 1–15. [CrossRef]

77. Strickling, H.; DiCarlo, M.F.; Shafiee, M.E.; Berglund, E. Simulation of Containment and Wireless Emergency Alerts within Targeted Pressure Zones for Water Contamination Management. *Sustain. Cities Soc.* **2019**, 101820. [CrossRef]

78. Kandiah, V.; Ar, B.; Berglund, E. An Empirical Agent-Based Model to Simulate the Adoption of Water Reuse Using the Social Amplification of Risk Framework. *Risk Anal.* **2017**, *10*, 2005–2022. [CrossRef]

79. Grimm, V.; Berger, U.; Bastiansen, F.; Eliassen, S.; Ginot, V.; Giske, J.; Goss-Custard, J.; Gr, T.; Heinz, S.K.; Huse, G.; et al. A standard protocol for describing individual-based and agent-based models. *Ecol. Model.* **2006**, *198*, 115–126. [CrossRef]

80. Willis, R.; Stewart, R.; Williams, P.; Hacker, C.; Emmonds, S.; Capati, G. Residential potable and recycled water end uses in a dual reticulated supply system. *Desalination* **2011**, *272*, 201–211. [CrossRef]

81. NAHB. Cost of Constructing a Home. 2010. Available online: https://www.nahbclassic.org/generic.aspx?genericContentID=248306 (accessed on 31 October 2020).

82. Duraplas. Available online: https://www.duraplas.com.au/ (accessed on 31 October 2020).

83. Gikas, G.D.; Tsihrintzis, V.A. Assessment of water quality of first-flush roof runoff and harvested rainwater. *J. Hydrol.* **2012**, *466–467*, 115–126. [CrossRef]

84. Texas Water Development Board. *The Texas Manual on Rainwater Harvesting*; Technical Report; Texas Water Development Board: Austin, TX, USA, 2005.

85. Lawson, S.; LaBranche-Tucker, A.; Otto-Wack, H.; Hall, R.; Sojka, B.; Crawford, E.; Crawford, D.; Brand, C. *Virginia Rainwater Harvesting Manual*; Technical Report; Cabell Brand Center: Salem, VA, USA, 2009.

86. Luke, S. *Multiagent Simulation and the MASON Library*; George Mason University: Fairfax, VI, USA, 2019.

87. Ramsey, L. RainwaterABM. 2020. Available online: https://github.com/evramsey/RainwaterABM (accessed on 31 October 2020).

88. Jacobs, H.; Haarhoff, J. Structure and data requirements of an end-use model for residential water demand and return flow. *Water SA* **2004**, *30*, 293–304. [CrossRef]

89. Rossman, L.A. *EPANET 2 Users Manual*, 2nd ed.; US Environmental Protection Agency: Cincinnati, OH, USA, 2000.

90. Marchi, A.; Salomons, E.; Ostfeld, A.; Kapelan, Z.; Simpson, A.R.; Zecchin, A.C.; Maier, H.R.; Wu, Z.Y.; Elsayed, S.M.; Song, Y.; et al Battle of the Water Networks II. *J. Water Resour. Plan. Manag.* **2014**, *140*, 04014009. [CrossRef]

91. Bhattacharya, A. *Harvesting Rainwater: Catch Water Where It Falls*; Working Papers id:2349; eSocialScience: New Delhi, India, 2009.

92. United States Geological Service. USGS 12113346 Springbrook Creek at Orillia, WA. 2020. Available online: https://waterdata.usgs.gov/usa/nwis/uv?12113346 (accessed on 31 October 2020).

93. PRISM Climate Group-Oregon State University. 2004. Available online: http://prism.oregonstate.edu (accessed on 31 October 2020).

94. Doorenbos, J.; Pruitt, W. Guidelines for predicting crop water requirements. *FAO Irrig. Drain. Pap.* **1977**, *24*, 144.

95. Beal, C.D.; Flynn, J. Toward the digital water age: Survey and case studies of Australian water utility smart-metering programs. *Util. Policy* **2015**, *32*, 29–37. [CrossRef]

96. Boryczko, K.; Rak, J. Method for Assessment of Water Supply Diversification. *Resources* **2020**, *9*, 87. [CrossRef]

97. Grubic, T.; Vara, L.; Hu, Y.; Tewari, A. Micro-generation technologies and consumption of resources: A complex systems' exploration. *J. Clean. Prod.* **2020**, *247*, 119091. [CrossRef]

98. Paudel, U.; Imteaz, M. Spatial Variability of Reasonable Government Rebates for Rainwater Tank Installations: A Case Study for Adelaide. In *Sustainability Perspectives: Science, Policy and Practice*; Khaiter, P., Erechtchoukova, M., Eds.; Springer: New York, NY, USA, 2019; Chapter 13.

99. Ramsey, E.; Berglund, E.; Goyal, R. The Impact of Demographic Factors, Beliefs, and Social Influences on Residential Water Consumption and Implications for Non-Price Policies in Urban India. *Water* **2017**, *9*, 844. [CrossRef]

100. Dolnicar, S.; Hurlimann, A.; Brun, B. What affects public acceptance of recycled and desalinated water? *Water Res.* **2011**, *45*, 933–943. [CrossRef] [PubMed]

101. Garcia-Cuerva, L.; Berglund, E.; Binder, A. Public Perceptions of Water Shortages, Conservation Behaviors, and Support for Water Reuse in the U.S. *Resour. Conserv. Recycl.* **2016**, *113*, 106–11. [CrossRef]

102. Shen, P.; Deletic, A.; Bratieres, K.; McCarthy, D. Real time control of biofilters delivers stormwater suitable for harvesting and reuse. *Water Res.* **2020**, *169*, 115257. [CrossRef]

Publisher's Note: MDPI stays neutral with regard to jurisdictional claims in published maps and institutional affiliations.

 *water*

Article

Is Clustering Time-Series Water Depth Useful? An Exploratory Study for Flooding Detection in Urban Drainage Systems

Jiada Li [1,*], Daniyal Hassan [1], Simon Brewer [2] and Robert Sitzenfrei [3]

[1] Department of Civil and Environmental Engineering, University of Utah, 201 Presidents Circle, Salt Lake City, UT 84112, USA; u6010713@umail.utah.edu

[2] Geography Department, University of Utah, 201 Presidents Circle, Salt Lake City, UT 84112, USA; simon.brewer@geog.utah.edu

[3] Unit of Environmental Engineering, University of Innsbruck, Innrain 52, 6020 Innsbruck, Austria; Robert.Sitzenfrei@uibk.ac.at

* Correspondence: jiada.li@utah.edu

Received: 26 July 2020; Accepted: 25 August 2020; Published: 30 August 2020

Abstract: As sensor measurements emerge in urban water systems, data-driven unsupervised machine learning algorithms have drawn tremendous interest in event detection and hydraulic water level and flow prediction recently. However, most of them are applied in water distribution systems and few studies consider using unsupervised cluster analysis to group the time-series hydraulic-hydrologic data in stormwater urban drainage systems. To improve the understanding of how cluster analysis contributes to flooding location detection, this study compared the performance of K-means clustering, agglomerative clustering, and spectral clustering in uncovering time-series water depth dissimilarity. In this work, the water depth datasets are simulated by an urban drainage model and then formatted for a clustering problem. Three standard performance evaluation metrics, namely the silhouette coefficient index, Calinski–Harabasz index, and Davies–Bouldin index are employed to assess the clustering performance in flooding detection under various storms. The results show that silhouette coefficient index and Davies–Bouldin index are more suitable for assessing the performance of K-means and agglomerative clustering, while the Calinski–Harabasz index only works for spectral clustering, indicating these clustering algorithms are metric-dependent flooding indicators. The results also reveal that the agglomerative clustering performs better in detecting short-duration events while K-means and spectral clustering behave better in detecting long-duration floods. The findings of these investigations can be employed in urban stormwater flood detection at the specific junction-level sites by using the occurrence of anomalous changes in water level of correlated clusters as flood early warning for the local neighborhoods.

Keywords: smart stormwater; machine learning; cluster analysis; data science; flooding detection

---

1. Introduction

Urban drainage systems (UDSs) are the infrastructures constructed to provide conveyance ability and storage capability for drainage overflow mitigation, surface inundation reduction, and pollutant removal. However, the existing UDSs, whose functionality can only serve for a limited number of years, might degrade and even deteriorate as time goes by [1]. In recent years, retrofitting the traditional UDSs with water-level sensors, velocity meters, and flow sensors has been widely adopted as an adaptive and cost-effective solution for flooding challenges [2,3]. The deployed sensors can measure the water quantity and quality data in a real-time way, which now makes it feasible for decision-makers and stakeholders to foresee the potential flood events and locate the vulnerable sites, which supports decision making. The need to understand the emerging data is crucial for forecasting flash floods, reducing sewer overflows, and detecting flooded sites [4–6]. Interpreting big water data for flood detection is attracting increasing attention from researchers [7–10] and can be employed to reduce potential flood damages.

In the last decade, many scholars have introduced several machine learning techniques to investigate the available water resources and hydrological datasets [11–13]. The major machine learning algorithms employed for flood detection are support vector machines [14,15], neuro-fuzzy [16], adaptive neuro-fuzzy inference systems, multilayer perceptron [17], random forest [18], and classification and regression trees [19]. Bowes et al. compared long short-term memory and recurrent neural networks by using a time-series of groundwater table data in the city of Norfolk, Virginia [20]. They explained that a long short-term memory neural network is better than the recurrent neural network in predicting groundwater level, but takes about three times longer to train the model. Hu et al. applied a boosted decision regression tree to detect drainage floods with over 90% accuracy in combined sewer systems of Detroit city, Michigan [21]. Li proposed a data-driven fuzzy neural method for reducing downstream urban flooding volume and showed that with an enhanced genetic algorithm optimization the regression deviations could be reduced from 0.22 to 0.07 [22]. However, the majority of these studies have focused on supervised learning (i.e., when a known outcome is used to train the model), and unsupervised machine learning algorithms (UMLA) are not commonly used in stormwater UDSs.

Clustering algorithms are a data-driven technology without considering the classification standard of different risk levels and thereby provide more objective and reasonable results [23]. Therefore, cluster analysis, one of the key unsupervised machine learning methods, has been applied in many fields, including pattern recognition, image analysis, data compression, and anomaly detection [24]. However, its applicability in urban flood detecting is yet to be fully investigated. In general, cluster analysis is based on identifying similarities between observations. If a water quantity or quality event happens in the water system, these observations are likely to be highly dissimilar to other observations [25]. The increment in dissimilarity would lead to these observations being considered as outliers, and thus detected as anomalies. Although cluster analysis has been extensively discussed in municipal topology classification and water distribution network simplification [26,27], the ability of UMLA methods to group time-series data at UDSs is still unknown, and the most appropriate methods to assess these algorithms are unclear. Keogh and Lin concluded that clustering time-series data is meaningless, but this argument does not cover the similarity-based clustering algorithms such as K-means and agglomerative clustering [28]. In contrast, Chen demonstrated that similarity-based cluster analysis could be successfully applied to sequence datasets by using different distance measures [29,30]. Wu et al. adopted the clustering algorithm [24], developed by Rodriguez and Laio [31], to detect the short-duration pipe burst with a 0.61% false positive in water distribution systems. Xing and Sela selected SCI (silhouette coefficient index) and CHI (Calinski–Harabasz index) as the metrics to evaluate K-mean clustering (KC) performance in clustering time-series water pressure data and they finally identified the number of clusters for the pressure sensor placement [32]. However, it was unclear why they chose these two indexes as the UMLA performance metrics. Previous studies from the computer science field have demonstrated the differences and similarities among the popular performance evaluation indices such as SCI, CHI, and DBI (Davies–Bouldin index) [33–35]. However, there is no systematic study of how these apply to time-series data from UDSs.

Floods are one of the most hazardous natural events in the world. The short response time against flood events makes them challenging for the hydrologists, and as a result, floods cause loss of life, economics, infrastructure, and property worldwide annually [36,37]. Researchers are trying to promote flooding indicators to identify flooding locations ahead of extreme storm events. There are several hydro-meteorological indicators, such as temperature, humidity, and precipitation, which are related to flood events. The most widely used indicator is hydraulic water level since it can be efficiently and continuously monitored and forecasted to facilitate floods early detection and warning [38]. To efficiently capture the flood events, the flooding water level should be well investigated.

In this study, clustering algorithms, including KC, agglomerative (AC), and spectral clustering (SC), are applied for the urban flood tracking. A storm water management model (SWMM) is established to represent the real-world stormwater urban drainage systems, located in Sugar House neighborhood, Salt Lake City, UT, USA. Three evaluation indices are used to test the performance analysis of the clustering algorithms, namely SCI, CHI, and DBI. The whole research is driven by the hypothesis that the clustering of time-series water level data has the potential to facilitate flooding location detection in the Sugar House Area. The investigations provide answers to various inter-related research questions: (1) What is the performance of different clustering algorithms in capturing the floods? (2) Which metrics are the most suitable for assessing cluster model performance based on hydraulic-hydrologic data in UDSs? (3) Which features of flood time-series data (length, volume and variability) are the most influential for flooding detection, and how does the choice of data feature affect the clustering performance in localizing the flooding sites?

To answer these questions, it is necessary to explore how UMLA groups time-series water depth data, and which assessment score can best represent UMLA performance. However, challenges to implement UMLA with time-series data still exist. Firstly, it is essential to re-format the time-series water depth datasets to make them suitable for clustering problem. This difficulty is associated with the second research question above since the features of datasets determine how we re-structure the data frame [39]. Secondly, the connection between the number of clusters and the clustering model performance is another obstacle. As it is still unknown how to correlate clustering performance and the number of clusters in the stormwater systems, it is necessary to build such a theoretical relationship for a practical application like the flooding detection herein [40]. Therefore, the study aims to improve the understanding of how UMLA facilitates detecting hydraulic anomaly according to the characteristics of water depth datasets in urban drainage networks.

The layout of the study is as follows: (1) build KC, AC, and SC algorithms to group the time-series water depth data; (2) use UMLA metrics, including SCI, CHI, and DBI, to evaluate these algorithms; (3) compare the best number of clusters obtained by each method; (4) investigate the relationship between model performance of flooding detection and water depth data characteristics (see Figure 1 for details). We start by describing the implementation of different UMLA methods, followed by the research methodology with an overview of the real-world case study, performance metrics, and simulation scenarios for cluster analysis. Then, we present the results, discussion, and finally, the conclusions.

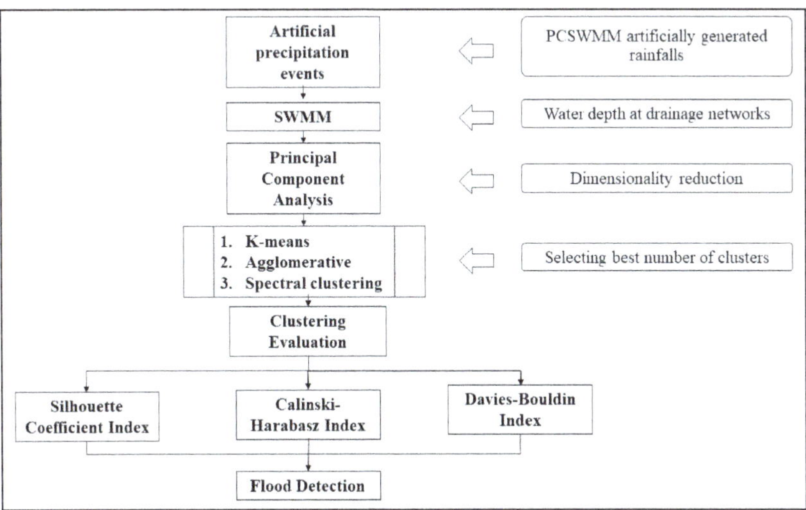

Figure 1. Representing the workflow of the whole study.

2. Materials and Methods

This study was organized in four steps: (i) time-series data preprocessing; (ii) clustering modeling implementation; (iii) clustering performance assessment; (iv) applications analysis of clustering results for urban floods detection. The workflow of the methods can be found in Figure 1.

2.1. Description of Unsupervised Machine Learning Algorithms

Current machine learning techniques mainly fall into two groups: supervised and unsupervised learning [41]. The UMLA is a self-organization method to find patterns in unlabeled data. Cluster analysis is, a subset of UMLA methods, and in general, is based on the principle of grouping similar observations and segmenting dissimilar observations [42]. Anomalous data points that differ from others may then be filtered [43]. A large number of clustering algorithms exist, including K-means, Affinity Propagation, and Mean Shift. In this research, we employed the SCI, CHI, and DBI to assess the performance of the cluster, because of their accuracy and wide applicability in a similar type of studies [44–46].

2.1.1. K-Means Clustering

K-means clustering (KC) is a centroid-based unsupervised clustering algorithm, originally designed for signal processing. It is the most widely applied method of cluster analysis in data mining [33]. K-means aims to partition the inputs into k clusters. Given a set of observations ($x_1, x_2, ...,$ x_i) for p variables, the algorithm runs as follows:

(1)   Choose k initial centroids, each defined by a value for each of the p variables. These are chosen randomly, often by simply choosing k observations.
(2)   Assign each observation to the centroid it is most similar to. The similarity is generally measured as the Euclidean distance between the observation and centroid in parameter space.
(3)   Once all observations are assigned, re-estimate the centroids location as the mean of the p variables of all observations assigned to that centroid.
(4)   Repeat until the algorithm stabilizes (minimize the within-cluster sum of squares).

The goal then is to minimize kC_ℓ the within-cluster sum of squares:

$$argmin_{\mu,C} \sum_{\ell=1}^{k} \sum_{x_i \in C_\ell}^{i} \|x_i - \mu_\ell\|^2 \tag{1}$$

where k is the number of cluster centers and $\{\mu_\ell\}$, $\ell = 1, \ldots, k$ are the cluster centroids $C_\ell\mu_\ell\mu_\ell C_\ell$. The total intra-cluster distance is the total squared Euclidean distance from each point to the center of its cluster, and this is a measure of the variance or internal coherence of the clusters [47]. This can be used to assess the stability of the solution. When this falls below a predefined threshold, the algorithm stops. The algorithm is often run multiple times with different random initialization of cluster centroids to avoid sub-optimal problems in convergence. The clustering solution with the lowest sum-of-squares is chosen as the final output.

However, the choice of k is challenging when model performance metrics are not available. Often, an initial value of k is chosen, then the algorithm is repeated for higher and lower values. To improve the efficiency of discovering the best k value, a score (SCI, CHI, DBI)-based performance assessment method is recommended in many prior studies [42].

2.1.2. Agglomerative Clustering

Agglomerative clustering (AC) is one of the main forms of hierarchical clustering. These algorithms do not provide a single partitioning of the data but instead provide a full hierarchy of cluster solutions from all observations in a single cluster (i.e., $k = 1$) to all observations in individual clusters (i.e., k

$= n$) [48]. In contrast to KC, hierarchical methods allow existing clusters to be split or merged, with the result that smaller clusters are related to large clusters in a hierarchy. The rules governing which clusters are again based on their distance or similarity. The AC algorithm consists of the following steps:

(1) Start with each data point as its own cluster.
(2) Select the distance metric and linkage criteria to calculate the dissimilarity between pairs of observations.
(3) Link together the two clusters with the minimum dissimilarity.
(4) Continue this process until there is only one cluster.

A key decision in the AC algorithm is the calculation of dissimilarity between clusters. In this study, we used Euclidean distance [47], and the Ward linkage, which measures the distance between the cluster centroids, similar to the K-means clustering method. The equations for Euclidean distance and Ward linkage are defined by Equations (2) and (3), respectively:

$$\|a - b\|_2 = \sqrt{\sum_I (a_i - b_i)^2} \tag{2}$$

where a and b mean the Euclidean vector; a_i and b_i are the point position for the Euclidean vector; i is the number of vectors.

$$d_{ij} = d\big(\{X_i\}, \{X_j\}\big) = \left\| X_i - X_j \right\|^2 \tag{3}$$

where d_{ij} is the squared Euclidean distance between point i and point j; X_i and X_j are Ward's vectors.

The resulting hierarchy of clusters can be represented using a dendrogram plot [48]. The detailed introduction of the dendrogram plot can be found in Section 2.3.5 below.

2.1.3. Spectral Clustering

Spectral clustering (SC) is an unsupervised learning technique based on graph theory, where SC takes advantage of graph information from the spectrum to find the number of clusters [49]. Unlike the previous methods that tend to prioritize clusters by proximity, SC aims to identify observations that are linked, and therefore may not form classical spherical groups in parameter space. The SC algorithm is as follows:

(1) Create a similarity matrix S between observations. This is the complement to the dissimilarity matrices used in other methods, and here is calculated as the negative Euclidean distance.
(2) Create an adjacency matrix A, representing the graph or connectivity between observations. This is a transformation of S, where for each observation, we find the k nearest neighbors (i.e., with the highest similarity). If observations i and j are considered to be neighbors, we set $A_{ij} = S_{ij}$. If not, we set $A_{ij} = 0$.
(3) Create a degree matrix D, where the diagonal values are the degree of connectivity for each observations, given as diag$\{D\} = \sum_{i,j}^{n} A_{ij}, \; i,j=1,\,2,3,\,...,\,n$
(4) Next, calculate the graph Laplacian matrix L. This can be normalized or unnormalized. Here, we use the unnormalized: $L = D - A$
(5) The clustering solution is then found by eigendecomposition of the Laplacian, and selecting the k smallest eigenvectors. Consequently, these result in a perfect separation of the observations. K-means is then run on these eigenvectors, to get the final cluster assignment of each observation: $L_{(N \times N)} = D - A$

As SC performs dimensionality reduction before clustering data points, it is a very flexible approach for complex data sets. However, the similarity matrix generated by SC may include negative values, which can be problematic for grouping time-series points.

2.1.4. Summary and Comparison of Clustering Algorithms

In general, it is difficult to recommend a single algorithm as being the most suitable for clustering, particularly with data that is uncertain and of poor quality, such as the features of pipe flow or water level data used here [41]. It is, therefore, advisable to use several algorithms and compare their performance for specific applications. Here, we use KC, SC, and AC to discover the unknown subgroups in simulated water depth data of UDSs' junctions. Table 1 summarizes the advantages and disadvantages of these algorithms from review papers [24,33,44].

Table 1. Clustering algorithm information summary.

Models	Definition	Pros	Cons
K-means Clustering	A kind of vector quantization, partition data points into clusters by minimizing the intra-cluster distance.	(1) Fast, easy-to-understand, and wide applications; (2) Stable for time series data; (3) Simple and efficient optimization performance; (4) Suitable for huge datasets.	(1) Number of clusters; (2) Spherical assumption.
Agglomerative Clustering	A kind of hierarchical clustering for merging clusters according to a measure of data dissimilarity.	(1) Stable runs (2) Reasonable dendrogram cut-off nodes; (3) Clusters growth without globular assumption; (4) Good performance for time-series data; (5) No need to know the correct clusters' number.	(1) Number of clusters; (2) Slow implementation; (3) Cluster with polluted noise.
Spectral Clustering	A kind of graph clustering based on the distances between points.	(1) Stable due to the data transformation; (2) No purely globular cluster assumption; (3) Easy to implement.	(1) Number of clusters; (2) Slow performance; (3) Cluster with polluted noise.

2.2. Clustering Model Implementation

The SWMM model was run six times, once with each of the rainfall scenarios described above. We collected the simulated time-series water depth from each node in the stormwater drainage network for cluster analysis. As there are 60 junctions in the SWMM model, this results in a matrix where each column represents a single time step with a 5-min interval, and each row (60 rows) stands for a junction or node in the network. We then used the principal component analysis (PCA) to reduce the dimensionality of this matrix. PCA uses the eigendecomposition of the correlation matrix to identify a small set of principal components that represent the majority of variance in the original data [50]. Here, we used correlations between the time-series at different nodes to reduce the column of matrix to 2, which means the number of timesteps is compressed to 2 principal components. Finally, the dataset matrix is configured with 60 rows and 2 columns under each modeling scenarios. The datasets used in this work are not large, and for computational costs are limited. While other techniques for data reduction exist (e.g., correspondence analysis, factor analysis, or non-metric multi-dimensional scaling), we used PCA due to the assumed linear response of the water depth values. Although the reduction of dimensionality might cause data loss or an undesirable relationship between score axes, PCA indeed helps reduce computation time and remove redundant data features in the following cluster analysis.

All clustering algorithms were then run using this set of two principal components shown in Figure 2, with the following set up:

(1)　K-means: We initially set the number of clusters (*k*) to 2 for each modeling scenarios. The algorithm was repeated ten times with different random initialization, and a maximum of 5 iterations was used to converge the algorithm.

(2) Agglomerative clustering model: We used Ward linkage, as this is robust to outliers and unequal variance in the data. As only 'Euclidean' supports 'Ward' linkage distance computation. If 'Ward' linkage is used for cluster distance computation, 'Euclidean' would be the best way to measure the data dissimilarity [51]. Thus, the cluster distance calculation method and dissimilarity metric among sample points are set to be 'Ward' and 'Euclidean' distance, respectively. The resulting hierarchy was cut to provide 2 clusters.

(3) Spectral clustering: The algorithm was used to identify 2 clusters, using the unnormalized graph Laplacian.

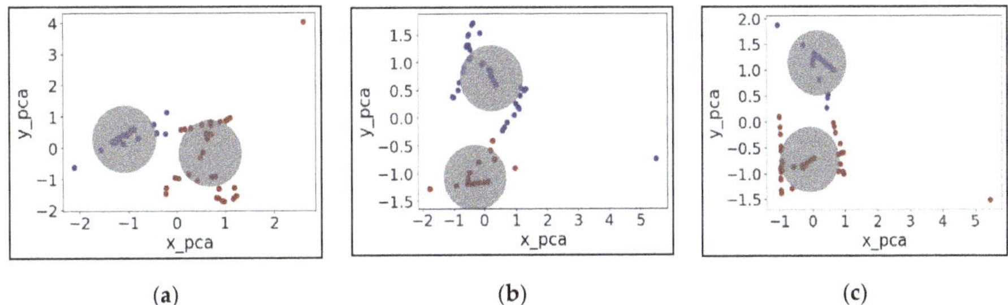

(a)   (b)   (c)

Figure 2. Principal component scores for the two components (x_pca means the first component score; y_pca means the second component score) by K-mean under varying rainfall scenarios: (**a**) 3 h' duration rainfall, (**b**) 12 h' duration rainfall, (**c**) 48 h' duration rainfall. The principal component scores are used to examine if these two clusters are reasonably distinguished from each other clustering (gray circles the blue and red dots assigned to the closest cluster).

In Figure 2 below, there is no sample marginal overlapping, which indicates the cluster classification is reasonable with respect to grouping the time-series water level data. Additionally, the isolated dots in the subplots of Figure 2 present the dissimilarity of the water depth datasets under this event, indicating these isolated dots might be the potential flooded junctions, which help the decision-makers to pre-screen the vulnerable sites in the drainage networks.

2.3. Clustering Model Evaluation and Validation

Unlike the supervised machine learning algorithms that compare the predicted and actual values to compute the model accuracy, the UMLA assess performance directly on the characteristics of the clusters that were obtained. The performance then depends on data features selected, data preprocessing, and parameter settings such as the distance function to use, a density threshold, or the number of expected clusters, which can be modified according to the varying datasets and object inputs. As a result, there is rarely a single obvious solution for clusters, and cluster analysis is an iterative process of knowledge discovery or interactive multi-objective optimization that involves trial and failure, aimed to obtain the desired results [52–55].

Several indices, including SCI, DBI and CHI, are employed to measure the relative performance of clustering algorithms. In general, these metrics provide an assessment of how the data variance is partitioned. An ideal cluster solution will have low intra-cluster variance (i.e., all observations should be similar within a cluster) and high inter-cluster variance (the clusters should be well separated).

2.3.1. Silhouette Coefficient Index

The silhouette coefficient index is an example of model-self-evaluation, where a higher SCI score relates to a model with better-defined clusters [56]. This score is bounded between −1 for incorrect clustering and +1 for well-formed clusters. Scores around zero indicate overlapping clusters. The SCI is defined for each observation, which can be calculated as Equation (4):

$$\text{SCI} = \frac{m - n}{\max(m, n)} \tag{4}$$

where the SCI is for a single observation; m is the mean distance between an observation and all other observations in the same class; n is the mean distance between the same observation and all observations in the next nearest cluster. The SCI has the advantage that it can be used to examine how well individual observation are clustered, or an estimate can be obtained for each cluster or for the whole cluster solution by averaging across a cluster or the entire dataset, respectively. An estimate can be obtained for each cluster or for the whole clusters solution. A set of samples is given as the mean of the SCI for each sample, and it would be relatively higher when clusters are dense and well separated [57].

2.3.2. Calinski-Harabasz Index

The CHI is calculated as the ratio of the between-clusters dispersion average and the within-cluster dispersion [58], penalized by the number of clusters (k). A higher CHI score indicates better-defined clusters (i.e., dense and well separated). CHI for a set of k clusters is calculated as:

$$\text{CHI} = \frac{T_r(B_k)}{T_r(W_k)} \times \frac{N - k}{k - 1} \tag{5}$$

where N is the number of points in our data; k is the number of the cluster; T_r represents dispersion matrix; B_k is the between-group dispersion matrix, and W_k is the within-cluster dispersion matrix. B_k and W_k are defined by the following equations:

$$W_k = \sum_{q=1}^{k} \sum_{x \in C_q} (x - c_q)(x - c_q)^T \tag{6}$$

$$B_k = \sum_{q}^{k} n_q (c_q - c)(c_q - c)^T \tag{7}$$

where C_q is the set of points in the cluster q, c_q is the center of the cluster q, c is the center of the whole data set which has been clustered into k clusters, n_q is the number of points in the cluster q.

2.3.3. Davies-Bouldin Index

The DBI can also be used to evaluate the model, where a lower DBI relates to a model with better separation between the clusters [59]. The index is defined as the average similarity (R_{ij}) between each cluster k and the next closest (i.e., most similar) cluster. The DBI is calculated as Equation (8):

$$\text{DBI} = \frac{1}{k} \sum_{i=1}^{k} \max_{i \neq j}(R_{ij}) \tag{8}$$

where DBI is the Davies–Bouldin index. Zero is the lowest possible score. Values closer to zero indicate a better partition. k is the number of the cluster. R_{ij} is the similarity measure which features per Equation (9):

$$R_{ij} = \frac{s_i + s_j}{d_{ij}} \tag{9}$$

where s_i is the average intra-distance between each point of cluster i and the centroid of that cluster representing as cluster diameter; d_{ij} is the inter-cluster distance between cluster centroids i and j; R_{ij}

is set to the trade-off between inter-cluster distance and intra-cluster distance. The computation of DBI is simpler than that of SC since this index is computed only with quantities and features inherent to the dataset [60]. However, a good value reported by DBI might not imply the best information retrieval [55].

2.3.4. Intra-Cluster Distance

Intra-cluster distance (ICD) is the distance between two samples belonging to the same cluster. Three types of intra-cluster distance, including complete diameter distance, average diameter distance, and centroid diameter distance, are popular in prior studies. As the number of clusters increase, individual clusters become more homogenous, and the ICD decreases. At a certain point, the decrease in distances becomes negligible. Plotting this distance against k usually results in an inflection point or elbow point where this occurs, and can be used to identify the optimal value of k [61]. The number of clusters is chosen at this point, hence the "elbow criterion." Here we use the centroid distance to represent ICD, given as double the average distance between all of the objects:

$$\Delta(S) = 2\left\{\frac{\sum_{x \in S} d(x, T)}{|S|}\right\} \tag{10}$$

$$T = \frac{1}{|S|}\sum_{x \in S} x \tag{11}$$

where $\Delta(S)$ is the centroid diameter distance of the formed cluster representative S; x is the samples belonging to cluster S; $d(x, T)$ is the distance between two objects, x and T; $|S|$ is the number of objects in cluster S.

2.3.5. Dendrogram

A dendrogram is a visualization in the form of a tree that shows the hierarchical relationship like the order and distance (dissimilarity) between samples [62]. The individual samples are located along the bottom of the dendrogram and referred to leaf nodes. The hierarchical clusters are formed by merging individual samples or existing lower-level clusters. In a dendrogram, the vertical axis is labeled distance and refers to a dissimilarity measure between individual samples or clusters. Generally, in a dendrogram, horizontal lines can be regarded as places where clusters merge, while vertical lines show the distance at which lower-level clusters were merged, forming a new higher-level cluster. The dissimilarity measure between two groups is calculated as Equation (12):

$$\text{Dis} = 1 - \text{C} \tag{12}$$

where Dis means the dissimilarity or distance among objects and C means the correlation degree between clusters.

If clusters are highly correlated to each other, they will have a correlation value close to 1. To that, Dis = 1 − C will be given a value close to zero. Therefore, highly related clusters are nearer to the bottom of the dendrogram. Those clusters that are not correlated have a correlation value close to zero. Clusters that are negatively correlated will give a distance value larger than 1 in the dendrogram. The dendrogram can be used to visually allocate correlated objects to clusters or to detect outliers and anomaly in a diagram [47]. In the dendrogram, each sample is treated as a single cluster and then successively combines pairs of clusters until all clusters have been merged into a single cluster. In this process, the dendrogram shows how the aggregations are performed from bottom to top of the dendrogram statically. This procedure allows the cut-off points to flexibly and efficiently represent the number of clusters. Therefore, this study used the number of cut-off points in the dendrogram to validate the cluster number of the agglomerative clustering.

2.4. Study Area and Data Description

A real-world urban stormwater system located in Salt Lake City, UT, U.S., was selected as the case study, shown in Figure 3. This study case, with an area of 81-ha, is semi-arid, and has soil composed of four primary types: alluvial fan, artificial fill, silt and clay, and sand and gravel deposits. The soil surrounding the study area is classified as hydrologic soil groups B and C, with low infiltration capacity, which has a relatively poorly draining surface. Due to climate change and urbanization, the studied area has suffered from floods more frequently than 1990s, and the increase in the magnitude and duration of the storm events has pushed the resulting stormwater system out of service. This urban drainage network was represented by a rainfall-runoff SWMM model. SWMM is a state-of-art tool developed to help support local, state, and national stormwater management objectives to reduce runoff, discharge, and improve stormwater quality [63,64]. It has been widely used all over the world in similar type of investigations including stormwater runoff, combined and sanitary sewers, and other drainage systems [65–67]. Figure 3 shows the components of this SWMM model, which includes one rain gauge, 60 junctions, 61 conduits, two outfalls, and seven sub-catchments, while the groundwater interflow, water evaporation, snowmelt, and manhole hydraulic loss are neglected during the simulation [68].

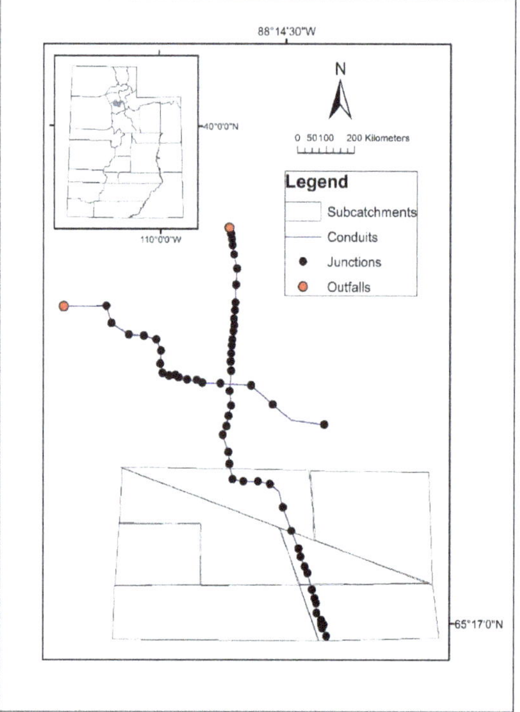

Figure 3. Study area located in the northern Utah state (left-top sub-figure: 1 degree roughly means 106 kilometer), the U.S. and the topological view of the stormwater urban drainage system model plotted by the PCSWMM v.7.3. (major sub-figure, scale unit is kilometer).

For this study, we created 6 artificial precipitation series according to the Chicago distribution method in PCSWMM v.7.3, and then imported them as modeling inputs. The distribution for the synthetic rains is shown in Figure 4. These rainfalls with durations of 3 h, 12 h, to 48 h and return periods ranging from 2-year to 5-year almost contain all typical features and characteristics of real storms in the study area. Additionally, rainfall measurements for two real rainfall events were collected to test the clustering algorithm. These rain records from 5 May 2015 rainfall event and 8 July 2015 rainfall event are representative for the typical real storms under average climatic conditions in the study area. Compared with water depth generated by the artificially designed rainfall data, the time-series water depth produced by the real-world storms contains more non-stationarity and noise. Nevertheless, the obtained findings are subsequent validated with real rain records.

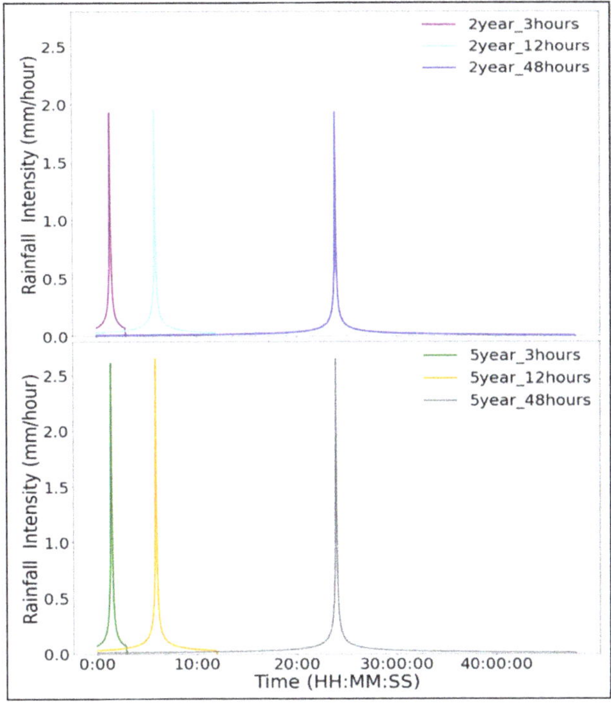

Figure 4. Distribution plots of artificially designed rainfalls with different return periods and rainfall duration.

3. Results

3.1. Clustering Performance Evaluation

3.1.1. K-Means

A detailed investigation was carried out to assess the performance of the clustering algorithms. Figure 4 shows how three performance metrics SCI, CHI and DBI change with different cluster numbers when using K-means to cluster the time-series water depth data. Values for the CHI value increase with higher cluster numbers, whereas the SCI and DBI values fluctuate. The SCI and DBI values show opposite trends, reflecting the different methods by which they are calculated (see Section 2.3 above). In particular, Figure 5b,c show that the best solution is with eight clusters, reflected in the largest SCI value and smallest DBI value. These results suggest that the SCI and DBI are more suitable to assess the performance of K-means, while any peak in the CHI related to cluster quality is eclipsed by the influence of increasing the number of clusters. Based on the SCI and DBI value in Figure 5a, the optimal number of clusters is six for the two year-3 h and five year-3 h rainfall scenarios. The differences in the optimal number of clusters in Figure 5a–c indicate that rainfall duration has impacts on the number of clusters when utilizing K-means to group time-series water depth datasets.

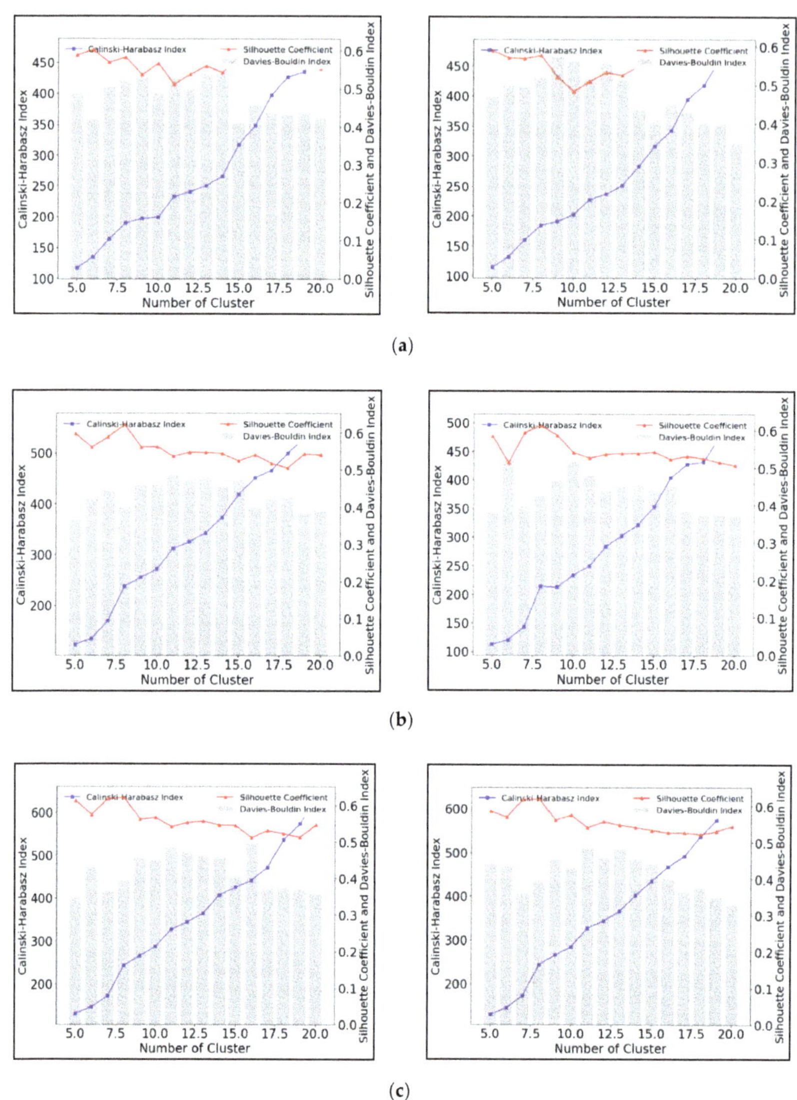

Figure 5. Performance evaluation for K-means Clustering with different cluster numbers under synthetic rainfall scenarios including (**a**) 3-h (left 2-year and right 5-year), (**b**) 12-h (left 2 year and right 5 year), and (**c**) 48-h duration (left 2 year and right 5 year).

3.1.2. Agglomerative Clustering

Figure 6 shows the same results but based on the use of Agglomerative Clustering (AC) to group the time-series water depth data. As with the K-means results (Figure 5), the CHI value increase with the number of clusters for all scenarios from short-duration to long-duration rainfall. Again, it is difficult to identify an optimal number of clusters, and this suggests that the CHI is not suitable for ascertaining the best clustering solution with these data. In contrast, the SCI and DBI show clear peaks in their values. Figure 6a shows that 16 clusters result in the maximum SCI close to 0.76 and minimum DBI with 0.38. Figure 5c shows a peak in SCI values (~0.6) for eight clusters, with a corresponding minimum in the DBI value (<0.4). However, Figure 6b shows that eight clusters could produce the largest SCI (~0.62) and the lowest DBI (~0.40) with the two year-12 h rainfall duration scenario (left subplot), but that 16 clusters are the optimal solution for the two year-12 h rainfall (SCI ~0.58 and DBI ~0.38; right subplot). In summary, the best cluster solutions AC algorithms are 16, eight, and eighteen under 3 h, 12 h, and 48-h duration rainfalls, respectively. Comparing the left subplots with the right subplots (Figure 6) provides evidence that the cluster number for the best AC performance remains the same, although the return period has been shifted from two-year to five-year. The rainfall return period (annual exceedance probability) was found to be less related to the number of clusters.

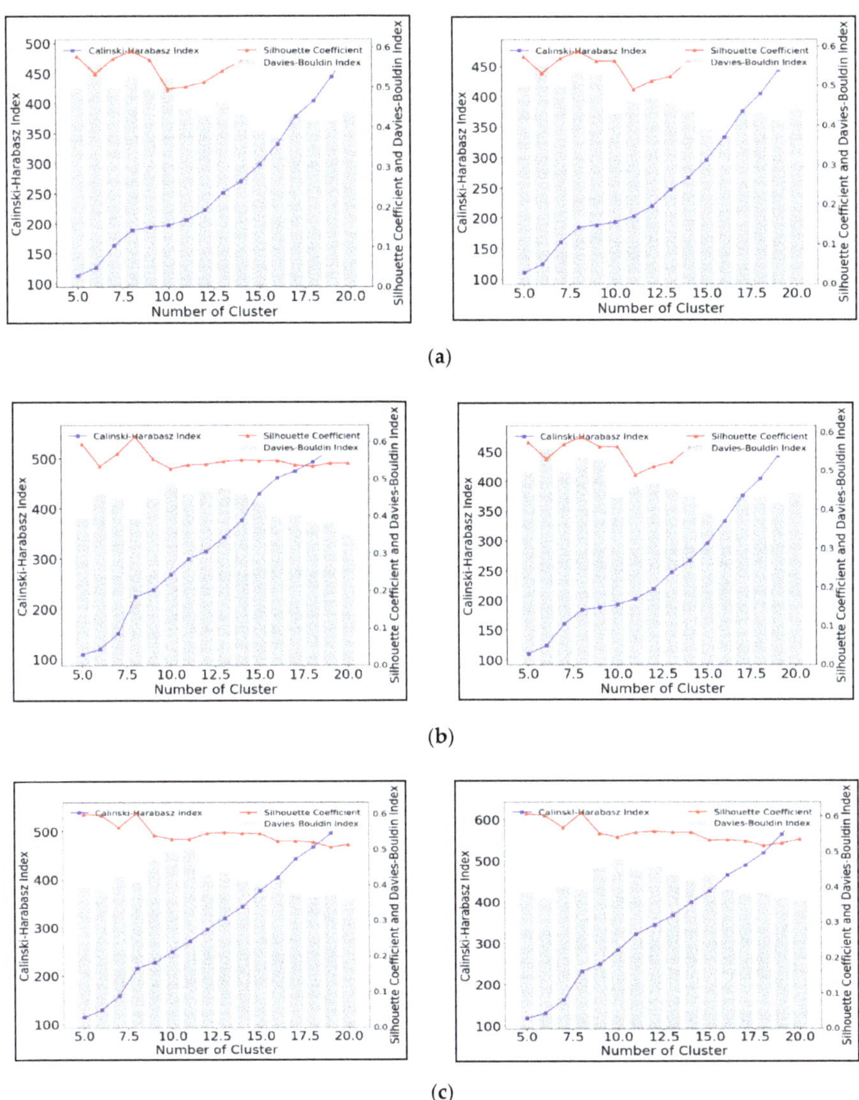

Figure 6. Performance evaluation for Agglomerative Clustering with different cluster numbers under synthetic rainfall scenarios including (**a**) 3-h (left 2-year and right 5-year), (**b**) 12-h (left 2-year and right 5-year), and (**c**) 48-h duration (left 2-year and right 5-year).

3.1.3. Spectral Clustering

Figure 7 shows the results obtained for different cluster numbers using Spectral Clustering to group the time-series water depth data. In contrast to the two previous methods, the SCI values decrease as the number of clusters increase. For the 12 and 48 h scenarios, this index identifies solutions at about 6–7 clusters, but no clear optimal solution is identified in the shorter scenarios (panel a). This suggests that this index is unsuitable for assessing this algorithm. The DBI values show greater variation as the number of clusters change, although minima can be observed at 6 to 7 clusters for most scenarios. The CHI values no longer show a linear increase, but show clear peaks, although usually for higher numbers of clusters than the DBI identifies. The highest CHI values (275 for 2 year-12 h and 190 for 5 year-12 h) are all generated by the SC with 13 clusters. For the for two year-48 h and five year-48 h scenarios, the largest CHI values are approximately 200 and 270, respectively, in both cases for 12 clusters.

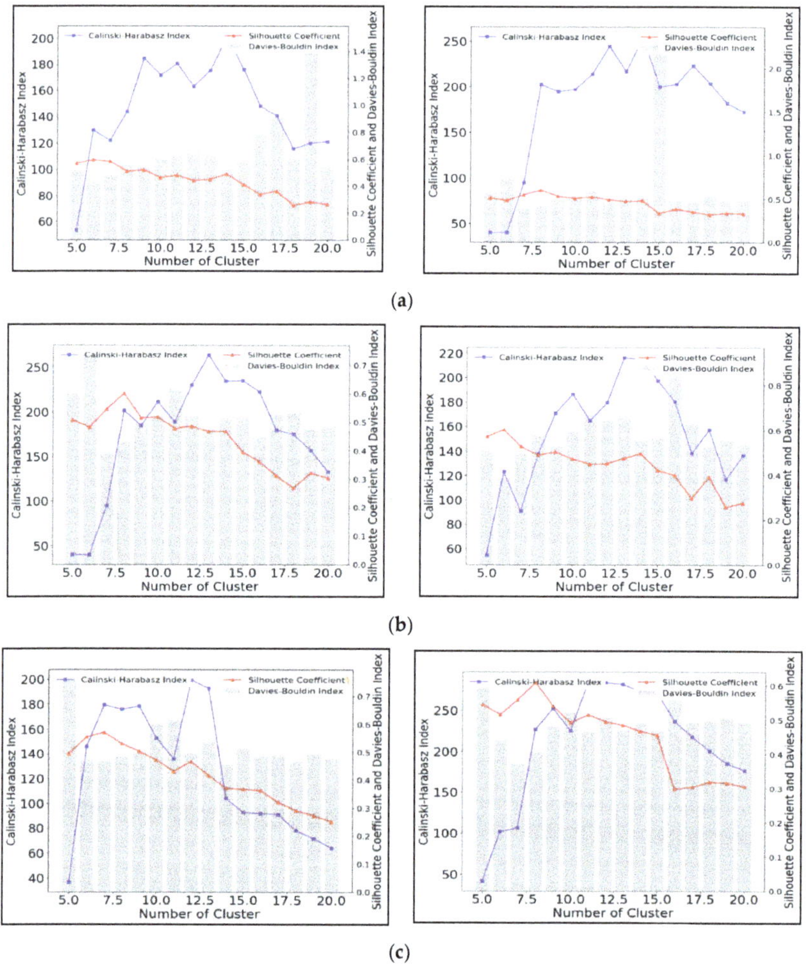

Figure 7. Performance evaluation for Spectral Clustering with different cluster numbers under synthetic rainfall scenarios including (**a**) 3-h (left 2-year and right 5-year), (**b**) 12-h (left 2-year and right 5-year), and (**c**) 48-h duration (left 2-year and right 5-year).

3.2. Clustering Performance Testing

The analysis of cluster performance in the previous section is based on synthetic rainfall datasets, due to lack of water depth data in the drainage network. However, the use of noise-free synthetic data may have a significant impact on the results obtained [69], and our results may not represent real storm situations or current climate conditions. In contrast, the trends identified here might be masked by time series noise, making it more difficult to identify optimal solutions. In order to validate that the results obtained from designed rainfalls can also be applied to non-stationary real-storms, we evaluate the performance of the clusters in grouping flooding water depth datasets generated by two real flood events described below.

The left plot in Figure 8 indicates that the best number of clusters for the 5 May 2015 event (Figure 8a) and 8 July 2015 event (Figure 8b) are five and four, respectively. Increasing the number of clusters beyond this causes both the SCI and the DBI to decline. The distribution of different clusters obtained is shown in the PCA plots in the right panel of Figure 7. These show that the cluster analysis resulted in a good separation of the storm events (indicated by the lack of overlap between the gray circles).It should be noted that both subplots 8a and 8b have an isolated cluster on the top. This is the only cluster composed of one sample, which means the water depth from the corresponding junction is significantly distinguishable to others. One possible reason for this phenomena is that the flooding or overflow events have occurred, triggering a very different signal in water depth at this location. Besides, as the rainfall duration increases from 3 h (the 5 May 2015 storm) to 24 h (the 8 July 2015 storm), the reduction in the number of clusters selected is in line with the results of Section 4, supporting the negative correlation between the number of clusters and event duration.

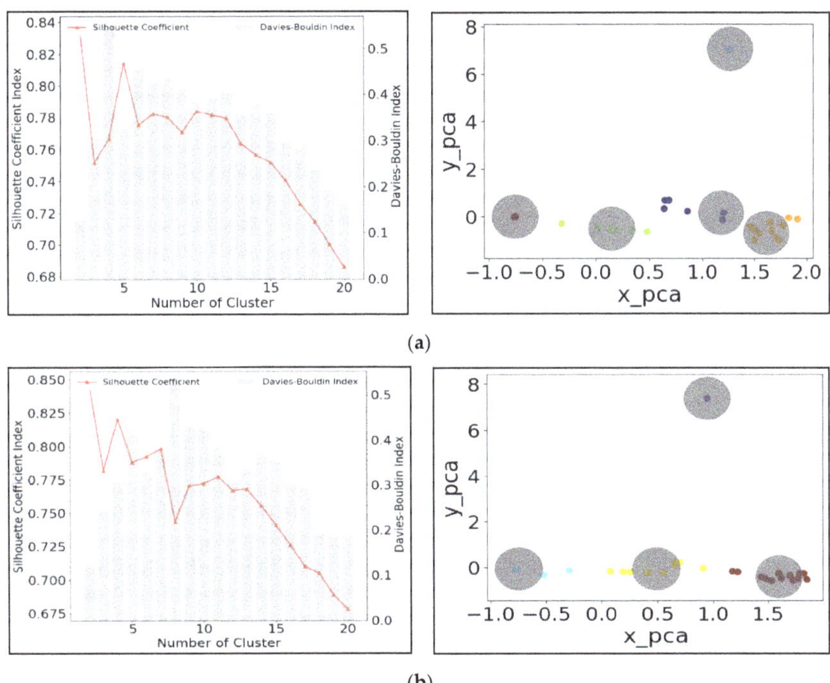

Figure 8. Cluster analysis test for time-series water depth generated by (**a**) 5 May 2015 flooding event; (**b**) 8 July 2015 flooding event (gray circles same to clusters), (x_pca means the first component score; y_pca means the second component score; The principal component scores are used to examine if these two clusters are reasonably distinguished from each other clustering).

3.3. Cluster Number Validation

The dendrogram plots are also used to validate the number of clusters. Figure 9 shows the dendrogram plots obtained from applying the AC algorithm to the flooding water depth data. Generally, the cut-off point should be at least 70% dissimilarity between two clusters or cutting where the dendrogram difference is most significant [69]. The number of clusters was selected by using a distance threshold of 0.9 distance or 90% dissimilarity, and this is plotted as a horizontal cut-off line in all dendrograms of Figure 9. The cross points (highlighted as green X in dendrogram) between the cut-off line and dendrogram leaves identify the accepted clusters. In Figure 9, one point identified by the cut-off line (junction 8; highlighted as red X in dendrogram) was considered as an outlier in the dendrogram and excluded. In practice, this algorithm might be helpful for anomaly detection in the sensor monitoring network. For instance, real-time monitoring is built to capture the varying different features of measurements as much as possible within a limited number of sensors [70,71]. Further, the clusters represent different parts of the hydrological network and can be used to help target locations for sensor deployment to observe overflow and flood events in the field.

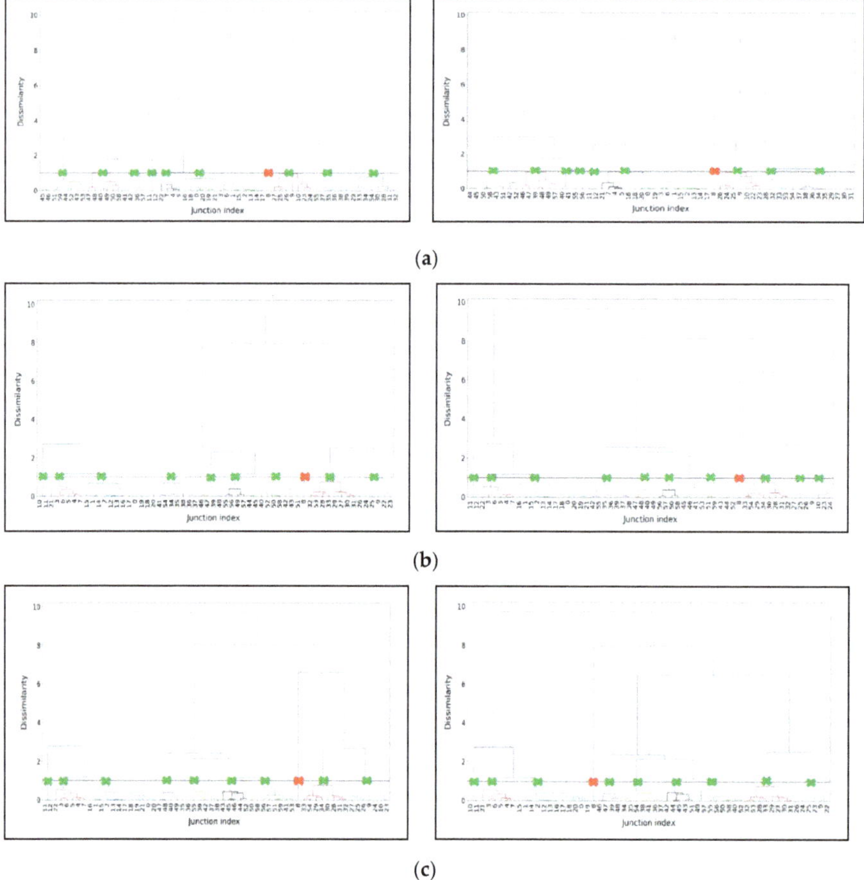

Figure 9. Dendrogram (green X representing acceptable cluster; red X representing unacceptable cluster) for comparing agglomerative cluster numbers between 2-year return period (the left subplots) and 5-year return period (the right subplots) rainfall scenarios. (**a**): left 2 year-3 h; right 5 year-3 h; (**b**): left 2 year-12 h; right 5 year-12 h; (**c**): left 2 year-48 h; right 5 year-48 h.

The vertical comparisons among the subplots of Figure 9a–c disclosed that the appropriate cluster numbers for 3 h, 12 h and 48 h rainfall scenarios are quite similar: eight, nine, and nine, respectively. Meanwhile, comparing cluster solutions for different time periods (e.g., left and right plot of Figure 9a, the number of clusters and their structure is remarkably similar, implying that the event return period has fewer impacts on AC model performance. This supports the conclusions reached with the synthetic time series, that the AC model performance noticeably depends on the flooding duration but not the event return period (exceedance probability).

This study adopted intra-cluster distance as the metric to assess the effects of flooding duration and return period (exceedance probability) on the performance of the K-means and Spectral Clustering algorithm. Figure 10 shows the results of this comparison, with the decay in the intra-cluster distance as the number of clusters increases. A notable elbow point (the cross between red dashed line and intra-distance curves) can be seen at the four clusters, as the decrease in distances becomes much smaller. Using the elbow criterion described in Section 2.3.4, this suggests that four clusters are the best solution. Increasing the number of clusters beyond this would result in a little additional gain for the extra complexity of the solution. Figure 10 shows that the intra-cluster distance changes in a similar way for all six rainfall scenarios, and that the intra-cluster distance is close in those rainfalls with the same duration. For example, the solid purple line with purple circle markers (representing two year-3 h rainfall scenario) overlaps the red dashed line with the red circle markers (representing five year-3 h rainfall scenario). However, there are still some differences between scenarios with different rainfall duration. Notably, the intra-cluster distance increases as the rainfall duration decreases (the distance for the '3 h' duration rainfall is the largest, followed by the '12 h' cases, and then the '48 h' scenarios). As a metric for clustering performance, intra-cluster distance is therefore useful in determining how well these algorithms group the water depth time-series. These results suggest that the K-means and spectral clustering algorithms work best with longer duration rainfalls, implying that the longer event duration produces greater similarity in the water depth at different junctions. This, coupled with the larger set of observations from a longer period, results in better formed individual clusters. Wu et al. have shown that these cluster methods work optimally when trained on massive datasets, which is supported by the results herein [72].

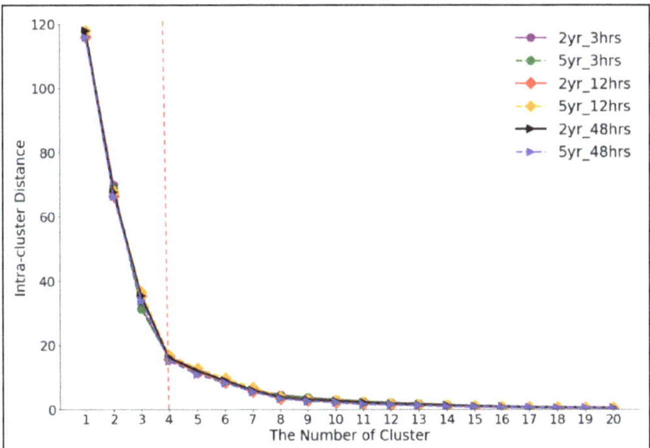

Figure 10. Cluster Intra-distance for comparing the effects of rainfall duration and return period on the performance of K-means and Spectral model (elbow point is the cross between the red dash-line and curves) under 6 synthetic rainfall scenarios ('yr' represents year while 'hrs' stands for h).

4. Discussions

4.1. Clustering Parametric Discussion

Previous cluster-based studies have mainly focused on detecting pressure, demand, pipe burst, infrastructure damage, and illicit intrusion in water distribution systems [71–73]. In the cluster analysis here, the features, such as the length of time-series water depth from UDSs, are found to be negatively correlated with the number of clusters. This finding has been validated by the dendrogram cut-off points in those designed rainfalls and also by the cluster center mapping based on real storm events. The similar results between the artificial (noise-free) and practical (noise-polluted) scenario infer that event duration (data length) overwhelms the event exceedance probability (data magnitude) in the cluster number identification, which agrees with the findings from [25,72]. Increasing the number of clusters often results in many more errors. One extreme case is that the zero error happens when each data point is equal to every cluster. Intuitively, the choice of the best number of clusters can be interpreted into a trade-off between the maximum reduction of complexity of the data with a single cluster and maximum accuracy by assigning each data point to its cluster. For long time series, we suggest starting with a small number of clusters and increasing the number, testing the performance at each increase.

In addition to the determination of the number of clusters, the structure of datasets may also affect the clustering model performance. KC and SC algorithms are able to robustly group water depth datasets from longer duration flood events. However, there is a limited relationship between algorithm performance and annual exceedance probability. The sharply rising trend (Figures 4–6) demonstrates that the CHI is not suitable to identify the best number of clusters in the KC and AC algorithms, but that the SCI and DBI work quite well and give comparable results (Figures 4–6). In contrast, the CHI works well in identifying the optimal cluster number with the SC algorithm. This difference reflects the different nature of the algorithms: KC and AC are based on simple dissimilarity measures between observations, whereas the SC is based on a graph representing connectivity. This is because that DBI evaluates intra-cluster similarity among every data point and inter-cluster differences among each group. Similarly, the SCI measures the distance between each data point and the centroid of the cluster it was assigned to. An SCI value close to 1 is always good, and a DBI value close to 0 is also good whatever clustering you are trying to evaluate. However, the CHI is not normalized, and it is difficult to compare two values of the CHI index from different data sets.

4.2. Implications of Clustering Application

This study provides an understanding of different clustering algorithms, applicability with different datasets, and an assessment of cluster solutions in flood detection strategies. For instance, as water level is one of the inferential indicators of local flood events, clusters with abnormal water level can be identified as early warning signals of flooding. As new data become availabel during monitoring, these can be assigned to the most similar cluster. Decreasing dissimilarity to abnormal cluster therefore indicates increasing likelihood of flooding. In Figure 8, we observed that there is one isolated dot for each subplot. These separated points represent the highly dissimilar water depth data, indicating the possibility of triggering flood events. These same cases are also captured in the dendrogram of Figure 9 which presents that the junction 8 highlighted with red cross might be the source of anomalous water level. One reasonable explanation for the anomalous cluster is the resultant flooding or overflow events occuring around the corresponding location. More attention are recommended to investigate if this location is flooded. Thus, it can be seen that classifying these points as anomalies is helpful for narrowing down the spatial searching domain from network-level to node-level, and consequently also reducing the timing and efforts in identifying the flooded locations in the complex network system [74–76]. We concluded that the occurrence of anomalous changes in water level in UDSs could be a timely reminder of the upstream or downstream overflow events for the neighborhoods. Our findings also explain how the characteristics of the dataset (notably length and magnitude) influence the number of clusters. This information could be employed to detect urban flood events using water depth datasets in other real drainage networks [66,67]. These clustering algorithms aim to efficiently capture the urban drainage flooding locations providing a basis for managing the existing drainage structures and developing sustainable urban drainage networks in urbanized areas [77].

4.3. Limitations and Future Work

Although this study has identified some clear differences in the application of cluster analysis, there are several limitations. Firstly, the majority of scenarios used time-series water depth datasets generated by model simulation. As these are smooth and noise-free, the results may not scale to field application. However, we found similarities between the results with the limited set of observed rainfall series used here, notably in the use of the different indices, but tend to result in a smaller number of clusters. Further work should apply these methods to a wider set of observed data to reduce the input (meteorological) uncertainties and meteorological variances if such data becomes available [36,37,78,79]. The possible integration of ensemble prediction system (EPS) and data assimilation techniques might be of interest for future work, which could provide help for estimating forecast uncertainty via a linear combination of suitable meteorological variances and uncertainties linked to the rainfall and hydraulics [80,81]. Secondly, as this paper only focuses on exploring usefulness of clustering model implementation and performance evaluation, analysis of errors and sensitivity analysis of water level datasets are recommended for to improve the reliability of results. Future work will concentrate on the application of these methods, including water-level sensor placement, combined sewer overflow detection, and urban flooding prediction. Since the dendrogram enables the AC algorithm to detect outliers in time-series water depth datasets, this can be used to help guide sensor deployment on vulnerable sites for observing overflow and flood events in the field [76]. It is planned to consider strengthening the connection between the theoretical results and field application by conducting a cluster analysis to optimize the sensor monitoring network for flooding detection at UDSs.

5. Summary and Conclusions

In the age of 'smart stormwater,' the increased deployment of sensors to monitor water level characteristics is resulting in rapidly accumulating data. It is becoming crucial to understand and promote methods to handle these big datasets to help in flood detection and control. This study aims to promote understanding of how cluster analysis facilitates the interpretation of the unlabeled time-series water depth data for flooding location detection at the stormwater urban drainage systems. In this work, three indexes, including silhouette coefficient index, Calinski–Harabasz index, and Davies–Bouldin index, were used to evaluate the performance of three popular unsupervised cluster analysis models namely K-means clustering, agglomerative clustering and spectral clustering. A real-world stormwater urban drainage systems SWMM model was applied to test the performance of clustering algorithms in capturing urban floods. Five conclusions were drawn below:

(1) Silhouette coefficient index and Davies–Bouldin index are suitable metrics to measure the performance of K-means and agglomerative clustering model when subject to identify the number of clusters for the best performance. However, the Calinski–Harabasz Index is found to be more favorable to assess the performance of the spectral clustering model in grouping time-series water depth datasets for urban drainage flooding detection.

(2) In K-means and spectral clustering models, the number of the clusters for maximizing model performance is highly related to the dataset length (flooding duration) but is slightly associated with the dataset magnitude. There is a negative correlation between the number of clusters and the length of datasets.

(3) The short-period water depth data can be well-grouped by the agglomerative clustering model. In contrast, K-means and spectral clustering models are better able to handle time-series water depth datasets from long-duration storm scenarios.

(4) This research work provides insight into unlabeled hydraulic data-driven techniques by conducting clustering experiments. The outcomes are useful for researchers to select the appropriate clustering model and to choose the corresponding performance metrics for specific urban flooding applications.

(5) The detailed analyses in this work provide guidance concerning how to use cluster solutions to isolate or prescreen vulnerable locations for flooded location detection strategies. The water level in isolated clusters can be considered as the floods early warning for the local residents. The occurrence of anomalous changes in water level in urban drainage systems could be a timely reminder of the upstream or downstream flood events for the surrounding neighborhoods.

Author Contributions: Conceptualization, J.L.; Data curation, J.L.; Formal analysis, J.L. and D.H.; Funding acquisition, J.L. and R.S.; Investigation, J.L., D.H., and S.B.; Methodology, J.L. and S.B.; Resources, R.S.; Software, J.L.; Validation, J.L.; Visualization, J.L. and D.H.; Writing—original draft, J.L.; Writing—review & editing, D.H., S.B., and R.S. All authors have read and agreed to the published version of the manuscript.

Funding: This research is funded by the Research Assistant Scholarship at the University of Utah. The contribution of the University Innsbruck is financially supported by the Austrian Climate and Energy Fund and by the programme "Smart Cities Demo - Living Urban Innovation 2018" (project 872123).

Acknowledgments: We would like to thank the Salt Lake City Department of Public Utilities (SLCDPU) for their efforts in developing the SWMM model. We also thank the Computational Hydraulics Int. (CHI) company for offering the research license of PCSWMM. Finally, we thank Steven Burian and the Student Engineering Association (SAE) of the University of Utah for sharing their relevant files and materials.

Conflicts of Interest: The authors declare no conflict of interest.

References

1. Li, J.; Tao, T.; Kreidler, M.; Burian, S.; Yan, H. Construction Cost-Based Effectiveness Analysis of Green and Grey Infrastructure in Controlling Flood Inundation: A Case Study. *J. Water Manag. Model.* **2019**, *27*, C466. [CrossRef]

2. Kerkez, B.; Gruden, C.; Lewis, M.; Montestruque, L.; Quigley, M.; Wong, B.; Bedig, A.; Kertesz, R.; Braun, T.; Cadwalader, O.; et al. Smarter stormwater systems. *Environ. Sci. Technol.* **2016**, *50*, 7267–7273. [CrossRef]

3. Li, J.; Yang, X.; Sitzenfrei, R. Rethinking the framework of smart water system: A review. *Water (Switzerland)* **2020**, *12*, 412. [CrossRef]

4. Morales, V.M.; Mier, J.M.; Garcia, M.H. Innovative modeling framework for combined sewer overflows prediction. *Urban Water J.* **2017**, *14*, 97–111. [CrossRef]

5. Norbiato, D.; Borga, M.; Degli Esposti, S.; Gaume, E.; Anquetin, S. Flash flood warning based on rainfall thresholds and soil moisture conditions: An assessment for gauged and ungauged basins. *J. Hydrol.* **2008**, *362*, 274–290. [CrossRef]

6. Wong, B.P.; Kerkez, B. Adaptivemeasurements of urban runoff quality. *Water Resour. Res.* **2016**, *52*, 8986–9000. [CrossRef]

7. Solomatine, D.P.; Ostfeld, A. Data-driven modelling: Some past experiences and new approaches. *J. Hydroinformatics* **2008**, *10*, 3–22. [CrossRef]

8. Henonin, J.; Russo, B.; Mark, O.; Gourbesville, P. Real-time urban flood forecasting and modelling—A state of the art. *J. Hydroinformatics* **2013**, *15*, 717–736. [CrossRef]

9. Koo, D.; Piratla, K.; Matthews, C.J. Towards Sustainable Water Supply: Schematic Development of Big Data Collection Using Internet of Things (IoT). *Procedia Eng.* **2015**, *118*, 489–497. [CrossRef]

10. Vojinovic, Z.; Abbott, M.B. Twenty-five years of hydroinformatics. *Water* **2017**, *9*, 59. [CrossRef]

11. Diao, K.; Farmani, R.; Fu, G.; Astaraie-Imani, M.; Ward, S.; Butler, D. Cluster analysis of water distribution systems: Identifying critical components and community impacts. *Water Sci. Technol.* **2014**, *70*, 1764–1773. [CrossRef] [PubMed]

12. Kang, O.Y.; Lee, S.C.; Wasewar, K.; Kim, M.J.; Liu, H.; Oh, T.S.; Janghorban, E.; Yoo, C.K. Determination of key sensor locations for non-point pollutant sources management in sewer network. *Korean J. Chem. Eng.* **2013**, *30*, 20–26. [CrossRef]

13. Mullapudi, A.; Lewis, M.J.; Gruden, C.L.; Kerkez, B. Deep reinforcement learning for the real time control of stormwater systems. *Adv. Water Resour.* **2020**, *140*, 103600. [CrossRef]

14. Tehrany, M.S.; Pradhan, B.; Jebur, M.N. Flood susceptibility mapping using a novel ensemble weights-of-evidence and support vector machine models in GIS. *J. Hydrol.* **2014**. [CrossRef]

15. Yu, P.S.; Yang, T.C.; Chen, S.Y.; Kuo, C.M.; Tseng, H.W. Comparison of random forests and support vector machine for real-time radar-derived rainfall forecasting. *J. Hydrol.* **2017**, *118*, 489–497. [CrossRef]

16. Shu, C.; Ouarda, T.B.M.J. Regional flood frequency analysis at ungauged sites using the adaptive neuro-fuzzy inference system. *J. Hydrol.* **2008**, *552*, 92–104. [CrossRef]

17. Zadeh, M.R.; Amin, S.; Khalili, D.; Singh, V.P. Daily Outflow Prediction by Multi Layer Perceptron with Logistic Sigmoid and Tangent Sigmoid Activation Functions. *Water Resour. Manag.* **2010**, *24*, 2673–2688. [CrossRef]

18. Wang, Z.; Lai, C.; Chen, X.; Yang, B.; Zhao, S.; Bai, X. Flood hazard risk assessment model based on random forest. *J. Hydrol.* **2015**, *527*, 1130–1141. [CrossRef]

19. Choubin, B.; Darabi, H.; Rahmati, O.; Sajedi-Hosseini, F.; Kløve, B. River suspended sediment modelling using the CART model: A comparative study of machine learning techniques. *Sci. Total Environ.* **2018**, *615*, 272–281. [CrossRef]

20. Bowes, B.D.; Sadler, J.M.; Morsy, M.M.; Behl, M.; Goodall, J.L. Forecasting groundwater table in a flood prone coastal city with long short-term memory and recurrent neural networks. *Water (Switzerland)* **2019**, *11*, 1098. [CrossRef]

21. Hu, Y.; Scavia, D.; Kerkez, B. Are all data useful? Inferring causality to predict flows across sewer and drainage systems using directed information and boosted regression trees. *Water Res.* **2018**, *145*, 697–706. [CrossRef] [PubMed]

22. Li, J. A data-driven improved fuzzy logic control optimization-simulation tool for reducing flooding volume at downstream urban drainage systems. *Sci. Total Environ.* **2020**, *732*, 138931. [CrossRef] [PubMed]

23. Yang, J.; Ye, M.; Tang, Z.; Jiao, T.; Song, X.; Pei, Y.; Liu, H. Using cluster analysis for understanding spatial and temporal patterns and controlling factors of groundwater geochemistry in a regional aquifer. *J. Hydrol.* **2020**, *583*, 124594. [CrossRef]

24. Jain, A.K.; Murty, M.N.; Flynn, P.J. Data clustering: A review. *ACM Comput. Surv.* **1999**, *31*, 264–323. [CrossRef]

25. Wu, Y.; Liu, S.; Wu, X.; Liu, Y.; Guan, Y. Burst detection in district metering areas using a data driven clustering algorithm. *Water Res.* **2016**, *100*, 28–37. [CrossRef] [PubMed]

26. Perelman, L.; Ostfeld, A. Topological clustering for water distribution systems analysis. *Environ. Model. Softw.* **2011**, *26*, 969–972. [CrossRef]

27. Sela Perelman, L.; Allen, M.; Preis, A.; Iqbal, M.; Whittle, A.J. Automated sub-zoning of water distribution systems. *Environ. Model. Softw.* **2015**, *65*, 1–14. [CrossRef]

28. Keogh, E.; Lin, J. Clustering of time-series subsequences is meaningless: Implications for previous and future research. *Knowl. Inf. Syst.* **2005**, *8*, 154–177. [CrossRef]

29. Chen, J.R. Making subsequence time series clustering meaningful. In Proceedings of the Fifth IEEE International Conference on Data Mining (ICDM'05), Houston, TX, USA, 27–30 November 2005; pp. 114–121.

30. Chen, J.R. Useful clustering outcomes from meaningful time series clustering. *Conf. Res. Pract. Inf. Technol. Ser.* **2007**, *70*, 101–109.

31. Rodriguez, A.; Laio, A. Clustering by fast search and find of density peaks. *Science* **2014**, *344*, 1492–1496. [CrossRef]

32. Xing, L.; Sela, L. Unsteady pressure patterns discovery from high-frequency sensing in water distribution systems. *Water Res.* **2019**, *158*, 291–300. [CrossRef] [PubMed]

33. Xu, D.; Tian, Y. A Comprehensive Survey of Clustering Algorithms. *Ann. Data Sci.* **2015**, *2*, 165–193. [CrossRef]

34. Aggarwal, C.C.; Zhai, C.X. A survey of text clustering algorithms. In *Mining Text Data*; Springer: Boston, MA, USA, 2012; pp. 77–128. ISBN 9781461432234.

35. Mosavi, A.; Ozturk, P.; Chau, K.W. Flood prediction using machine learning models: Literature review. *Water (Switzerland)* **2018**, *10*, 1536. [CrossRef]
36. Mel, R.A.; Viero, D.P.; Carniello, L.; D'Alpaos, L. Optimal floodgate operation for river flood management: The case study of Padova (Italy). *J. Hydrol. Reg. Stud.* **2020**, *30*, 100702. [CrossRef]
37. Mel, R.A.; Viero, D.P.; Carniello, L.; D'Alpaos, L. Multipurpose use of artificial channel networks for flood risk reduction: The case of the waterway Padova-Venice (Italy). *Water (Switzerland)* **2020**, *12*, 1609. [CrossRef]
38. Hsu, M.H.; Chen, S.H.; Chang, T.J. Inundation simulation for urban drainage basin with storm sewer system. *J. Hydrol.* **2000**, *234*, 21–37. [CrossRef]
39. Yaseen, Z.M.; Sulaiman, S.O.; Deo, R.C.; Chau, K.W. An enhanced extreme learning machine model for river flow forecasting: State-of-the-art, practical applications in water resource engineering area and future research direction. *J. Hydrol.* **2019**, *569*, 387–408. [CrossRef]
40. Fotovatikhah, F.; Herrera, M.; Shamshirband, S.; Chau, K.W.; Ardabili, S.F.; Piran, M.J. Survey of computational intelligence as basis to big flood management: Challenges, research directions and future work. *Eng. Appl. Comput. Fluid Mech.* **2018**, *12*, 411–437. [CrossRef]
41. Kubat, M. *An Introduction to Machine Learning*; Publisher: New York City, NY, USA, 2017; ISBN 9783319639130.
42. Xu, R.; Wunsch, D. Survey of clustering algorithms. *IEEE Trans. Neural Networks* **2005**, *16*, 645–678. [CrossRef]
43. Shannon, W.D. 11 Cluster Analysis. *Handb. Stat.* **2007**, *27*, 342–366.
44. Celebi, M.E.; Kingravi, H.A.; Vela, P.A. A comparative study of efficient initialization methods for the k-means clustering algorithm. *Expert Syst. Appl.* **2013**, *40*, 200–210. [CrossRef]
45. Lloyd, S.P. Least Squares Quantization in PCM. *IEEE Trans. Inf. Theory* **1982**, *28*, 129–137. [CrossRef]
46. Stanford, M. Chapter 7 Hierarchical cluster analysis. *Stat. Med.* **2012**, *2*, 1–11.
47. Danielsson, P.E. Euclidean distance mapping. *Comput. Graph. Image Process.* **1980**, *14*, 227–248. [CrossRef]
48. Forina, M.; Armanino, C.; Raggio, V. Clustering with dendrograms on interpretation variables. *Anal. Chim. Acta* **2002**, *454*, 13–19. [CrossRef]
49. Von Luxburg, U. A tutorial on spectral clustering. *Stat. Comput.* **2007**, *17*, 395–416. [CrossRef]
50. Bro, R.; Smilde, A.K. Principal component analysis. *Anal. Methods* **2014**, *6*, 2812–2831. [CrossRef]
51. Maier, H.R.; Kapelan, Z.; Kasprzyk, J.; Kollat, J.; Matott, L.S.; Cunha, M.C.; Dandy, G.C.; Gibbs, M.S.; Keedwell, E.; Marchi, A.; et al. Evolutionary algorithms and other metaheuristics in water resources: Current status, research challenges and future directions. *Environ. Model. Softw.* **2014**, *62*, 271–299. [CrossRef]
52. Aghabozorgi, S.; Seyed Shirkhorshidi, A.; Ying Wah, T. Time-series clustering—A decade review. *Inf. Syst.* **2015**, *53*, 16–38. [CrossRef]
53. Rokach, L.; Maimon, O. Clustering Methods. *Data Min. Knowl. Discov. Handb.* **2006**, *14*, 321–352. [CrossRef]
54. Hastie, T.; Tibshirani, R.; Friedman, J. *The Elements of Statistical Learning: The Elements of Statistical Learning Data Mining, Inference, and Prediction*, 2nd ed.; Publisher: New York City, NY, USA, 2009; ISBN 978-0-387-84858-7.
55. Pedregosa, F.; Varoquaux, G.; Gramfort, A.; Michel, V.; Thirion, B.; Grisel, O.; Blondel, M.; Prettenhofer, P.; Weiss, R.; Dubourg, V.; et al. Scikit-learn: Machine learning in Python. *J. Mach. Learn. Res.* **2011**, *12*, 2825–2830.
56. Maulik, U.; Bandyopadhyay, S. Performance evaluation of some clustering algorithms and validity indices. *IEEE Trans. Pattern Anal. Mach. Intell.* **2002**, *24*, 1650–1654. [CrossRef]
57. Al-Zoubi, M.B.; Al Rawi, M. An efficient approach for computing silhouette coefficients. *J. Comput. Sci.* **2008**, *4*, 252–255. [CrossRef]
58. Aranganayagi, S.; Thangavel, K. Clustering categorical data using silhouette coefficient as a relocating measure. In Proceedings of the Proceedings—International Conference on Computational Intelligence and Multimedia Applications, Sivakasi, Tamil Nadu, India, 13–15 December 2007; Volume 2, pp. 13–17.
59. Caliñski, T.; Harabasz, J. A Dendrite Method Foe Cluster Analysis. *Commun. Stat.* **1974**, *3*, 1–27. [CrossRef]
60. Davies, D.L.; Bouldin, D.W. A Cluster Separation Measure. *IEEE Trans. Pattern Anal. Mach. Intell.* **1979**, *PAMI-1*, 224–227. [CrossRef]
61. Petrovic, S. A Comparison Between the Silhouette Index and the Davies-Bouldin Index in Labelling IDS Clusters. In Proceedings of the 11th Nordic Workshop of Secure IT Systems, Linköping, Sweden, 19–20 October 2006; pp. 53–64.
62. Xiao, J.; Lu, J.; Li, X. Davies Bouldin Index based hierarchical initialization K-means. *Intell. Data Anal.* **2017**, *21*, 1327–1338. [CrossRef]
63. Thorndike, R.L. Who belongs in the family? *Psychometrika* **1953**, *18*, 267–276. [CrossRef]

64. Rossman, L.A. *Storm Water Management Model User's Manual Version 5.1*; EPA/600/R-14/413b; Natl. Risk Manag. Lab. Off. Res. Dev. United States Environ. Prot. Agency: Cincinnati, OH, USA, 2015.

65. Li, J.; Burian, S.; Oroza, C. Exploring the potential for simulating system-level controlled smart stormwater system. In Proceedings of the World Environmental and Water Resources Congress 2019: Water, Wastewater, and Stormwater; Urban Water Resources; and Municipal Water Infrastructure—Selected Papers from the World Environmental and Water Resources Congress, Pittsburgh, Pennsylvania, 19–23 May 2019; pp. 46–56.

66. Kroll, S.; Weemaes, M.; Van Impe, J.; Willems, P. A methodology for the design of RTC strategies for combined sewer networks. *Water (Switzerland)* **2018**, *10*, 1675. [CrossRef]

67. Rinaldo, A.; Rodriguez-Iturbe, I. Geomorphological theory of the hydrological response. *Hydrol. Process.* **1996**, *10*, 803–829. [CrossRef]

68. Moazenzadeh, R.; Mohammadi, B.; Shamshirband, S.; Chau, K.W. Coupling a firefly algorithm with support vector regression to predict evaporation in northern iran. *Eng. Appl. Comput. Fluid Mech.* **2018**, *12*, 584–597. [CrossRef]

69. Suzuki, R.; Shimodaira, H. Pvclust: An R package for assessing the uncertainty in hierarchical clustering. *Bioinformatics* **2006**, *22*, 1540–1542. [CrossRef] [PubMed]

70. Sambito, M.; Di Cristo, C.; Freni, G.; Leopardi, A. Optimal water quality sensor positioning in urban drainage systems for illicit intrusion identification. *J. Hydroinform.* **2020**, *22*, 46–60. [CrossRef]

71. Shende, S.; Chau, K.W. Design of water distribution systems using an intelligent simple benchmarking algorithm with respect to cost optimization and computational efficiency. *Water Sci. Technol. Water Supply* **2019**, *19*, 1892–1898. [CrossRef]

72. Wu, Y.; Liu, S. Burst Detection by Analyzing Shape Similarity of Time Series Subsequences in District Metering Areas. *J. Water Resour. Plan. Manag.* **2020**, *146*, 04019068. [CrossRef]

73. Mel, R.; Sterl, A.; Lionello, P. High resolution climate projection of storm surge at the Venetian coast. *Nat. Hazards Earth Syst. Sci.* **2013**, *13*, 1135–1142. [CrossRef]

74. Flowerdew, J.; Horsburghb, K.; Wilson, C.; Mylne, K. Development and evaluation of an ensemble forecasting system for coastal storm surges. *Q. J. R. Meteorol. Soc.* **2010**, *136*, 1444–1456. [CrossRef]

75. Chang, L.C.; Shen, H.Y.; Wang, Y.F.; Huang, J.Y.; Lin, Y.T. Clustering-based hybrid inundation model for forecasting flood inundation depths. *J. Hydrol.* **2010**, *385*, 257–268. [CrossRef]

76. Guo, X.; Zhao, D.; Du, P.; Li, M. Automatic setting of urban drainage pipe monitoring points based on scenario simulation and fuzzy clustering. *Urban Water J.* **2018**, *15*, 700–712. [CrossRef]

77. Mel, R.; Viero, D.P.; Carniello, L.; Defina, A.; D'Alpaos, L. Simplified methods for real-time prediction of storm surge uncertainty: The city of Venice case study. *Adv. Water Resour.* **2014**, *71*, 177–185. [CrossRef]

78. Sitzenfrei, R.; Rauch, W. Optimizing small hydropower systems in water distribution systems based on long-time-series simulation and future scenarios. *J. Water Resour. Plan. Manag.* **2015**, *141*, 04015021. [CrossRef]

79. Lionello, P.; Sanna, A.; Elvini, E.; Mufato, R. A data assimilation procedure for operational prediction of storm surge in the northern Adriatic Sea. *Cont. Shelf Res.* **2006**. [CrossRef]

80. Buizza, R.; Milleer, M.; Palmer, T.N. Stochastic representation of model uncertainties in the ECMWF ensemble prediction system. *Q. J. R. Meteorol. Soc.* **2007**, *26*, 539–553. [CrossRef]

81. Panganiban, E.B.; Cruz, J.C.D. Rain water level information with flood warning system using flat clustering predictive technique. In Proceedings of the IEEE Region 10 Annual International Conference, Penang, Malaysia, 5–8 November 2017; Volume 2017, pp. 727–732.

Article

Comparative Analysis of Reliability Indices and Hydraulic Measures for Water Distribution Network Performance Evaluation

Gimoon Jeong and Doosun Kang *

Department of Civil Engineering, Kyung Hee University, 1732 Deogyeong-daero, Giheung-gu, Yongin-si, Gyeonggi-do 17104, Korea; gimoon1118@gmail.com
* Correspondence: doosunkang@khu.ac.kr

Received: 5 August 2020; Accepted: 25 August 2020; Published: 26 August 2020

Abstract: The performance of water distribution networks (WDNs) can be quantified by several types of hydraulic measure. In design and operation of a WDN, sufficient consideration should be given to system performance, and it would be inefficient to separately consider individual characteristics of hydraulic measures. Instead, various reliability indices have been developed and utilized to evaluate the performance of WDNs; however, deciding which index to use according to a particular WDN situation has not been investigated in sufficient depth. In this regard, this study analyzes the correlation between representative reliability indices and hydraulic measures to propose the most adequate reliability index according to the desired system performance in various situations. Specifically, six hydraulic measures representing system performance were selected from the viewpoint of redundancy, robustness, and serviceability. In addition, nine indices for estimating system reliability were classified based on theoretical backgrounds such as hydraulic, topological, entropic, and mixed approaches. The correlations between the nine indices and six measures were analyzed using 17 sample hypothetical networks with different layouts, under three water supply scenarios, and the overall evaluation results for each reliability index are presented through multi-criteria decision analysis.

Keywords: comparative analysis; hydraulic measure; multi-criteria decision analysis (MCDA); reliability index; water distribution network (WDN)

---

1. Introduction

The expected performance of an infrastructure system can be interpreted through the concept of "system reliability", which quantifies marginal capacity to fulfil the users' requirements. In a water distribution network (WDN), the system reliability indicates the stable performance of supplying required water with adequate service pressure. Here, the specific performance of WDN could be assessed by representative hydraulic measures.

Wildavsky [1] defined "resilience", one of the most important performance parameters of WDNs, as the capacity to cope with unanticipated dangers after they have become manifest and learning to bounce back. Subsequently, Comfort [2] defined resilience as the capacity to adapt existing resources and skills to new situations and operating conditions. For theoretical concepts of reliability, Maier et al. [3] suggested first-order estimators such as reliability, vulnerability, and resilience of water quality service in rivers, and Bruneau et al. [4] summarized the seismic resilience of an infrastructure system into 4 R's: robustness, redundancy, resourcefulness, and rapidity. For reliability assessment of WDNs, several studies [5–7] compared the performance of different WDNs using simple types of representative hydraulic measure such as average surplus head, minimum surplus head, and supplied demand. Moreover, Marlim et al. [8] divided and formulated the reliability objectives of a WDN's user service into social, economic, hydraulic, and water quality, and Markov et al. [9] also found that the performance of a WDN could be measured via users' satisfaction and proposed a serviceability indicator to quantify this.

However, any individual hydraulic measurement is too fragmentary to be applied as the objective for WDN design and operation; hence, a large number of studies have been attempting to formulate a single "synthetic" index using various theoretical approaches. Wagner et al. [10] were the first to introduce and apply the concepts of mechanical and hydraulic reliability approaches to WDNs. Mays [11] also defined mechanical reliability as network topology evaluating system connectivity, given failure conditions, and hydraulic reliability as the ability of a system to meet the required water demand and pressure under normal and abnormal conditions. Later, Ostfeld [12] categorized WDN reliability evaluations into topological, hydraulic, and entropic backgrounds.

With regard to the hydraulic approach, Todini [13] developed the resilience index (RI), which represents the surplus and required energy in a WDN, whereas Jayaram and Srinivasan [14] developed a modified resilience index (MRI) with a different energy composition. Later, Liu et al. [15] and Jeong et al. [16] identified that a topographical relationship alters the reliability of the network performance, and proposed mixed reliability indices, a pipe hydraulic resilience index (PHRI) and a revised resilience index (RRI) by incorporating hydraulic and topographical approaches.

Within topological methods, research using a geometric approach [17–19] was performed, leading to different measures for estimating network reliability such as network efficiency (NE), average degree (AD), and link density. Creaco et al. [7] found that network performance is represented by the uniformity of pipe diameters in loop structures and developed a uniformity coefficient as the topological index. Moreover, Prasad and Park [6] also proposed a mixed reliability index, namely network resilience index (NRI), considering diameter uniformity along with the existing resilience index.

Regarding entropic reliability approaches, Awumah et al. [20] proposed an entropy reliability index by formulating water supply diversity in a WDN, and Tanyimboh and Templeman [21] developed and applied flow entropy (FE) into a WDN study based on the entropy concept of Shannon [22]. Raad et al. [23] suggested another mixed reliability index incorporating hydraulic and entropic approaches and compared four different reliability indices using performance measures in a benchmark network. Moreover, Jeong and Kang [24] suggested a hydraulic uniformity index (HUI), which is a mixed reliability index considering uniformity of the hydraulic gradients of pipes within a WDN.

However, the previously mentioned reliability indices have a bias towards certain system performances as influenced by their theoretical background. For example, in a recent study by Paez et al. [25], the correlation between different indices was analyzed through five arbitrary network designs. In addition, Tanyimboh et al. [26] investigated the correlations between surrogate reliability/redundancy measures (e.g., FE, RI, NRI) and surplus power factor with hydraulic reliability in hypothetical WDNs. In the most recent study of Sirsant and Reddy [27], the correlation between a reliability index and hydraulic and mechanical performance was also analyzed based on an optimally designed network and multipurpose functions of design cost, entropy, resiliency, and combined indices.

Eventually, it is necessary to appropriately examine which reliability index best reflects each type of system performance according to various situations and purposes required in the design and operation of WDNs. To that end, in this study, the correlations between representative reliability indices and hydraulic measures under three abnormal conditions (pipe failure; demand increase; fire flow) are analyzed for various types of application networks. Through these simulations, it was intended to determine the most adequate reliability index for evaluation of WDN performance in various abnormal water supply conditions.

The rest of this paper is organized as follows. The following section provides an overview of the proposed correlation analysis, and details of the hydraulic measures and various reliability indices are also described. Section 3 explains the design process of application networks and three application scenarios, and the application results and analyses are provided in Section 4. Finally, the conclusions of the study are summarized in Section 5.

2. Methodology

2.1. Overview of Correlation Analysis

In this study, representative reliability indices and hydraulic measures were classified and estimated for application networks following the procedure shown in Figure 1. First, nine reliability indices are selected and estimated through a base scenario representing normal operating conditions. Meanwhile, hydraulic measures show the change in performance of the system under abnormal conditions (as compared with normal conditions), and thus six hydraulic measures according to performance properties were selected and simulations were performed according to three scenarios representing "abnormal" conditions. The overall framework can be summarized as follows: (1) categorization of hydraulic measures and reliability indices, (2) establishing synthetic application networks, (3) calculation of the measures and indices in normal/abnormal water supply conditions for each application network, (4) correlation analysis between the measures and indices, and (5) multi-criteria decision analysis (MCDA) for evaluation of reliability indices.

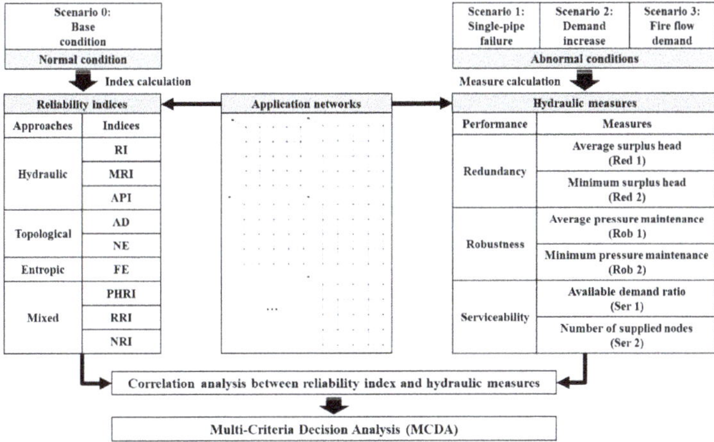

Figure 1. Correlation analysis between reliability indices and hydraulic measures. RI: resilience index, MRI: modified resilience index, API: available power index, AD: average degree, NE: network efficiency, FE: flow entropy, PHRI: pipe hydraulic resilience index, RRI: revised resilience index, NRI: network resilience index.

2.2. Hydraulic Measures

Hydraulic measures are an indication of the water supply's status in the system, which directly represents the performance of a WDN. In this study, the properties that can ensure the performance of a WDN were classified as redundancy, robustness, and serviceability, and change in performance was quantified based on two hydraulic measures of each property. Detailed descriptions of the concept of each performance property and the selected hydraulic measures are presented below.

2.2.1. Redundancy Measures

Redundancy used to be defined as substitutable capacity such as excessive backup [4] and the extent captured by the loops [28] in WDN design. In this study, network average surplus head (Red1) and minimum surplus head (Red2) under abnormal conditions were used as measures to indicate the redundancy of the system. Here, the surplus head refers to the head supplied in excess of the required head at each node (i.e., total head, subtracting elevation and minimum required pressure head); it can be shown that the higher the values of Red1 and Red2, the higher the network redundancy. The detailed calculation of redundancy measures can be presented as follows:

$$\text{Red1} = \frac{\sum_{j=1}^{nnode}\left(H_j - H_{req,j}\right)}{nnode} \tag{1}$$

$$\text{Red2} = \min\left\{H_j - H_{req,j}\right\}, j = 1, 2, \ldots, nnode \tag{2}$$

where *nnode* is the number of nodes; H_j is the total head at node *j*; and $H_{req,j}$ is the minimum required head at node *j*.

2.2.2. Robustness Measures

Robustness represents the capacity of a system to withstand a given level of stress or demand without suffering degradation or loss of function [4]. Hence, robustness performance keeps the variability of losses within a narrow band [29]. In this study, network average pressure maintenance (Rob1) and minimum pressure maintenance (Rob2) under abnormal conditions (as opposed to normal conditions) were used as measures to reflect the robustness of the system. Herein, pressure maintenance refers to the ratio between nodal pressure under abnormal and normal conditions (i.e., pressure under abnormal conditions divided by the pressure under normal conditions), and it can be understood that the higher the values of Rob1 and Rob2, the stronger the robustness of the network. The detailed calculation of robustness measures can be expressed as follows:

$$\text{Rob1} = \frac{1}{nnode}\sum_{j=1}^{nnode}\frac{H_{j,\,abnormal}}{H_{j,\,normal}} \tag{3}$$

$$\text{Rob2} = \min\left\{\frac{H_{j,\,abnormal}}{H_{j,\,normal}}\right\}, j = 1, 2, \ldots, nnode \tag{4}$$

where $H_{j,normal}$ and $H_{j,abnormal}$ are the total heads at node *j* in normal and abnormal water supply conditions, respectively.

2.2.3. Serviceability Measures

Serviceability represents users' satisfaction with a system's functionality, depending on several factors such as the vulnerability of system components, topology, and operation scenarios [9]. In this study, the available demand ratio (Ser1) and number of supplied nodes (Ser2) under abnormal conditions were used as measures to represent the serviceability of the system. Herein, the available demand ratio indicates the ratio between the nodal supplied demand under abnormal and normal conditions (i.e., supplied demand under abnormal conditions divided by the desired demand), and supplied node refers to a node in which all desired demands are satisfied. Therefore, it can be understood that the higher the values of Ser1 and Ser2, the better the serviceability of the network. The detailed calculation of serviceability measures can be presented as follows:

$$\text{Ser1} = \frac{\sum_{j=1}^{nnode} Q_{j,\,avl}}{\sum_{j=1}^{nnode} Q_j} \tag{5}$$

$$\text{Ser2} = \sum_{j=1}^{nnode} A_j \text{ where } A_j = \begin{cases} 1 & if\ Q_{j,\,avl} = Q_j \\ 0 & \text{otherwise} \end{cases} \tag{6}$$

where Q_j is the water demand at node *j*; $Q_{j,avl}$ is the supplied water demand at node *j*; and A_j is the water availability indicator of node *j*.

2.3. Reliability Assessment Index

A reliability index can be defined as an indicator calculated by system states such as elements, measurements, and structural characteristics. So far, various indices have been proposed through hydraulic, topological, and entropic approaches in search of a comprehensive understanding of the reliability of a system, and furthermore, these indices have been combined with each other or developed in numerous other forms. In this study, nine representative reliability indices were selected, and they were analyzed as to how well they reflect changes in hydraulic measures according to abnormal conditions. The following is a description of the concept and specific calculation for each reliability index.

2.3.1. Hydraulic Reliability Index

The resilience index (RI) quantifies WDN reliability based on system power such as input, dissipated, and surplus power. In a looped network, the goal is to provide more power (energy per unit time) at each node than is required, in order to have a sufficient surplus to be dissipated internally in case of failures. This surplus can be used to characterize the resilience of the looped network, i.e., its intrinsic capability for overcoming sudden failures [13]. In other words, RI is the ratio of surplus power excluding the minimum required power at the node with the total system power supplied; the detailed calculation method can be expressed as

$$RI = \frac{\gamma \sum_{j=1}^{nnode} Q_j \left(H_j - H_{req,j} \right)}{\gamma \sum_{s=1}^{nsource} Q_s H_s + \gamma \sum_{p=1}^{npump} Q_p H_p - \gamma \sum_{j=1}^{nnode} Q_j H_{req,j}} \tag{7}$$

where γ is the specific weight of water; Q_s is the inflow at source s; Q_p is the pumping flow at pump p; H_s is the total head at source s; H_p is the pumping head at pump p; *nsource* is the number of sources; and *npump* is the number of pumps.

The modified resilience index (MRI) quantifies the reliability of a WDN based on system power, similarly to RI. In particular, MRI uses only required and surplus power to calculate the index and could be used to compare the uncertainty handling of one network relative to another, which is essential in design and rehabilitation problems [14]. A detailed calculation method for MRI is given as follows:

$$MRI = \frac{\gamma \sum_{j=1}^{nnode} Q_j \left(H_j - H_{req,j} \right)}{\gamma \sum_{j=1}^{nnode} Q_j H_{req,j}} \tag{8}$$

The available power index (API) quantifies the WDN reliability using total available power and input power. Here, the available power represents the output power at demand nodes; while the unavailable power includes the power dissipated due to pipe friction losses and various minor losses in the network [15]. The detailed calculation method of API is given as

$$API = \frac{\gamma \sum_{j=1}^{nnode} Q_j H_j}{\gamma \sum_{s=1}^{nsource} Q_s H_s + \gamma \sum_{p=1}^{npump} Q_p H_p + \gamma \sum_{t=1}^{ntank} Q_t H_t} \tag{9}$$

where Q_t denotes the inflow at tank t; H_t is the total head at tank t; and *ntank* is the number of tanks.

2.3.2. Topological Reliability Index

The average degree (AD) is a convenient geometric index for quantifying system reliability based on the number of node and pipe elements. It implies that network reliability is proportional to the diversity of link (i.e., pipe) elements and paths. If a network has too few pipes, there will be many isolated nodes and clusters with a small number of nodes. As more pipes are added to the network, the small clusters are connected to larger clusters [18]. The calculation method of AD can be presented as follows:

$$AD = \frac{2 \times npipe}{nnode} \tag{10}$$

While many topological indices, including AD, quantify connectivity of networks using only the number of nodes and pipes without distinction of their layouts, Latora and Marchiori [17] suggested an index, called network efficiency (NE), that quantifies the average efficiency of paths. The NE can be calculated as an average distance between two generic nodes and is expressed by Equation (11).

$$NE = \frac{1}{nnode(nnode - 1)} \sum_{j=1}^{nnode} \sum_{\substack{j^* = 1 \\ j \neq j^*}}^{nnode} \frac{1}{d_{jj^*}} \tag{11}$$

where d_{jj^*} is the shortest path length from node j to node j^*.

2.3.3. Entropic Reliability Index

Shannon [22] derived the informational entropy function as a statistical measure of the amount of uncertainty that a probability distribution represents. The flow entropy (FE) proposed by Tanyimboh and Templeman [21] is another representative entropic reliability index available for WDNs. Prasad and Tanyimboh [30] suggested that this measure can better represent multi-source networks, and it was observed that, as the FE increases, the network becomes more reliable. The detailed calculation method of FE can be presented by Equations (12)–(15).

$$FE = E_0 + \sum_{j=1}^{nnode} P_j E_j \tag{12}$$

$$P_j = \frac{T_j}{T} \tag{13}$$

$$E_0 = - \sum_{s=1}^{nsource} \frac{Q_s}{T} \ln\left(\frac{Q_s}{T}\right) \tag{14}$$

$$E_j = -\frac{Q_j}{T_j} \ln\left(\frac{Q_j}{T_j}\right) - \sum_{ji \, \in \, ND_j} \frac{Q_{ji}}{T_j} \ln\left(\frac{Q_{ji}}{T_j}\right) \tag{15}$$

where T_j is the total flow reaching node j; T is the sum of the nodal demands; ND_j denotes the set of all pipe flows emanating from node j; and Q_{ji} represents the flow rate at pipe i from node j.

2.3.4. Mixed Reliability Index

The pipe hydraulic resilience index (PHRI) focuses on nodal water head and simultaneously considers the hydraulic gradient between upstream and downstream nodes; hence, it can be categorized as a mixed reliability index based on hydraulic and topographical aspects. The detailed calculation method for PHRI can be presented as Equations (16)–(19).

$$PHRI = \frac{\sum_{i=1}^{npipe} (S_i)}{\sum_{i=1}^{npipe} (A_i + S_i)} \tag{16}$$

$$S_i = \frac{1}{2}\left(H_{ds,i} - H_{ds,req,i}\right)L_{pro,i} \tag{17}$$

$$S_i + A_i = \frac{1}{2}\left(H_{us,i} - H_{ds,req,i}\right)L_{pro,i} \tag{18}$$

$$L_{pro,i} = \sqrt{L_i^2 - \left(Z_{us,i} - Z_{ds,i}\right)^2} \tag{19}$$

where $H_{ds,i}$ is the total head at the downstream node of pipe i; $H_{us,i}$ is the total head at the upstream node of pipe i; $H_{ds,req,i}$ is the minimum required head at the downstream node of pipe i; L_i is the length of pipe i; $Z_{ds,i}$ is the elevation at the downstream node of pipe i; and $Z_{us,i}$ is the elevation at the upstream node of pipe i.

The revised resilience index (RRI) is the mixed reliability index based on hydraulic and topographical approaches. Although the calculation method for RRI is identical with MRI, RRI applies the hydraulic gradient representing network topography when calculating the minimum required head at downstream nodes [16]. The calculation method for RRI is as shown in Equation (20).

$$\text{RRI} = \frac{\gamma \sum_{j=1}^{nnode} Q_j \left(H_j - H^*_{req,j}\right)}{\gamma \sum_{j=1}^{nnode} Q_j H^*_{req,j}} \tag{20}$$

where $H^*_{req,j}$ denotes the actual minimum required head at node j.

The network resilience index (NRI) is another mixed reliability index incorporating network topology into the formulation of RI. In WDNs, reliable loops can be ensured, if the pipes connected to a node are not widely varying in diameter. NRI incorporates the effects of both surplus power and reliable loops [6]. The detailed calculation method of NRI is as shown in Equations (21) and (22):

$$\text{NRI} = \frac{\gamma \sum_{j=1}^{nnode} C_j Q_j \left(H_j - H_{req,j}\right)}{\gamma \sum_{s=1}^{nsource} Q_s H_s + \gamma \sum_{p=1}^{npump} Q_p H_p - \gamma \sum_{j=1}^{nnode} Q_j H_{req,j}} \tag{21}$$

$$C_j = \frac{\sum_{i=1}^{npipe_j} D_i}{npipe_j \times \max\{D_i\}} \tag{22}$$

where $npipe_j$ is the number of pipes connected with node j; and D_i denotes the diameter of pipe i.

2.4. Multi-Criteria Decision Analysis

Since each index may fit different scenarios and performance aspects, it can be seen that comprehensive comparative analysis is difficult to achieve. Thus, multi-criteria decision analysis (MCDA) can be applied for drawing quantitative and comprehensive conclusions in complex decisions with multiple criteria. In urban water infrastructure, MCDA techniques such as weighted sum model (WSM), weighted product model (WPM), analytic hierarchy process (AHP), technique for order of preference by similarity to ideal solution (TOPSIS), and 'elimination et choix traduisant la realité' (ELECTRE) are used [31]. Gheisi et al. [32] applied MCDA methods such as WSM, WPM, and TOPSIS to five reliability indices as a way of determining the optimal pipe layout and diameter designs in a hypothetical WDN and claimed that WSM can be easily applied to MCDA of WDN.

3. Applications

3.1. Application Networks

In this study, to consider various types of network, application networks of various layouts were designed for the same region. Specifically, as shown in Figures 2 and 3, for a region with 25 demand nodes distributed in a grid format, application networks were configured with 17 different layouts, which ranged from the most-looped P-41 layout with all of the 41 pipes arranged to the most-branched P-25 layout with only 25 pipes arranged. All 17 networks supply the 25 demand nodes from a single source with a total head of 45 m.

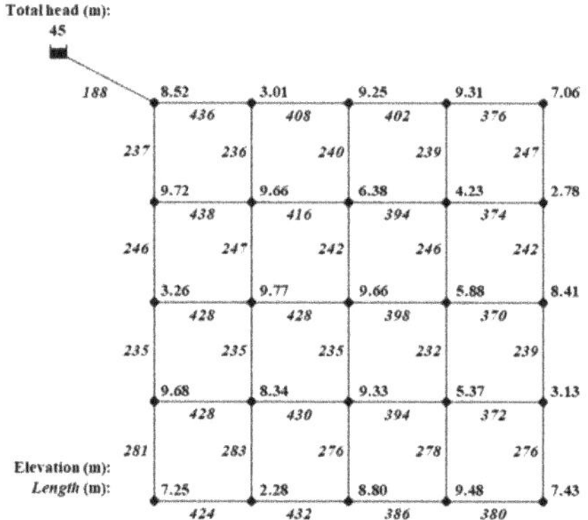

Figure 2. Configuration of application grid network (P-41).

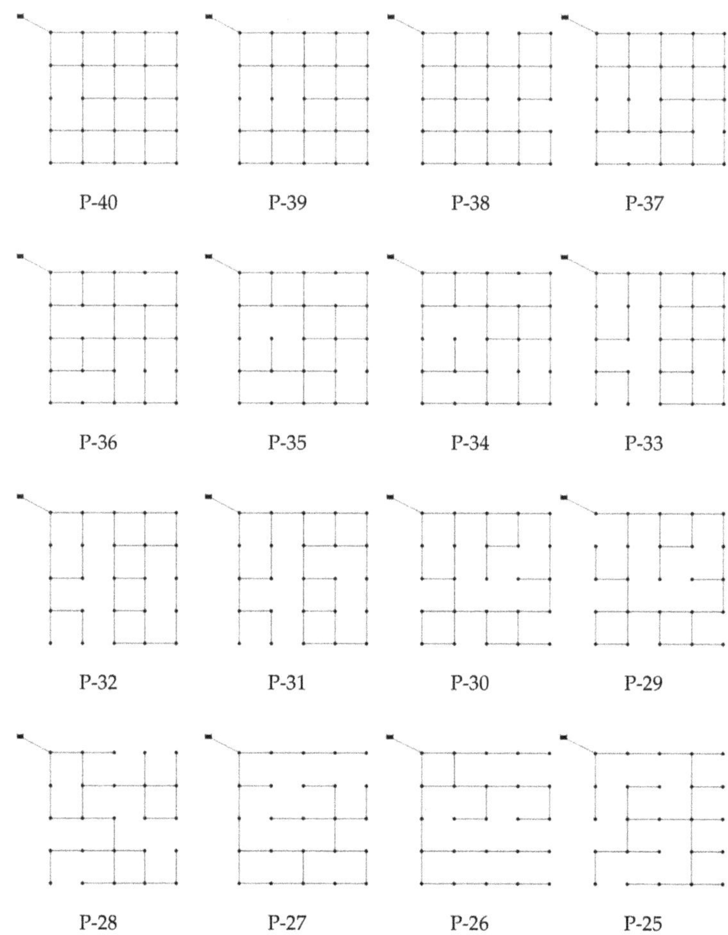

Figure 3. Optimally designed application network layout.

Base demand at all nodes was equal at 9.49 m^3/h, and 17 networks were designed by applying an hourly peaking factor (HPF) of 2.25 to produce design flow. Here, a multi-objective genetic algorithm (MOGA) was applied with decision variables of pipe layout and diameters, objective functions of design cost (minimization) and reliability indices (maximization), and a constraint of satisfying the minimum required pressure (15 m). Finally, 17 optimally designed networks were selected from the P-41 to P-25 layout considering distribution of reliability-index values. Figure 2 shows the node elevation and pipe length of the P-41 network, and the optimal layouts of the P-40–P-25 networks can be comprehensively viewed in Figure 3.

3.2. Application Scenarios

In this study, four scenarios, including a normal condition, were constructed to simulate performance change under abnormal conditions. The simulation of each scenario was performed using the pressure driven analysis (PDA) module provided by EPANET 3 [33], a WDN analysis software. The specific water supply conditions according to the four scenarios applied in this study are described in the following subsections.

3.2.1. Base Scenario

In the base scenario corresponding to a normal operating condition, a daily peaking factor (DPF) of 1.80 was applied to the base demand to simulate daily water usage. The hydraulic measures under the base scenario serve as the reference values of performance changes according to the subsequent abnormal scenarios and are used for calculation of the nine reliability indices as summarized in Table 1. Note that the higher reliability index value indicates a higher system reliability.

Table 1. Reliability index values of 17 application networks in the base scenario.

Network Layout	Reliability Index								
	RI	MRI	API	AD	NE	FE	PHRI	RRI	NRI
P-25	0.31	0.32	0.65	2.00	0.59	1.18	0.77	0.06	0.26
P-26	0.40	0.41	0.70	2.08	0.54	1.55	0.84	0.07	0.36
P-27	0.40	0.42	0.70	2.16	0.60	2.26	0.81	0.04	0.35
P-28	0.41	0.43	0.70	2.24	0.60	2.14	0.82	0.15	0.35
P-29	0.55	0.57	0.77	2.32	0.64	3.74	0.85	0.16	0.46
P-30	0.62	0.64	0.81	2.40	0.70	3.04	0.88	0.17	0.52
P-31	0.50	0.52	0.75	2.48	0.71	4.16	0.90	0.19	0.41
P-32	0.51	0.53	0.75	2.56	0.77	4.99	0.90	0.20	0.41
P-33	0.56	0.58	0.77	2.64	0.78	5.33	0.92	0.27	0.44
P-34	0.81	0.84	0.90	2.72	0.80	5.93	0.96	0.59	0.66
P-35	0.82	0.85	0.91	2.80	0.91	5.92	0.96	0.58	0.64
P-36	0.84	0.87	0.92	2.88	0.86	6.21	0.97	0.61	0.66
P-37	0.80	0.82	0.90	2.96	0.94	6.81	0.96	0.53	0.61
P-38	0.73	0.75	0.86	3.04	0.89	8.82	0.94	0.41	0.57
P-39	0.88	0.91	0.94	3.12	0.97	8.93	0.98	0.60	0.69
P-40	0.89	0.92	0.94	3.20	0.98	8.98	0.98	0.67	0.69
P-41	0.89	0.92	0.94	3.28	0.99	9.22	0.98	0.69	0.74

RI: resilience index, MRI: modified resilience index, API: available power index, AD: average degree, NE: network efficiency, FE: flow entropy, PHRI: pipe hydraulic resilience index, RRI: revised resilience index, NRI: network resilience index.

3.2.2. Scenario 1: Single-Pipe Failure

In Scenario 1, based on the base scenario flow condition, a single-pipe failure, in which each pipe is sequentially closed one by one, is constructed. Here, since all the pipes are closed once each, the probability of failure can be assumed to be the same. Therefore, changes in the performance of the entire system were calculated by averaging the measured changes of individual nodes over all failure cases.

3.2.3. Scenario 2: Water Consumption Increase

In Scenario 2, a scenario of increasing water consumption according to climate change and population growth was assumed. A study by Pachauri and Meyer [34] predicted that the global average temperature would increase by about 2 °C by 2050 according to the RCP (Representative Concentration Pathway) 8.5 climate change scenario of the Intergovernmental Panel on Climate Change (IPCC). Kenney et al. [35] found that daily water consumption increased by 2% for average temperature increases of 0.56 °C. Furthermore, Hoornweg and Pope [36] predicted that the population of major metropolitan cities in the world will increase by about 43.1% from 2025 to 2050, according to the shared socioeconomic pathways (SSP) scenario. Therefore, in this scenario, it is assumed that the base demand will increase by 41.5%, from 9.49 to 13.43 m^3/h, according to the climate change and population growth prediction. Here, a DPF of 1.80 was also applied to establish the daily water usage condition.

3.2.4. Scenario 3: Fire Flow

In Scenario 3, adding to the base scenario flow condition, fire flow demands are sequentially generated at each node. It is assumed that fire occurs once for each node, and the probability of fire outbreak is the same in all cases; therefore, the performance change according to Scenario 3 can be identified through the average of the measured changes according to the simulation of individual fire flows. The Ontario Ministry of the Environment and Climate Change (OMOECC) [37] estimated fire flow demand according to the supply population of a WDN as shown in Table 2. Since the equivalent population of each node according to the base demand of the application network is approximately 1000, the fire flow demand at each node was applied as 230.4 m^3/h. Table 3 presents a brief summary of each design condition, the base scenario and the three abnormal scenarios introduced above.

Table 2. Estimation of fire flow requirements (OMOECC [37]).

Equivalent Population	Nodal Base Demand (m3/h)	Suggested Fire Flow (m3/h)
500–1000	4.6	136.8
1000	9.2	230.4
1500	13.8	284.4
2000	18.3	342.0
3000	27.5	396.0
4000	36.7	450.0
5000	45.8	518.4
6000	55.0	572.4

Table 3. Application scenarios comparison.

Scenario	Base Demand per Node (m^3/h)	Peaking Factor	Peak Demand per Node (m^3/h)	Description
Design	9.49	2.25	21.35	Network design condition
Base	9.49	1.80	17.08	Base condition for normal operation
Scenario 1	9.49	1.80	17.08	Single-pipe failure
Scenario 2	13.43	1.80	24.17	Water consumption increase
Scenario 3	9.49	1.80	17.08	Fire flow at single node

4. Results

4.1. Correlation between Reliability Indices and Hydraulic Measures

For 17 application networks, a total of 54 correlations were analyzed through the results of nine reliability indices calculated in the base scenario and the results of six hydraulic measures collected in Scenarios 1–3. As all reliability indices and hydraulic measures indicate that the higher the value, the more superior the reliability and performance of the network, it can be concluded that the larger the derived correlation coefficient, the better the index is at reflecting system performance.

4.1.1. Results for Scenario 1: Single-Pipe Failure

Figure 4 and Table 4 show the correlation between reliability indices and hydraulic measures according to the application of Scenario 1. In Figure 4, the points in the scatter plot represent the reliability index values and hydraulic measures derived from each of the 17 networks (*x*-axis—Reliability index; *y*-axis—Hydraulic measure), and Table 4 summarizes the correlation coefficients calculated from each scatter plot. The plots and table values indicated in yellow are index-measure combinations with relatively high correlation coefficients of 0.95 or higher.

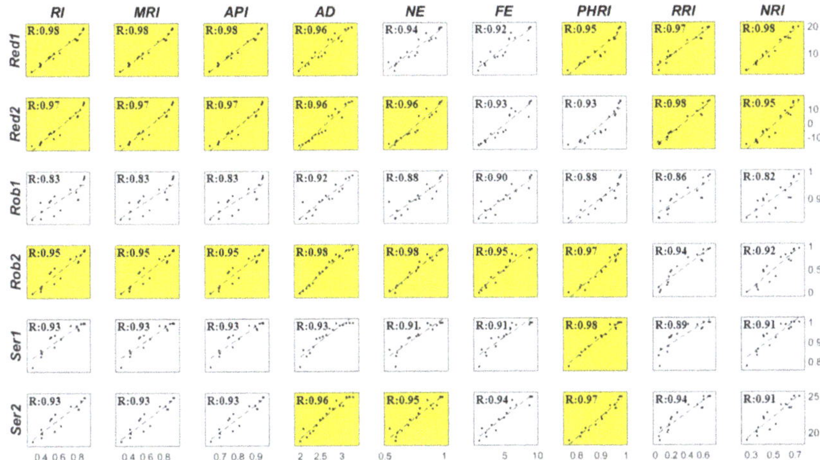

Figure 4. Scatter plots of Scenario 1 (*x*-axis—Reliability index; *y*-axis—Hydraulic measures). The plots highlighted in yellow show high correlation coefficient of 0.95 or higher.

Table 4. Correlation coefficients for Scenario 1.

Hydraulic Measure		Reliability Index								
		RI	MRI	API	AD	NE	FE	PHRI	RRI	NRI
Redundancy	Red1	0.98	0.98	0.98	0.96	0.94	0.92	0.95	0.97	0.98
	Red2	0.97	0.97	0.97	0.96	0.96	0.93	0.93	0.98	0.95
Robustness	Rob1	0.83	0.83	0.83	0.92	0.88	0.90	0.88	0.86	0.82
	Rob2	0.95	0.95	0.95	0.98	0.98	0.95	0.97	0.94	0.92
Serviceability	Ser1	0.93	0.93	0.93	0.93	0.91	0.91	0.98	0.89	0.91
	Ser2	0.93	0.93	0.93	0.96	0.95	0.94	0.97	0.94	0.91

Note: The values highlighted in yellow indicate high correlation coefficient of 0.95 or higher.

In Scenario 1, a broken pipe causes increment of flow and head loss to substitute paths. The hydraulic indices (RI, MRI, and API) are proportional to excessive nodal pressure in normal water demand conditions. When the pipe is broken, the network which has bigger excessive pressure remains more backup pressure after the event, thus the redundancy measures (Red1 and Red2) have relatively high correlation with the hydraulic indices (*R* values are 0.98 and 0.97, respectively).

The topological indices (AD and NE) quantify the diversity of water paths. For the pipe failure scenario, therefore, these indices show relatively high correlation with all measures indicating good representation of system performance under the specific condition (*R* values are higher than 0.95 for Red2, Rob2, and Ser2).

The entropic index (FE) quantifies the diversity and relative uniformity of flow distribution. When a pipe is isolated, the higher flow diversity leads to better adaptation against rapid increase of water flow. However, increased flow occurs in limited local paths, thus the overall correlation with the measures is relatively low compared to other indices.

Among the mixed indices (PHRI, RRI, and NRI), PHRI shows good correlation, especially with the serviceability measures (*R* values are 0.98 for Ser1 and 0.97 for Ser2). For a pipe isolation condition, a network with a lower hydraulic gradient gets better service. RRI and NRI are calculated in a similar way to RI, thus, they also show high correlation with the redundancy measures, but have a relatively low correlation with other measures.

4.1.2. Results for Scenario 2: Water Consumption Increase

Figure 5 and Table 5 show the correlation between reliability indices and hydraulic measures according to the application of Scenario 2. In Scenario 2, system-wide increased water consumption causes an increment of flow and head loss in all pipes; that is, a system-wide stress is produced under this scenario. This mainly affects the branch-type networks (P-25–P-33) and significantly reduces the measures of Red2 and Rob2 that are related to the minimum nodal pressure as seen in Figure 5 (see scatter plots in second and fourth rows).

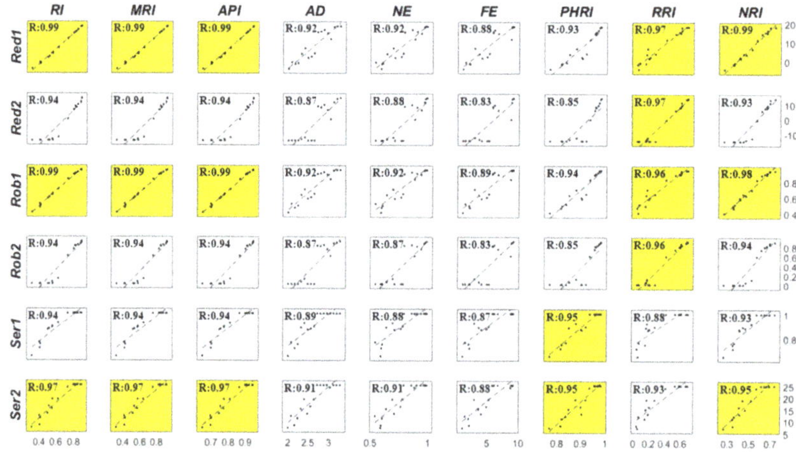

Figure 5. Scatter plots of Scenario 2 (*x*-axis—Reliability index; *y*-axis—Hydraulic measures). The plots highlighted in yellow show high correlation coefficient of 0.95 or higher.

Table 5. Correlation coefficients for Scenario 2.

Hydraulic Measure		Reliability Index								
		RI	MRI	API	AD	NE	FE	PHRI	RRI	NRI
Redundancy	Red1	0.99	0.99	0.99	0.92	0.92	0.88	0.93	0.97	0.99
	Red2	0.94	0.94	0.94	0.87	0.88	0.83	0.85	0.97	0.93
Robustness	Rob1	0.99	0.99	0.99	0.92	0.92	0.89	0.94	0.96	0.98
	Rob2	0.94	0.94	0.94	0.87	0.87	0.83	0.85	0.96	0.94
Serviceability	Ser1	0.94	0.94	0.94	0.89	0.88	0.87	0.95	0.88	0.93
	Ser2	0.97	0.97	0.97	0.91	0.91	0.88	0.95	0.93	0.95

Note: The values highlighted in yellow indicate high correlation coefficient of 0.95 or higher.

The hydraulic indices (RI, MRI, and API) show great correlations with the measures overall (in range of 0.94–0.99) under Scenario 2. The higher value of the hydraulic indices implies that the networks have excessive pressure, thus providing redundancy, guaranteeing maintenance of pressure, and allowing better serviceability.

In contrast, the topological indices (AD and NE) and entropic index (FE) show relatively low correlation with the performance measures under Scenario 2 in which overall flows increase without changes in network connectivity/layout and flow paths.

For the mixed indices, PHRI quantifies the serviceability well, while RRI represents the redundancy and robustness measures well. It is noteworthy that RRI has a strong correlation with Red2 and Rob2 (*R* values are 0.97 and 0.96, respectively), which cannot be seen in other indices under Scenario 2.

4.1.3. Results for Scenario 3: Fire Flow

Figure 6 and Table 6 show the correlation between reliability indices and hydraulic measures according to the application of Scenario 3. In Scenario 3, node-assigned fire flow causes increment of flow and head loss along a specific flow path from the source to the node at which the fire occurred. As seen in Figure 6 and Table 6, overall, the correlation coefficients in Scenario 3 are less than those of the other two abnormal scenarios.

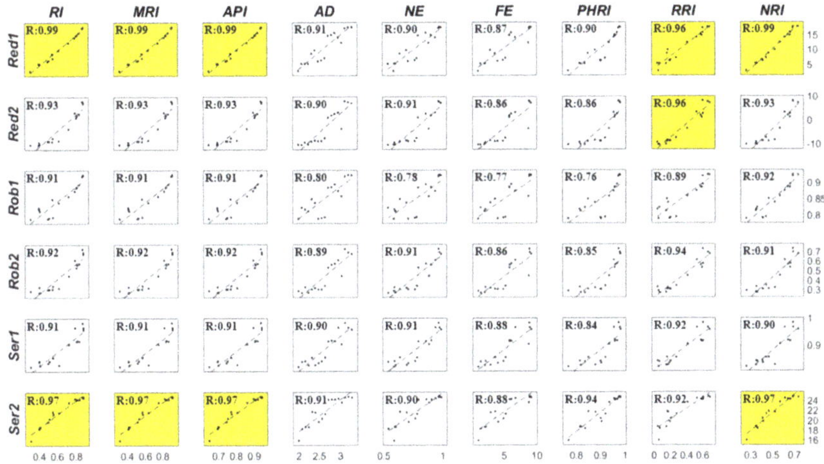

Figure 6. Scatter plots of Scenario 3 (*x*-axis—Reliability index; *y*-axis—Hydraulic measure). The plots highlighted in yellow show high correlation coefficient of 0.95 or higher.

Table 6. Correlation coefficients for Scenario 3.

Hydraulic Measure		Reliability Index								
		RI	MRI	API	AD	NE	FE	PHRI	RRI	NRI
Redundancy	Red1	0.99	0.99	0.99	0.91	0.90	0.87	0.90	0.96	0.99
	Red2	0.93	0.93	0.93	0.90	0.91	0.86	0.86	0.96	0.93
Robustness	Rob1	0.91	0.91	0.91	0.80	0.78	0.77	0.76	0.89	0.92
	Rob2	0.92	0.92	0.92	0.89	0.91	0.86	0.85	0.94	0.91
Serviceability	Ser1	0.91	0.91	0.91	0.90	0.91	0.88	0.84	0.92	0.90
	Ser2	0.97	0.97	0.97	0.91	0.90	0.88	0.94	0.92	0.97

Note: The values highlighted in yellow indicate high correlation coefficient of 0.95 or higher.

Overall, the hydraulic indices (RI, MRI, and API) and mixed hydraulic index (NRI) show higher correlations (*R* values are in the range of 0.90–0.99) compared to other indices (in the range of 0.76–0.91).

It is interesting to observe that RRI shows strong correlation with Red2 and Rob2 under Scenarios 1–3 (*R* values are always ranged in 0.94–0.98). It should be noted that Red2 and Rob2 measure the minimum pressure at critical nodes and RRI represents the resilience of the network while maintaining the minimum required head.

The topological indices (AD and NE), entropic index (FE), and a hydraulic-gradient based mixed index (PHRI) show relatively low correlation with the measures under the fire-flow scenario; none of them yield a correlation coefficient of greater than 0.95.

4.2. Multi Criteria Decision Analysis (MCDA)

In this study, MCDA was performed using WSM by assigning equal weights to all six performance measures. Here, two different MCDAs are conducted: a scenario-based MCDA and a performance-based MCDA. The scenario-based MCDA aims to find the optimal indices representing the individual abnormal conditions; while the performance-based MCDA intends to find the best fit indices for individual performance such as redundancy, robustness, and serviceability. Finally, a comprehensive evaluation of individual indices is achieved by combining both MCDA results.

4.2.1. Scenario-Based MCDA

Table 7 shows the MCDA results of nine indices according to different scenarios. In Scenario 1, AD and PHRI showed high correlation coefficients from MCDA, which indicates that these indices appropriately quantify the system performance under the single pipe failure condition. In a similar way, under Scenario 2 (system-wide water consumption increase), hydraulic indices (RI, MRI, and API) show great performance; while for Scenario 3 (single-node fire flow), the hydraulic indices still show better performance than other indices but, overall, correlation coefficients were lower than those of the other scenarios.

Table 7. Correlation coefficients obtained by scenario-based MCDA.

Abnormal Scenario	Reliability Index								
	RI	MRI	API	AD	NE	FE	PHRI	RRI	NRI
Scenario 1	0.93	0.93	0.93	0.95	0.94	0.93	0.95	0.93	0.92
Scenario 2	0.96	0.96	0.96	0.90	0.90	0.86	0.91	0.95	0.95
Scenario 3	0.94	0.94	0.94	0.89	0.89	0.85	0.86	0.93	0.94
Average	0.94	0.94	0.94	0.91	0.91	0.88	0.91	0.94	0.94

Note: The values highlighted in yellow indicate high correlation coefficient of 0.95 or higher.

4.2.2. Performance-Based MCDA

Table 8 shows the MCDA results of nine indices according to the system performance. For system redundancy quantification, the hydraulic indices (RI, MRI, and API) and two mixed indices (RRI, and NRI) showed great performance in all scenarios. Regarding the robustness context, the overall correlation coefficients are relatively low, which indicates that the applied indices are inadequate to represent the system robustness under the applied scenarios. For measuring system serviceability, the hydraulic indices and PHRI, showed relatively good results.

Finally, the combined MCDA results can be obtained via the averages of Tables 7 and 8 (assuming equal weights for WSM). As seen, among the applied nine indices, the hydraulic indices (RI, MRI, and API) and the hydraulic-based mixed indices (RRI, and NRI) showed high correlation coefficients (i.e., 0.94) and are considered as the optimal indices to quantify the system performance under the applied abnormal scenarios and hydraulic measures.

Table 8. Correlation coefficients obtained by performance-based MCDA.

System Performance	Reliability Index								
	RI	MRI	API	AD	NE	FE	PHRI	RRI	NRI
Redundancy	**0.97**	**0.97**	**0.97**	0.92	0.92	0.88	0.90	**0.97**	**0.96**
Robustness	0.92	0.92	0.92	0.90	0.89	0.87	0.88	0.93	0.92
Serviceability	0.94	0.94	0.94	0.92	0.91	0.89	0.94	0.91	0.93
Average	0.94	0.94	0.94	0.91	0.91	0.88	0.91	0.94	0.94

Note: The values highlighted in yellow indicate high correlation coefficient of 0.95 or higher.

5. Conclusions

In this study, correlations between hydraulic measures and reliability indices were analyzed to evaluate the ability of reliability indices to reflect system performance. To that purpose, six (hydraulic) measures of redundancy, robustness, and serviceability and nine reliability indices based on hydraulic, topological, entropic, and mixed approaches were selected. Correlation analyses were performed using the 17 optimally designed hypothetic WDNs under three abnormal operation scenarios. To obtain a comprehensive evaluation of individual indices, an MCDA based on WSM was performed as a post-analysis. Based on the results obtained in the study, the following conclusions can be drawn:

1. In the single-pipe-failure scenario, the topological indices (AD and NE) and mixed index (PHRI) were found to be the best at the quantification of network performance.
2. In the demand increase scenario, the hydraulic or mixed hydraulic indices (RI, MRI, API, RRI, and NRI) were found to be the best at representation of network performance.
3. In the fire flow scenario, the hydraulic indices (RI, MRI, and API) were found to be best at representation of network performance in terms of redundancy and serviceability. It was found that NRI represents very close correlation trends with the hydraulic indices in all scenarios.
4. For redundancy quantification, the hydraulic or mixed hydraulic indices (RI, MRI, API, RRI, and NRI) showed the best performance.
5. For robustness quantification, all indices showed relatively low correlation with the measures, indicating that the applied indices do not sufficiently reflect system robustness.
6. For serviceability quantification, the hydraulic indices (RI, MRI, and API) and PHRI were found to best reflect network performance.
7. From the combined MCDA, the hydraulic or mixed hydraulic indices (RI, MRI, API, RRI, and NRI) were found to be the optimal indices to quantify the system performance under the applied abnormal scenarios.

Prior to using any reliability index for the design and operation of a WDN, it is essential to select an adequate reliability index suitable for the goals of the designer/operator, and the results of this study can contribute to that purpose. In addition, the proposed comparative analyses are expected to be useful in research on index development that can better reflect various network performances. Future study should consider the needs of water managers for scenario development reflecting real operational conditions and the selection of reliability indices and hydraulic measures. Further comparative analyses, including various reliability indices and hydraulic measures for a number of WDNs with diverse sizes and layouts, are a fruitful area to pursue as a future study. Through the proposed comparative analysis, it is anticipated that the best fit reliability indices can be suggested and developed for quantification of WDN performance.

Author Contributions: Conceptualization, G.J. and D.K.; methodology, G.J. and D.K.; software, G.J.; validation, D.K.; formal analysis, G.J. and D.K.; writing—original draft preparation, G.J.; writing—review and editing, D.K.; supervision, D.K.; funding acquisition, D.K. All authors have read and agreed to the published version of the manuscript.

Funding: This work was supported by the National Research Foundation of Korea (NRF) grant funded by the Korea government (MSIT—Ministry of Science and ICT) (No. NRF-2020R1A2C2009517).

Conflicts of Interest: The authors declare no conflict of interest.

References

1. Wildavsky, A. *Searching for Safety*; Transaction Books: New Brunswick, NJ, Canada, 1988.
2. Comfort, L. *Shared Risk: Complex Systems in Seismic Response*; Pergamon: New York, NY, USA, 1999.
3. Maier, H.R.; Lence, B.J.; Tolson, B.A.; Foschi, R.O. First-order reliability method for estimating reliability, vulnerability, and resilience. *Water Resour. Res.* **2001**, *37*, 779–790. [CrossRef]
4. Bruneau, M.; Chang, S.E.; Eguchi, R.T.; Lee, G.C.; O'Rourke, T.D.; Reinhorn, A.M.; Von Winterfeldt, D. A framework to quantitatively assess and enhance the seismic resilience of communities. *Earthq. Spectra* **2003**, *19*, 733–752. [CrossRef]
5. Xu, C.; Goulter, C. Reliability-based optimal design of water distribution networks. *J. Water Resour. Plan. Manag.* **1999**, *125*, 352–362. [CrossRef]
6. Prasad, T.D.; Park, N.S. Multiobjective genetic algorithms for design of water distribution networks. *J. Water Resour. Plan. Manag.* **2003**, *1*, 73–82. [CrossRef]
7. Creaco, E.; Franchini, M.; Todini, E. The combined use of resilience and loop diameter uniformity as a good indirect measure of network reliability. *Urban Water J.* **2016**, *13*, 167–181. [CrossRef]
8. Marlim, M.S.; Jeong, G.; Kang, D. Identification of Critical Pipes Using a Criticality Index in Water Distribution Networks. *Appl. Sci.* **2019**, *9*, 4052. [CrossRef]
9. Markov, I.; Grigoriu, M.; O'Rourke, T.D. *An Evaluation of Seismic Serviceability of Water Supply Networks with Application to the San Francisco Auxiliary Water Supply System*; NCEER: Buffalo, NY, USA, 1994.
10. Wagner, J.M.; Shamir, U.; Marks, D.H. Water distribution reliability: Simulation methods. *J. Water Resour. Plan. Manag.* **1998**, *114*, 276–294. [CrossRef]
11. Mays, L.W. *Reliability Analysis of Water Distribution Systems*; ASCE: Reston, VA, USA, 1989.
12. Ostfeld, A. Reliability analysis of water distribution systems. *J. Hydroinform.* **2004**, *6*, 281–294. [CrossRef]
13. Todini, E. Looped water distribution networks design using a resilience index based heuristic approach. *Urban Water* **2000**, *2*, 115–122. [CrossRef]
14. Jayaram, N.; Srinivasan, K. Performance-based optimal design and rehabilitation of water distribution networks using life cycle costing. *Water Resour. Res.* **2008**, *44*, W01417. [CrossRef]
15. Liu, H.; Savić, D.A.; Kapelan, Z.; Creaco, E.; Yuan, Y. Reliability surrogate measures for water distribution system design: Comparative analysis. *J. Water Resour. Plan. Manag.* **2017**, *143*, 04016072. [CrossRef]
16. Jeong, G.; Wicaksono, A.; Kang, D. Revisiting the resilience index for water distribution networks. *J. Water Resour. Plan. Manag.* **2017**, *143*, 04017035. [CrossRef]
17. Latora, V.; Marchiori, M. Efficient behavior of small-world networks. *Phys. Rev. Lett.* **2016**, *87*, 198701. [CrossRef] [PubMed]
18. Costa, L.D.F.; Rodrigues, F.A.; Travieso, G.; Villas Boas, P.R. Characterization of complex networks: A survey of measurements. *Adv. Phys.* **2007**, *56*, 167–242. [CrossRef]
19. Jamakovic, A.; Uhlig, S. On the relationships between topological measures in real-world networks. *Netw. Heterog. Media* **2008**, *3*, 345–359. [CrossRef]
20. Awumah, K.; Goulter, I.; Bhatt, S.K. Assessment of reliability in water distribution networks using entropy based measures. *Stoch. Hydrol. Hydraul.* **1990**, *4*, 309–320. [CrossRef]
21. Tanyimboh, T.T.; Templeman, A.B. Calculating maximum entropy flows in networks. *J. Oper. Res. Soc.* **1993**, *44*, 383–396. [CrossRef]
22. Shannon, C.E. A mathematical theory of communication. *Bell Syst. Tech. J.* **1948**, *27*, 379–423, 623–656.
23. Raad, D.N.; Sinske, A.N.; Van Vuuren, J.H. Comparison of four reliability surrogate measures for water distribution systems design. *Water Resour. Res.* **2010**, *46*, W05524. [CrossRef]
24. Jeong, G.; Kang, D. Hydraulic Uniformity Index for Water Distribution Networks. *J. Water Resour. Plan. Manag.* **2020**, *146*, 04019078. [CrossRef]
25. Paez, D.; Filion, Y.; Suribabu, C.R. Correlation analysis of reliability surrogate measures in real size water distribution networks. In Proceedings of the WDSA/CCWI Joint Conference Proceedings, Kingston, ON, Canada, 23–25 July 2018; Volume 1.

26. Tanyimboh, T.T.; Siew, C.; Saleh, S.; Czajkowska, A. Comparison of surrogate measures for the reliability and redundancy of water distribution systems. *Water Resour. Manag.* **2016**, *30*, 3535–3552. [CrossRef]

27. Sirsant, S.; Reddy, M.J. Assessing the Performance of Surrogate Measures for Water Distribution Network Reliability. *J. Water Resour. Plan. Manag.* **2020**, *146*, 04020048. [CrossRef]

28. Yazdani, A.; Otoo, R.A.; Jeffrey, P. Resilience enhancing expansion strategies for water distribution systems: A network theory approach. *Environ. Model. Softw.* **2011**, *26*, 1574–1582. [CrossRef]

29. Cimellaro, G.P.; Reinhorn, A.M.; Bruneau, M. Framework for analytical quantification of disaster resilience. *Eng. Struct.* **2010**, *32*, 3639–3649. [CrossRef]

30. Prasad, T.D.; Tanyimboh, T.T. Entropy based design of Anytown water distribution network. *Water Distrib. Syst. Anal.* **2008**, *2008*, 1–12.

31. Tscheikner-Gratl, F.; Egger, P.; Rauch, W.; Kleidorfer, M. Comparison of multi-criteria decision support methods for integrated rehabilitation prioritization. *Water* **2017**, *9*, 68. [CrossRef]

32. Gheisi, A.; Shabani, S.; Naser, G. Flexibility ranking of water distribution system designs under future mechanical and hydraulic uncertainty. *Procedia Eng.* **2015**, *119*, 1202–1211. [CrossRef]

33. EPANET3 in Github. Available online: https://github.com/OpenWaterAnalytics/epanet-dev/blob/master/src/epanet3.h (accessed on 15 August 2016).

34. Pachauri, R.; Meyer, L. *Climate Change 2014: Synthesis Report. Fifth Assessment Report of the Intergovernmental Panel on Climate Change*; IPCC: Geneva, Switzerland, 2014.

35. Kenney, D.S.; Goemans, C.; Klein, R.; Lowrey, J.; Reidy, K. Residential water demand management: lessons from Aurora, Colorado 1. *J. Am. Water Resour. Assoc.* **2008**, *44*, 192–207. [CrossRef]

36. Hoornweg, D.; Pope, K. Population predictions for the world's largest cities in the 21st century. *Environ. Urban.* **2017**, *29*, 195–216. [CrossRef]

37. OMOECC Design Guidelines for Drinking Water Systems, Ontario Ministry of the Environment and Climate Change. Available online: http://www.ontario.ca/document/design-guidelines-drinking-water-systems (accessed on 22 May 2020).

Article

Spatial Aggregation Effect on Water Demand Peak Factor

Giuseppe Del Giudice [1], Cristiana Di Cristo [1] and Roberta Padulano [2,*]

[1]   Department of Civil, Environmental and Architectural Engineering, Università degli Studi di Napoli Federico II, Via Claudio 21, 80125 Naples, Italy; delgiudi@unina.it (G.D.G.); cristiana.dicristo@unina.it (C.D.C.)

[2]   Regional Models and geo-Hydrological Impacts, Centro Euro-Mediterraneo sui Cambiamenti Climatici, Via Thomas Alva Edison, 81100 Caserta (CE), Italy

*   Correspondence: roberta.padulano@cmcc.it

Received: 28 May 2020; Accepted: 14 July 2020; Published: 16 July 2020

Abstract: A methodological framework for the estimation of the expected value of hourly peak water demand factor and its dependence on the spatial aggregation level is presented. The proposed methodology is based on the analysis of volumetric water meter measurements with a 1-h time aggregation, preferred by water companies for monitoring purposes. Using a peculiar sampling design, both a theoretical and an empirical estimation of the expected value of the peak factor and of the related standard error (confidence bands) are obtained as a function of the number of aggregated households (or equivalently of the number of users). The proposed methodology accounts for the cross-correlation among consumption time series describing local water demand behaviours. The effects of considering a finite population is also discussed. The framework is tested on a pilot District Metering Area with more than 1000 households equipped with a telemetry system with 1-h time aggregation. Results show that the peak factor can be expressed as a power function tending to an asymptotic value greater than one for the increasing number of aggregated households. The obtained peak values, compared with several literature studies, provide useful indications for the design and management of secondary branched pipes of water distribution systems.

Keywords: cross-correlation; data spatial aggregation; finite population effect; metering; sample mean; sampling design; standard error; stochastic analysis; water demand peak factor; water distribution networks

---

1. Introduction

In the last decades, the understanding and prediction of water consumption have become a focal point of EU policies and directives, with the general aim of supporting safe access to drinking water and basic sanitation services to the people. In this context, the estimation of water demand in a distribution system is a key issue when applying management strategies to reduce costs and preserve the resource [1].

The water demand of a single user exhibits a random and pulsing behaviour; however, the aggregation of a large number of consumers is able to highlight trends, seasonal cycles, and the possible existence of peaks. Such quantities usually have different values and features according to the scales of the measured or aggregated data (hourly, daily, weekly, monthly, seasonally, yearly). The estimate of peak values is crucial to design drinking water distribution networks, in order to obtain reliable systems, able to provide a good level of service in terms of demands and pressures [2]. The knowledge of water consumptions and of the relative peak values is also required in many applications where the simulation of the system functioning is needed, whose results are strongly affected by demand uncertainty [3–6].

The estimation of hourly or sub-hourly peak demand due to residential uses has been widely studied adopting different methods and techniques. Top-Down Deterministic Approaches (TDAs) provide empirical relationships based on the number of users for the estimation of the hourly or sub-hourly demand peak factor, defined as the ratio between the maximum and mean flow. TDAs usually focus on the whole network, analysing the water consumption of the total served population. The first relationships [7,8] estimated the dependence on the population of the instantaneous peak factor in sewer systems. Some research found that the hourly peak factor can be considered constant when the population is lower than a fixed threshold, while it decreases when the users exceed the threshold value [9]. Those empirical equations for wastewater peak factors tend to be restricted to a minimum population of one thousand and a maximum population of one million. More recently, a formula was proposed for characterizing the mean value of the peak water demand for small towns through statistical inferences on a large database [10], providing a lower estimate compared to the Babbitt's formula [7]. Moreover, investigated the effect of the data time sampling interval on the evaluation of the peak factor was investigated [10]. The dependence of peak factors on the number of users was also the subject of investigations [11], to provide empirical relationships for the estimation of the parameters of the Gumbel probability distribution, able to represent the stochastic behaviour of peak water demand.

Bottom-Up Approaches (BUAs) try to reconstruct nodal demands generating a large number of synthetic realizations of individual users' consumptions described by a stochastic variable. It has been proved that at the fine temporal scale the nodal demand takes the shape of a pulse [12]. In this context, temporal trends of instantaneous nodal consumptions are reconstructed aggregating demands produced by stochastic pulse generation methods, such as the Poisson Rectangular Pulse (PRP) (e.g., [13–18]) or the cluster Neyman-Scott Rectangular Pulse (NSRP) (e.g., [4,19,20]). A single pulse is associated with each demand event, whose arrival time is described through a Poisson process. In the proposed methods, pulse duration and intensity have been generated assigning different specific probability distributions: Normal [15], exponential [4,19–21], log-normal [12] for the duration; exponential [4,15,19,20], Weibull [21], log-normal [12] for the intensity. More recently, a method was proposed to account for the correlation between pulse duration and intensity, which led to some improvement in pulse consistency [22].

To apply these methods, model parameters need to be assigned. The parameters' values can be obtained using measured pulse features obtained by monitoring consumptions with an ultra-high time resolution [12,17,18] or reproducing statistical properties of aggregated consumptions, when they are known at a higher temporal step (1 min or larger) [19,23].

In this context, Blokker E.J.M., et. al., [24,25] proposed the SIMDEUM model for the reconstruction of water consumptions starting from the micro-components of water demand. The PRP model was used, but different distributions were adopted to generate the pulses produced by the different household fixtures and users. Then, for its parametrization, knowledge is required about the occupants' habits and about the end uses of the fixtures obtained from a survey of the considered households. This can be done, for example, by analysing the water end-users that drive peak daily demand and examining their diurnal demand patterns using data obtained from high resolution smart meters [26]. The PRP and SIMDEUM models have similar performances [27], with the former prevailing at the single household scale and the latter prevailing in case of multiple households. In all cases, BUAs require a significant computational effort and, for their parametrization, a detailed knowledge of the consumptions at a small spatial scale is required.

More recently, proposed a probabilistic approach was proposed for a reliable estimation of the maximum residential water demand represented by a single variable [28], showing the reliability of the log-normal and Gumbel distributions in representing peak water demand during the day. The authors suggested practical equations for the estimation of the expected value and coefficient of variation of the daily peak factor and investigated time scaling effects.

Many studies investigated the influence of the acquisition time step in water demand modelling [23,29,30]. In this context, analysing water consumption data recorded at time intervals from 5 min to one hour, a significant effect of the sampling time step was observed [31] and new equations were derived for the evaluation of the peak value. A comparison of the instantaneous estimate of the maximum demand obtained at a 1 s time step through a BUA with the one computed from hourly average estimates using a TDA was also performed [32]. As expected, results showed that the latter gives small demand values, especially at small spatial aggregation scales, while at increasing aggregation levels the difference decreases, because the random fluctuations tend to be smoothed with consequently smaller peak values.

The effect of spatial aggregation is less studied. First attempts investigated the effect of both time step and spatial aggregation on the cross-correlation between nodal demands, however limiting the analysis to a group of five and ten houses [33]. Results highlighted an increase in correlation for increasing spatial aggregation, while a decrease of the standard deviation was observed.

In the last decades, the rising development of smart meters systems for household water consumption monitoring provided new modelling perspectives [34,35]. Smart metering can provide data recordings at different levels of accuracy, from 1 s to hours, depending on the characteristic of the system and on the objective of the investigations [36,37]. With a reasonable economic impact, water companies started with the installation of smart water meters, usually placed in a large number of households and collecting hourly measurements. In fact, water companies are mainly interested in controlling and understanding aggregated consumptions in order to make decisions on pricing strategies, on future interventions, and on consumption reduction. Some approaches have been recently proposed for modelling demand patterns using measurements at large time steps [38–41].

A first objective of the present study is to understand how water companies can obtain information about the estimate of the peak factor, starting from measurements realized for different purposes on large networks with an hourly temporal scale. The paper presents a methodology for performing a statistical analysis of hourly data in order to analyse the behaviour of the hourly peak demand values as a function of spatial data aggregation using a high number of measurements. The considered test-case is a large-size District Metering Area of the water distribution network of Naples (Italy) equipped with a smart metering system, which provided water demand measurements performed with a one hour time aggregation on more than 1000 households for one year [39,40]. The main novelty of the study lies in the complex sampling design adopted, which allows treating hourly peak factors as stochastic variables for each fixed number of aggregated meters, accounting for possible cross-correlation and finite population effects. In this way, the main statistics (including expected values and variability) of the peak factor can be obtained as a function of the size of the considered group of users, and compared with other literature indications adequately scaled to account for different time scales. The main goal of the research is to provide the operators with a procedure for understanding the reliability of the network in terms of demand and pressure at different levels of users' aggregation using available data. This information is particularly useful to analyse the behaviour of old water networks, where the operating conditions may differ from the ones considered at the design stage, or to design future measures to improve the system management, such as the creation of District Meter Areas.

The paper is structured as follows. Section 2 describes the District Metering Area under investigation and the collected measurements, the main objectives and features, and the methodological framework of the analysis. Section 3 reports the outcomes, while Section 4 provides a discussion about the applicability of results. Finally, significant conclusions are drawn in Section 5.

2. Materials and Methods

2.1. The District Metering Area

The area under investigation is "North Soccavo" (Figure 1), which is one of the administrative neighbourhoods of the Municipality of Naples (Italy), counting about 20,000 inhabitants. The area was selected by the local water company as the pilot case for the implementation of a District Metering Area. The reason for this particular choice lies in the observation that this area is connected by a single branch to the water network of the City, which makes it particularly prone to be distrectualized. In recent years, the local water company connected all the existing water meters with a telemetry system, allowing for the automatic radio collection of consumed water volumes at a 1-h time step. The connection involved water meters related to both single and multiple households, as well as commercial activities and public buildings.

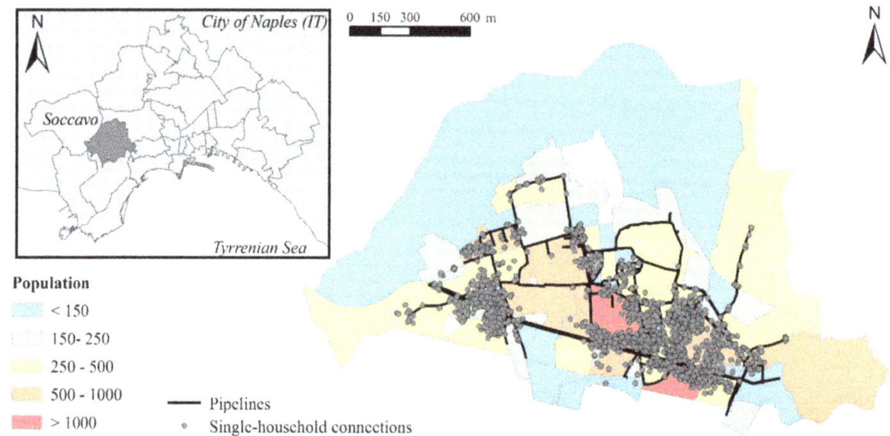

Figure 1. Pilot area: Water distribution network, water meters, and census particles.

Data referring to the year 2016 have recently been the subject of a multi-purpose research focusing on the prediction of water demand patterns, useful for the local water company to define management and leakage detection strategies [39,40,42]. The findings, based on hourly consumption data collected from 1 January to 31 December 2016, highlighted the following issues:

1. The neighbourhood is mainly residential. Of the total $K = 4253$ water meters, about 86% serve flats, apartments, and other inhabited buildings, whereas 14% have a non-residential purpose (commercial activities, public offices, and schools).

2. Not all the consumption time series collected by the telemetry system were suitable for the analyses. An anomaly/outlier detection procedure was applied, based on the use of the Completeness-Continuity Triangle (CCT) and on the application of the MAD criterion at different time aggregation levels. Such a strategy enabled the identification and subsequent removal of unreliable hourly data or the entire time series having a large number of outliers and/or missing values, as shown in [42]. Focusing on residential water meters serving single households, a number of 1162 passed the proposed anomaly/outlier detection procedure. Those data are used in the present research.

3. Focusing on the connections serving single households, a reduced number of significant patterns showing the annual cycle of water demand was detected [39]. Specifically, in the pilot area five patterns were identified representing different clusters of consumption behaviours. Figure 2 shows that those patterns are significantly similar at large aggregation levels (e.g., monthly), only differing because of the consumption in August. This occurs because most of the people in Italy spend their holidays in August, and this produces a decrease in consumptions. However, the trends are different depending on the number of vacation days that is, in turn, proportional to the income level; a cluster without reduction, corresponding to users taking no summer holidays, can also be observed. Such a behaviour is expected to repeat cyclically every year. Removing the

August consumption, no further seasonal cycle can be observed, and the five patterns in Figure 2 show no significant differences in the remaining months. As a consequence, if August data is discarded, daily discharges can be considered a random variable with no deterministic dependence. As far as the daily cycle is concerned, three different non-dimensional patterns were identified corresponding to Sundays, Saturdays, and Mondays–Fridays, respectively [40].

4. For residential connections, scaling laws were proposed [40] providing the mean hourly discharges and related standard deviations as a function of the number of aggregated households. The regression parameters depend on the characteristics of the specific dataset in terms of single-user behaviour and cross-correlation structure.

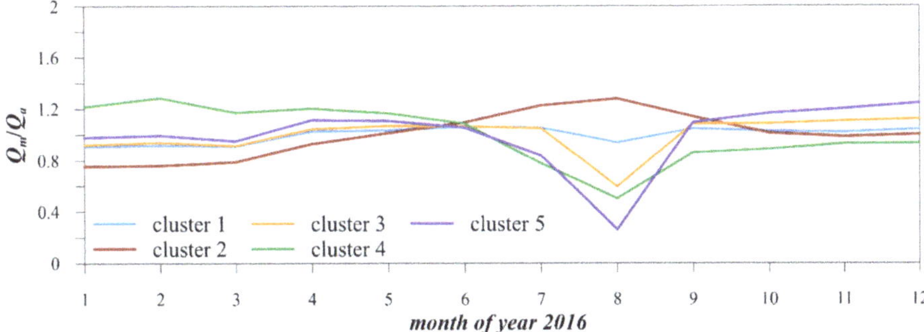

Figure 2. Average nondimensional patterns for the five main different clusters of consumption behaviours observed in the pilot area (Q_m and Q_a are the monthly and annual discharges, respectively).

The scaling laws proposed for the pilot area [40] are a function of the number of aggregated households, instead of the number of consumers, because the information about the number of people "hiding" behind each water meter was known only for a few cases. In the present paper, however, this information was derived by intersecting the spatial distribution of residential water meters (Figure 1), with the number of inhabitants at the census scale. As a result, the average number of users per connection ranges from 2.8 to 3. This uncertainty is caused by the delay between the date of the census survey (2011) and that of data collection (2016), and to a small number of water meters that still missed their connection with the telemetry system by the end of 2016.

2.2. Rationale and Structure of the Analysis

In the present paper the hourly water consumption database collected within the pilot area is used to obtain a comprehensive sample of hourly peak factors. Such a database can be potentially used to draw significant information allowing for the prediction of fundamental statistics of hourly peak demand such as central values, variability, and probability distribution. This research particularly focuses on the first issue, namely the sample mean of peak factors and related statistics.

As mentioned in the previous sections, peak factors in water networks can be deeply influenced by the amount and behaviour of consumers. Any statistical analysis should comply with the fact that the peak factors' values and the related statistics could be affected by the number and quality of the aggregated time series. For instance, if the network serves a small number of users, there is a large possibility that those consumers will highlight similar behaviours, resulting in higher peak factor values. On the contrary, if the network serves a large number of consumers, different behaviours are expected and this translates in a global water demand more homogeneously distributed within the day, with smaller values of peak factors.

The peak factors evaluation in water networks usually consists of understanding how peak factors change under a progressive aggregation of the users, namely in finding a mathematical or statistical dependence of peak factors on the number of users N_u. Synchronicity of consumption behaviours is usually accounted for by means of the cross-correlation among consumption time series. Those considerations imply that, when N_u is small, a dependence can be found not only on *how many* but also on *which* time series is going to be aggregated. In other words, results may deeply vary according to the specific performed selection of consumers. On the contrary, when N_u is large, results are expected not to be significantly altered whichever time series is selected.

To overcome this issue and to investigate the statistical structure of peak factors in a way that is reliable, rigorous, and robust, the following sampling design is proposed. A discretized number N of households is set and, for each of them, the N time series (each corresponding to a water meter) with size D (corresponding to the considered monitored days) are extracted from the consumption database of the pilot area and aggregated. For each N, the operation is repeated M times, allowing the same water meters to be extracted in different samples, whereas, for each sample, extraction is performed without replacement. In this way, N artificial populations are obtained (one for each aggregation level) and M representative samples with size D are available. Finally, for each N, the main focus concerns the analysis of the following quantities assumed as the most important when using the concept of peak factors for the design or verification of water networks:

- Expected value of the sample mean of hourly peak factors;
- Standard error of the sample mean of hourly peak factors.

To correctly address the above-mentioned items, the usual sampling theory (e.g., [43]) cannot be adopted straightforwardly. The first reason lies in the observation that each random sample consists of a time series made up of a number D of independent realizations of the variable of interest (hourly peak factor), but there could be a non-negligible cross-correlation among the M samples that has to be taken into account. In this perspective, literature provides suggestions about including cross-correlation in the analyses [44].

The second reason is that the effect of a finite population must be taken into account. In this perspective, literature suggests that the classic sampling theory should be adopted when the population fraction ψ (namely the ratio of the amount of extracted data to the maximum number of available data, or, in other words, the ratio of sample size to finite population size) is small [45]. Indeed, in this condition, sample sizes comparable with the population size provide unnaturally small variabilities, since different samples will contain the same elements when $\psi \rightarrow 1$, with a progressive degeneration of the variance [45]. In turn, this could result in the need for very large and expensive databases to investigate large aggregation levels. For large ψ values, in case of sampling without replacement, suitable correction factors should be applied when estimating standard errors from the population variance, whereas the effect of a finite population on central values is usually considered negligible [45]. Especially concerning the variance of sample means, a correction factor, usually referred to as the Finite Population Correction Factor (*FPCF*) [45,46], a function of the population fraction, should be used when relating this quantity to the population variance. In the present research, the investigated population is characterized by two different dimensions, namely the number of monitored days D and the number of aggregated households N. For the adopted sampling design, D is the sample size, directly affecting computations, but the scientific interest mainly lies in understanding the effect of N, which, in turn, acts as a hidden variable with no explicit mathematical effect.

2.2.1. Parameter Definition

Let $q_{h,i}(d)$ be defined as a random variable which describes the water volume consumed by a single household i within a specific hour h of a specific day d; if D is the number of days with hourly registrations, the recorded sample for the hour h is made up of a maximum of D data. M random samples of N households are drawn from the database of N_{max} households ($1 \leq N \leq N_{max}$) so that each household can belong to different samples, but every household can only be extracted once within each sample. The aggregated water demand for each day d at hour h of the random sample m is:

$$Q^m_{h,N}(d) = \sum_{i=1}^{N} q_{h,i}(d) \qquad h = 1, \ldots, 24 \tag{1}$$

For a group of N households, the hourly peak water demand $Q^m_{p,N}(d)$ of the random sample m for each day d is defined as:

$$Q^m_{p,N}(d) = \max_{h=1,\ldots,24}\left[Q^m_{h,N}(d)\right] \tag{2}$$

where $Q^m_{\mu,N}(d)$ is the daily mean water demand of the random sample m for a group of N households for each day d, expressed as:

$$Q^m_{\mu,N}(d) = \frac{\sum_{h=1}^{24} Q^m_{h,N}(d)}{24} \tag{3}$$

Then, for a group of N households, the dimensionless hourly peak water demand factor $CP_{m,N}(d)$ of the random sample m for each day d is defined as:

$$CP_{m,N}(d) = \frac{Q^m_{p,N}(d)}{Q^m_{\mu,N}(d)} \tag{4}$$

By the adopted notation, $CP_{m,N}(d)$ stands for a CP value belonging to the m-th sample of size N and referring to day d. According to the purpose of the analysis, it could be either seen as part of a sub-sample of size D made up of all the daily observations of CP within one specific sample m, or, alternatively, it can be considered as part of a sub-sample of size M made up of all the observations referring to one specific day d across all the extracted samples. In all cases, $CP_{m,N}(d)$ is a single realization drawn from the population of the random variable CP_N with expected value μ_N.

2.2.2. Expected Value, Variance, and Distribution of the Sample Mean

For a group of N households, the sample mean of the hourly peak water demand factor related to a sample m of size D is:

$$\overline{CP}_{m,N} = \frac{\sum_{d=1}^{D} CP_{m,N}(d)}{D} \tag{5}$$

If CP_N is an independent random variable, the mean (i.e., the expected value) of the sample means $\overline{CP}_{m,N}$ coincides with the population mean μ_N:

$$\mu_N = \frac{\sum_{m=1}^{M} \overline{CP}_{m,N}}{M} \tag{6}$$

Literature suggests an empirical relationship between μ_N and the number N of aggregated households in the following form [7,10,11,16,28]:

$$\mu_N = \frac{a}{N^b} + c \tag{7}$$

where, a, b, and c are the estimated regression coefficients. c is the horizontal asymptote of the function, representing the expected value of the sample mean of peak factor CP_N for a very large N.

According to the classic sampling theory, the standard deviation of the sample mean, usually referred to as, "standard error of the sample mean" [47], $ES_{D,N}$, is directly related to the population variance and to sample size D:

$$ES_{D,N}^2 = Var\{\overline{CP}_{m,N}\} = Var\left\{\frac{\sum_{d=1}^{D} CP_{m,N}(d)}{D}\right\} = \frac{1}{D^2} \sum_{d=1}^{D} Var\{CP_{m,N}(d)\} = \frac{\sigma_N^2}{D} \tag{8}$$

where σ_N^2 is the population variance of CP_N.

If the random variable CP_N is normally distributed, the sample mean will be normally distributed too, with $\overline{CP}_{m,N} \sim N(\mu_N, ES_{D,N})$ independently of sample size D. Otherwise, based on the central limit theorem, when the dimension of the random sample becomes sufficiently large ($D \geq 30$), the distribution of the sample mean can be approximated by a normal distribution independently of the specific distribution of the random variable CP_N. To verify the normality of the sample mean $\overline{CP}_{m,N}$, well-known statistical tests can be adopted such as the Kolmogorov-Smirnov (KS) test [48].

If the random variable CP_N is not independent (as will be demonstrated in the present paper), Equation (6) is still valid, whereas the standard error of the sample mean can be estimated according to the following Equation [45] that explicitly accounts for the covariance matrix:

$$
\begin{aligned}
ES_{D,N}^2 &= Var\left\{\frac{\sum_{d=1}^{D} CP_{m,N}(d)}{D}\right\} \\
&= \sum_{d=1}^{D} \frac{Var\{CP_{m,N}(d)\}}{D^2} + \sum_{i=1}^{D} \sum_{\substack{j=1 \\ j \neq i}}^{D} \frac{Cov\{CP_{m,N}(i), CP_{m,N}(j)\}}{D^2} \\
&= \frac{1}{D^2}\left[\sum_{d=1}^{D} Var\{CP_{m,N}(d)\} + \sum_{i=1}^{D} \sum_{\substack{j=1 \\ j \neq i}}^{D} Cov\{CP_{m,N}(i), CP_{m,N}(j)\}\right]
\end{aligned}
\tag{9}
$$

where the first term sums up the cross-sample variance for each day d, and the second term sums up the cross-correlation among pairs of samples.

Equation (9) estimates the standard error of sample means. When sample data are extracted from a finite population, as in the present paper, the values of the standard error can be influenced and underestimated, because there is a high probability that the same elements are extracted from the total population. Indeed, for $N = N_{max}$ Equation (9) gives a null value for the standard error, which is a degeneration caused by the fact that the M samples are made up of exactly the same CP_N values. Instead, for an infinite population, a finite, although small, value for the standard error should be expected even for very high N values.

In case there is no spatial correlation among water demands, the covariance term in Equation (9) is null and the variance collapses back to Equation (8), with an inverse dependence on sample size D. In any other case, also accounting for the finite population effect (i.e., a null asymptotical value for ES) the dependence of ES on D can be formulated for each N in the generic form:

$$ES_D = \alpha_1 \times D^{\beta_1} \tag{10}$$

where the coefficients depend on the structure of the spatial correlation [30,41]. Since Equation (10) can be applied for each fixed N, the following general equation is proposed to consider the additional dependence of the variance on N:

$$ES_{D,N} = \frac{\alpha_1 \times D^{\beta_1}}{(\alpha_2 + N)^{\beta_2}} \tag{11}$$

When the distribution of the sample mean $\overline{CP}_{m,N}$ for a group of N households is (at least approximately) normal, the lower and upper limits $\left[\overline{CP}_{m,N}\right]_p$ of a confidence interval centered on the mean μ_N, for a predefined probability p, are:

$$\left[\overline{CP}_{m,N}\right]_p = \mu_N \pm \xi_p \times ES_{D,N} \tag{12}$$

where ξ_p is the normal p-th quantile; the standard error $ES_{D,N}$ can be estimated as the square root of either Equation (8) or Equation (9), or directly by its empirical approximation provided by Equation (11), based on the probability distribution of random variable CP_N. If the sample mean $\overline{CP}_{m,N}$ is normally distributed, substituting in Equation (12) the 2.5-th and 97.5-th normal percentile values $\xi_p = \pm 1.96$, the 95% confidence interval is obtained.

2.2.3. Variability of Peak Coefficient among Weekdays

The water demand can show different trends between working days and weekends, and this can affect the maximum daily water demand. For the investigated dataset, water consumption exhibits a significant weekly cycle, and water demand was clustered in three groups: weekdays, Saturdays, and Sundays [40].

To verify if the sample mean of hourly peak factors has a day-to-day variability, the ANOVA test is used, which is able to identify significant differences in the central values of different groups [49]. In the present study, seven groups, one for each day of the week, are defined and the related sample means are estimated by Equation (6). For each group, summation in Equation (6) is only extended to the days D_i with $i = 1, 2, \ldots , 7$ (1 = Mondays, 2 = Tuesdays, $\ldots$, 7 = Sundays). In other words, for each value of N, seven groups of M sample means are evaluated. It is worth noting that the mean of all the $7 \times M$ sample means coincides with the population mean μ_N.

As highlighted in the previous sections, the statistical behaviour of peak factors is influenced by the number N of aggregated households; thus, it is expected that the outcomes of ANOVA show a similar dependence. Specifically, for small N values any differences of the peak factor values among the days of the week could be covered by the high peak demand variability. In turn, those differences could become more evident for higher N values, when peak demand variability is lower due to the stabilization of the aggregated water demand pattern.

3. Results

The methodology proposed in the previous sections was applied to the water consumption database of the pilot area, made up of N_{max} = 1162 connections, each corresponding to a registered consumption time series. The analysis was initialized by setting 50 different values of N to be tested, ranging between 1 and 1162 (Table 1). For each N, M = 150 samples of N time series were randomly extracted from the consumption database and aggregated. Then, for each aggregated series, hourly peak demand factors $CP_{m,N}(d)$ were computed for the m-th sample by means of Equation (5). The total number of available monitored days D_{max} in the 2016 database is equal to 322; thus, for each day of the week, the maximum number of monitored days is 46 (Table 1).

The computation of sample means $\overline{CP}_{m,N}$ by means of Equation (6) was performed gathering $CP_{m,N}(d)$ values in seven groups according to the day of the week. Then, the ANOVA test was performed to highlight possible differences in the behaviour of peak factors during the week. Figure 3 shows the results of the ANOVA test as box-plots of peak factor sample means for two different values of aggregated households N = 10 and N = 1000. ANOVA outcomes highlight that there are significant differences in terms of expected values of sample means between the weekdays, the Saturdays, and the Sundays, so that three clusters can be identified, coherently with findings shown in [40]. Moreover, as expected, those differences are more and more evident the higher the N value and can be considered statistically significant starting from N = 5–10.

Table 1. Cluster definition and relevant parameters.

Cluster		1	2	3	4
Day		Saturdays	Sundays	Weekdays	All Days
D_{min}				30	
D_{max}		46	46	230	322
N_{min}				1	
N_{max}				1162	
Equation (7)	a			1.763	
	b			0.670	
	c	1.573	1.611	1.536	1.552
	R^2	0.998	0.997	0.997	0.998
Equation (11)	α_1	0.274	0.293	0.300	0.274
	β_1			0	
	α_2	-0.726	-0.632	-0.665	-0.695
	β_2			0.5	
	R^2	0.999	0.997	0.999	0.999

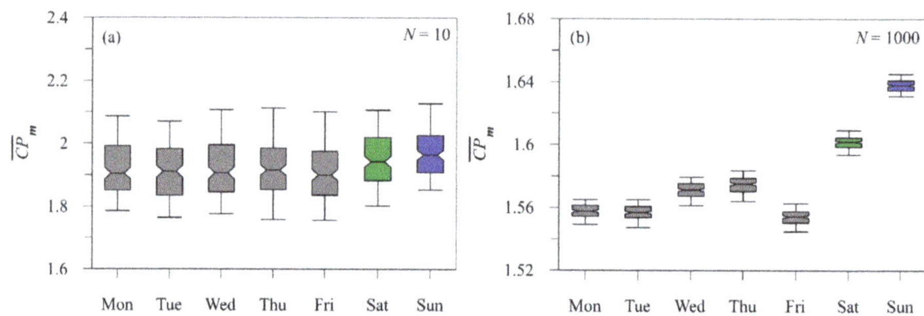

Figure 3. Box-plot of hourly peak factor sample means for the different days of the week and for two different numbers of aggregated households: (**a**) $N = 10$ and (**b**) $N = 1000$.

Figure 3 also shows that weekends are characterized by an expected value of peak factors higher than the weekdays. This could be explained considering that, during the weekend, people tend to adopt predictable schedules, translating in more homogeneous consumption behaviours, leading to more synchronous water uses and, therefore, producing more coherent water demand diurnal patterns. Finally, Figure 3 demonstrates that the differences among the three clusters are statistically significant and should be accounted for in further analyses. As a consequence, in the following sections four clusters will be investigated separately: Cluster 1, made up of Saturdays (D_{max} = 46); Cluster 2, made up of Sundays (D_{max} = 46); Cluster 3, made up of the remaining weekdays (D_{max} = 230); Cluster 4, made up of all the days of the week (D_{max} = 322). This last cluster is considered in order to better understand the significance of cluster separation in evaluating the statistics of interest.

3.1. Sample Mean: Expected Value, Standard Error, and Scaling Laws

According to the proposed methodology, the analysis of hourly peak factor sample means consists of the estimation of the expected value, associated standard deviation, and confidence band.

For each N value, M sample means $\overline{CP}_{m,N}$ were computed by means of Equation (5) and the corresponding expected values μ_N were estimated by means of Equation (6); then, the empirical relation between N and μ_N was found by calibrating parameters in Equation (7). Sample means and expected values are shown, for each Cluster, in Figure 4 as a function of the number of aggregated households. Table 1 shows the estimated values of the regression coefficients a, b, and c and the value for the coefficient of determination, which is very high for all Clusters. Figure 5a shows the comparison between the observed expected values, computed by means of Equation (6), and the predicted expected values, obtained from Equation (7), for all Clusters. It is evident that points gather almost perfectly along the 1:1 line, showing a high accordance between the observed and the predicted values, with just a slight deviation for the highest mean values, corresponding to $N = 1$.

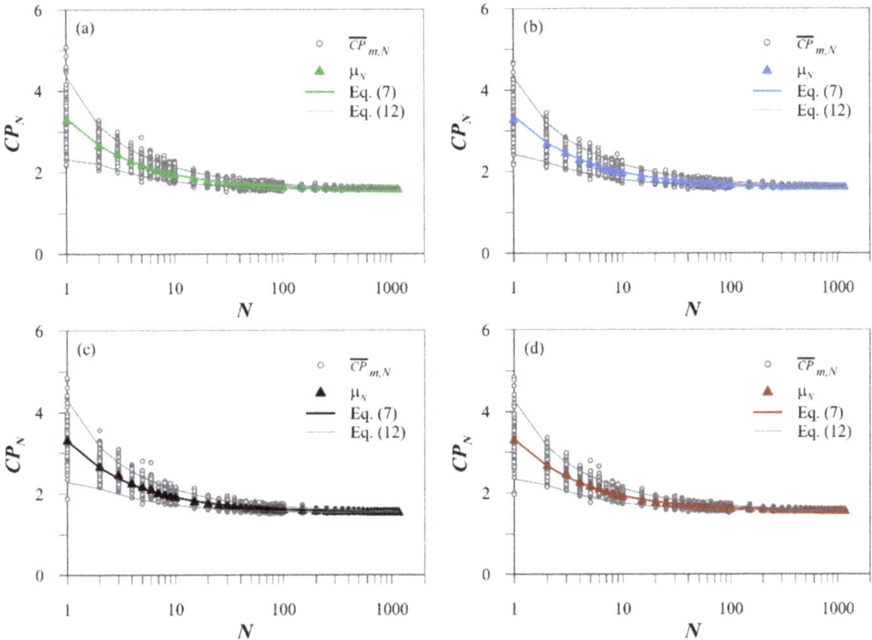

Figure 4. Sample means, expected values, and confidence bands as a function of the number of aggregated households for: (**a**) Cluster 1; (**b**) Cluster 2; (**c**) Cluster 3; (**d**) Cluster 4.

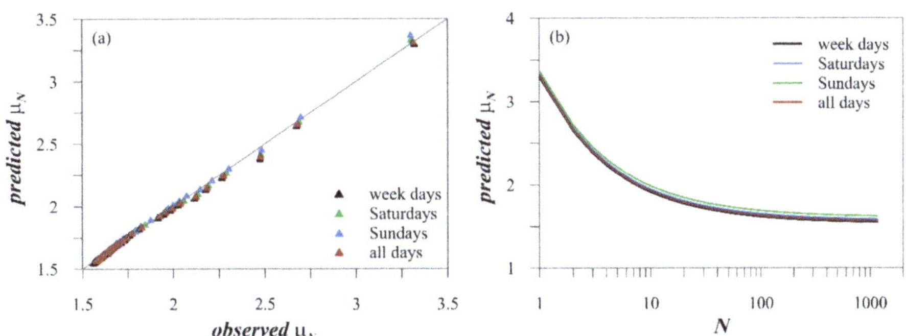

Figure 5. (**a**) Accordance between expected values of sample means estimated by Equations (6) and (7) for all the Clusters. (**b**) Comparison among calibrations of Equation (7) performed on the different Clusters.

Table 1 and Figure 5b show that the regression curves of the four Clusters are very similar, with only a different value for the c coefficient, which represents the expected value of hourly peak demand factor for a large number of households. As Figure 4 shows, this asymptotic value can be considered attained for $N > 100$–200 for every Cluster. Figure 5b and Table 1 also show that the highest asymptotic expected value is observed for the Sundays Cluster, followed by the Saturdays, and the Weekdays Clusters. Cluster 4 shows intermediated values.

As Figure 4 shows, for a fixed N, the M sample means $\overline{CP}_{m,N}$ show a non-negligible variability, which can be quantified by means of the standard error $ES_{D,N}$. In order to compute standard errors, the regression coefficients in Equation (11) were calibrated for each Cluster by using the estimate of $ES_{D,N}$ provided by Equation (9), and their values are shown in Table 1 along with the very high coefficient of determination. To capture the dependence of $ES_{D,N}$ on both N and D, different values of D were tested in the range D_{min}–D_{max}, where $D_{min} = 30$ was set to ensure normality, as previously mentioned. However, in all cases the dependence on D resulted to be negligible with respect to the aggregation level, with very small values for the exponent β_1, that was approximated to zero for all Clusters (Table 1). Moreover, Table 1 shows that for all the Clusters β_2 resulted equal to 0.5, with a simplification in the proposed regression equation.

Figure 6 shows, for each Cluster, the regression curve provided by Equation (11) as well as the standard errors estimated as the square root of Equation (9) for three different values of D (D_{min}, D_{max}, and intermediate value depending on D_{max}). Coherently with the approximation $\beta_1 = 0$, no effect of the number of recording days can be observed, with all the points gathering along the regression curve, with just a slight deviation for $N = 1$.

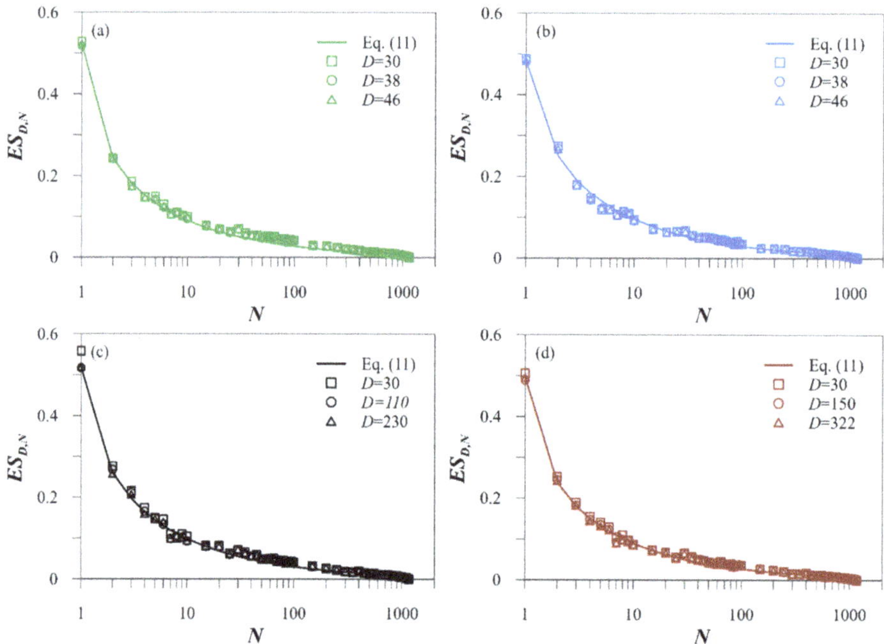

Figure 6. Standard errors of the sample mean estimated by Equation (9) and predicted by Equation (11) for three different D for: (**a**) Cluster 1; (**b**) Cluster 2; (**c**) Cluster 3; (**d**) Cluster 4.

As a goodness-of-fit measure, Figure 7a shows a comparison between the squared standard error estimated by means of Equation (9) and the values predicted by Equation (11) for all the Clusters, with regression coefficients shown in Table 1. The points in Figure 7a gather almost perfectly along the 1:1 line, ensuring an extremely satisfying prediction of the sample mean standard deviation by Equation (11). Figure 7 shows a comparison among the prediction curves of the standard error as a function of N for the different Clusters for $D = D_{max}$. It can be observed that the four curves show small differences for small values of N, which become negligible for $N > 100$–200. Coherently, in the same range of N, the prediction curve for μ_N reaches its asymptotic value for all the Clusters, which suggests an extreme accuracy in the estimation of the expected value of the sample mean for $N > 100$–200. On the other hand, this can be regarded as an effect of investigating a finite population. Indeed, if the same N values were analysed based on a more extended database (i.e., if a higher number of recorded households were monitored), higher values for the standard error would possibly be expected. Moreover, for $N = N_{max}$, each of the M samples is made up of the same elements, so that the M estimates of the sample mean are equal, and the standard error of the sample mean is equal to zero.

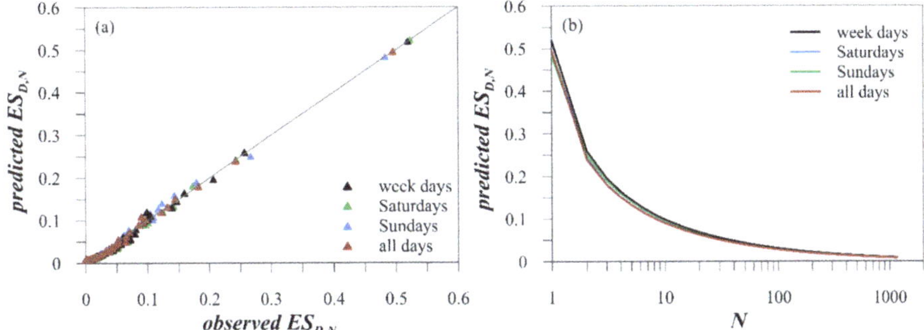

Figure 7. (**a**) Accordance between the standard error estimated for $D = D_{max}$ by Equations (9) and (11) for all the Clusters. (**b**) Comparison among calibrations of Equation (11) performed on the different Clusters for $D = D_{max}$.

The empirical estimates of the standard error were adopted in Equation (12) to obtain the 95% confidence band centred on the expected value of the sample mean, as shown in Figure 4. Confirming the previous evidence, the confidence band reduces as N increases, with an amplitude that can be considered negligible for $N > 100$–200.

3.2. Sample Mean: Probability Distribution and Final Considerations

The assumption of normality was verified for sample sizes $D \geq 30$ independently on the distribution of the original sample variable CP_N. However, in order to highlight the possible effect of a finite population, the normality assumption was checked for each Cluster and each value $N < N_{max}$ by means of the Kolmogorov-Smirnov (KS) test [48]. For $N = N_{max}$ no probability distribution can be defined since the variance is null.

The KS test was run under two different assumptions for the distribution of the sample mean:

- normal distribution with unknown mean and variance parameters m and s ("assumption 1");
- normal distribution with $m = \mu_N$, estimated by means of Equation (6), and $s = ES_{D,N}$, estimated as the square root of Equation (9) ("assumption 2").

Figure 8 shows the results of the Kolmogorov-Smirnov test for the two considered assumptions, in terms of percentage of samples passing/not passing the KS test for the four Clusters. It can be observed that under assumptions 1 and 2 the KS test is passed for all the Clusters for all the tested N values. This proves that the sample means are rigorously distributed by means of a normal model with the mean and variance correctly estimated by Equations (6) and (9), respectively. This also confirms that the estimation of the probability distribution of the sample mean is not affected by any finite population effect.

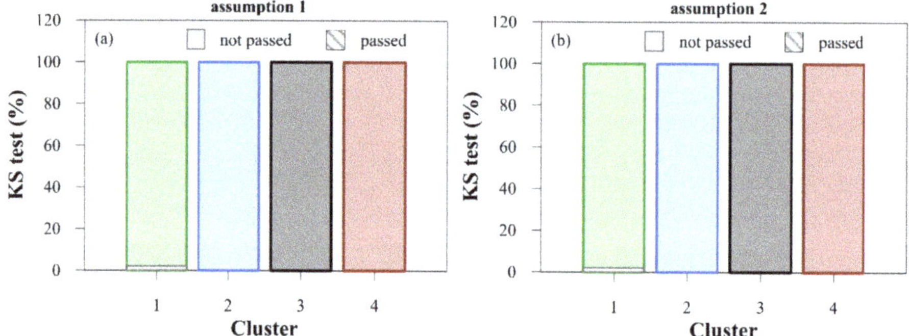

Figure 8. Percentage of samples passing/not passing the Kolmogorov-Smirnov test for the four tested Clusters under (**a**) assumption 1 and (**b**) assumption 2 for the underlying normal distribution of sample means.

Finally, for Cluster 2, Figure 9 shows a comparison between the empirical frequency and the normal probability models under the two assumptions, for the values $N = 60$ and $N = 500$; for these values, the KS test is passed under both assumptions. It can be noted that the CDF curves representing assumptions 1 and 2 are overlapped, highlighting the accuracy of the theoretical estimators adopted for the expected value and the standard deviation. Those results are shown for Cluster 2 but can be extended to all the Clusters.

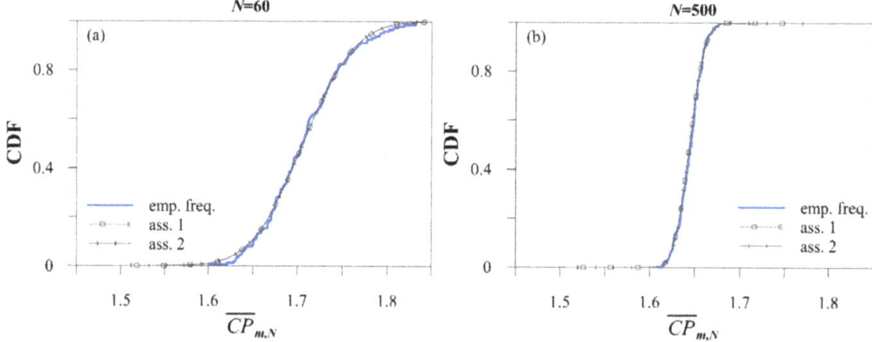

Figure 9. Empirical vs. theoretical probability distributions of sample means for Cluster 2 under two different assumptions for the underlying normal model: (**a**) $N = 60$ and (**b**) $N = 500$.

4. Discussion

4.1. Comparison with Literature

In this paragraph the results obtained from the presented analysis are compared with previous literature analyses. Different literature deterministic relationships for peak factor evaluation have been inspired by the following well known Babbitt's formula [7] deduced for domestic wastewater:

$$C_{P,B} = \frac{5}{\left(\dfrac{N_u}{1000}\right)^{0.2}} \tag{13}$$

where N_u is the number of users, usually ranging between one thousand and one million in a population, as previously mentioned. The Babbitt's relationship was successively reformulated [28] as:

$$\mu_{N_u}(\Delta t) = K_{CP}(\Delta t) \times \frac{10}{N_u^{0.2}} \tag{14}$$

Equation (14) is valid for $250 < N_u < 1250$, and was originally obtained analysing data measured with a 1-min frequency ($K_{CP} = 1$). K_{CP} is a reduction coefficient that takes into account the effect of the time aggregation scale for time steps higher than 1 min. The results of the above relationships are compared with Equation (7), herein rewritten in terms of number of users N_u assuming that each meter serves 2.9 inhabitants on average, and considering the parameters corresponding to all the days of the week (Cluster 4 in Table 1):

$$\mu_{N_u} = \frac{3.60}{N_u^{0.67}} + 1.552 \tag{15}$$

Differently from Equation (15), Equations (13) and (14) do not exhibit any asymptote.

For $N_u = 500$ and $N_u = 1000$, Table 2 reports: (i) The values of the peak factor estimated by Equation (15); (ii) the values obtained by adopting Equation (13), and (iii) the values obtained by adopting Equation (14). In Equation (14), considering the experimental field data reported in [28], K_{CP} is assumed to be equal to 0.65 for a sampling time step of 60 min. Table 2 highlights that the Babbitt's formula overestimates the peak factor [10], while the prediction obtained with the present analysis is comparable with the estimate of the formula proposed by [28]. In particular, the values obtained with Equation (14) are within the uncertainty range of Equation (7).

Table 2. Hourly peak factor values estimated with different relationships.

N_u	Equation (15)	Equation (13)	Equation (14)
500	1.61	6.60	1.87
1000	1.59	5.00	1.63

Forcing Equation (15) to assume a structure similar to Equation (14), it can be approximated by the following expression:

$$\mu_{N_u} = \frac{4.3}{N_u^{0.18}} \tag{16}$$

where the exponent for the number of users is very similar to the one in the empirical relationship in Equation (14) proposed by [28].

As previously mentioned, differently from the empirical literature relationships, the proposed Equation (7) for the evaluation of the hourly peak coefficient tends to an asymptotic value as the number of household increases. A similar result was also obtained by [16], who derived an estimation of the instantaneous peak factor using a probabilistic approach to describe the residential water use based on the Poisson Rectangular Pulse (PRP) model and adopting the Gumbel distribution for the extreme values. The asymptotic value can be assumed to be equal to the asymptotical hourly peak factor for a growing population [11]. Analyses performed on different towns in Italy showed that the asymptotic value ranges between 1.5 and 1.7 [11], similarly to the one deduced herein.

For a number of users varying between 3 and 3000, Equation (15) predicts a peak factor ranging between 3.3 and 1.55, which is the asymptotic value. Those values are also comparable with the range 1–5 reported in [26] considering the results of recent studies in different countries. The obtained values, smaller than the one provided by the empirical Babbitt's relationship, may be ascribed to a different kind of analysis and/or to a change in consumption behaviours compared to 30–50 years ago.

4.2. Applicability Example

The proposed procedure helps the operators in understanding the reliability of a network in terms of demand and pressure at a different level of the users' aggregation using hourly meter data. It can be adopted for understanding if peak values are changed with respect to the ones considered at the design stage, for planning DMAs and for verifying the behavior of existing networks in case of problems in the branched pipes where a lower number of household is served.

As noted above, Equation (15) tends to an asymptotic value as the number of households increases ($N > 100$–200). This means that for looped networks, which serve more than about 600 inhabitants, the peak value can be considered equal to the asymptotic value. Conversely, when considering a single mainline serving different small groups of households, the variability of the peak factor should be accounted for. The synthetic following example shows an application of the proposed formulation in verifying a branched pipe serving different groups of households. Figure 10 shows the main line with six nodes and Table 3 reports the number of households (each represented by a water meter) assumed connected to each node, and the corresponding number of users under the assumption that each meter serves 2.9 inhabitants.

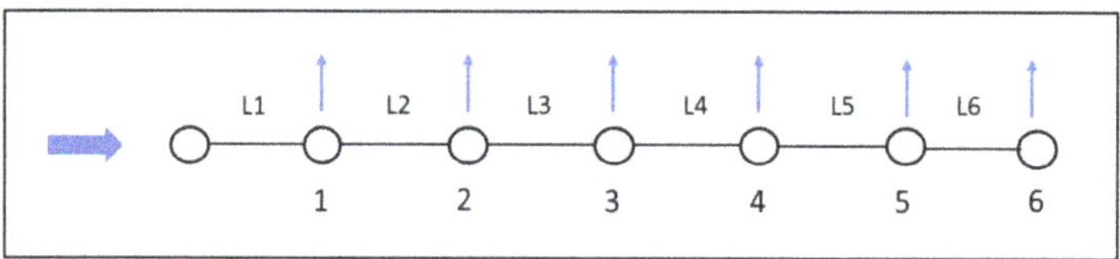

Figure 10. Sketch of a schematic mainline with six nodes serving different groups of households.

Table 3. Example data.

Node	N	N_u	Link	CP	Q_m (L/s)	Q_p (L/s)
1	60	174	L1	1.60	36	59
2	32	93	L2	1.62	24	40
3	40	116	L3	1.64	18	29
4	30	87	L4	1.68	9.7	16
5	15	43	L5	1.81	3.6	6.6
6	3	9	L6	2.41	0.61	1.5

For a total number of 180 served households, equivalent to 522 users, Equation (15) provides a peak factor (CP) equal to 1.60, which is the value that should be considered for designing the pipe L1. Indeed, while link L1 serves 180 households, L6 serves only three of them. Assuming a water supply of 0.07 L/s per inhabitant, Table 3 reports, for each link, the peak value obtained by means of Equation (15), μ_{N_u}, as well as the corresponding mean, Q_m, and peak, Q_p, discharge. A correct evaluation of the hourly peak factor is important for designing the trunks of branch pipes, where an underestimation of the discharge may produce situations of pressure deficit. Conversely, an overestimation of the pipe diameter may produce low velocity and an increase of the water age with a consequent decay of water quality [50]. Concluding, the performed study highlights that the peak factor changes drastically in the interval $1 < N < 100$, and this change has to be carefully considered for a correct design of branch pipes.

5. Conclusions

The proposed analysis provides a methodological framework to investigate the main features of water demand hourly peak factors based on hourly consumption data. The main objective is the estimation of the sample mean of hourly peak factors, the associated standard error (allowing for the definition of confidence bands), and its probability distribution. Those quantities are investigated in a perspective of spatial aggregation: For each considered aggregation level, artificial populations are created by aggregating multiple consumption time series and analysing the related statistics.

Theoretical expressions for the sample mean and for the standard error are provided (Equations (6) and (9), respectively), where the standard error expression accounts for the cross-correlation among samples. Moreover, empirical relations of the sample mean and standard error as a function of the number of aggregated households or meters (or users) are also provided (Equations (7) and (11), respectively). Concerning the probability distribution, sample means can be considered normally distributed, with model parameters effectively estimated by Equations (6) and (9).

The outcomes of the research in terms of mean peak factor are consistent with previous literature analyses focusing on similar or higher-resolution consumption datasets. In addition, the confidence band suggests a high accuracy of its estimation. The structure of the dependence on the aggregation level suggests the presence of an asymptotic value for a high number of users, as also suggested by some recent literature works.

The research confirms the possibility of using 1 h-aggregation consumption datasets for the analysis of water demand peak factors and provides a general framework to perform the stochastic analysis for aggregated consumption data. The empirical relation for the estimation of the expected value of the hourly peak factor has a general validity, although regression parameters' values are a reflection of the specific consumptions of the pilot area. General validity can be also extended to Equation (11) for the estimation of the standard deviation if the effect of a finite population is neglected. Indeed, results showed that the finite population condition does not affect the probability distribution of sample means, which remains normal, but it may affect the amplitude of the confidence bands, which could be underestimated. The proposed methodology will be further applied on other distribution systems. Moreover, additional investigations about the effect of spatial correlation on the coefficient of variation of peak discharges, as well as the quantification of the peak factor variance, will be the object of future research.

As a final remark, the structure and the coefficients of the empirical relationship described by Equation (7) for the expected value of the hourly peak water demand factor allows formulating the following general considerations, that can be of significant aid in the design and verification of water distribution networks.

1. Several relationships provided by the literature asymptotically tend to zero as the number of households increase; conversely, in the present research the peak factor asymptotically tends to a constant value greater than one.
2. The asymptotic value is reached for values of the number of households N of about 100–200 (approximately corresponding to a number of users N_u of about 300–600); conversely, the increase in the peak factor mainly affects only the secondary pipe networks of the urban centres which serve a reduced number of users.
3. The secondary pipe networks generally consist of a branched pipe structure, which is more sensitive to the flow variation than looped networks; in this case, the peak factor must be adequately considered for design purposes.

Author Contributions: Conceptualization, G.D.G., C.D.C., and R.P.; data curation, G.D.G., C.D.C., and R.P.; formal analysis, G.D.G., C.D.C., and R.P.; investigation, G.D.G., C.D.C., and R.P.; methodology, G.D.G., C.D.C., and R.P.; software, G.D.G., C.D.C., and R.P.; supervision, G.D.G., C.D.C., and R.P.; validation, G.D.G., C.D.C., and R.P.; visualization, G.D.G., C.D.C., and R.P.; writing—original draft, G.D.G., C.D.C., and R.P.; writing—review and editing, G.D.G., C.D.C., and R.P. All authors have read and agreed to the published version of the manuscript.

Funding: This research received no external funding.

Conflicts of Interest: The authors declare no conflict of interest.

References

1. Cominola, A.; Spang, E.S.; Giuliani, M.; Castelletti, A.; Lund, J.R.; Loge, F.J. Segmentation analysis of residential water-electricity demand for customized demand-side management programs. *J. Clean. Prod.* **2018**, *172*, 1607–1609. [CrossRef]
2. National Research Council of the National Academies. *Drinking Water Distribution Systems: Assessing and Reducing Risks*; National Academies Press: Washington, DC, USA, 2007.

3. Babayan, A.; Kapelan, Z.; Savic, D.; Walters, G. Least-Cost design of water distribution networks under demand uncertainty. *J. Water Resour. Plan. Manag.* **2005**, *131*, 375–382. [CrossRef]

4. Alcocer-Yamanaka, V.H.; Tzatchkov, V.G.; Arreguin-Cortes, F.I. Modeling of drinking water distribution networks using stochastic demand. *Water Resour. Manag.* **2012**, *26*, 1779–1792.

5. Di Cristo, C.; Leopardi, A.; de Marinis, G. Assessing measurement uncertainty on trihalomethanes prediction through kinetic models in water supply systems. *J. Water Supply Res. Technol. AQUA* **2015**, *64*, 516–528. [CrossRef]

6. Marquez Calvo, O.O.; Quintiliani, C.; Alfonso, L.; Di Cristo, C.; Leopardi, A.; Solomatine, D.; de Marinis, G. Robust optimization of valve management to improve water quality in WDNs under demand uncertainty. *Urban Water J.* **2018**, *15*, 943–952. [CrossRef]

7. Babbitt, H.E. *Sewerage and Sewage Treatment*, 3rd ed.; Wiley: New York, NY, USA, 1928.

8. Johnson, C.F. Relation between average and extreme sewage flow rates. *Eng. News Rec.* **1942**, *129*, 500–501.

9. Metcalf, L.; Eddy, H.P. *American Sewerage Practice, Volume III: Design of Sewers*, 3rd ed.; McGraw-Hill: New York, NY, USA, 1935.

10. Tricarico, C.; de Marinis, G.; Gargano, R.; Leopardi, A. Peak residential water demand. *Proc. Inst. Civ. Eng. Water Manag.* **2007**, *160*, 115–121. [CrossRef]

11. Balacco, G.; Gioia, A.; Iacobellis, V.; Piccinni, A.F. At-site assessment of a regional design criterium for water-demand peak factor evaluation. *Water* **2019**, *11*, 24. [CrossRef]

12. Buchberger, S.G.; Carter, J.T.; Lee, Y.H.; Schade, T.G. *Random Demands, Travel Times and Water Quality in Dead-Ends*; AWWARF Report No. 294; American Water Works Association Research Foundation: Denver, CO, USA, 2003.

13. Buchberger, S.G.; Wu, L. Model for instantaneous residential water demands. *J. Hydraul. Eng.* **1995**, *121*, 232–246. [CrossRef]

14. Buchberger, S.G.; Wells, G.J. Intensity, duration and frequency of residential water demands. *J. Water Resour. Plan. Manag.* **1996**, *122*, 11–19. [CrossRef]

15. Guercio, R.; Magini, R.; Pallavicini, I. Instantaneous residential water demand as stochastic point process. *WIT Trans. Ecol. Environ.* **2001**, *48*, 129–138.

16. Zhang, X.; Buchberger, S.; Van Zyl, J. A theoretical explanation for peaking factors. In Proceedings of the ASCE EWRI Conferences, Anchorage, AK, USA, 15–19 May 2005.

17. Creaco, E.; Alvisi, S.; Farmani, R.; Vamvakeridou-Lyroudia, L.; Franchini, M.; Kapelan, Z.; Savic, D. Preserving duration–intensity correlation on synthetically generated water-demand pulses. *Procedia Eng.* **2015**, *119*, 1463–1472.

18. Creaco, E.; Farmani, R.; Kapelan, Z.; Vamvakeridou-Lyroudia, L.; Savic, D. Considering the mutual dependence of pulse duration and intensity in models for generating residential water demand. *J. Water Resour. Plan. Manag.* **2015**, *141*, 04015031.

19. Alvisi, S.; Franchini, M.; Marinelli, A. A stochastic model for representing drinking water demand at residential level. *Water Resour. Manag.* **2003**, *17*, 197–222.

20. Alcocer-Yamanaka, V.H.; Tzatchkov, V.; Buchberger, S.G. Instantaneous water demand parameter estimation from coarse meter readings. In Proceeding of the 8th Water Distribution Systems Analysis Symposium, Cincinnati, OH, USA, 27–30 August 2006; ASCE: Reston, VA, USA; pp. 1–14.

21. García, V.J.; García-Bartual, R.; Cabrera, E.; Arregui, F.; García-Serra, J. Stochastic model to evaluate residential water demands. *J. Water Resour. Plan. Manag.* **2004**, *130*, 386–394.

22. Creaco, E.; Kossieris, P.; Vamvakeridou-Lyroudia, L.; Makropoulos, C.; Kapelan, Z.; Savic, D. Parameterizing residential water demand pulse models through smart meter readings. *Environ. Model. Softw.* **2016**, *80*, 33–40.

23. Alvisi, S.; Franchini, M.; Marinelli, A. Generation of synthetic water demand time series at different temporal and spatial aggregation levels. *Urban Water J.* **2014**, *11*, 297–310.

24. Blokker, E.J.M.; Vreeburg, J.H.G.; van Dijk, J.C. Simulating residential water demand with a stochastic end-use model. *J. Water Resour. Plan. Manag.* **2010**, *136*, 19–26.

25. Blokker, E.J.M.; Pieterse-Quirijns, E.J.; Vreeburg, J.H.G.; van Dijk, J.C. Simulating Nonresidential Water Demand with a Stochastic End-Use Model. *J. Water Resour. Plan. Manag.* **2011**, *137*, 511–520.

26. Beal, C.; Stewart, R.A. Identifying Residential Water End-Uses Underpinning Peak Day and Peak Hour Demand. *J. Water Resour. Plan. Manag.* **2014**, *140*, 04014008.

27. Creaco, E.; Pezzinga, G.; Savic, G. On the choice of the demand and hydraulic modeling approach to WDN real-time simulation. *Water Resour. Res.* **2017**, *53*, 6159–6177.

28. Gargano, R.; Tricarico, C.; Granata, F.; Santopietro, S.; de Marinis, G. Probabilistic Models for the Peak Residential Water Demand. *Water* **2017**, *9*, 417.

29. Filion, Y.R.; Adams, B.; Karney, B. Cross correlation of demands in water distribution network design. *J. Water Resour. Plan. Manag.* **2007**, *133*, 137–144.

30. Magini, R.; Pallavicini, I.; Guercio, R. Spatial and temporal scaling properties of water demand. *J. Water Resour. Plan. Manag.* **2008**, *134*, 276–284.

31. Gato-Trinidad, S.; Gan, K. Characterizing maximum residential water demand. *Water* **2012**, *122*, 15–24.

32. Creaco, E.; Signori, P.; Papiri, S.; Ciaponi, C. Peak demand assessment and hydraulic analysis in WDN design. *J. Water Resour. Plan. Manag.* **2018**, *144*, 04018022.

33. Moughton, L.J.; Buchberger, S.G.; Boccelli, D.L.; Filion, Y.R.; Karney, B.W. Effect of time step and data aggregation on cross correlation of residential demands. In Proceeding of the 8th Annual Water Distribution Systems Analysis Symposium, Cincinnati, OH, USA, 27–30 August 2006; ASCE: Reston, VA, USA; pp. 1–14.

34. Boyle, T.; Giurco, D.; Mukheibir, P.; Liu, A.; Moy, C.; White, S.; Stewart, R. Intelligent metering for urban water: A review. *Water* **2013**, *5*, 1052–1081.

35. Cominola, A.; Giuliani, M.; Piga, D.; Castelletti, A.; Rizzoli, A.E. Benefits and challenges of using smart meters for advancing residential water demand modeling and management: A review. *Environ. Model. Softw.* **2015**, *72*, 198–214.

36. Willis, R.; Stewart, R.A.; Panuwatwanich, K.; Capati, B.; Giurco, D. Gold Coast domestic water end use study. *Water J. Aust. Water Assoc.* **2009**, *36*, 79–85.

37. Britton, T.; Stewart, R.; O'Halloran, K. Smart metering: Enabler for rapid and effective post meter leakage identification and water loss management. *J. Clean. Prod.* **2013**, *54*, 166–176.

38. Gargano, R.; Tricarico, C.; Del Giudice, G.; Granata, F. A stochastic model for daily residential water demand. *Water Sci. Technol. Water Supply* **2016**, *16*, 1753–1767.

39. Padulano, R.; Del Giudice, G. A mixed strategy based on Self-Organizing Map for water demand pattern profiling of large-size smart water grid data. *Water Resour. Manag.* **2018**, *32*, 3671–3685.

40. Padulano, R.; Del Giudice, G. Pattern Detection and Scaling Laws of Daily Water Demand by SOM: An Application to the WDN of Naples, Italy. *Water Resour. Manag.* **2019**, *33*, 739–755.

41. Creaco, E.; De Paola, F.; Fiorillo, D.; Giugni, M. Bottom-up generation of water demands to preserve basic statistics and rank cross-correlations of measured time series. *J. Water Resour. Plan. Manag.* **2020**, *146*, 06019011.

42. Padulano, R.; del Giudice, G. A nonparametric framework for water consumption data cleansing: An application to a smart water network in Naples (Italy). *J. Hydroinf.* Available online: https://iwaponline.com/jh/article/doi/10.2166/hydro.2020.133/72559/A-nonparametric-framework-for-water-consumption (accessed on 6 March 2020).

43. Frankel, M. *Handbook of Survey Research*; Academic Press: San Diego, CA, USA, 1983.

44. Kay, S. *Intuitive Probability and Random Processes Using MATLAB®*; Springer Science & Business Media: Berlin, Germany, 2006.

45. McCharty, P.J.; Snowden, C.B. *The Bootstrap and Finite Population Sampling*; Vital and Health Statistics Series 2, No. 95. DHHS Pub. No. (PHS) 85–1 369; Public Health Service, U.S. Government Printing Office: Washington, DC, USA, 1985.

46. Starsinic, M. Incorporating a Finite Population Correction Factor into American Community Survey Variance Estimates. In Proceedings of the Section on Survey Research Methods; American Statistical Association: Alexandria, VA, USA, 2011; pp. 3621–3631.

47. Payton, M.E.; Greenstone, M.H.; Schenker, N. Overlapping confidence intervals or standard error intervals: What do they mean in terms of statistical significance? *J. Insect Sci.* **2003**, *3*, 1–6.

48. Kolmogorov, A.N. Confidence limits for an unknown distribution function. *Ann. Math. Stat.* **1941**, *12*, 461–463.

49. Kottegoda, N.T.; Rosso, R. *Applied Statistics for Civil and Environmental Engineers*, 2nd ed.; Blackwell: Malden, MA, USA, 2008.
50. Quintiliani, C.; Marquez-Calvo, O.; Alfonso, L.; Di Cristo, C.; Leopardi, A.; Solomatine, D.P.; de Marinis, G. Multiobjective Valve Management Optimization Formulations for Water Quality Enhancement in Water Distribution Networks. *J. Water Resour. Plan. Manag.* **2019**, *145*, 04019061.

Article

Hybrid Model for Short-Term Water Demand Forecasting Based on Error Correction Using Chaotic Time Series

Shan Wu [1], Hongquan Han [1], Benwei Hou [1,*] and Kegong Diao [2]

[1] College of Architecture and Civil Engineering, Beijing University of Technology, Beijing 100124, China; wushan@bjut.edu.cn (S.W.); hanhongquan_2018@163.com (H.H.)
[2] Faculty of Computing, Engineering, and Media, De Montfort University, The Gateway, Leicester LE1 9BH, UK; kegong.diao@dmu.ac.uk
* Correspondence: benweihou@bjut.edu.cn

Received: 13 April 2020; Accepted: 9 June 2020; Published: 12 June 2020

Abstract: Short-term water demand forecasting plays an important role in smart management and real-time simulation of water distribution systems (WDSs). This paper proposes a hybrid model for the short-term forecasting in the horizon of one day with 15 min time steps, which improves the forecasting accuracy by adding an error correction module to the initial forecasting model. The initial forecasting model is firstly established based on the least square support vector machine (LSSVM), the errors time series obtained by comparing the observed values and the initial forecasted values is next transformed into chaotic time series, and then the error correction model is established by the LSSVM method to forecast errors at the next time step. The hybrid model is tested on three real-world district metering areas (DMAs) in Beijing, China, with different demand patterns. The results show that, with the help of the error correction module, the hybrid model reduced the mean absolute percentage error (MAPE) of forecasted demand from (5.64%, 4.06%, 5.84%) to (4.84%, 3.15%, 3.47%) for the three DMAs, compared with using LSSVM without error correction. Therefore, the proposed hybrid model provides a better solution for short-term water demand forecasting on the tested cases.

Keywords: water demand forecasting; hybrid model; error correction; chaotic time series; least square support vector machine

---

1. Introduction

One critical factor in planning, design, operation, and management of water distribution system (WDS) is satisfying quality water demand at reasonable pressure [1–3]. An accurate hydraulic model of WDS will help water utilities to improve their operation ability and management effectively. Because the WDS hydraulics are driven by consumer demands, it is necessary to estimate consumer demands prior to performing hydraulic evaluation [4]. Water demand at a given time in the future is usually related to historical water consumption and meteorological factors such as humidity, air temperature, and wind velocity [5]. Water demand forecasting plays an important role in activities of the WDS such as water production, pump station operation, real-time modeling, and other strategic decisions of water management [1,6].

The water demand forecasting models can be categorized into long-term and short-term models according to the forecast horizon (i.e., the time period that the water demand will be forecasted) and forecast frequency (i.e., the time step that the water demand forecasts are performed within the time period) [7]. The long-term forecasting model (1 to 10 years' forecast horizon) pays more attention to the plan and design of WDSs. The short-term forecasting model (1 day to 1 month's forecast horizon) targets the real-time water demands of the existing WDSs, which is generally used for daily operation of water plants and pump stations [8]. In this study we focus on the short-term model. The accurate model for short-term water demand forecasting with a forecast frequency ranging from daily to sub-hourly is an essential support for optimal scheduling and better decision marking for WDS management [9].

Many studies have proposed forecasting models for short-term water demand forecasting, which can be generally classified into traditional methods and learning algorithms [9]. Early works used traditional statistical models to settle this problem, such as liner regression, exponential smoothing, and auto regressive integrated moving average (ARIMA) [7]. These models have been widely applied in practice because they are simple to understand and implement. Whereas, the traditional models are not always able to accurately predict the nonlinear changes of water demands. Recently, more sophisticated models that use machine learning algorithms and artificial intelligence have been utilized to address this problem. The models utilizing machine learning algorithms are typical data-driven nonlinear models, which are mainly based on historical data to establish the relationships between water demand and related variables (e.g., previous water consumption, air humidity, and temperature).

A number of data-driven models that use machine learning algorithms have been developed for short-term water demand forecasting, such as artificial neural networks (ANN) models [10–12], support vector machine models (SVM) [13–16], project pursuit regression models [1,17], and random forests [18]. Herrera et al. [1] conducted a comparison of these aforementioned models, and found that the SVM model has the most accurate results. Khan and Coulibaly [15] performed a comparison between SVM, ANN, and seasonal autoregressive model in forecasting lake water levels, and the results indicated the SVM model outperforms the other two. The main reason is because the SVM exhibits inherent advantages in formulating cost functions by using structural risk minimization principle instead of the empirical risk minimization of ANN [19].

SVM maps the nonlinear trends of input space to linear trends in a higher dimensional space and recognizes the subtle patterns in complex datasets by using a learning algorithm [20]. The least squares support vector machine (LSSVM) is an extension of SVM which involves equality constraints instead of inequality constraints and works with a least squares cost function [21,22]. Due to the equality constraints, the LSSVM reduces the computational complexity by solving a set of linear equations rather than the quadratic programming problem in standard SVM. Chen and Zhang [14], Herrera et al. [1], and Praveen and Bagavathi [23] established an LSSVM-based model to forecast hourly water demand; it was found that the LSSVM model has better generalization ability than ANN. Other examples of LSSVM applications include river flow estimation [24], discharge-suspended sediment estimation [25], and pipeline network failure estimation [26]. When forecasting water demand with the LSSVM-based model, Chen and Zhang [13] utilized the Bayesian framework to determine the model parameters (namely, the regularization constant and the width of the RBF kernel). Their case study showed that parameter determination by Bayesian method is faster than that of cross-validation [26,27].

Both the traditional models and the learning algorithms have achieved promising results in their own linear or nonlinear domains, whereas, none of them are universally suitable for all circumstances. To improve the performance of the forecasting models, the hybrid models combining two or more different algorithms/models are developed by some studies. Zhang [28] established the hybrid model with ANN and ARIMA to forecast time series, in which the ARIMA model was firstly used to predict the linear part of the data, then ANN was performed to model the errors between the linear part and the observed data (i.e., the nonlinear part of the data). The application results of three benchmark time series data showed that the hybrid model improved forecasting accuracy more than the independent models. Odan and Reis [7] associated the Fourier series (FS) to ANN for hourly water demand forecasting. ANN were used to model the errors of the FS forecast (i.e., the difference between the FS model and the observed data). Brentan et al. [29] proposed a hybrid model based on SVM and adaptive FS, where SVM firstly provided the initial forecasting and then the adaptive FS was utilized to model the errors between the initial forecasting and the observed data. Thus, the nonlinear and periodical behavior of water demand can be captured by the SVM and FS model, respectively.

In addition to FS, the chaotic time series method gives the possibility of detecting instability phenomena hidden behind random-looking phenomena, which has been widely used in short-term time series forecasting of rainfall, traffic, and other fields. For example, Dhanya et al. [30] examined the chaotic characteristics of daily rainfall data of the Malaprabha basin, India, and they established a daily rainfall prediction model based on the theory of chaotic time series. Liu et al. [31] combined chaos theory with SVM to perform short-term prediction of network traffic. Yang et al. [32] proposed an improved fuzzy neural system based on chaotic reconstruction technology for short-term load forecasting of electric power systems, and the application showed that the chaotic technology-based model performs better than the conventional neural network model. So far, chaotic time series has rarely been implemented to forecast water demand, and its performance in this field is unknown.

As aforementioned, with the help of error correction of the initial forecasting, hybrid models could perform better than any individual model [7,28,29]. Therefore, it is worthwhile to integrate the chaotic time series method in the hybrid forecasting model and investigate their performance. This paper aims to achieve better predictions of short-term water demand by presenting a hybrid forecasting model which couples the chaotic time series with LSSVM in the error correction module. Specifically, it will:

- Present the framework, methods, and performance indicators of the hybrid forecasting model,
- Test the hybrid model's accuracy based on case studies of three real-world DMAs in Beijing WDSs,
- Verify the effectiveness of the model by comparing it with the results of other models, including ARIMA, LSSVM without error correction, and LSSVM using Fourier series for error correction.

2. Methodology

2.1. Research Framework

The historical water consumption and calendar data are used as the model inputs in this study, as many researchers have proved that the hourly and 15-min forecasting model only considering historical water consumption data is able to achieve reliable forecast results [9,33,34]. Further, this study tests the model's capability of forecasting without real-time meteorological (e.g., temperature, humidity, and wind speed) data which is usually unavailable in real-time or highly uncertain. Admittedly, there are studies considering meteorological data for hourly water demand forecasting (e.g., Al-Zahrani et al. [35] and Brentan et al. [29]), but there is no proof that use of meteorological data can significantly improve the prediction accuracy without increasing the complexity of the method.

This study addresses the problem of short-term water demand forecasting with the prediction horizon of 24 h with time intervals of 15 min. Firstly, historical water demand data from DMA cases are collected, and the features of the historical data are extracted to select valuable information as the inputs of the forecasting model. Then the forecasting model is trained and tested using the historical water demand data and will be rebuilt every 24 h on the basis of an updated data set. When applying the forecasting model, the newly observed water demand data are collected at 15-min intervals. The historical data set always maintains the same size and is updated once a day by adding the newly observed data and deleting the earliest data.

There are 96 time steps in the water demand forecasting for one day ahead. The water demand forecasting for each time step in one day ahead is performed as follows: (1) Establish the forecasting model by LSSVM according to the historical water demand data (see Section 2.2, Section 2.3, and Section 3.2). (2) Predict the water demand at the first future time step (15 min) on the forecasting day by the forecasting model; the model inputs for the 15-min prediction are provided by the historical data. (3) Predict the water demand at the second future time step (30 min) on the forecasting day; the model inputs for the 30-min prediction are obtained from the newly observed data at 15 min and the historical data. (4) The input data for the 45-min prediction is obtained from the newly observed data at 30 min, the observed data at 15 min, and the historical data, and so on. This stepwise data updating procedure is shown Figure 1. It should be noted that the model parameters of the forecasting model remain unchanged for the 96 time steps, but the model inputs for different time steps are updated as illustrated in Figure 1.

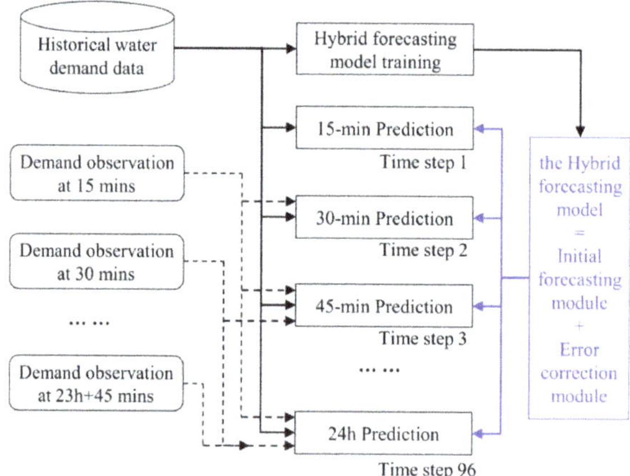

Figure 1. Water demand data processing procedure.

The hybrid forecasting model is mainly constituted of two parts, namely the initial forecasting module and the error correction module. The framework of the hybrid model is shown in Figure 2. The outline of the initial forecasting module is actually similar to the traditional water demand forecasting model. The difference between the hybrid model and the traditional one is the error correction module.

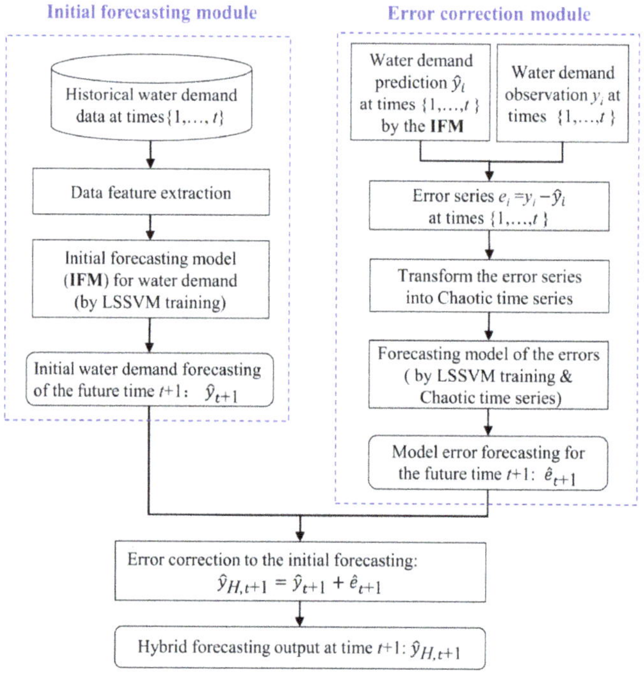

Figure 2. Hybrid framework for water demand forecasting.

In the initial forecasting module, historical water demand data and other relevant information are firstly collected into a data set with the time step of 15 min. After identification and processing of abnormal data, data features are extracted to provide valuable information to the forecasting model inputs. Furthermore, the nonlinear relationship between the historical water demand data and the demand at the next time step is constructed by LSSVM training, which provides the initial forecasting model $F(y)$ of water demand. Then, the forecasted water demand $\hat{y}_{t+1}$ at the future time (target time) $t + 1$ is obtained by the initial forecasting model. The errors of the initial forecasting model on the training data at historical time steps $(1, \ldots , t)$ is expressed as:

$$e_i = y_i - \hat{y}_i \tag{1}$$

where e_i is the error of the initial forecasting model at the time step i $(i = 1, \ldots , t)$; y_i is the observed water demand at time step i; $\hat{y}_i$ is the output value of the initial forecasting model at time step i. Note that, $t + 1$ is the first target time step at which the water demand is unknown and needs forecasting.

The error correction module has three steps. Firstly, the error time series $(e_1, e_2, \ldots , e_i, \ldots , e_t)$ from the initial forecasting model is transformed into a chaotic time series. Secondly, the LSSVM is adopted to establish the relationship between the errors of the initial forecasting at next time step and the chaotic time series at current and previous time steps, which provides the error forecasting model $f(e)$. Thirdly, the forecasted error for the target time $t + 1$ is obtained and used to correct the initially forecasted demand value as follows:

$$\hat{y}_{H,t+1} = \hat{y}_{t+1} + \hat{e}_{t+1} \tag{2}$$

where $\hat{y}_{H,t+1}$ is the water demand forecasting by the hybrid model, in other words, the final output of water demand forecasting at the target time $t + 1$; $\hat{y}_{t+1}$ is the forecasted water demand by the initial forecasting model $F(y)$; and $\hat{e}_{t+1}$ is the forecasted error by the error forecasting model $f(e)$.

2.2. Initial Forecasting Model by LSSVM

SVM has been widely applied in several areas including pattern recognition, regression, nonlinear classification, and function estimation. LSSVM is originated from SVM and first proposed by Suykens and Vandewalle [21], which is believed, takes a computational advantage over standard SVM by converting quadratic optimization problem into linear equations. In the field of water demand forecasting, the LSSVM is used to establish the nonlinear relationship between model inputs and outputs.

Consider a given training set of N samples $(\mathbf{X}_i; y_i)(i = 1, \ldots , N)$, where $\mathbf{X}_i$ denotes the ith input vector in n-dimensional space $(\mathbf{X}_i = (X_{1i}, \ldots , X_{ni})\in\mathbf{R}^n)$ and y_i is the corresponding desired output value (i.e., the observed value) of the ith sample. The nonlinear function between the inputs and outputs can be given as below [19,26,36]:

$$\hat{y}_i(\mathbf{X}_i) = \boldsymbol{\omega}^T \varphi(\mathbf{X}_i) + b \tag{3}$$

where $\hat{y}_i$ is the model output corresponding to the sample i, the nonlinear transformation function $\varphi(*)$ maps the $\mathbf{X}_i$ to the m-dimensional feature space, $\boldsymbol{\omega}$ is the m-dimensional weight parameter vector, and b is the bias parameter $(\boldsymbol{\omega}\in\mathbf{R}^m, b\in\mathbf{R})$.

Equation (3) provides the initial forecasting model of water demand, in other words, the relationship between the model input and output, where the input data is $\mathbf{X}_i = (Q_t, Q_{t-1}, Q_{t-2}, Q_{t-95}, Q_{t-191}, Q_{t-671})$ and the output $\hat{y}_i$ is the forecasted water demand Q_{t+1} at the target time $t + 1$. Detailed description of model input data selection is presented the Section 3.1.

Considering the complexity of minimizing the model errors between y_i and $\hat{y}_i$, in the LSSVM, the parameters ω and b in equation (3) can be estimated according to the structural risk minimization principle [19,36]:

$$\min J(\omega, \xi) = \frac{1}{2}\omega^T\omega + \frac{1}{2}\gamma \sum_{i=1}^{N} \xi_i^2 \tag{4}$$

where γ is the regularization constant determining the tradeoff between the training error and the generalization performance, ξ_i is a slack variable denotes model error.

The solution of the optimization problem (Equation (4)) can be obtained by Lagrange function [19,36]. Then the LSSVM model for the non-learner function in Equation (3) is finally turned into:

$$\hat{y}(\mathbf{X}) = f(X) = \sum_{i=1}^{N} \alpha_i K(\mathbf{X}_i, \mathbf{X}) + b \tag{5}$$

where α_i ($i = 1, \dots, N$) is the Lagrange multiplier and can be evaluated by γ, $K(\mathbf{X}_i, \mathbf{X})$ is the kernel function. The radial basis function (RBF) kernel is one of the most popular kernel functions, and is used in this study as below:

$$K(\mathbf{X}_i, \mathbf{X}) = \exp\left(\frac{-\|\mathbf{X}_i - \mathbf{X}\|^2}{2\sigma^2}\right) \tag{6}$$

where σ is the width parameter that reflects the radius enclosed by the boundary closure.

It is worth mentioning that, at this point, Equation (3) is transformed into Equation (5) which can be directly established though the training samples $(\mathbf{X}_i; y_i)$ ($i = 1, \dots, N$) and model parameters σ and γ. Therefore, establishing an LSSVM model with RBF kernel involves the selection of RBF kernel width σ and the regularization constant parameter γ. Among the available methods for parameter tuning of LSSVM such as the cross-validation method [19], the grid search method [26], and Bayesian framework-based inferring [13,37], the Bayesian approach with three levels of inference is chosen for parameter tuning of LSSVM in this study.

2.3. Error Forecasting Model Based on Chaotic Time Series

Chaos is a quasi-stochastic irregular motion possibly appearing in deterministic nonlinear dynamic systems [38]. Since various nonlinear systems exhibit chaotic features, chaos theory is widely used in nonlinear system analysis to detect deterministic relationships hidden behind random-looking phenomena, and has been increasingly used in time series analysis [30,31]. According to the delay coordinate embedding technique, the underlying dynamical system can be faithfully reconstructed from stochastic time series under fairly general conditions [39]. Therefore, a one-to-one correspondence can be established between the reconstructed and the true but unknown dynamical systems [40].

Given a scalar time series of model errors $e = (e_1, e_2, \dots, e_N)$ with time step Δt, and N is the number of elements in the time series, the element e_i ($i = 1, \dots, N$) is computed by Equation (1). According to the procedure of phase space reconstruction, the scalar time series e is transformed in phase space as follows:

$$\begin{aligned}
\mathbf{E}_1 &= \left(e_1,\ e_{1+\tau},\ e_{1+2\tau},\ \dots, e_{1+(m-1)\tau}\right) \\
\mathbf{E}_2 &= \left(e_2,\ e_{2+\tau},\ e_{2+2\tau},\ \dots, e_{2+(m-1)\tau}\right) \\
&\qquad\qquad \cdots \\
\mathbf{E}_M &= \left(e_M,\ e_{M+\tau},\ e_{M+2\tau},\ \dots, e_{M+(m-1)\tau}\right)
\end{aligned} \tag{7}$$

where τ is the delay time, it could be several times of Δt; $\mathbf{E}_i$ ($i = 1, \dots, M$) is a chaotic vector in the phase space, m is the embedding dimension of the phase space, $M = N-(m-1)\tau$ is the number of phase point.

Takens [39] has proved that the chaotic attractor of a time series would be revealed in the phase space if the parameters τ and m are properly selected. The dimension parameter m is usually larger than three, to entirely reveal the underlying information of the time series [31]. Among existing methods for determining parameters τ and m, the coupled-cluster (C-C) method [41] is used in this study.

In the case of chaotic systems, the Lyapunov exponent (λ) gives a system the sensitivity to initial conditions and determines the total predictability of the system, and a positive λ indicates the system is chaotic [42]. Therefore, the reconstructed time series ($\mathbf{E}_1$, $\mathbf{E}_2$, ... , $\mathbf{E}_M$) is tested for the chaotic signature through the maximum Lyapunov exponent which is evaluated by Wolf's algorithm [43].

In the phase space of a chaotic system, the dynamic information could be interpreted in the form of m-dimensional mapping as [30]:

$$\mathbf{E}_{M+1} = f(\mathbf{E}_M) \tag{8}$$

where $\mathbf{E}_M$ is the state at current time, $\mathbf{E}_{M+1} = (e_{M+1}, e_{M+1+\tau}, e_{M+1+2\tau}, \cdots , e_{M+1+(m-1)\tau})$ is the state at future time. Note that, the last element $e_{M+1+(m-1)\tau}$ of $\mathbf{E}_{M+1}$ is exactly the next element e_{k+1} of the error series e which needs to be forecasted. Therefore, the phase point $\mathbf{E}_i(i = 1, 2, \ldots , M)$ further evolves into $\mathbf{E}_{i+1}$, and there is a determinism mapping function between $e_{i+1+(m-1)\tau}$ (i.e., the last element of $\mathbf{E}_{i+1}$) and $\mathbf{E}_i$ as follows:

$$e_{i+1+(m-1)\tau} = f(\mathbf{E}_i) = f\left(e_1, e_{1+\tau}, e_{1+2\tau}, \ldots, e_{1+(m-1)\tau}\right) \tag{9}$$

According to the properties shown in Equations (8) and (9), the chaotic time series can be utilized for prediction, and then the LSSVM approach described in Section 2.3 can be used to establish the nonlinear functions in Equation (9). The model input data and output data for LSSVM training are shown as follows:

$$\mathbf{X}_{\text{error}} = \begin{bmatrix} \mathbf{E}_1 \\ \mathbf{E}_2 \\ \mathbf{E}_{M-1} \end{bmatrix}; \; \mathbf{Y}_{\text{error}} = \begin{bmatrix} e_{2+(m-1)\tau} \\ e_{3+(m-1)\tau} \\ e_{M+(m-1)\tau} \end{bmatrix} \tag{10}$$

where $\mathbf{X}_{\text{error}}$ is the input data with the dimension of $(M{-}1) \times m$, $\mathbf{Y}_{\text{error}}$ is the output data with the dimension of $(M{-}1) \times 1$.

Note that, due to $M = N{-}(m{-}1)\tau$, the last element of $\mathbf{Y}_{\text{error}}$ is actually e_N, in other words, the last element of the error time series e. After the nonlinear function of Equation (9) is established by LSSVM, one can predict the future element of e at next time step through $e_{N+1} = f(\mathbf{E}_M)$.

2.4. Performance Indicators of Forecasting Models

In terms of accuracy evaluation of water demand forecasting models, variety of measures are available to characterize the performance of the models [1,7,9]. This study adopts four widely used indicators as evaluation criteria, including the mean absolute error (MAE), the mean absolute percentage error (MAPE), the root means square error (RMSE), and the coefficient of determination (R^2). The equations of these aforementioned indicators are shown as follows:

$$\text{MAE} = \frac{1}{N_f} \sum_{i=1}^{N_f} |y_i - \hat{y}_i| \tag{11}$$

$$\text{MAPE} = \frac{1}{N_f} \sum_{i=1}^{N_f} \frac{|y_i - \hat{y}_i|}{y_i} \times 100\% \tag{12}$$

$$\text{RMSE} = \sqrt{\frac{1}{N_f} \sum_{i=1}^{N_f} (y_i - \hat{y}_i)^2} \tag{13}$$

$$R^2 = 1 - \frac{\sum_{i=1}^{N_f} (y_i - \hat{y}_i)^2}{\sum_{i=1}^{N_f} (y_i - \bar{y})^2} \tag{14}$$

where y_i and $\hat{y}_i$ are the observed value and the predicted value of water demand at time i, respectively; $\bar{y}$ and $\bar{\hat{y}}$ are the corresponding mean values; N_f is the number of forecasted time steps, which is equal to 96 for the water demand forecasting problem with a one day horizon and a frequency of 15 min.

3. Case Study

3.1. Data Feature Extraction and Model Inputs

The historical water demand data from three actual DMAs (namely, DMA1, DMA2, and DMA3) in Beijing, China, were collected and used to train and test the forecasting model. On the inlet of the DMA, the water demand data were metered with the unit of m^3 and recorded every 15 min; then the data were transferred to the database of the Beijing Water Works in real time. The water consumption pattern and the composition of customers in DMA1 is very different from that in DMA2 and DMA3; DMA1 includes more than 10,000 residential customers, 168 business customers, and 68 industrial customers. The number of water customers in DMA2 and DMA3 are 1822 and 1936, respectively; water customers in DMA2 and DMA3 are mostly residential and there are also some business customers. The statistics of the three DMAs' water consumption data are show in Table 1. The three DMAs' water consumptions at different times in one week are shown in Figure 3. From the weekly curves of water demands in Figure 3, one can see the different demand patterns of the three DMAs, for example, there is no obvious peak hour in the evening for DMA1, and there are no obvious morning peak hours on weekends for DMA3.

Table 1. Characteristics of water demand data in 2018 for the three case study district metering areas (DMAs).

DMAs	Date of Data	Minimum (m^3/h)	Maximum (m^3/h)	Mean (m^3/h)	Standard Deviation (m^3/h)	Coefficient of Variation
DMA1	1 November–26 December	120.00	2224.00	1192.03	467.81	0.39
DMA2	1 November–26 December	16.88	67.04	35.85	19.97	0.30
DMA3	17 June–11 August	28.48	95.12	63.45	15.14	0.24

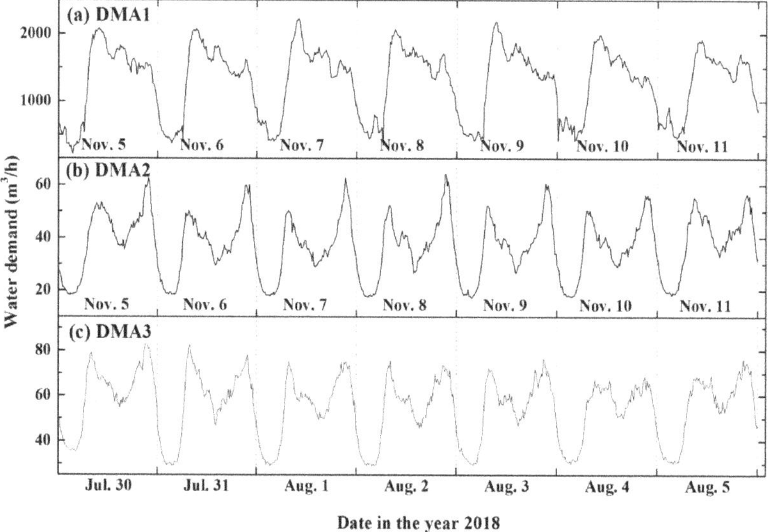

Figure 3. One-week water consumption curves of the case study DMAs. (**a**) DMA1; (**b**) DMA2, and (**c**) DMA3.

In total, 8 weeks' data were collected from the water demand record in 2018 for training and testing the forecasting model. The data set contains 5376 observations for each DMA. Seven weeks' data were used as training data, while the last week's data were used for model testing. When using the hybrid framework to predict the water demand at 96 time steps on the next day, the water demand data of the current day and previous days were used for model training, for example, the historical water demand data of the previous 49 days were used for model training to predict the demand on the 50th day, and the water demand data of the previous 50 days were used for model training to predict the demand on day 51, and so on.

When selecting the input data for the forecasting model from the historical water demand data, Guo et al. [9] categorized the historical data into three fragments, namely, recent time, near time, and distant time, and selected five time-steps in each time fragment as the input data. Herrera et al. [1] selected the historical water demand data at three time-steps including the current time, the previous time, and the target time in the previous week as the input data. Ordan and Reis [7] selected six time-steps including four continuous time-steps before the target time, the target time on the previous day, and previous week. According to these literatures, the historical water demand at the current time, the previous time, the target time on the previous day, and the previous week are usually adopted as the model input data in the short-term water demand forecasting. In this study, to better model the characteristics of the water demand time series, a correlation analysis [7] is performed based on the data of three DMAs to find the data that is highly related to the water demand data at the target time from the historical water demand data. Furthermore, various combinations of the related data are tested as the input for the forecasting model, and the following combination is identified as having the best performance, in other words, three continuous time-steps before the target time (Q_t, Q_{t-1}, Q_{t-2}), the target time on the previous one day and two days (Q_{t-95} and Q_{t-191}), and the target time on the previous week (Q_{t-671}). Therefore, the historical data set (Q_t, Q_{t-1}, Q_{t-2}, Q_{t-95}, Q_{t-191}, Q_{t-671}) is adopted as the input data for the initial forecasting model in this study.

3.2. Model Setup

In addition to the hybrid forecasting model proposed in this study, two other forecasting models are established to make comparisons with and to validate the performance of the proposed hybrid forecasting approach. As summarized in Table 2, the hybrid model H_LSSVM_Chaos is the one established by the hybrid framework of this study (see Figure 2), and the other two are a single forecasting (S_LSSVM) and a hybrid forecasting model (H_LSSVM_FS), respectively. The single forecasting model S_LSSVM uses the traditional prediction procedure without error correction module, in other words, only the initial forecasting module is used. The hybrid forecasting model (H_LSSVM_Chaos and H_LSSVM_FS) adopts both the initial forecasting module and the error correction module. The model inputs of the initial forecasting module are the feature data extracted from the historical water demand data, while the model inputs of the error correction module are the error series of the initial forecasting model. The error series can be evaluated according to Equation (1) and the flowchart in Figure 2. In the hybrid forecasting model, the initial forecasting module is the same one applied in the single forecasting model.

Table 2. Characteristics of forecasting models.

Models	Forecasting Category	Model Inputs
S_LSSVM	Single forecasting	Feature values of historical water demand data
H_LSSVM_Chaos	Hybrid forecasting	Chaotic time series of the errors of the initial forecasting
H_LSSVM_FS	Hybrid forecasting	Scalar time series of the errors of the initial forecasting

The hybrid model H_LSSSVM_FS uses the Fourier series as the forecasting model of the error time series in the error correction module, which is similar to the approach used by Brentan et al. [29] and Ordan and Reis [7]. Model inputs of the hybrid models' error correction modules are based on the errors of the initial forecasting by the S_LSSVM model.

For the error correction module in the H_LSSVM_FS model, the error time series of the previous seven days (i.e., 672 values) is used to compute the coefficients of the Fourier series; the number of harmonics of FS is set to 336. The LS-SVMlab Toolbox developed by Brabanter et al. [44] is used to train the forecasting models by LSSVM, and the three-Level Bayesian inferring method is adopted for parameter tuning of the LSSVM. Table 3 displays the model parameters for the application of LSSVM and chaos methods. Parameters γ and δ^2 in Table 3 were obtained by Bayesian method for the LSSVM model training. In addition, m and τ are the essential parameters for chaotic time series construction.

Table 3. Model parameters for the application of LSSVM and chaos methods.

Models	DMA ID	γ	δ^2	m	τ
	1	0.1431	5.7407	-	-
S_LSSVM	2	0.0378	12.5104	-	-
	3	0.0457	13.4042	-	-
	1	0.5827	5.7680	4	9
H_LSSVM_Chaos	2	2.7872	4.5586	4	8
	3	3.1269	6.1330	3	11

3.3. Application Results

3.3.1. Overall Performance

Figure 4 compares the observed water demand with the forecasted water demand using the S_LSSVM, H_LSSVM_Chaos, and H_LSSVM_FS models at 15 min steps for one day ahead. It can be seen that the predicted water demand by the three models is consistent with the trend of the observations, and the hybrid models perform better than the single forecasting models (S_LSSVM) during the periods of water demand fluctuations. As quantified below by the model performance indicators, the H_LSSVM_Chaos models provide the closest estimates to the corresponding observed water demand during most of the peak periods.

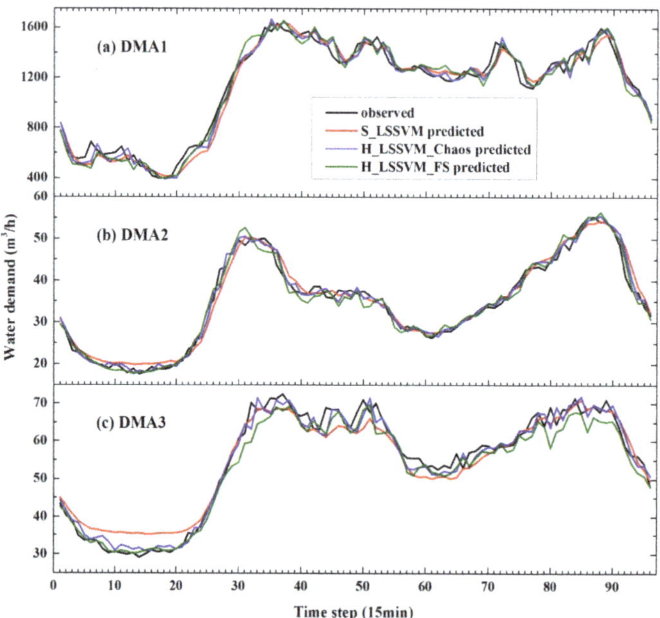

Figure 4. Water demand forecasting for one day ahead with the time step of 15 min. (**a**) Water demand of DMA1 on 26 December; (**b**) water demand of DMA2 on 26 December; and (**c**) water demand of DMA3 on 11 August.

Table 4 gives the overall performance of the different forecasting models for the three DMAs in Beijing. It can be seen that the H_LSSVM_Chaos provides a higher accuracy than the other two models according to the performance indicators R^2, MAE, MAPE, and RMSE. The single forecasting model S_LSSVM is the least accurate.

Table 4. Performance indicators of forecasting models on testing data.

Models	DMA ID	R^2	MAE (m^3/h)	MAPE (%)	RMSE (m^3/h)
	1	0.9654	54.43	5.64	68.61
S_LSSVM	2	0.9722	1.31	4.06	1.76
	3	0.9447	2.70	5.84	3.31
	1	0.9711	47.92	4.84	62.66
H_LSSVM_Chaos	2	0.9817	1.08	3.15	1.43
	3	0.9701	1.86	3.47	2.44
	1	0.9626	56.35	5.44	71.30
H_LSSVM_FS	2	0.9782	1.18	3.33	1.56
	3	0.9533	2.20	3.72	3.05

Among the three DMAs, the prediction accuracy to DMA1 is slightly worse than to DMA2 and DMA3, for example, the MAPEs of (DMA1, DMA2, DMA3) of the H_LSSVM_Chaos models and the H_LSSVM_FS models are (4.84%, 3.15%, 3.47%) and (5.44%, 3.33%, 3.72%), respectively. The reason is that the composition of the water customers in DMA1 is relatively complex, not only including residential users, but also a large number of commercial and industrial users. The statistical parameter COV of DMA1's water demand data is 0.39, which is the largest one among the three DMAs. Larger COV indicates a high level of water demand floating and makes the demand pattern more difficult to capture. As a result, even using the error correction module, the hybrid model H_LSSVM_Chaos only reduced the MAPE of DMA1 from 5.64% to 4.84%, which is less than the reductions for the other DMAs. Moreover, because the water consumptions in DMA2 are mostly residential demands which thus lead to a simple water demand pattern, the prediction results for DMA2 give the highest accuracy. Therefore, as for the error correction module performance on short-term water demand forecasting, the DMAs with simple customer composition have better prediction accuracy when using error correction module.

3.3.2. Comparisons Between the Hybrid Forecasting Models

Figure 5 shows the error forecasting by the error correction module in the hybrid models. Compared to the water demand data in Figure 4, the errors of initial forecasting in Figure 5 have a large number of fluctuations, in other words, the value of errors has a greater frequency of change. In addition, the complex and disorderly change in the peak values of the error data are also shown in Figure 5; there is no obvious rule on the occurrence time of the peak value, such as peaks at the time steps (7, 45, 71, 75) in Figure 5a. The results in Figure 5 can be summarized as follows:

- The error forecasting models based on the chaos method and the FS method can both obtain more reasonable prediction results in some periods where the error data changes mildly, such as time steps 5 to 23 and 60 to 72 in DMA2, and 10 to 24 in DMA3.
- The prediction accuracy of the two methods is relatively low in the periods where the error data change frequently, such as time steps 33 to 55 in DMA1, 24 to 34 in DMA2, and 35 to 53 in DMA3. It should be noted that even in these hardly predictable time steps, however, the predictions from the chaos method is closer to the errors of the initial prediction than the FS model, e.g., for the error predictions at time steps 30 to 55 in the three DMAs, the MAEs obtained by chaos method and FS model are (47.54, 1.17, 2.40) and (64.53, 1.84, 3.15), respectively.
- At some time steps, the error predictions from the FS method are larger than the errors of initial prediction, which leads to misleading corrections to the initial forecasting, such as the time steps

32, 33, and 62 to 64 in DMA1; time steps 32, 46, 55 and 50 in DMA2; time steps 30 to 35, 80 and 86 in DMA3. While this kind of misleading correction is not much in the chaos method.

In general, the chaos method performs better than the FS method in predicting such a complex fluctuated error time series, and the practice also proves that the errors predicted by the chaotic method are closer to the initial errors in the three DMAs.

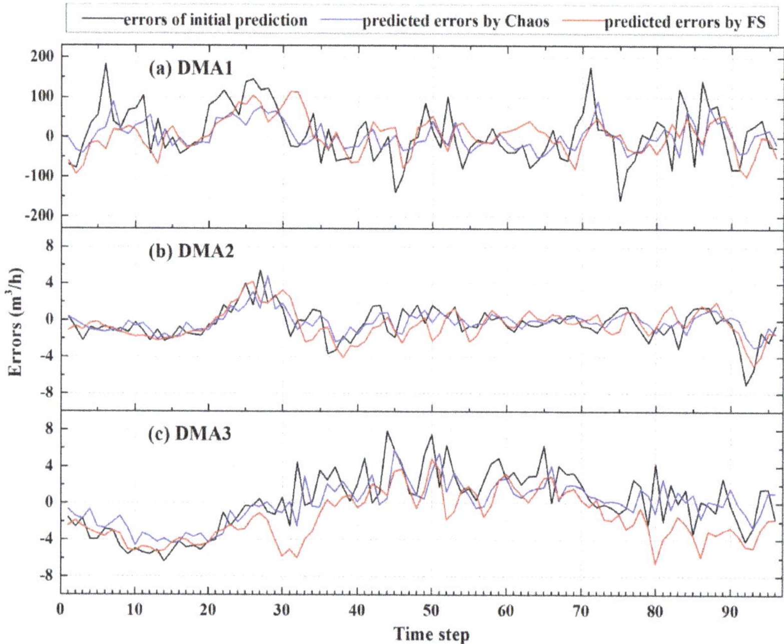

Figure 5. Comparison between the errors of the initial forecasting and the predicted errors. (**a**) DMA1 on 26 December; (**b**) DMA2 on 26 December; and (**c**) DMA3 on 11 August.

The statistics of absolute percentage errors (APE) between the single forecasting model S_LSSVM and the hybrid models are provided in Figure 6. From the mean, median, maximum, and minimum values of APEs of the predictions for the three DMAs in Beijing, the H_LSSVM_Chaos models perform better than that of the S_LSSVM models. Therefore, the hybrid framework using the LSSVM and chaotic time series gives more accurate predictions. The hybrid models using LSSVM and Fourier series did not always perform as well as the H_LSSVM_Chaos. The MAPEs of the H_LSSVM_FS model for DMA1 is 5.44%, which is better than that of the single forecasting model S_LSSVM 5.68%. Whereas, other statistics of the H_LSSVM_FS model in DMA1, such as the 75-percentile value and the maximum value of the APE, are similar or even worse than that of the S_LSSVM. The reason is that the H_LSSVM_FS model performs a misleading correction for the severely fluctuated time steps, as shown in Figure 5a. For DMA2, although the mean and median APEs of the H_LSSVM_FS models are similar to that of the H_LSSVM_Chaos models, the overestimates of the errors during the time steps 38 to 58 in Figure 5b by the FS method are still notable. Therefore, more attention should be paid when using the error correction module in short-term water demand forecasting.

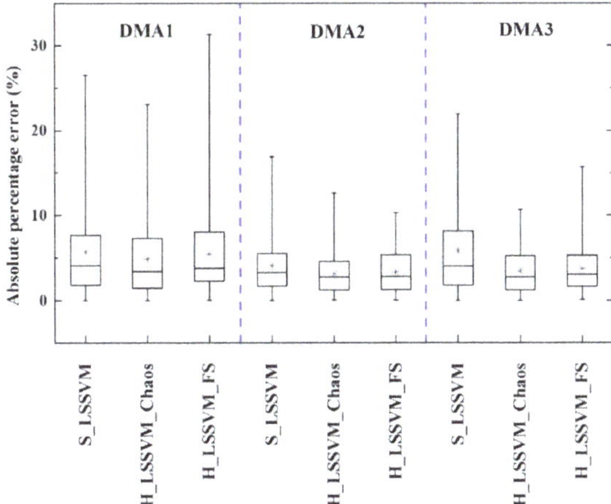

Figure 6. Statistics of the absolute relative errors for different forecasting models.

3.4. Discussion

In the initial forecasting module and error correction module of the hybrid forecasting framework, the forecasting models are established by LSSVM. The successful implementation of the LSSVM model depends on the precision of model parameters (i.e., γ and δ^2). In this study, the three-level Bayesian evidence inferring method is adopted to infer LSSVM model parameters. To investigate the influence of model parameters on the performance of LSSVM models, the application of the S_LSSVM model on DMA2 is taken as an example. With the same model input data, Table 5 shows the model performances to different model parameters which are obtained by the 1-level Bayesian inferring, 3-level Bayesian inferring, and the grid search algorithm. These parameters are computed by the LS-SVMlab Toolbox [45]. As Table 5 shows, after 3-level inferring, the Bayesian evidence method catches reasonable model parameters with moderate computation burden. The grid search algorithm provides the best performance, but it takes the longest computation time. As shown in Table 4, the hybrid model H_LSSVM_Chaos model using 3-level Bayesian inferred parameters performs even better than the grid search algorithm built S_LSSVM model. The computation time of the H_LSSVM_Chaos model is about 1 time (including initial forecasting and error correction) longer than the 3-level Bayesian built S_LSSVM model, which is much shorter than that of the grid search algorithm built S_LSSVM model (Table 5). Therefore, the hybrid framework using 3-level Bayesian built LSSVM for initial forecasting and error time series forecasting is suitable for the short-term water demand forecasting.

Table 5. Performances of the S_LSSVM model with different parameters with application to DMA2.

Methods	Model Parameters			Model Performance Indicators			Computation Time (s)
	γ	δ^2	R^2	MAE (m³/h)	MAPE (%)	RMSE (m³/h)	
1-level Bayesian inferring	0.0183	13.6763	0.9632	1.56	4.98	2.03	179
3-level Bayesian inferring	0.0378	12.5104	0.9722	1.31	4.06	1.76	1253
Grid search	0.1097	0.7515	0.9809	1.13	3.38	1.46	3685

The hybrid model (H_LSSVM_Chaos) is also compared to the traditional ARIMA model, and Table 6 shows the results on the three DMAs. The development of the ARIMA models follows the procedure described by Adamowski [45]. The parameters of the ARIMA are trained and tested based on different combinations, the number of autoregressive parameters (p), the number of difference (d) and the number of moving average parameters (q) are set as (3, 1, 1). Note that, the same set of historical water demand data are used to build the H_LSSVM_Chaos and ARIMA forecasting models; the historical data before the forecasting day are used to establish the forecasting models.

Table 6. Performance comparison between the auto regressive integrated moving average (ARIMA) and the hybrid forecasting models.

DMA ID	Forecasting Date	Forecasting Models	R^2	MAE (m^3/h)	MAPE (%)	RMSE (m^3/h)
1	26 December	ARIMA	0.9656	55.43	5.53	68.34
		H_LSSVM_Chaos	0.9711	47.92	4.84	62.66
2	26 December	ARIMA	0.9723	1.31	3.83	1.76
		H_LSSVM_Chaos	0.9817	1.08	3.15	1.43
3	11 August	ARIMA	0.9687	1.92	3.44	2.50
		H_LSSVM_Chaos	0.9701	1.86	3.47	2.44
	8–10 August	ARIMA	0.9687	1.90	3.49	2.44
	(mean values)	H_LSSVM_Chaos	0.9772	1.64	3.00	2.08

As shown in Table 6, the H_LSSVM_Chaos model perform better than the ARIMA model on DMA1 and DMA2, for example, the MAPEs (DMA1, DMA2) of the H_LSSVM_Chaos model and the ARIMA model are (4.84%, 3.15%) and (5.53%, 3.83%), respectively. Whereas, the application results of DMA3 show some variations: (i) on the forecasting day August 11, the H_LSSVM_Chaos has a similar result to the ARIMA, for example, the R^2 and MAPEs of the two models are (0.9701, 0.9687) and (3.47%, 3.44%), respectively; (ii) on the forecasting days from August 8 to 10, the H_LSSVM_Chaos perform better than the ARIMA, for example, the three days' MAPEs of the H_LSSVM_Chaos and the ARIMA are (3.48%, 2.81%, 2.71%) and (4.03%, 3.10%, 3.35%), respectively. The reason for the variations is that August 11 is Saturday while August 8 to 10 are weekdays. As shown in Figures 3c and 4c, for DMA3, the water consumption curve on Saturday is different and more complex than that of weekdays. The distinctive water consumption curve on Saturday results in fewer training samples for establishing the forecasting model, which affects the forecasting accuracy for Saturday. However, the overall performance of the H_LSSVM_Chaos model is still better than the ARIMA model, despite the variations in the forecasting accuracy on Saturday. These comparisons verified the validity of the H_LSSVM_Chaos model.

Generally, one single model could not identify the underlying patterns for every case, and the hybrid framework including different models is able to capture different aspects of the available information for prediction [5,46]. The LSSVM method in the initial prediction module captures nonlinear relationships between the discontinuous feature data (Q_t, Q_{t-1}, Q_{t-2}, Q_{t-95}, Q_{t-191}, Q_{t-671}) of the historical water demand data set and the water demand Q_{t+1} on the forecasting day; the chaotic time series method in the error correction module captures the continuous and periodic changes from the errors of the initial forecasting module.

4. Conclusions

Short-term water demand forecasting with the horizon ranges from sub-hourly to daily plays an important role in the field of optimal operation of pump stations and online hydraulic simulation of water distribution systems. To obtain more accurate predictions, this study proposes a hybrid framework with the error correction module which uses the chaotic time series, and investigates the performance of the framework in the short-term water demand forecasting with one day ahead and a 15-min time step. The hybrid framework is developed by integrating two modules, namely, the initial forecasting module and the error correction module. The initial forecasting model is established by the least squares support vector machines (LSSVM). In the error correction module the errors forecasting model is established by LSSVM using chaotic time series of error data from initial forecasting.

The hybrid model is implemented in the water demand forecasting of three actual district metering areas (DMAs) in Beijing, China, and the application results of the hybrid model are comparable to that of other two models including the forecasting model without error correction and the hybrid model using Fourier series for error correction. From the case study results, the following conclusions could be drawn:

- In most instances, the hybrid models perform better than the forecasting model without error correction. The error correction module performs better in the short-term water demand forecasting than the DMAs whose composition of customers is simple. A simple composition of customers indicates a simple water consumption pattern and less peak fluctuations in the water consumption curves.

- Due to the capability of detecting the underlying instability characteristics of time series, the error correction module using chaotic time series performs better than the Fourier series in predicting a complex disordered time series of errors.

- For the periods of frequent and disordered peak fluctuations in the error time series, the performance of the error correction module is not good, and the error forecasting model based on Fourier series may lead to unreasonable forecasts by misleading the corrections to the initial forecasting. As a result, more attention should be paid to the features of the error time series when using the error correction module.

In the presented study, the hybrid forecasting framework is tested by three actual DMAs in Beijing with different characteristics. Further work on other DMAs are needed to test and verify the robustness of the hybrid forecasting framework, and much more effort is needed to test the performance of chaotic methods in mining the characteristics of the disordered peak fluctuated data. This study only tested the proposed model for the 24 h forecast horizon, whereas, the hybrid forecasting framework is not limited to the forecast horizon of one day, there is a potential to implement the model to a much longer forecast horizon and frequency, such as one week ahead with a time step of 6 h. Then the feature data for model training obtained from the historical data set should be adjusted accordingly.

Author Contributions: Conceptualization, S.W. and B.H.; methodology, H.H. and B.H.; validation, S.W., H.H., B.H., and K.D.; formal analysis, H.H., B.H., and K.D.; investigation, H.H., B.H., and K.D.; resources, S.W., H.H., B.H., and K.D.; writing—original draft preparation, H.H. and B.H.; writing—review and editing, S.W. and K.D.; visualization, B.H. and K.D.; supervision, S.W.; project administration, S.W. and B.H.; funding acquisition, S.W. All authors have read and agreed to the published version of the manuscript.

Funding: This research was funded by the Major Science and Technology Program for Water Pollution Control and Treatment, grant number 2017ZX07108-002.

Acknowledgments: The authors would like to thank the editors and reviewers for bringing the paper to a scientific standard for inclusion in the journal.

Conflicts of Interest: The authors declare no conflict of interest.

References

1. Herrera, M.; Torgo, L.; Izquierdo, J.; Pérez-García, R. Predictive models for forecasting hourly urban water demand. *J. Hydrol.* **2010**, *387*, 141–150. [CrossRef]
2. Anele, A.; Hamam, Y.; Abu-Mahfouz, A.; Todini, E. Overview, comparative assessment and recommendations of forecasting models for short-term water demand prediction. *Water* **2017**, *9*, 887–898. [CrossRef]
3. Chen, J.; Boccelli, D.L. Demand forecasting for water distribution systems. *Procedia Eng.* **2014**, *70*, 339–342. [CrossRef]
4. Qin, T.; Boccelli, D.L. Estimating distribution system water demands using Markov chain Monte Carlo. *J. Water Resour. Plan. Manag.* **2019**, *145*, 04019023. [CrossRef]
5. Donkor, E.A.; Mazzuchi, T.A.; Soyer, R.; Alan Roberson, J. Urban water demand forecasting: Review of methods and models. *J. Water Resour. Plan. Manag.* **2014**, *140*, 146–159. [CrossRef]
6. Chen, J.; Boccelli, D.L. Forecasting hourly water demands with seasonal autoregressive models for real-time application. *Water Resour. Res.* **2018**, *54*, 879–894. [CrossRef]
7. Odan, F.K.; Reis, L.F.R. Hybrid water demand forecasting model associating artificial neural network with Fourier series. *J. Water Resour. Plan. Manag.* **2012**, *138*, 245–256. [CrossRef]

8. Pacchin, E.; Gagliardi, F.; Alvisi, S.; Franchini, M. A comparison of short-term water demand forecasting models. *Water Resour. Manag.* **2019**, *33*, 1481–1497. [CrossRef]

9. Guo, G.; Liu, S.; Wu, Y.; Li, J.; Zhou, R.; Zhu, X. Short-term water demand forecast based on deep learning method. *J. Water Resour. Plan. Manag.* **2018**, *144*, 04018076. [CrossRef]

10. Crommelynck, V.; Duquesne, C.; Mercier, M. Daily and Hourly Water Consumption Forecasting Tools Using Neural Networks. In Proceeding of the AWWA's Annual Computer Specialty Conference, Nashville, TN, USA, 12–15 April 1999; pp. 665–676.

11. Jain, A.; Ormsbee, L. Short-term water demand forecast modeling techniques—Conventional Methods Versus AI. *Am. Water Work. Assoc.* **2002**, *94*, 64–72. [CrossRef]

12. Bougadis, J.; Adamowski, K.; Diduch, R. Short-term municipal water demand forecasting. *Hydrol. Process.* **2005**, *19*, 137–148. [CrossRef]

13. Chen, L.; Zhang, T.Q. Hourly water demand forecast model based on Bayesian least squares support vector machine. *J. Tianjin Univ.* **2006**, *39*, 1037–1042.

14. Chen, L.; Zhang, T.Q. Hourly water demand forecast model based on least squares support vector machine. *J. Harbin Inst. Technol.* **2006**, *38*, 1528–1530.

15. Khan, M.S.; Coulibaly, P. Application of support vector machine in lake water level prediction. *J. Hydrol. Eng.* **2006**, *11*, 199–205. [CrossRef]

16. Braun, M.; Bernard, T.; Piller, O.; Sedehizade, F. 24-Hours demand forecasting based on SARIMA and support vector machines. *Procedia Eng.* **2014**, *89*, 926–933. [CrossRef]

17. Dahl, C.M.; Hylleberg, S. Flexible regression models and relative forecast performance. *Int. J. Forecast.* **2004**, *20*, 201–217. [CrossRef]

18. Chen, G.; Long, T.; Xiong, J.; Bai, Y. Multiple random forests modelling for urban water consumption forecasting. *Water Resour. Manag.* **2017**, *31*, 4715–4729. [CrossRef]

19. Tripathi, S.; Srinivas, V.V.; Nanjundiah, R.S. Downscaling of precipitation for climate change scenarios: A support vector machine approach. *J. Hydrol.* **2006**, *330*, 621–640. [CrossRef]

20. Ghalehkhondabi, I.; Ardjmand, E.; Young, W.A.; Weckman, G.R. Water demand forecasting: Review of soft computing methods. *Env. Monit Assess.* **2017**, *189*, 313. [CrossRef]

21. Suykens, J.A.K.; Vandewalle, J. Least squares support vector machine classifiers. *Neural Process. Lett.* **1999**, *9*, 293–300. [CrossRef]

22. Suykens, J.A.K.; De Brabanter, J.; Lukas, L.; Vandewalle, J. Weighted least squares support vector machines robustness and sparse approximation. *Neurocomputing* **2002**, *48*, 85–105. [CrossRef]

23. Vijai, P.; Bagavathi Sivakumar, P. Performance comparison of techniques for water demand forecasting. *Procedia Comput. Sci.* **2018**, *143*, 258–266. [CrossRef]

24. Samsudin, R.; Saad, P.; Shabri, A. River flow time series using least squares support vector machines. *Hydrol. Earth Syst. Sci.* **2011**, *15*, 1835–1852. [CrossRef]

25. Kisi, O. Modeling discharge-suspended sediment relationship using least square support vector machine. *J. Hydrol.* **2012**, *456–457*, 110–120. [CrossRef]

26. Aydogdu, M.; Firat, M. Estimation of failure rate in water distribution network using fuzzy clustering and LS-SVM methods. *Water Resour. Manag.* **2014**, *29*, 1575–1590. [CrossRef]

27. Cherkassky, V.; Ma, Y. Practical selection of SVM parameters and noise estimation for SVM regression. *Neural Netw.* **2004**, *17*, 113–126. [CrossRef]

28. Zhang, G.P. Time series forecasting using a hybrid ARIMA and neural network model. *Neurocomputing* **2003**, *50*, 159–175. [CrossRef]

29. Brentan, B.M.; Luvizottom, E.; Herrera, M.; Izquierdo, J.; Pérez-García, R. Hybrid regression model for near real-time urban water demand forecasting. *J. Comput. Appl. Math.* **2017**, *309*, 532–541. [CrossRef]

30. Dhanya, C.T.; Nagesh Kumar, D. Multivariate nonlinear ensemble prediction of daily chaotic rainfall with climate inputs. *J. Hydrol.* **2011**, *403*, 292–306. [CrossRef]

31. Liu, X.; Fang, X.; Qin, Z.; Ye, C.; Xie, M. A Short-term forecasting algorithm for network traffic based on Chaos theory and SVM. *J. Netw. Syst. Manag.* **2010**, *19*, 427–447. [CrossRef]

32. Yang, H.Y.; Ye, H.; Wang, G.; Khan, J.; Hu, T. Fuzzy neural very-short-term load forecasting based on chaotic dynamics reconstruction. *Chaos Solitons Fractals* **2006**, *29*, 462–469. [CrossRef]

33. Bakker, M.; Vreeburg, J.H.G.; van Schagen, K.M.; Rietveld, L.C. A fully adaptive forecasting model for short-term drinking water demand. *Environ. Model. Softw.* **2013**, *48*, 141–151. [CrossRef]

34. Cutore, P.; Campisano, A.; Kapelan, Z.; Modica, C.; Savic, D. Probabilistic prediction of urban water consumption using the SCEM-UA algorithm. *Urban. Water J.* **2008**, *5*, 125–132. [CrossRef]

35. Al-Zahrani, M.A.; Abo-Monasar, A. Urban residential water demand prediction based on artificial neural networks and time series models. *Water Resour. Manag.* **2015**, *29*, 3651–3662. [CrossRef]

36. Van Gestel, T.; Suykens, J.A.; Baesens, B.; Viaene, S.; Vanthienen, J.; Dedene, G. Benchmarking least squares support vector machine classifiers. *Mach. Learn.* **2004**, *54*, 5–32. [CrossRef]

37. Van Gestel, T.; Suykens, J.A.; Baestaens, D.E.; Lambrechts, A.; Lanckriet, G.; Vandaele, B. Financial time series prediction using least squares support vector machines within the evidence framework. *IEEE Trans. Neural Netw.* **2001**, *4*, 809–821. [CrossRef]

38. Li, D.; Han, M.; Wang, J. Chaotic time series prediction based on a novel robust echo state network. *Trans. Neural Netw. Learn. Syst.* **2012**, *23*, 787–799. [CrossRef]

39. Takens, F. Detecting strange attractors in turbulence. In *Lecture Notes in Mathematics*; Rand, D.A., Young, L.S., Eds.; Springer: Berlin, Germany, 1981; Volume 898, pp. 366–381. [CrossRef]

40. Lai, Y.C.; Ye, N. Recent developments in chaotic time series analysis. *World Sci. Publ. Co.* **2003**, *13*, 1383–1422. [CrossRef]

41. Kim, H.S.; Eykholt, R.; Salas, J.D. Nonlinear dynamics, delay times, and embedding windows. *Phys. D Nonlinear Phenom.* **1999**, *127*, 48–60. [CrossRef]

42. Dhanya, C.T.; Nagesh Kumar, D. Nonlinear ensemble prediction of chaotic daily rainfall. *Adv. Water Resour.* **2010**, *33*, 327–347. [CrossRef]

43. Wolf, A.; Swift, J.B.; Swinney, H.L.; Vastano, J.A. Determining Lyapunov exponents from a time series. *Phys. D Nonlinear Phenom.* **1985**, *16*, 285–317. [CrossRef]

44. De Brabanter, K.; Karsmakers, P.; Ojeda, F. *LS-SVMlab Toolbox User's Guide Version 1.8*; Katholieke Universiteit Leuven: Leuven, Belgium, 2011; p. 115.

45. Adamowski, J.F. Development of a short-term river flood forecasting method for snowmelt driven floods based on wavelet and cross-wavelet analysis. *J. Hydrol.* **2008**, *353*, 247–266. [CrossRef]

46. Wang, X.; Sun, Y.; Song, L.; Mei, C. An eco-environmental water demand based model for optimising water resources using hybrid genetic simulated annealing algorithm. Part II: Model application and results. *J. Environ. Manag.* **2009**, *19*, 2612–2619. [CrossRef] [PubMed]

Article

On the Use of an IoT Integrated System for Water Quality Monitoring and Management in Wastewater Treatment Plants

Ramón Martínez [1], Nuria Vela [2], Abderrazak el Aatik [2], Eoin Murray [3], Patrick Roche [3] and Juan M. Navarro [1,*

[1] Research Group in Advanced Telecommunications (GRITA), Universidad Católica de Murcia (UCAM), 30107 Guadalupe, Spain; rmcarreras@ucam.edu

[2] Applied Technology Group to Environmental Health, Universidad Católica de Murcia (UCAM), 30107 Guadalupe, Spain; nvela@ucam.edu (N.V.); aaatik@ucam.edu (A.e.A.)

[3] Research & Development, T.E. Laboratories Ltd. (TelLab), Tullow, Carlow R93 N529, Ireland; emurray@tellab.ie (E.M.); proche@tellab.ie (P.R.)

* Correspondence: jmnavarro@ucam.edu

Received: 10 February 2020; Accepted: 3 April 2020; Published: 12 April 2020

Abstract: The deteriorating water environment demands new approaches and technologies to achieve sustainable and smart management of urban water systems. Wireless sensor networks represent a promising technology for water quality monitoring and management. The use of wireless sensor networks facilitates the improvement of current centralized systems and traditional manual methods, leading to decentralized smart water quality monitoring systems adaptable to the dynamic and heterogeneous water distribution infrastructure of cities. However, there is a need for a low-cost wireless sensor node solution on the market that enables a cost-effective deployment of this new generation of systems. This paper presents the integration to a wireless sensor network and a preliminary validation in a wastewater treatment plant scenario of a low-cost water quality monitoring device in the close-to-market stage. This device consists of a nitrate and nitrite analyzer based on a novel ion chromatography detection method. The analytical device is integrated using an Internet of Things software platform and tested under real conditions. By doing so, a decentralized smart water quality monitoring system that is conceived and developed for water quality monitoring and management is accomplished. In the presented scenario, such a system allows online near-real-time communication with several devices deployed in multiple water treatment plants and provides preventive and data analytics mechanisms to support decision making. The results obtained comparing laboratory and device measured data demonstrate the reliability of the system and the analytical method implemented in the device.

Keywords: smart city; water quality monitoring; Internet of Things; wireless sensor networks; water treatment plant; data analytics; nitrate; nitrite

---

1. Introduction

Water is a scarce and precious resource that is being put under pressure due to the fast-growing population that is extracting too much water and polluting our rivers, lakes, and groundwater with municipal, agricultural, and industrial wastes. Climate change, loss of biodiversity, unsustainable use of natural resources, and environmental pressures have a negative impact on water quality and quantity which are inextricably linked, with over extraction causing low river flows, low ground water levels, and drying up of wetlands. The deteriorating water environment, accelerating the shortage of water and affecting human health, has become an important problem that restricts the development of cities.

One of the most important environmental problems today is, undoubtedly, the contamination of water by nitrates, especially in areas with significant agricultural activity, as occurs in the southeast of Spain [1,2]. The nitrates are natural components of soil and water, both surface and underground, which come, in part, from the decomposition of nitrogenous organic matter, although their presence in the soil and in aquifers increases with the use of nitrogenous fertilizers and manure in areas with a high level of agricultural activity. Farmers invest large amounts of nitrogenous fertilizers in the fields to maintain adequate production and increase yields. Most of these are not absorbed by plants, so they settle in the soil and gradually filter through it, reaching groundwater. Similarly, these compounds can circulate through surface runoff and cause contamination problems in surface, fresh, or marine waters [3].

An excessive contribution of nutrients in surface waters, especially nitrogen and phosphorus, gives rise to a rapid proliferation of aquatic vegetation, as a consequence of oxygen depletion on the surface, which favors the appearance of eutrophication processes [4]. The Mar Menor (Region of Murcia, southeast of Spain) has been in the news in recent years due to the eutrophication, which refers to the processes of the ecosystem originated by the enrichment of nutrients of the water, especially nitrogen and/or phosphorus [5,6]. This situation, added to by the fact that most of the effluents from the wastewater treatment plants (WWTP) in this area are used for irrigation in agriculture, implies an increase in responsibility of the water industry to adopt a more sustainable management of urban water systems for this type of compound [7]. One of the most effective approaches to address this challenge of sustainability is wastewater treatment, in which water quality monitoring (WQM) plays a key role.

WQM can be described as a method for periodically sampling and analyzing water conditions and characteristics [8]. This method forms the basis for water environmental management, as it is vital to monitor source waters and the aquatic systems that receive inputs from industrial waste and sewage treatment plants, stormwater systems, and runoff from urban and agricultural lands [9]. Similarly, domestic sewage and water flows resulting from chemical processes and waste in industry and sanitation should be monitored in wastewater treatment plants that purify the water to decontaminate it before releasing it into the sea (or other large bodies of water), or be used for other applications such as irrigation, and to detect possible toxic or radioactive discharges [10]. Wastewater, also known as sewage, contains more than 99% water and is characterized by volume or rate of flow, physical condition, chemical constituents, and the bacteriological organisms that it contains. The quality of treated wastewater is defined by physical-chemical parameters such as pH, temperature, conductivity, turbidity, Biological Oxygen Demand (BOD), Chemical Oxygen Demand (COD), Total Organic Carbon (TOC), Total Suspended Solids (TSS), and nitrogen and phosphorus compounds [11,12]. From an environmental perspective, the concentrations of phosphate, nitrate and nitrite in water are crucial due to their role in eutrophication. They are important analytes for environmental, food and human health monitoring and thus their detection and quantification is essential [13]. The sensor implemented in this paper within the developed integrated system for water quality monitoring is a low-cost device that consists of a nitrate and nitrite analyzer based on a novel ion chromatography detection method [14].

Wastewater treatment is an important component in the water cycle, as it ensures that the environmental impact of human usage of water is significantly reduced. Wastewater treatment plants (WWTPs) use a series of treatment stages to clean up the contaminated water so that the treated effluent is safely discharged to inland water, estuaries and the sea. Wastewater treatment consists of several processes (physical, biological, and chemical) that aim to reduce nitrogen, phosphorous, organic matter, and suspended solids content [15]. The purpose of WQM is to support the control of these processes by accurately monitoring water parameters (e.g., nitrate, nitrite, phosphate, and pH) mainly in the influent and effluent of each WWTP. Specifically, WQM performs (i) the detection and quantification of these parameters in the influent wastewater that could affect the treatment processes, providing the plant operator with valuable information to foresee such effects, and (ii) the analytical control of the effluent to verify that the treated waters comply with the standards required by the current regulations [16], ensuring the environmental sustainability of water. In the European context, environmental legislation requires improvements in water quality and effluent discharged to waterways due to the Water Framework Directive [17] and related Directives, e.g., the Urban Wastewater Treatment Directive [18] and the Nitrates Directive [19]. The need for compliance with these Directives has created a demand among Government Monitoring Agencies and legislative bodies throughout Europe for frequent monitoring, both temporally and spatially. Traditional WQM methods involve the manual collection of water samples at different locations, followed by laboratory analytical techniques in order to characterize the water quality. Such methods take a long time and are no longer considered efficient. Although these methodologies analyze physical, chemical, and biological agents, they have several drawbacks: (i) poor spatiotemporal coverage [20], (ii) they are labor intensive and high cost (labor, operation, and equipment), and (iii) the lack of near-real-time water quality information to enable critical decisions for public health and environment protection [21]. Therefore, there is a need for WQM systems that enable reliable performance of WWTPs through effective data management and the online near real-time monitoring capability. The WQM system presented in this work is tested in a wastewater treatment real scenario and reported results are compared with analytical techniques values.

In the recent years, the vision of the Internet of Things (IoT) [22] augmented with advances in software technologies, such as service-oriented architecture (SOA), software as a service (SaaS), cloud computing, and others, has stimulated the development of smart water quality monitoring systems (SWQMSs) [23,24]. These systems combine technologies and components from microsystems (miniaturized electric, mechanical, optical, and fluid devices) with knowledge, technology, and functionality from disciplines like biology, chemistry, nanosciences, and cognitive sciences. Fortunately, the use of IoT software platforms helps to overcome the challenges associated with the broad set of technologies, systems, and design principles of the IoT [25,26]. SWQMSs are a new generation of systems architecture (hardware, software, network technologies, and managed services) that provides near-real-time awareness based on inputs from machines, people, video streams, maps, news feeds, sensors, and more that integrate people, processes, and knowledge to enable collective awareness and decision making where devices can offer more advanced access to their functionality [27]. As such, event-based information can be acquired, and then processed on-device and in-network. This capability provides new ground for approaches that can be more dynamic and highly sophisticated and that can take advantage of the available context (readings of water quality parameters). For this reason, SWQMSs allow to optimize the performance of the WWTP in particular and the treatment system in general achieving a smart wastewater management. Wired SWQMSs are still the main approach to monitor the parameters in existing wastewater treatment plants. However, this type of system has the drawbacks of high cost, poor expansion capability and difficult maintenance due to inefficient operating environment [28]. In order to overcome these previously mentioned drawbacks, a cost-effective decentralized SWQMS is designed in this work, using a low-cost water quality monitoring device that is integrated in an IoT software platform and in a Wireless Sensor Network (WSN) [29].

Wireless Sensor Networks have proven to be a very effective technology for numerous environmental monitoring applications. WSNs currently enable the automatic monitoring of air pollution [30], noise pollution [31–33], forest fires [34], climatological conditions [35], and much more over wide areas, something previously impossible. The use of WSNs for WQM is particularly appealing due to the low cost of the sensor nodes and hence the cost-effectiveness of this solution. These simple and low-cost networks allow monitoring of processes remotely, in near-real-time and with minimal human intervention. Considerable research has been conducted to monitor water quality through the development of WSNs. Adu-Manu and Pule [36,37] study and analyze recent developments in the sensor devices, data acquisition procedures, communication and network architectures, and power management schemes of WSNs to maintain a long-lived operational SWQMSs. Adamo [38] presents a SWQMS that supports to strategic decisions concerning critical environment issues of the marine ecosystem by implementing an smart buoy prototype designed for in situ and in continuous space-time monitoring of water temperature, salinity/conductivity, turbidity, and chlorophyll-a concentration as biological indicators of water eutrophication. Jiang [39] developed a WSN based on ZigBee technology for online auto-monitoring of the water temperature and pH value of an artificial lake.

In the field of wastewater treatment, WSNs represent a promising technology because of their rapid deployment and their ability to acquire, process and transmit data at a number of distributed sampling points. The application of WSNs to WQM has opened up a new avenue of research towards the development of decentralized SWQMSs that evolve with the changing wastewater infrastructure to meet the water requirements of smart cities [40–42]. These decentralized SWQMSs (i) offer great potential for cost reduction, (ii) allow for precise matching of growing wastewater capacity requirements, (iii) take advantage of the relative homogeneity of wastewater streams at their point of origin, (iv) do not need large sewer systems nor require extensive networks for the distribution of treated water, and (v) present probability of failure significantly lower than that of failure of centralized system. The advent of WSNs allows the replacement of traditional WQM methods or the expansion of existing wired SWQMS. Tadokoro and Wang [43,44] describe the design of SWQMSs using wired and wireless technologies for online near-real-time supervisory, control, and data acquisition (SCADA) of wastewater treatment processes. The designs conceived support many functions directed at multiple wastewater treatment plants, such as decentralized control, centralized management, remote diagnosis and fault early warning. Regarding WSNs sensor nodes, there is research work focused on the design of devices to monitor diverse parameters. In this sense, the work presented by Geetha [24] is based on the single-chip TI CC3200 microcontroller to monitor pH, conductivity, water level and turbidity and upload them to the Ubidots cloud. Reference [45] is based on Arduino to monitor pH, conductivity and dissolved oxygen and upload them to the ThingSpeak cloud, whereas the work presented by Saravanan [46] is based on Arduino to monitor flow, temperature, color, and turbidity, and upload them to the SWQMS cloud server. However, the prototypes cited previously are far from the close-to-market stage.

This paper presents the integration to a WSN and a preliminary validation in a wastewater treatment plant scenario of a low-cost water quality monitoring device in the close-to-market stage. This device consists of a nitrate and nitrite analyzer based on a novel ion chromatography detection method. The analytical device is integrated using an Internet of Things software platform and tested under real conditions in a wastewater treatment plant scenario. By doing so, a decentralized SWQMS conceived and developed for wastewater quality monitoring and management is accomplished. This investigation is part of an ongoing research project, referred to as LIFE EcoSens Aquamonitrix [47], which aims to validate and optimize this solution to achieve a low-cost, fully automated in situ analyzer for environmental water monitoring ready to be launched in the market after the project.

The paper is structured as follows. After this introduction, Section 2 describes the analytical device, the IoT software platform, the developed SWQMS called the EcoSens Aquamonitrix System, and the methodology followed for the validation of the system implemented in a real wastewater treatment scenario. In Section 3, the features of the EcoSens Aquamonitrix System are shown and the results obtained from the experiments to validate the analytical device are discussed. Finally, Section 4 presents the general conclusions of this study and proposes future work.

2. Materials and Methods

The proposed integrated system encompasses the analytical device that was connected to an IoT software platform, henceforth IoT platform, via a wireless sensor network. The description of the device design is highlighted in Section 2.1. Section 2.2 presents the IoT platform and discusses the middleware that supports the system. Section 2.3 describes the implementation of the IoT water quality monitoring system in a wastewater treatment plant scenario. Finally, Section 2.4 details the methodology followed in this investigation for a preliminary validation of the system.

2.1. Device

The core of the analytical device is a portable ion chromatography (IC) system based on the method previously reported by Murray [14]. This system employs a novel design of a ultraviolet (UV) light-emitting diode (LED)-based optical detector which enables cost-effective direct in situ detection of nitrite and nitrate in natural waters. The automated portable IC system is described in detail by Murray [48]. The main functional blocks of the system are depicted in Figure 1 and each component is subsequently described.

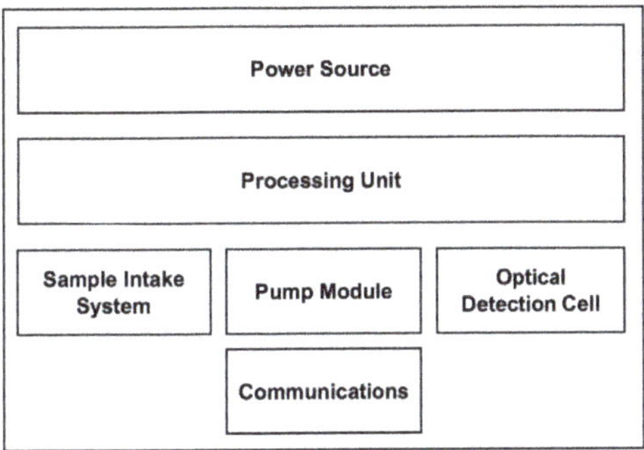

Figure 1. Functional block diagram highlighting core components of analytical system.

The sample intake system is comprised of a 12 V high flow pump and a reservoir. The pump draws sample from the water source filling the reservoir. The sample intake pump runs at the beginning of each analysis cycle for 30 s. The pump module is responsible for loading sample into the system and pumping the eluent through the ion exchange column and detector for analysis. The pump module consists of four 3D printed syringe pumps, three of which are used for eluent delivery, whereas the fourth syringe pump draws sample from the reservoir. Once full, the syringe pumps empty at a set flow rate enabling chromatographic analysis and analyte detection. The optical detection cell consists of a low-cost, UV absorbance detector which incorporates a 235 nm LED and photodiode [14]. The photodiode is coupled with an ADS1115 analogue to digital converter (ADC). The signal from the photodiode (0–3.3 V) is sampled every 50 ms by the ADC which communicates with the systems microcontroller via I2C protocol. A HTU21D-F temperature and humidity sensor is used to measure the internal parameters of the analyzer. The sensor communicates over I2C and the readings are logged once at the beginning of every cycle.

The system is powered from a portable battery (Voltaic V88), which has a capacity of 24 Ampere-hour and supplies 12 Volts to the embedded system. The battery is charged from an alternating current source with a supplied adapter. The system can operate on battery alone for short periods of time, however, for long term deployments the unit runs from the battery while it is charging. This setup allows the system to function for up to 5 days on battery alone at an hourly sampling frequency, if the main power supply fails.

The analyzer is housed within a Peli 1510M Mobility Case (see Figure 2a). The case is water resistant, crushproof, dust resistant, and features a pressure equalization valve to balance interior pressure. The modular design of the system facilitates maintenance and exchange of components, without affecting the functioning of the other modules (see Figure 2b).

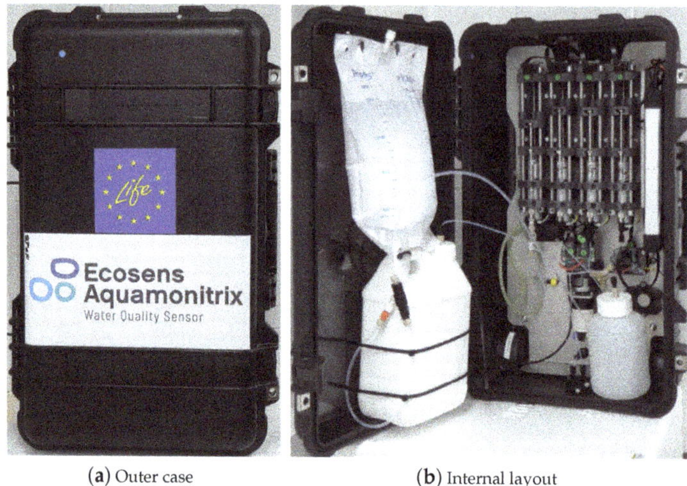

(**a**) Outer case (**b**) Internal layout

Figure 2. Design of the device.

An embedded system based on the Teensy 3.6 microcontroller unit is used to automate all functionality of the system. The firmware allows the unit to operate independently without user interaction once set up. A real-time clock wakes the system at a defined interval upon which system sampling and analysis functions begin to execute.

In this work, to add the connectivity with the wireless sensor network, an IoT solution associated with the system is implemented using a Raspberry Pi Zero W (Rpi) connected to a SimCom SIM800 Quad-Band GSM/GPRS integrated component. Raw signal transmission is acquired in real-time via an RS232 serial connection between the microcontroller and Rpi. The raw data is processed at the end of each IC run and used to calculate retention time and peak area of nitrate and nitrite. These values are transmitted via the SIM800 module. A small buffer of processed data are stored on the RPi unit in the event of a data transmission failure. The IoT devices attempt to transmit the buffered data at the end of subsequent IC runs.

2.2. IoT Platform

The IoT platform applied in this research is based on the middleware called thethings.io [49]. As shown in the Figure 3, the architecture of the IoT platform has been structured in five layers. The aspects of IoT middleware which are relevant for the understanding of the developed system are described below.

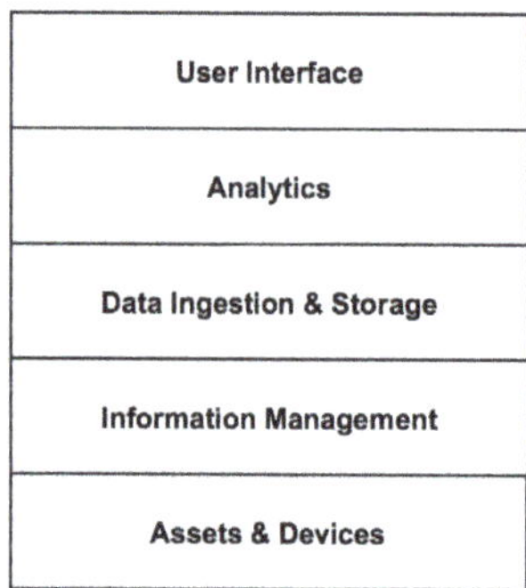

Figure 3. Architecture of the IoT platform.

At the lowest layer are the *assets* and *devices*. Assets are the reason for the development of IoT applications. The assets of interest are the real-world objects and *entities* subject to monitoring, as well as having digital representations and identities. Assets are instrumented with embedded technologies that bridge the digital realm with the physical world, and that provide the capability to monitor the assets as well as providing identities to the assets. Sensors and actuators in various devices, e.g., Wireless Sensor Networks, provide the main functional capabilities of sensing and embedded identities.

The second layer corresponds to the information management middleware, which is based on the specification of an information model and an Application Program Interface (API) based on the Representational State Transfer (REST) software architectural style (see Zhou [27]), as well as the adoption of communication protocols to manage and exchange information. All the information handled by the IoT platform is modeled by defining an information model composed of seven types of entities. Figure 4 illustrates these entities, as well as the relationships between them.

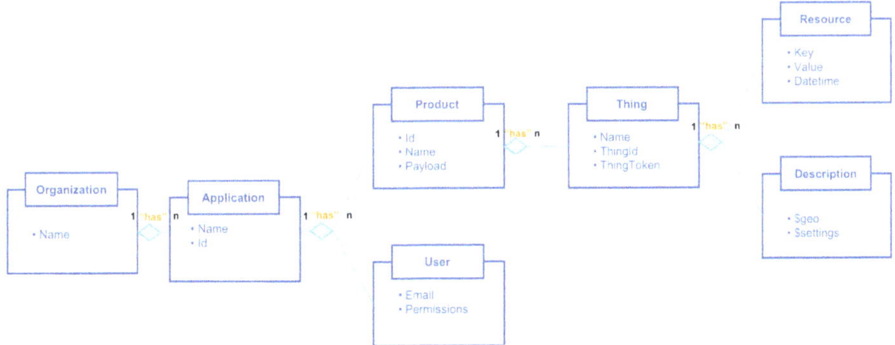

Figure 4. Entity relationship diagram. All the information handled by the platform is modelled by defining an information model composed of seven types of entities.

An entity can refer to hardware, that is called *thing*, e.g., IoT device, or software, that is called *resource*, e.g., hosted in the device, and to high-level abstractions, that are called *product*, that group entities of different types, e.g., product can have different types of things. An *organization*-type entity represents an account associated with the IoT platform and is uniquely identified by a name. An organization has *applications*. Applications are characterized by a pair <name_app, id>. Applications have *products* and *users*. Users are characterized by a pair <email, permissions>. Products are identified by a triplet <id, name_product, payload>. Each product has *things* also characterized by triplets <name_thing, thingid, thingtoken>. Finally, things have *resources* and *descriptions*. Descriptions are semantic data (metadata) characterized by a pair <$geo, $settings>. The attributes of the resources are data associated with a date and time grouped in triplets <key, value, datetime>. The values of these attributes become the information of the system.

The IoT platform provides a complete backend solution to develop IoT applications through an easy and flexible REST API, which allows the mediation between a large number of services. Table 1 shows the operations at thing level supported by information management middleware. Moreover, the IoT platform is agnostic to hardware, being possible to integrate any hardware platform. To do this, it is necessary to use the supported protocols [27,50]: Hypertext Transfer Protocol (HTTP), Websockets, Message Queue Telemetry Transport (MQTT), User Datagram Protocol (UDP), Transmission Control Protocol (TCP), or Constrained Application Protocol (CoAP). Regarding serialization formats, such as JavaScript Object Notation (JSON), Messagepack, and Protocol Buffers, are supported.

The third layer of the platform architecture is responsible for data ingestion and storage. Ingestion consists of getting data from producers, e.g., IoT devices, and making them available to consumers, e.g., IoT applications. For this purpose, a component called *message broker* is used. This component is based on Redis technology [51]. Producers send data to the message broker using the information management middleware. The data is pushed to the temporary storage of the message broker. The temporary storage consists of a cache that allows later stages (e.g., analytics) a simple and quick access to the incoming data. In doing so, producers can publish messages and consumers can consume them quickly using Redis database engine, which means real-time data management. The message broker persists the information published by the producers in two data warehouses called *DB Time series* and *object repository* that use NoSQL Cassandra [52] and MongoDB [53] technologies, respectively. The time series built from the attributes of the system resources, <key, value> pairs, are stored in the DB Time series. The rest of the information is stored in the object repository. The biggest challenge for the message broker is to be able to dispatch the received requests. For this purpose, a memory and CPU cluster are available. The management of these resources is automated thanks to the application of auto-scaling tasks [54].

Table 1. Description of the thing operations used in this research.

Operation	Description
Thing Activate	Activates a thing with an activation code. The result is a thing_token, which can be requested at any time from the control panel.
Thing Write	Writes the data records from the thing to the specified thing_token.
Thing Read	Returns the resource values by specifying the resource_key of the corresponding thing_token.
Get resources	Return the names of the resources from the thing.

At the fourth layer, the IoT middleware allows the modeling of data by scientists who perform analytical tasks to discover valuable insights hidden in the big data stored in the platform. In this analytical architecture, three types of mechanisms can be distinguished: *functions, jobs,* and *triggers.* Functions are fragments of code executed using a call from the RESTful API. Functions can be invoked by triggers and jobs. This mechanism is useful for encapsulating logic hosted in the cloud. The execution of functions is limited to 15 s. Jobs are executed automatically each predefined time period. Jobs are used to process data in order to generate *key performance indicators (kpis)* and analytics by aggregating event data. The execution of the jobs is limited to 5 min. Triggers are executed when an event occurs after using the RESTful API, e.g., thingWrite request. Triggers enable alarms to be sent through various methods (e-mail, short message service (SMS), twitter or voice calls). Triggers also allow the creation of aggregated resources or events for the calculation of kpis and analytics using jobs. Triggers execution is limited to 2 s.

The IoT platform provides the cloud code API (see Table 2) and the *Cloud Code Sandbox* to execute JavaScript code associated with triggers and jobs. The cloud code sandbox uses Jailed [55]. Thanks to this library, it is possible to launch an independent and secure sandbox for each request (trigger or job). In addition, Jailed allows to export a set of external functions into the sandbox.

Table 2. Description of the cloud code API operations used in this research.

Operation	Description
analytics.events	Allows to create and retrieve events.
analytics.kpis	Allows to create custom kpis.
thethingsAPI	Allows to read and write from/to an IoT device resource, call functions from triggers and jobs and get the things from a product.

At the top of the architecture is the user interface, which provides to the user data monitoring and management capabilities. For this purpose, both a dashboard system and a global online management panel are accessible from anywhere. The dashboard system incorporates widgets from libraries and proprietary that allow their customization to get the best possible response in the visualization of information. There are three levels of dashboard: main, application, and insight. The main dashboard displays the metrics of the applications and IoT devices, and their activity. The applications show the measurements of a subset of devices and their activity. Finally, the insight dashboard displays the measurements and activity of an IoT device related to an application. Therefore, the dashboard system can monitor the information associated with IoT applications: data from resources, kpis, device status, alarms, and other information.

After presenting the device (see Section 2.1) and the IoT platform, the integration process that results in a smart water quality monitoring system is addressed in the following section by describing a decentralized use case in the field of wastewater treatment.

2.3. EcoSens Aquamonitrix System in a WasteWater Treatment Domain

A distributed scenario is assumed in the wastewater treatment domain, e.g., WWTPs in many regions or countries. This is a complex IoT scenario, as it involves the management of a large number of sensors. The sensors are hosted in the *devices*, see Section 2.1, developed with the aim of collecting, processing and transmitting data associated with *assets* (nitrate and nitrite concentrations). The proposed scenario demands a solution that allows to integrate, manage and scale a large number of IoT devices and users. In order to meet these requirements, the EcoSens Aquamonitrix System has been implemented in a wastewater treatment plant scenario. This smart water quality monitoring system (SWQMS) depicted by Figure 5 is powered by the IoT platform described in the Section 2.2 to leverage the potential of wireless sensor networks and the sensor, see Section 2.1.

In the EcoSens Aquamonitrix System, the integration of the device with the IoT platform is achieved thanks to the information management middleware (see layer 2 of IoT architecture). Specifically, it uses an operation, henceforth *thingSend*, as a result of versioning the RESTful API to send the water quality data collected by the devices to the platform. This feature is useful because it allows to customize the body of the request and provides a special URL (endpoint) to access the *HTTP parser*, a Web service in charge of decoding and preprocessing the information. In this way, this innovative system takes advantage of the new functionalities offered by the Internet of Things based on the architecture studied in the Section 2.2. The devices are deployed in the influent and effluent of each WWTP. The devices communicate with the cloud via the mobile network, using the General Packet Radio Service (GPRS). At the top of the Figure 5 is the cloud, where instances of the components *message broker, DB Time series, Object repository,* and *Cloud Code SandBox* have been deployed. As a starting point, it is supposed the creation of an *organization* entity in the system. This entity has associated a *user* with administrator permissions. There are five different processes involved in the operation of the SWQMS: system initialization, capture and storage of information, information modeling, data analytics and visualization, and management of information. These processes are described below.

System initialization

1. The administrator sends a request to the *message broker* from the platform's user interface (things manager) to create a *product*. The message broker creates a product entity in the *object repository*. This entity is used to group the *resources* and *things* of the system.
2. Each device sends a thingActivate request, see Table 1, to the message broker. As a result, a unique identifier (thingToken) is created and transmitted to the device. For each request received, the message broker creates a thing entity in the object repository.
3. The administrator sends a request to the message broker from the user interface of the platform for the creation of an IoT application. The message broker creates an *application* entity in the object repository.
4. The administrator sends as many requests to the message broker as users to be registered in the system. As a result, a number of user entities is created in the object repository.

Note: step 2 can also be performed by the administrator using the thingActivate operation of the RESTful API or by sending a request from the user interface to create a thing.

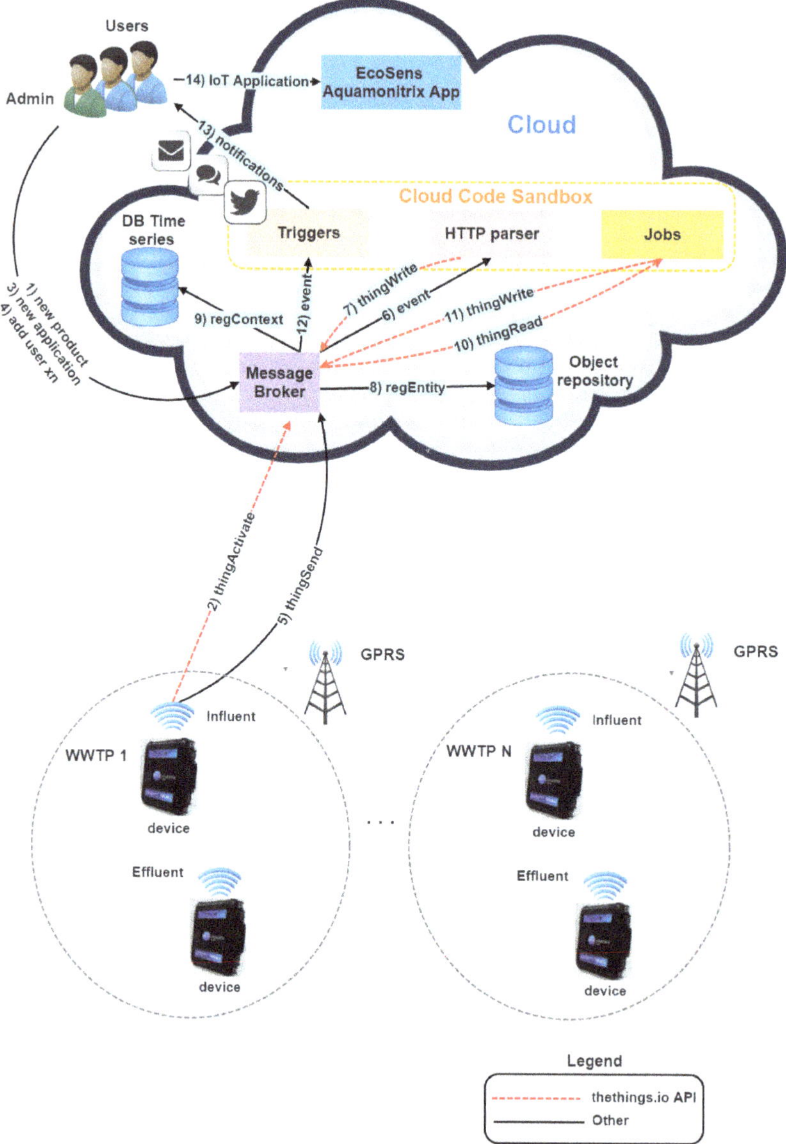

Figure 5. EcoSens Aquamonitrix Smart Wastewater Quality Monitoring System.

Capture and storage of information

5. Water quality data is acquired by sensors and collected by IoT devices. The IoT devices add spatial and temporal information to the data. The information is transmitted to the message broker using the *thingSend* operation.

6. When the information of an IoT device is published in the message broker, there is an event that notifies the consumer, in this case an *HTTP parser*. The HTTP parser decodes and preprocesses the message broker information to obtain the nitrate and nitrite measurements as well as the alarms. The alarms are generated if nitrate and nitrite concentrations exceed the allowed limits of 50 mg/L and 5 mg/L, respectively [56]. Note that other customized alarm levels are allowed.

7. The thingWrite operation, see Table 1, is used to store in the two system databases the information decoded and preprocessed: <key, value> pairs and associations of the nitrate, nitrite, and alarms resources.

Information modeling

8. The platform provides the RESTful API (see Table 1) and and other services, e.g., thingSend operation, used by both the administrator and the IoT devices to create or update entities. Entities represent different levels of abstraction (see Section 2.2). The mapping is done in the object repository, which records the associations between the system entities such as resource, thing, product, application, etc.

9. The information preprocessed by the HTTP parser, see step 6, is stored in the time series database of the system. This data warehouse provides the perfect infrastructure for mission-critical data.

Data analytics

10. In the Cloud Code SandBox, several *jobs*, such as hourly, daily, weekly, and monthly, have been implemented to get the system information using thingRead and exploit it through operations of the cloud code API, see Table 2. As a result, different *kpis*, e.g., mean, maximum, minimum, and standard deviation, related to nitrate and nitrite concentrations are obtained.

11. In a similar way to the alarms generated by the HTTP parser, the kpis are modeled as resources and stored using thingWrite in the time series database of the system.

12. *Triggers* are released when events occur after the thingWrite operations of the HTTP parser.

13. The use of triggers allows to notify alarms by e-mail, short message service, twitter or voice calls.

Visualization and management of the information

14. Finally, the visualization and management of the information is achieved through a user interface, which consists of a web service that takes advantage of the information stored in the system. The data sent by IoT devices and the alarms generated by the HTTP parser are published instantaneously, while key performance indicators calculated during the data analytics process are shown according to the job frequency (hourly, daily, weekly, or monthly). In this way, the interface provides to the user with valuable knowledge about water quality for decision making. Note that the interface has a functionality that allows exporting the data sets in csv, jpeg, and excel formats.

In summary, the EcoSens Aquamonitrix System is water quality aware: the information collected from the sensors is decoded, preprocessed, and modeled for processing, analysis, and knowledge extraction. This knowledge supports decision making by government agencies responsible for environmental protection and wastewater plant operators. The achieved behavior is possible thanks to the orchestration of the IoT services provided by the IoT platform, which allows the resources associated with the IoT devices to be searchable, accessible, and usable, thus maximizing their interaction with the user interface.

2.4. Integrated System Demonstration and Validation

Once the general integrated system has been applied to a IoT use case scenario in the wastewater treatment domain (see Section 2.3), a system customization has been developed for its demonstration and validation in real conditions. For these purposes, the experiments detailed in Sections 2.4.1 and 2.4.2 have been designed and carried out.

2.4.1. IoT System Demonstration in a Wastewater Treatment Scenario

The first experiment consisted of demonstrating the features implemented at the IoT application level. To do this, eight devices have been deployed in four WWTP located in the Region of Murcia (Spain): Alcantarilla, Molina de Segura, Los Alcázares, and San Pedro del Pinatar. A device has been placed at influent and effluent of each plant. During the test, devices have collected data related to nitrate and nitrite concentrations during the month of May 2019, twice a day, regardless of their location. This sampling frequency is appropriate for water quality monitoring applications, hence the term *near-real-time*. Note that the sampling frequency for traditional manual water quality monitoring techniques is usually once a week. The data transmitted by the devices are captured, stored, analyzed, and finally represented through the user interface by means of different customizations that compose the IoT application hosted in the system platform. In Section 3.1, all the customizations implemented in the integrated system are presented.

2.4.2. Detection Method Validation

The second experiment has focused on the preliminary validation in real conditions of the ion chromatography detection method in which the analytical device is based. To do this, nitrate and nitrite values collected by a device deployed in a WWTP and available on the IoT platform have been compared with experimental data analyzed in laboratory tests carried out with the usual procedures.

Sampling collection

The WWTP is located in Alcantarilla (Murcia, SE Spain) at the coordinates 37°55′39″ N 1°14′28″ W. This locality has an estimated population of 41,622 people in 2018. The plant has a design flow that exceeds 13,000 m^3/day corresponding to 151,667 population equivalent. The treatment included the following steps; (i) mechanical pretreatment, (ii) aerobic biological process as secondary treatment with active sludge with double stage, (iii) coagulation and flocculation, and (iv) sand filter and disinfection with ultraviolet in closed pipes. The device was placed in the wastewater effluent (see Figure 6).

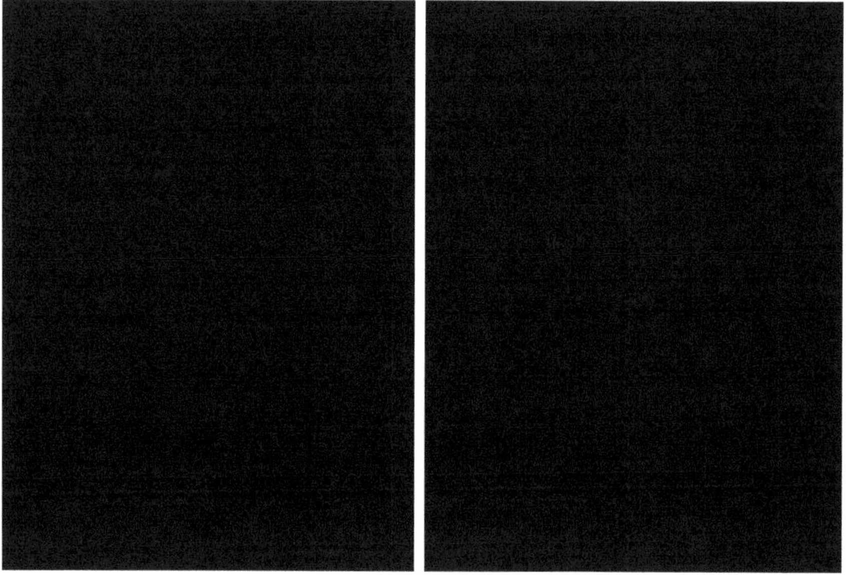

(**a**) Installation of the device   (**b**) Device deployed in the effluent

Figure 6. Device deployment in Alcantarilla wastewater treatment plant (WWTP).

The sampling procedure involved the acquisition of the analytical data obtained by the sensor daily (average value of 3 daily readings) and the analytical data obtained from 3 daily samples of effluent that were integrated and subsequently analyzed in the laboratory. The samples were stored at 4 °C until further analysis. The sampling was carried out during the month of March 2019, and eight samples were compared in the following days 1, 4, 8, 12, 14, 18, 20, and 22.

Analytical determination by portable IC

The prototype analytical device is a portable ion chromatography (IC) system based on the method previously reported by [14] as mentioned in the Section 2.1. Therefore, the study parameters have been extracted from this paper. Concretely, these parameters are Limit of Detection (LOD), Limit of Quantification (LOQ), Range, Linearity, Repeatability, and Accuracy.

Analytical determination by standardized method

Nitrate and nitrite were analyzed using ion chromatography. Water samples of effluent were filtered through a 0.2 μm Minisart® Plus syringe membrane filter (NY, USA). The available analytical standards for sodium nitrite, sodium nitrate, and sodium carbonate provided by Panreac (Barcelona, Spain). Distilled, de-ionized water (DDW, 18 Ω cm^{-1}) was used in all sample preparation procedures.

The samples were injected into the IC system consisting of a Metrohm pump (Herisau, Switzerland) coupled by a transfer line to a Metrohm conductimetric detector (Herisau, Switzerland). The column used was Metrosep A Supp 7, Metrohm (Herisau, Switzerland). The injection volume was 20 μL and the mobile phase flow (Na_2CO_3, 3.6 mM) was 0.7 mL.min^{-1}. The anions were identified and quantified according to their retention times (13.90 min $\pm$ 0.4 min and 21.19 min $\pm$ 0.5 min for nitrite and nitrate, respectively).

Statistical analysis

The statistical software SigmaPlot version 13 (Systat, Software Inc., San Jose, CA, USA) [57] was used to verify the correlation between the results obtained by the sensor and the experimental data, the average were compared by an hypothesis test ($p > 0.05$).

3. Results and Discussion

In this section, the proposed solution for water quality monitoring and management in distributed, dynamic, and complex scenarios such as water distribution systems (WDSs) in cities is analyzed. The evaluation will be detailed in two sections. In Section 3.1, all the customizations implemented in the integrated system are presented. In Section 3.2, the results obtained from the experiment to validate the detection method employed by the integrated system are shown and discussed.

3.1. Integrated System Customization

The general integrated system has been instantiated for a IoT use case scenario in the wastewater treatment domain. To achieve this, it was necessary to carry out tasks including deployment of devices in WWTP, integration with the IoT platform, implementation of data analysis algorithms, and alarm calculation functions, as well as customization of widgets for the representation of information. This section describes the features of the implemented IoT application that allow to visualize the information of the customized system and, therefore, to demonstrate the appropriateness of this smart water quality monitoring and management solution.

Before starting the presentation of the results, it is necessary to mention that eight devices transmit water quality data to the system cloud where it is decoded, stored, and processed (see Section 2.4.1). Therefore, eight thing-type entities (one per device) have been created in the system (see Section 2.3, step 2 of the system initialization process). In the customized system, a thing has four types of resource: nitrate, nitrite, alarms, and kpis. These resources and the values of their associated attributes, see Section 2.2, become the information of the developed system.

The IoT application provides customized features for online and near-real-time monitoring and management of water quality data associated with the influent and effluent of each WWTP. These customized features are arranged in two levels of dashboard: application and insight. The application level shows the measurements of all devices (things) deployed and their activity, whereas the insight level displays the measurements and activity of a single device. To begin with, the main implemented features of the application dashboard are explained. From this dashboard, it is available the product map to visualize the geographical location of the devices deployed. Figure 7a illustrates the four clusters of devices related to the WWTPs instrumented. Moreover, if one of these clusters is clicked, that area is automatically zoomed. For instance, if the Los Alcázares WWTP cluster is clicked, the two devices deployed in this facility are displayed. Figure 7b shows the thing detail window, which appears if an device on the product map is clicked. The detail window displays the latest nitrate and nitrite concentration values stored in the system as well as the alarms and kpis computed by the HTTP parser and jobs executed on the Cloud Code Sandbox.

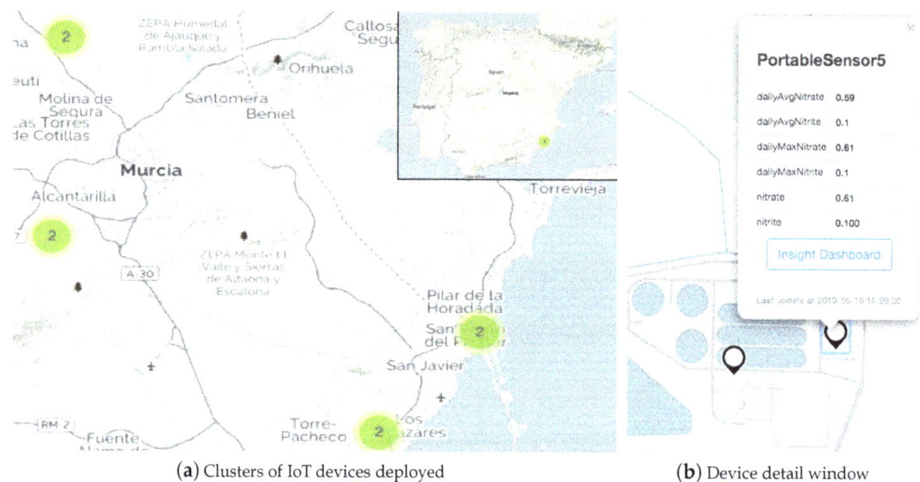

(**a**) Clusters of IoT devices deployed   (**b**) Device detail window

Figure 7. Product map.

Another widget implemented in the application dashboard is the general table, see Figure 8. This feature provides information about the WSN deployed. Specifically, for a given IoT device, e.g., PortableSensor1, it shows information about the last transmission of the associated device (3 h ago), the latest values of the attributes related to nitrate and nitrite resources (0.52 mg/L and 0.15 mg/L, respectively), the location of the IoT device (Alcantarilla WWTP), and a higher level of detail regarding its deployment (influent final pretreatment).

⇵ Name		⇵ Last Seen	⇵ Nitrate (mg/L)	⇵ Nitrite (mg/L)	⇵ Location	⇵ Description
PortableSensor8	3 hours ago		0.920	0.110	San Pedro WWTP	Effluent tertiary treatment
PortableSensor7	3 hours ago		0.570	0.050	San Pedro WWTP	Influent final pretreatment
PortableSensor6	3 hours ago		1	0.010	Los Alcázares WWTP	Effluent tertiary treatment
PortableSensor5	3 hours ago		0.61	0.100	Los Alcázares WWTP	Influent final pretreatment
PortableSensor4	3 hours ago		0.810	0.030	Molina de Segura WWTP	Effluent tertiary treatment
PortableSensor3	3 hours ago		0.430	0.130	Molina de Segura WWTP	Influent final pretreatment
PortableSensor2	3 hours ago		0.760	0.020	Alcantarilla WWTP	Effluent tertiary treatment
PortableSensor1	3 hours ago		0.520	0.150	Alcantarilla WWTP	Influent final pretreatment

Figure 8. General table. This feature provides information about the Wireless Sensor Network (WSN) deployed.

Figure 9 depicts the implementation of two widgets that simplify the tasks for monitoring and maintenance of the system. The first widget, Figure 9a, shows the alarms associated with all the things. Note that the data shown has been simulated as no alarms were generated during the system test. The second widget is called an actuator as it allows to modify the location (latitude and longitude) and metadata associated with a thing (thing name, thing description, serial number, and other metadata). Figure 9b shows the use of the actuator to set the location and the name associated with the PortableSensor8 thing.

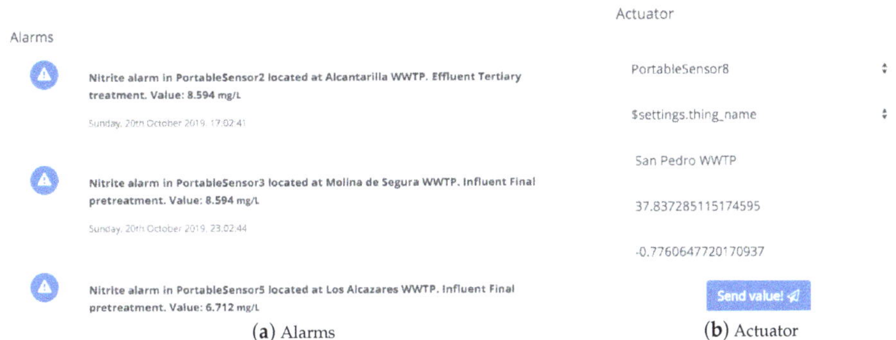

(**a**) Alarms   (**b**) Actuator

Figure 9. Widgets developed to ease monitoring and maintenance tasks.

The IoT application provides different mechanisms implemented to navigate towards the insight dashboard associated with a thing. For instance, this second level of dashboard can be visualized by clicking on a thing name in the general table (see Figure 8) or through the product map detail window (see Figure 7b). The insight dashboard provides a set of customized widgets that display information associated with the thing: metadata, alarms for high levels of nitrate and nitrite, the latest value of nitrate and nitrite concentrations acquired as well as graphs. By default, each graph shows the latest 14 values of at least one resource between nitrate, nitrite, alarms, and kpis. Among these kpis, information is displayed about the average, maximum, minimum, and standard deviation processed at different frequencies (hourly, daily, weekly, and monthly). In addition, it is possible to zoom in and out of the graphs to display a specific range of data representation or to filter by date to visualize or download a certain range of values. Figure 10 depicts a widget layout to monitor water quality in terms of nitrate concentrations of the Los Alcázares WWTP. The water quality data shown in Figure 10 corresponds to the week from 13 to 19 of May 2019 by the PortableSensor5 and PortableSensor6 devices deployed in the influent (Figure 10a,b) and the effluent (Figure 10c,d) of the plant, respectively. Note that in the current set-up, devices collect data a maximum of three times a day, but a more frequent sampling can be configured remotely.

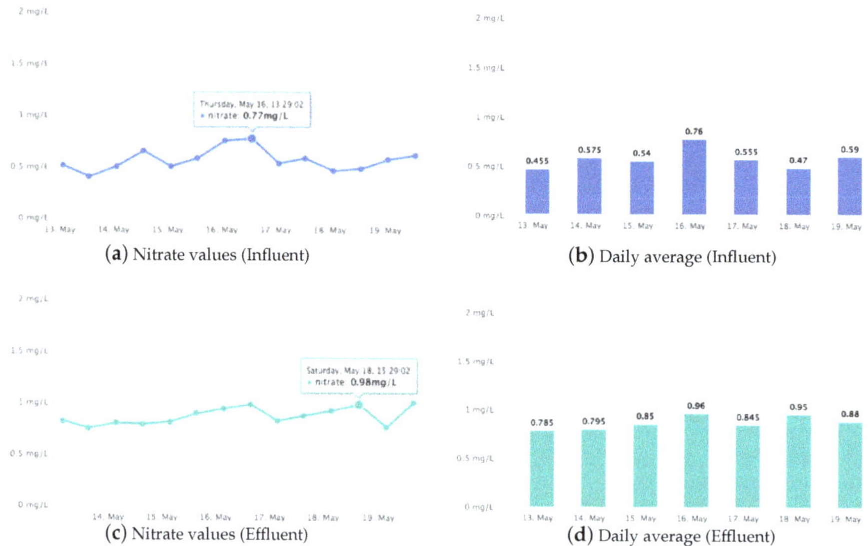

(a) Nitrate values (Influent)

(b) Daily average (Influent)

(c) Nitrate values (Effluent)

(d) Daily average (Effluent)

Figure 10. Water quality monitoring of the Los Alcázares WWTP.

Figure 10a,c shows the measurements of nitrate parameter acquired in the influent and effluent during the mentioned time period. In these customized widgets, it is possible to display accurate information about the values by placing the cursor over each point on the graph. In general, the nitrate values obtained in the influent are lower than those of the effluent. This fact can be appreciated in more detail in Figure 10b,d, which illustrate the daily averages of nitrate concentrations for the two measurement sites. Note that the maximum values obtained for the average kpi in the influent and effluent are 0.76 mg/L and 0.96 mg/L, respectively. In both cases, these are low values. Moreover, the weekly average obtained is 0.56 mg/L (influent) and 0.86 mg/L (effluent). These values are far from the alarm threshold of 50 mg/L specified by European regulations [56]. The deployment of the network has been extended in other facilities located around Europe (Spain, Finland, Ireland, and Portugal) showing its scalability and good performance in the collection and processing of data in near-real-time.

3.2. Device Measurements Validation

Nitrate concentrations in wastewater obtained from the standardized experimental method and automatically through the device are shown in Figure 11a. In spite of the observed variations between nitrate concentrations detected by the sensor and those analyzed by applying the experimental method, from a statistical level, the test performed indicates that there are no significant differences between the two methods ($p = 0.432$). In this sense, note that the average concentration in the eight days of sampling was approximately 4.43 mg/L and 4.58 mg/L for the sensor and the experimental method, respectively.

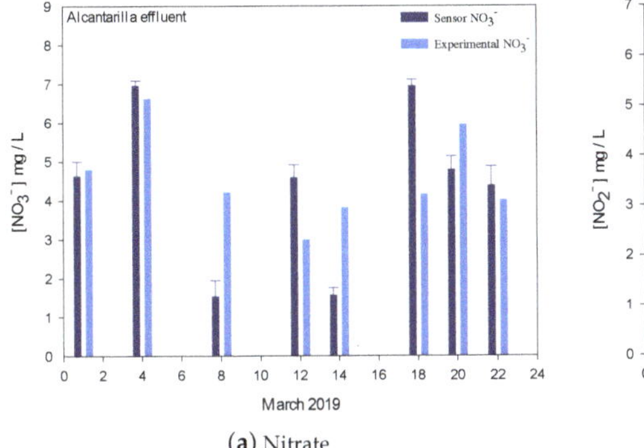
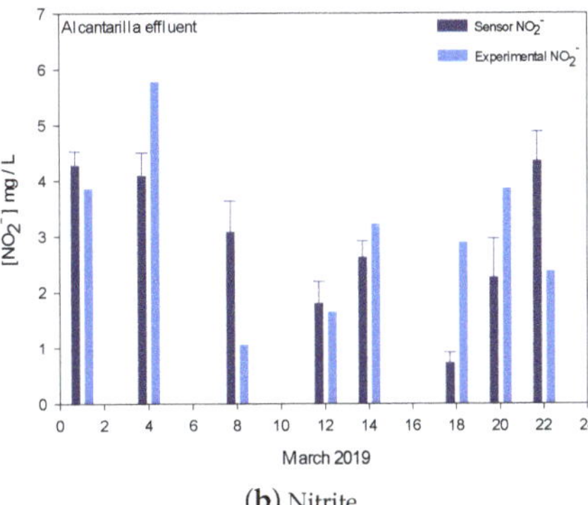

(**a**) Nitrate   (**b**) Nitrite

Figure 11. Concentration in wastewater obtained from standardized experimental method (blue sky) and automatically from the enhanced portable sensor (blue navy).

A similar behaviour is observed for the case of nitrite. In Figure 11b, the concentrations of this parameter obtained from the sensor and those analyzed by applying the experimental method in the eight days of sampling are shown. From a statistical level, there are also no significant differences between the two methods ($p = 0.395$). The average concentration is approximately 2.90 mg/L and 3.09 mg/L for the sensor and the experimental method, respectively.

The study of the analytical validation parameters of both methods (LD (mg/L), LQ (mg/L), Range (mg/L), Linearity (determination coefficient (R^2)), Repeatability (relative standard deviation (RSD%) of peak area and retention time), and Accuracy (%)) shows that the results achieved in the sensor validation [14] are very similar to those obtained from the standardized experimental method (Table 3). In this context, Figure 11 shows that the differences observed between the sensor data and the experimental data (0.2–2.8 mg/L and 0.2–2.2 mg/L for nitrate and nitrite, respectively) are within what is expected, taking into account the working ranges of each of the parameters for each of the methods of analysis. Therefore, the concentrations analyzed by the device are valid and demonstrate the potential of the system for portable analysis under real conditions, as the validation has been conducted in a wastewater treatment plant effluent, with the influence that the matrix effect can have on this type of waters.

Table 3. Analytical validation parameters for nitrate and nitrite obtained of device and standardized experimental method.

Parameters	Device		Experimental	
	Nitrate	Nitrite	Nitrate	Nitrite
LOD (mg/L)	0.04	0.007	0.1	0.003
LOQ (mg/L)	0.07	0.010	1.0	0.030
Range (mg/L)	0.07–75	0.01–15	1.0–100	0.030–100
Linearity (R^2)	≥0.995	≥0.995	≥0.995	≥0.995
Repeatability (RSD)	3.06–4.19	0.75–1.10	2.04–4.20	0.84–2.23
Accuracy (%)	91.2	92.2	93.4	91.6

4. Conclusions

In this paper, an integrated IoT system for water quality monitoring is conceived and customized for its demonstration and preliminary validation in wastewater treatment use case. The proposed system leverages an innovative low-cost analytical device at the close-to-market stage. The device consists of a nitrate and nitrite analyzer based on a novel ion chromatography detection method and equipped with IoT communication capabilities to build a WSN. An IoT software platform is used to integrate the analytical device. By doing so, a decentralized SWQMS adaptable to the dynamic and heterogeneous WDSs of cities is achieved. This SWQMS is composed of a wireless sensor network and an open cloud-based middleware.

A thoroughgoing analysis of the different layers of the conceived system is applied to proper design of the customized system in the field of wastewater treatment. The implemented platform provides near-real-time communication with devices and incorporates preventive functions and data analytics that support decision-making. To achieve these features, five different processes for the management and administration of the system by different organizations are implemented: system initialization, capture and storage of information, information modeling, data analytics and visualization, and management of information.

The results obtained from a real conditions wireless sensor network deployment in Murcia, Spain, as part of the Ecosens Aquamonitrix project, show that the implemented system provides features for online and near-real-time monitoring and management of wastewater quality parameters. The system architecture is extensible to include other features. Moreover, the scalability of the IoT ecosystem enables to increase both the number of sensor nodes and the storage and processing resources of the IoT platform. Regarding the preliminary validation of the device, the developed method was used to determine the content of nitrite and nitrate in the effluent of a WWTP. The results achieved show that this method is reliable and fast working in a wide range for nitrates and nitrites determination as well as avoiding the use of many reagents, some of which can be hazardous.

The advances features of the developed IoT integrated system will enable massive sensor deployments in the water distribution systems of smart cities allowing end users to detect pollution events and adverse trends in near real-time. Thus, the private or public entities in charge of water quality monitoring and management will be able to act in a more efficient and effective way tackling the problems detected (i.e., pollution sources), reacting to the problems more quickly and minimizing the negative environmental impact.

Author Contributions: Conceptualization, R.M. and J.M.N.; Formal analysis, R.M., J.M.N., N.V., and A.e.A.; Investigation, R.M., N.V., A.e.A., E.M., P.R., and J.M.N.; Methodology, R.M., J.M.N., N.V., and A.e.A.; Resources, R.M., J.M.N., N.V., and A.e.A.; Software, R.M., J.M.N., N.V., and A.e.A.; Visualization, R.M.; Supervision, J.M.N.; Validation, R.M., N.V., A.e.A., E.M., P.R., and J.M.N.; Writing—original draft preparation, R.M., J.M.N., N.V., and A.e.A.; Writing—review & editing, R.M., N.V., A.e.A., E.M., P.R., and J.M.N. All authors have read and agreed to the published version of the manuscript.

Funding: This paper is the result of the research carried out under the project Enhanced Portable Sensor for Water Quality Monitoring, moving to genuinely integrated Water Resource Management—ECOSENS AQUAMONITRIX—LIFE is co-funded by the LIFE Programme of the European Union under contract number LIFE17 ENV/IE/000237.

Acknowledgments: This work has been partially supported by the Entity of Sanitation and Wastewater Treatment of the Region of Murcia (ESAMUR).

Conflicts of Interest: The authors declare no conflict of interest.

Abbreviations

The following abbreviations are used in this manuscript:

WWTP	Wastewater treatment plant
WQM	Water Quality Monitoring
BOD	Biological Oxygen Demand
COD	Chemical Oxygen Demand
TOC	Total Organic Carbon
TSS	Total Suspended Solids
IoT	Internet of Things
SOA	Service-Oriented Architecture
SaaS	Software as a Service
SWQMS	Smart Water Quality Monitoring System

WSN	Wireless Sensor Network
SCADA	Supervisory Control And Data Acquisition
IC	Ion Chromatography
UV	Ultraviolet
LED	Light-Emitting Diode
ADC	Analogue to Digital Converter
Rpi	Raspberry Pi Zero W
API	Application Program Interface
REST	Representational State Transfer
HTTP	Hypertext Transfer Protocol
MQTT	Message Queue Telemetry Transport
UDP	User Datagram Protocol
TCP	Transmission Control Protocol
CoAP	Constrained Application Protocol
JSON	JavaScript Object Notation
kpis	key performance indicators
SMS	Short Message Service
GPRS	General Packet Radio Service
LD	Limit of Detection
LQ	Limit of Quantification
WDS	Water Distribution System
RSD	Relative Standard Deviation

References

1. Arauzo, M.; Martínez-Bastida, J.J. Environmental factors affecting diffuse nitrate pollution in the major aquifers of central Spain: Groundwater vulnerability vs. groundwater pollution. *Environ. Earth Sci.* **2015**, *73*, 8271–8286. [CrossRef]
2. Martínez-Bastida, J.J.; Arauzo, M.; Valladolid, M. Intrinsic and specific vulnerability of groundwater in central Spain: The risk of nitrate pollution. *Hydrogeol. J.* **2010**, *18*, 681–698. [CrossRef]
3. García-Garizabal, I.; Causapé, J.; Abrahao, R. Nitrate contamination and its relationship with flood irrigation management. *Hydrol. J.* **2012**, *442*, 15–22. [CrossRef]
4. Le Moal, M.; Gascuel-Odoux, C.; Menesguen, A.; Souchon, Y.; Etrillard, C.; Levain, A.; Moatar, F.; Pannard, A.; Souchu, P.; Lefebvre, A.; et al. Eutrophication: A new wine in an old bottle? *Sci. Total Environ.* **2019**, *651*, 1–11. [CrossRef] [PubMed]
5. Nixon, S.W. Coastal marine eutrophication: A definition, social causes, and future concerns. *Ophelia* **1995**, *41*, 199–219. [CrossRef]
6. Pérez-Ruzafa, A.; Pérez-Ruzafa, I.M.; Newton, A.; Marcos, C. Coastal lagoons: Environmental Variability, Ecosystem Complexity and Goods and Services Uniformity. In *Coasts and Estuaries*; Elsevier: New York, NY, USA, 2019; pp. 253–276.
7. Ramos, H.M.; McNabola, A.; López-Jiménez P.A.; Pérez-Sánchez, M. Smart Water Management towards Future Water Sustainable Networks. *Water* **2020**, *12*, 58. [CrossRef]
8. Farrel-Poe, K.; Payne, W.; Emanuel, R. *Arizona Watershed Stewardship Guide: Water Quality and Monitoring*; College of Agriculture and Life Sciences, University of Arizona: Tucson, AZ, USA, 2005; pp. 1–2.
9. Mentzafou, A.; Panagopoulos, Y.; Dimitriou, E. Designing the National Network for Automatic Monitoring of Water Quality Parameters in Greece. *Water* **2019**, *11*, 1310. [CrossRef]
10. Sempere-Payá, V.; Todolí-Ferrandis, D.; Santonja-Climent, S. ICT as an enabler to smart water management. In *Smart Sensors for Real-Time Water Quality Monitoring*; Springer: Berlin/Heidelberg, Germany, 2013; pp. 239–258.
11. Thomas, O.; Théraulaz, F.; Cerdà, V.; Constant, D.; Quevauviller P. Wastewater quality monitoring. *TRAC Trend Anal. Chem.* **1997**, *16*, 419–424. [CrossRef]
12. Bourgeois, W.; Burgess, J.; Stuetz, R. On-line monitoring of wastewater quality: A review. *J. Chem. Technol. Biotechnol.* **2001**, *76*, 337–348. [CrossRef]

13. Korostynska, O.; Mason, A.; Al-Shamma'a, A. Monitoring of nitrates and phosphates in wastewater: Current technologies and further challenges. *S2IS* **2012**, *5*. [CrossRef]

14. Murray, E.; Roche, P.; Harrington, K.; McCaul, M.; Moore, B.; Morrin, A.; Diamond, D.; Paull, B. Low cost 235 nm ultra-violet light-emitting diode-based absorbance detector for application in a portable ion chromatography system for nitrite and nitrate monitoring. *J. Chromatogr. A* **2019**, *1603*, 8–14. [CrossRef] [PubMed]

15. Mason, A.; Korostynska, O.; Al-Shamma'a, A.I. *Microwave Sensors for Real-Time Nutrients Detection in Water*; Springer: Berlin/Heidelberg, Germany, 2013; pp. 197–216.

16. Quevauviller, P.; Thomas O.; Derbeken A.V. In *Wastewater Quality Monitoring and Treatment*; John Wiley & Sons: Hoboken, NJ, USA, 2006; pp. 1–22.

17. European Commission, EC. Directive 2000/60/EC of the European Parliament and of the Council of 23 October 2000 establishing a framework for Community action in the field of water policy. *Off. J. Eur. Commun.* **2000**, *L327*, 1–73.

18. European Commission, EC. Council Directive 91/271/EEC of 21 May 1991 concerning urban wastewater treatment. *Off. J. Eur. Commun.* **1991**, *L135*, 40–52.

19. European Commission, EC. Council Directive 91/676/EEC of 12 December 1991 concerning the protection of waters against pollution caused by nitrates from agricultural sources. *Off. J. Eur. Commun.* **1991**, *L375*, 1–8.

20. Katsriku, F.; Wilson, M.; Yamoah, G.G.; Abdulai, J.; Rahman, B.; Grattan, K. Framework for time relevant water monitoring system. In *Computing in Research and Development in Africa*; Springer: Basel, Switzerland, 2015; pp. 3–19.

21. Vijayakumar, N.; Ramya, A.R. The real time monitoring of water quality in IoT environment. In Proceedings of the 2015 International Conference on Innovations in Information, Embedded and Communication Systems (ICIIECS), Nagercoil, India, 19–20 March 2015; pp. 1–5.

22. Borgia, E. The Internet of Things vision: Key features, applications and open issues. *Comput. Commun.* **2014**, *54*, 1–31. [CrossRef]

23. Dong, J.; Wang, G.; Yan, H.; Xu, J.; Zhang, X. A survey of smart water quality monitoring system. *ESPR* **2015**, *22*, 4893–4906. [CrossRef]

24. Geetha, S.; Gouthami, S. Internet of things enabled real time water quality monitoring system. *Smart Water* **2016**, *2*, 1. [CrossRef]

25. Martínez, R.; Pastor, J.; Álvarez, B.; Iborra, A. A testbed to evaluate the fiware-based IoT platform in the domain of precision agriculture. *Sensors* **2016**, *16*, 1979. [CrossRef]

26. Mineraud, J.; Mazhelis, O.; Su, X.; Tarkom, S. A gap analysis of Internet-of-Things platforms. *Comput. Commun.* **2016**, *89*, 5–16. [CrossRef]

27. Zhou, H. Four Pillars of IoT. In *The Internet of Things in the Cloud. A Middleware Perspective*; CRC Press: Boca Raton, USA, 2012; pp. 59–66.

28. Li, D.; Liu, S. WSN in Sewage Treatment. In *Water Quality Monitoring and Management: Basis, Technology and Case Studies*; Academic Press: London, UK, 2018; pp. 92–93.

29. Akyildiz, I.; Su, W.; Sankarasubramaniam, Y.; Cayirci, E. Wireless sensor networks: A survey. *Comput. Netw.* **2002**, *38*, 393–422. [CrossRef]

30. Yi, W.Y.; Lo, K.M.; Mak, T.; Leung, K.; Leung, Y.; Meng, M. A Survey of Wireless Sensor Network Based Air Pollution Monitoring Systems. *Sensors* **2015**, *15*, 31392–31427. [CrossRef] [PubMed]

31. Noriega-Linares, J.; Navarro, J. On the application of the Raspberry PI as an advanced acoustic sensor network for noise monitoring. *Electronics* **2016**, *5*, 74. [CrossRef]

32. Segura-Garcia, J.; Felici-Castell, S.; Perez-Solano, J.; Cobos, M.; Navarro, J. Low-Cost Alternatives for Urban Noise Nuisance Monitoring using Wireless Sensor Networks. *J. Sens.* **2014**, *99*, 9.

33. Noriega-Linares, J.E.; Rodriguez-Mayol, A.; Cobos, M.; Segura-Garcia, J.; Felici-Castell, S.; Navarro, J. A Wireless Acoustic Array System for Binaural Loudness Evaluation in Cities. *IEEE Sens. J.* **2017**, *17*, 7043–7052. [CrossRef]

34. Jadhav, P.S.; Deshmukh, V.U. Forest fire monitoring system based on ZIG-BEE wireless sensor network. *IJETAE* **2012**, *2*, 187–191.

35. Keshtgari, M.; Deljoo, A. A wireless sensor network solution for precision agriculture based on zigbee technology. *SCIRP* **2012**, *4*, 25–30. [CrossRef]

36. Adu-Manu, K.S.; Tapparello, C.; Heinzelman, W.; Katsriku, F.A.; Abdulai, J.D. Water quality monitoring using wireless sensor networks: Current trends and future research directions. *TOSN* **2017**, *13*, 4. [CrossRef]

37. Pule, M.; Yahya, A.; Chuma, J. Wireless sensor networks: A survey on monitoring water quality. *JART* **2017**, *15*, 562–570. [CrossRef]

38. Adamo, F.; Attivissimo, F.; Carducci, C.G.C.; Lanzolla, A.M.L. A smart sensor network for sea water quality monitoring. *IEEE Sens. J* **2014**, *15*, 2514–2522. [CrossRef]

39. Jiang, P.; Xia, H.; He, Z.; Wang, Z. Design of a water environment monitoring system based on wireless sensor networks. *Sensors* **2009**, *9*, 6411–6434. [CrossRef]

40. König, M.; Jacob, J.; Kaddoura, T.; Farid, A.M. The role of resource efficient decentralized wastewater treatment in smart cities. In Proceedings of the 2015 IEEE First International Smart Cities Conference (ISC2), Guadalajara, Mexico, 25–28 October 2015; pp. 1–5.

41. Massoud, M.A.; Tarhini, A.; Nasr, J.A. Decentralized approaches to wastewater treatment and management: Applicability in developing countries. *J. Environ. Manag.* **2009**, *90*, 652–659. [CrossRef] [PubMed]

42. O'Donovan, P.; Coburn, D.; Jones, E.; Hannon, L.; Glavin, M.; Mullins, D.; Clifford, E. A cloud-based distributed data collection system for decentralised wastewater treatment plants. *Procedia Eng.* **2015**, *119*, 464–469. [CrossRef]

43. Tadokoro, H.; Jp, P.E.; Nakamura, N.; Nishimura, T.; Uemura, K.; Kikuchi, N.; Hatayama, M. Monitoring and Control Systems for the IoT in the Water Supply and Sewerage Utilities. *Hitachi Rev.* **2017**, *66*, 704–711.

44. Wang, W.; Yao, J.; Qiu, S. Design and implementation of sewage cloud platform monitoring system. In *MATEC Web of Conferences*; EDP Sciences: Les Ulis, France, 2018; Volume 232, p. 4005.

45. Zakaria, Y.; Michael, K. An integrated cloud-based wireless sensor network for monitoring industrial wastewater discharged into water sources. *SCIRP* **2017**. [CrossRef]

46. Saravanan, K.; Anusuya, E.; Kumar, R. Real-time water quality monitoring using Internet of Things in SCADA. *Environ. Monit. Assess* **2018**, *190*, 556. [CrossRef]

47. Life EcoSens Aquamonitrix Project. Available online: https://www.ecosensaquamonitrix.eu/ (accessed on 3 January 2020).

48. Murray, E.; Roche, P.; Briet, M.; Moore, B.; Morrin, A.; Diamond, D.; Paull, B. Fully automated, low-cost ion chromatography system for in-situ analysis of nitrite and nitrate in natural waters. *Talanta* **2020**, *216*, 120955. [CrossRef]

49. thethings.iO IoT Platform. Available online: https://thethings.io (accessed on 3 January 2020).

50. Bahga, A.; Madisetti, V. *Internet of Things: A Hands-On Approach*; Vpt: Plymouth, MA, USA, 2014.

51. Redis. Available online: https://redis.io (accessed on 14 January 2020).

52. Apache Cassandra. Available online: http://cassandra.apache.org (accessed on 14 January 2020).

53. MongodB. Available online: https://www.mongodb.com (accessed on 14 January 2020).

54. Lorido-Botran, T.; Miguel-Alonso, J.; Lozano, J. A review of auto-scaling techniques for elastic applications in cloud environments. *J. Grid Comput.* **2014**, *12*, 559–592. [CrossRef]

55. Jailed, Flexible JavaScript Sandbox. Available online: https://github.com/asvd/jailed (accessed on 14 January 2020).

56. European Commission (EC). Directive 2006/118/EC of the European Parliament and of the Council of 12 December 2006 on the protection of groundwater against pollution and deterioration. *Off. J. Eur. Commun.* **2006**, *L372*, 19–31.

57. SigmaPlot Product Overview. Available online: https://systatsoftware.com/products/sigmaplot/ (accessed on 29 January 2020).

Article

Optimal Placement of Pressure Sensors Using Fuzzy DEMATEL-Based Sensor Influence

Jorge Francés-Chust [1,2], Bruno M. Brentan [2,3], Silvia Carpitella [2,4] and Joaquín Izquierdo [2,*] and Idel Montalvo [2,5]

[1]   Business Development, Aigües Bixquert, S.L., Xàtiva, 46800 Valencia, Spain; jorge@abxat.com
[2]   Fluing—Institute for Multidisciplinary Mathematics, Universitat Politècnica de València, 46022 Valencia, Spain; brunocivil08@gmail.com (B.M.B.); silvia.carpitella@gmail.com (S.C.); imontalvo@ingeniousware.net (I.M.)
[3]   School of Engineering, Hydraulic Engineering and Water Resources Department, Federal University of Minas Gerais, 31270-901 Belo Horizonte, MG, Brazil
[4]   Dipartimento di Ingegneria, Università degli Studi di Palermo, 90128 Palermo, Italy
[5]   Business Development, IngeniousWare GmbH, Jollystrasse 11, 76137 Kahlsruhe, Germany
*   Correspondence: jizquier@upv.es

Received: 30 December 2019; Accepted: 6 February 2020; Published: 12 February 2020

Abstract: Nowadays, optimal sensor placement (OSP) for leakage detection in water distribution networks is a lively field of research, and a challenge for water utilities in terms of network control, management, and maintenance. How many sensors to install and where to install them are crucial decisions to make for those utilities to reach a trade-off between efficiency and economy. In this paper, we address the where-to-install-them part of the OSP through the following elements: nodes' sensitivity to leakage, uncertainty of information, and redundancy through conditional entropy maximisation. We evaluate relationships among candidate sensors in a network to get a picture of the mutual influence among the nodes. This analysis is performed within a multi-criteria decision-making approach: specifically, a herein proposed variant of DEMATEL, which uses fuzzy logic and builds comparison matrices derived from information obtained through leakage simulations of the network. We apply the proposal first to a toy example to show how the approach works, and then to a real-world case study.

Keywords: water distribution network; leakage; optimal sensor placement; sensitivity; uncertainty; entropy; multi-criteria decision-making; DEMATEL

---

1. Introduction and Literature Review

Optimal sensor placement (OSP) for leakage detection in water distribution networks (WDNs) currently represents an exciting and lively field of research, aimed at optimising processes of network control, management, and maintenance [1].

With the explosion of the sensors market, and the consequent access to pressure sensors, various technical questions appear for water utilities. The most important are how many sensors to install, and, given a number of sensors, where to install them. There is a clear trade-off analysis to be performed that aims to answer the first question. Having more sensors in the network means more data that can be used to get more complete knowledge about the system. However, having more sensors also means more money spent. As a result, economical reasons make the choice of a monitoring strategy a crucial decision. The selection of good monitoring points may bring more and better information about the system, for less money.

Considering the complex task of defining the position and number of sensors in a water network, it is reasonable that there are many works in the literature presenting different approaches for the OSP problem. In this introduction we just focus on some of those most directly related to the approach we present in this paper.

With the aim of finding leaks in water systems, a methodology for pressure sensor placement based on sensitivity analysis is presented in [2]. The authors generate a sensitivity matrix based on simulations of leaks, from which a signature matrix may be extracted. Genetic algorithms are then used to maximise the isolation of leaks. In [3], a multi objective technique to find the optimal monitoring points for leak detection, given a number of sensors, is applied. The sensitivity matrix is calculated based on the percentage variation of pressure from the normal scenario to an abnormal one.

Hydraulic simulations are performed based on a body of input information, including pipe roughness, nodal water demand, etc. The accuracy of the hydraulic state derived depends on the quality of the input information. Since many inputs are not directly measured, uncertainty analysis is often needed. For example, the authors of [4] modify the work of [2], by adding nodal demand uncertainty analysis to build the sensitivity matrix.

Additionally, in [5], uncertainties coming from water demand in the network are included. The authors use genetic algorithms to investigate optimal sensor placement based on the sensitivity matrix and residual vectors.

To simulate small leaks in water distribution systems, a demand driven approach (DDA) can be used, modelling the leak as a function of only the pressure. Of course, for large leaks and pipe bursts, a DDA has a set of limitations. If the anomaly leads the system's pressure under the minimal operational pressure, the total demand cannot be supplied. In [6], an optimal pressure sensor placement methodology based on nodal entropy is presented. The authors simulate anomalies using a pressure driven approach (PDA) and compare the results with DDA simulations. Using the entropy method, the authors rank the nodes with high entropy as the best monitoring points.

The sensitivity matrix is widely used in works of OSP. However, the use of that matrix without considering other parameters can lead to the concentration of sensors in a reduced region of the network, which leads to significant coverage reduction. To cope with it, in [7], it is combined the sensitivity matrix with the maximisation of the entropy related to the sensor network. The entropy's maximisation guarantees better spread of the sensors, according to the authors. Optimisation approaches are also widely applied to locate optimal monitoring points. In [8], an approach to minimise the distance of localised leaks based on sensors' data is developed. A pre-defined number of sensors is used as a constraint for the optimisation problem. A genetic algorithm is used to find the optimal number of sensors and their strategic monitoring position. Aiming to maximise the number of failures detected, the authors in [9] present an optimal sensor placement using a minimum test cover (MTC) with approximated solutions. The authors develop a new augmented greedy algorithm for solving the MTC problem.

The main purpose of this paper is to evaluate relationships among pressure sensors in a network to get as complete as possible an understanding of their mutual influence, and thus identify those candidate nodes to host sensors that may have bigger impact on the network information. The ultimate aim is to guarantee better network control by optimising the number of sensors and their location in the network. Identifying those network nodes that capture bigger influence can be strategic, since variations on those nodes may directly reflect variations on other nodes in the system, and this will eventually reduce the investment in sensors.

We claim that a multi-criteria decision-making (MCDM) approach may effectively support the problem being faced. A methodology that appears to be best suited to such an aim is the decision making trial and evaluation laboratory (DEMATEL), first implemented by Fontela and Gabus [10,11].

DEMATEL is helpful when dealing with complex systems, such as water networks, since they are characterised by many aspects/elements directly or indirectly interdependent with each other, and this condition makes hard many decision-making tasks. As asserted, for example, in [12], the use of DEMATEL supports the visualisation of interferences existing among the relevant aspects of a given problem, thereby helping comprehensive understanding of the intensity and direction of direct and indirect relations for each pair of factors under study. This technique deals with interactions through a step-by-step approach [13]; it has been widely applied in the literature for management problems characterised by the presence of heavy interdependence among elements [14–17]; and many developments of its application have been proposed in a wide number of fields (see [18–20], among others).

To address the stated problem, we herein propose a new approach within the framework of the fuzzy DEMATEL method. The fuzzy DEMATEL represents a development of the traditional crisp DEMATEL, extended by Wu and Lee [21], and makes use of elements of the fuzzy set theory [22] for better managing uncertainty affecting input evaluations.

As stated in [23], criteria should be analysed under uncertain conditions when working in vague contexts. Additionally, after stating that decision-making processes are human activities mainly accomplished in uncertain environments, in [24] it is emphasised as the crisp DEMATEL can reflect information only in a partial way. The authors consider the usefulness of applying fuzzy theory to extend the traditional method, so that judgements of preference can be translated into fuzzy numbers, after having been expressed by decision-makers through the adoption of a specific fuzzy linguistic scale.

From that angle, the author of [25] agrees with the fact that making use of fuzzy numbers minimises subjective bias, and for this reason, the fuzzy DEMATEL has to be preferred to the traditional crisp version when it comes to real-world applications. After presenting a literature review related to the various fields of fuzzy DEMATEL application, the author applies this methodology to determine those critical aspects having a major impact on local sustainable development through adaptive reuse projects. The authors in [26] also highlight difficulties in making decisions in a fuzzy environment, especially when complex selection criteria are involved. The authors propose fuzzy DEMATEL to determine the most influential factors when evaluating/selecting suppliers, finding that the aspect of financial stability has the highest impact on project implementation. Additionally, in [27], use is made of fuzzy DEMATEL to design a formal framework to use as a driver during the process of business strategy formulation. The authors also stress that the integration with other methodologies is useful to overcome subjectivity of evaluations, and to generally optimise final results of analyses. With regard to the field of critical infrastructures, in [28], it is proposed a hybrid MCDM approach based on fuzzy DEMATEL for failure risk assessment to capture the dynamic nature of opinions provided by a team of experts.

To the best of the authors' knowledge, fuzzy DEMATEL has been scarcely applied in the sector of water network management so far. Water networks are really complex systems made of many interconnected elements, such as tanks, pumps, valves, treatment facilities, and hundreds or even thousands of kilometres of underground pipes [29]. Critical components of networks are characterised by the presence of strong degrees of interdependence, which have a huge impact on the quality of the final service, especially when it comes to minimising operation failures. For this main reason, not only does a fuzzy DEMATEL-based application appear suitable to dealing with the type of stated problem, but it also may represent a powerful approach to fuel the process of OSP.

This paper suggests a novel way to face the OSP problem, based on a new modified version of the traditional fuzzy DEMATEL. Our proposal addresses two main issues: (1) Reducing the huge amount of time often spent during the stage of collection of expert evaluations; and (2) making evaluations as objective as possible, despite that they are represented by fuzzy numbers. To pursue this twofold objective, we herein propose to replace the input matrices of expert linguistic assessments with a single input matrix of linguistic assessments related to a suitable quantitative parameter that expresses the degree of influence between pairs of elements. Even though such a new development is herein applied to the OSP process in a WDN, we claim that it can be extended to other kinds of complex problems.

The paper is structured as follows. After this introduction, including some literature reviewing and stating of the significance of the problem for the water supply field, Section 2, devoted to materials and methods, provides a concise description of the elements involved in the problem under analysis and presents the novel approach we propose, aimed at getting the final ranking of nodes showing those most convenient for hosting sensors. Section 3 provides a numerical example in which the proposed approach is applied, for exemplification purposes, to a very small WDN of the benchmark literature, whereas Section 4 shows the results for a larger network, including comparisons with two optimisation-based OSP methods. Lastly, Section 5 gives the conclusions and raises likely future developments of research.

2. Materials and Methods

This section describes the methodologies and the elements involved we propose to apply in water network monitoring. The section is divided into three sub-sections: sensitivity matrix for leak detection, redundancy analysis for optimal sensor placement, and DEMATEL-based approach to establish interdependence among sensors without the need for reliance on expert judgements. In this last regard, we propose a simple but effective modification within the framework of the traditional fuzzy DEMATEL procedure, specifically related to the step of input data collection, with the purpose of achieving much more objective results by reducing the uncertainty derived from the collection of human judgements.

2.1. Sensitivity Matrix for Leak Detection

Installing sensors in the network to find anomalies requires the identification of strategical points, which should be as sensitive as possible to anomalies. Considering normal operation, the pressure at a node i at a time step t is denoted P_j^N. If an anomaly (e.g., a leak) is simulated as an increase of the demand at a given node i, the sensitivity at node j related to that anomaly at node i can be written as

$$s_{i,j} = \frac{P_j^N - P_j^i}{q_i},$$

(1)

where P_j^j is the pressure at node j at time step t, under the anomaly occurring at node i; q_i is the leakage flow at node i.

The sensitivity matrix is calculated by simulating leaks at all nodes in the network. Row i of this matrix expresses the sensitivity of column nodes j to leaks q_i at node i.

Several works have been proposed in the literature to model leakage in water networks [30,31]. In general, leakage can be modelled as a nonlinear function of pressure. The software Epanet2.0 [32], used in this research, models a leak through an emitter, and the flow is written as:

$$q_i = \beta \cdot P_i^\alpha,$$

(2)

where β is the emitter coefficient, and α is the emitter exponent. The values of α and β depend on the leakage geometry, external environment, and other parameters. In this work, based on the the well-known orifice equation, the value of α is defined as 0.5. The emitter coefficient is discussed in the case study section.

Given a comprehensive representation of the network, especially focusing on its nodes, we assume that each node can potentially host a sensor, and proceed by quantitatively calculating the degree of interdependence between sensors in pairs. This approach aims to identify those nodes exhibiting great interdependence with the others, and thus, of most strategic value. Placing sensors in those nodes rather than in others would actually increase the control capability of the entire network.

Within this perspective, we propose a MCDM approach based on a modified fuzzy DEMATEL technique, presented later on.

2.2. Redundancy Analysis for Optimal Sensor Placement

For optimal sensor placement, not only the most sensitive nodes should be monitored, as it is also important to maximise the coverage of the sensor network. In general, the more spread-out the sensors are, the higher the coverage. In this sense, a joint analysis of sensitivity and entropy can help improve the final sensor network.

From a physical approach, entropy is a property that measures the order/disorder level in a system. Mathematically, the entropy $H(X)$ can be calculated as the product of the mass probability function $p(x)$ of a variable X times the logarithm of its inverse:

$$H(X) = \sum_{x \in X} p(x) \cdot \ln \frac{1}{p(x)}. \tag{3}$$

Considering the sensitivity matrix S composed by the elements $s_{i,j}$ (1), and following the proposal of [7], the function $p(x)$ is written as:

$$p(x) = \frac{a_i}{\sum_{i=1}^{n} a_i}, \tag{4}$$

where

$$a_i = \max s_i, \tag{5}$$

and s_i is the ith row of matrix S.

In this sense, the entropy is calculated based on maximal sensitivity for a given leakage level.

Anomalies occurring in the network can be observed by one sensor and not by others. This is an important point for optimal sensor placement to optimise the coverage of the sensor network. The conditional entropy $H(Y|X)$ has been used to measure the redundancy of data and was applied to sensor placement as presented by [33]. The conditional entropy represents the remaining entropy of a variable Y given the entropy of another variable X.

For sensor placement, in [33], it is shown that the increase of the total entropy leads to a wider coverage of the network. The increase of the total entropy can be reached by maximising the conditional entropy, expressed as:

$$H(Y|X) = \sum_{x \in X, y \in Y} p(x,y) \cdot \ln \frac{p(x)}{p(x,y)}, \tag{6}$$

$$H(Y|X) = - \sum_{x \in X, y \in Y} p(x,y) \cdot \ln p(x,y) + \sum_{x \in X, y \in Y} p(x,y) \cdot \ln p(x), \tag{7}$$

$$H(Y|X) = H(X,Y) + \sum_{x \in X} p(x) \cdot \ln p(x) = H(X,Y) - H(X). \tag{8}$$

A new matrix can be written, where the maximal sensitivity is used to calculate the probability function $p(x)$ (Equation(4)) and then the conditional entropy (Equation (8)). The new matrix of conditional entropy is used to measure the influence of setting a new sensor in the network. Or, in other words, the influence of a monitoring node on the others.

2.3. DEMATEL-Based Approach to Establish Interdependencies Among Sensors

In [34], the DEMATEL procedure is defined as an ad hoc approach transforming relations existing among causes and effects of elements into an intelligible system model. Of course, as in the solution of most problems, it is necessary to accomplish a previously detailed study of the problem under analysis, to clearly define the general objective and the main elements to be taken into account. This accomplished, the traditional (crisp) DEMATEL method follows a procedure consisting on a number of steps. These steps are conceptually recalled here (readers can study them further, for example, in [12], among many other sources).

- Collecting in non-negative matrices the judgements provided by the experts about the influence of one element over another; one matrix per expert.
- Aggregating those matrices into a single one, called direct relation matrix (DRM).
- Calculating the total relation matrix (TRM) by normalising first and then suitably manipulating the DRM so as to aggregate indirect influences.
- Drawing an influential relation chart to visually identify causal relationships among the considered elements.
- Deriving the final ranking of elements according to their prominence, which gives the global impact each element has over the others.

In the next subsections we shortly describe first the fuzzy DEMATEL procedure, and then introduce the modification we propose within the structure of the method, and justify the associated advantages.

2.3.1. Fuzzy DEMATEL

As already underlined, the fuzzy version of DEMATEL is more suitable than the crisp version to reduce uncertainty, and get more reliable results. We describe now the steps to implement the method as it exists in the literature. After having highlighted the general objective of the decision-making problem and the elements to be evaluated, and properly chosen the group of experts, the procedure is the following.

1. Defining the fuzzy linguistic scale that will be used to assess the elements belonging to the system. Judgments must be collected by pairwise comparing all the elements to express the influence of one element, i, over another, j, and vice versa. To such an aim, in [21], it is defined the linguistic variable "influence" through five linguistic terms of evaluation, each one associated to a positive triangular fuzzy number (TFN) (a_{ij}, b_{ij}, c_{ij}). TFNs expressing those evaluations are given in Table 1.

Table 1. Fuzzy linguistic scale for the linguistic variable "influence".

Linguistic Evaluation	Corresponding TFN
No Influence (NI)	$(0, 0, 0.25)$
Low Influence (LI)	$(0, 0.25, 0.5)$
Medium Influence (MI)	$(0.25, 0.5, 0.75)$
High Influence (HI)	$(0.5, 0.75, 1)$
Extreme Influence (EI)	$(0.75, 1, 1)$

2. Aggregating judgements attributed by decision makers, and defuzzifying the collected assessments to get the crisp DRM. Among the wide range of defuzzification methods in the literature, in [21] it is suggested making use of the converting fuzzy data into crisp scores (CFSC) algorithm, introduced by [35]. This is a five-step procedure for deriving a single and aggregated crisp DRM, $D = (d_{ij})$ (a squared $n \times n$ matrix), from the k matrices of input (k being the number of involved experts), each one containing the TFNs expressing the linguistic assessments provided by the experts.

3. Normalising the obtained DRM and calculating the TRM. The normalised DRM, $Z = (z_{ij})$, can be obtained by means of Equations (9) and (10):

$$Z = s \times D, \tag{9}$$

s being a positive number slightly smaller than

$$\min\left(\frac{1}{\max_{1\leq i\leq n}\sum_{j=1}^{n}d_{ij}}, \frac{1}{\max_{1\leq j\leq n}\sum_{i=1}^{n}d_{ij}}\right). \tag{10}$$

After deriving matrix Z, it is possible to proceed to the calculation of the TRM, $T = (t_{ij})$, as follows:

$$T = Z(I - Z)^{-1}, \tag{11}$$

I being the identity matrix. This matrix represents the build-up of mutual direct and indirect effects among elements, since T, being the sum of all the powers of Z, reflects both direct and indirect effects among elements (note that the series of powers of Z is convergent (see, for example, [36]), since, because of Equation (10), the spectral radius of Z is smaller than 1).

4. Building the relational chart on the plane "prominence" (horizontal axis) versus "relation" (vertical axis). Values of prominence $(A + B)$ and relation $(A - B)$ can be derived from matrix T, by calculating the sums of the rows, A, and the sums of the columns, B:

$$A = \sum_{j=1}^{n} t_{ij}, \tag{12}$$

$$B = \sum_{i=1}^{n} t_{ij}. \tag{13}$$

The resulting mapping represents the core of the methodology, since just by observing the positions of the elements in the four quadrants of the plane, it is possible to establish which elements have: (i) high prominence and high relation; (ii) low prominence and low relation; (iii) high prominence and low relation; and (iv) low prominence and high relation. This distinction is very useful to understand how interdependence among the elements is organised, and thus to establish future lines of intervention. Moreover, relations among elements are visually represented by means of arrows: two elements are linked by an arrow if the corresponding value of the TRM overcomes a given threshold, herein calculated by averaging all the values of the TRM [37].

5. Ordering in a decreasing way the elements of the decision-making problem, according to their corresponding values of prominence, to obtain a structured ranking. The "prominence" of an element indicates how much it influences the others, thereby providing a global measurement of its importance. Values of "relation" are instead useful to cluster factors into groups of causes or effects. If the "relation" value corresponding to an element is positive, that means that it has to be considered as a cause, while as an effect otherwise.

2.3.2. Modified Fuzzy DEMATEL: Improvements and Advantages

After the above preliminaries, this subsection presents the novel modified fuzzy DEMATEL, which will be applied to solve a simple case by example, and then to address a real-world case study.

As already stressed throughout the paper, we apply the method to the specific OSP problem. However, the application can be extended to use cases of other nature.

The modification proposed regards the first two steps of the procedure sketched in the previous section, while keeping invariant the remaining last three steps, which are actually shared also by the crisp version of the method. Specifically, we propose to bond the linguistic variable "influence" with a measurable parameter, quantitatively expressing the degree of interconnection among the network elements. In other terms, once defining and numerically calculating this parameter for each pair of elements, we propose to fix five numerical intervals, corresponding to the linguistic assessments and related TFNs of Table 1.

The flowchart of Figure 1 provides a detailed description of the steps of the new procedure. We summarise next the main advantages derived from our modified fuzzy DEMATEL.

First of all, the first step of the traditional procedure requires one to undertake a long process of feedback exchange with as many experts as possible, in order to accomplish a reliable acquisition of data. Each expert is asked to fill in a non-negative input matrix, providing subjective evaluations about the degree of influence between pairs of elements. It is evident as this stage may be highly time-consuming and scarcely precise. Moreover, experts can pairwise compare just a limited number of elements because they may doubt some evaluations. Obviously, it is nonsense to ask someone to pairwise compare hundreds of elements. This is the case with many real, complex problems involving a plethora of factors, which cannot be reduced if effective decisions have to be made. The traditional DEMATEL-based approach cannot be applied in such cases. Instead, our modified fuzzy DEMATEL may take into account very large sets of elements, since linguistic assessments are directly correlated to the numerical values taken by the chosen parameter of interest, according to which the input matrix can be easily compiled.

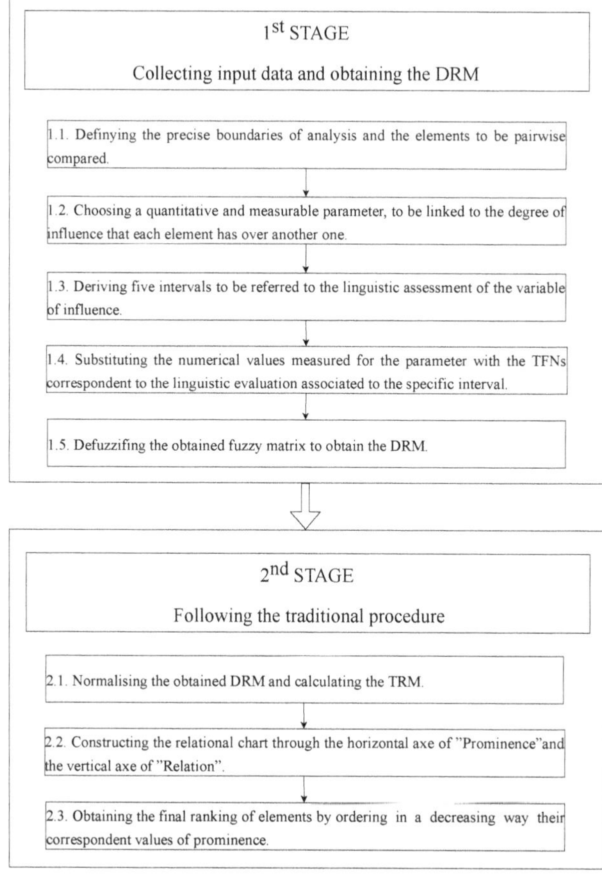

Figure 1. Flowchart of the modified fuzzy DEMATEL procedure.

Moreover, our solution permits one to better manage vagueness, since the collected data refer to a measurable parameter, not to personal opinions of experts, and the consequent use of fuzzy numbers further reduces uncertainty due to measurement errors.

Lastly, our proposal gives back directly a single fuzzy DRM, which will require just a simple operation of defuzzification, without aggregating data coming from many matrices of input issued by many experts.

Once this single matrix is defuzzified and normalised, the application continues through the same steps of the traditional method (from the TRM till the final ranking of elements and their graphical representation).

3. Numerical Example

In this section we apply the proposed procedure, as an example, to a very small WDN. Using this small network means indeed to deal with a small number of elements to be evaluated, and then with a small number of matrices. This enables us to show the calculations of our procedure step-by-step.

The small network used as the numerical example is known as a two-loop network [38] (Figure 2). This network has six junctions, eight pipes, and one reservoir. Classically, the network is used as a benchmark for optimal design in water distribution systems. For this example, the optimally designed network is used, and a demand curve with residential features has been added. This allows the simulation of the network for 24-h. Leaks are created using emitters node by node. An emitter coefficient equal to 1 has been used, in accordance with [39], which investigated the effects of the emitter coefficient on different geometries and hydraulic head loads to simulate leaks. Observe that the authors established an interval to simulate single leaks varying from 0.5 to 8.

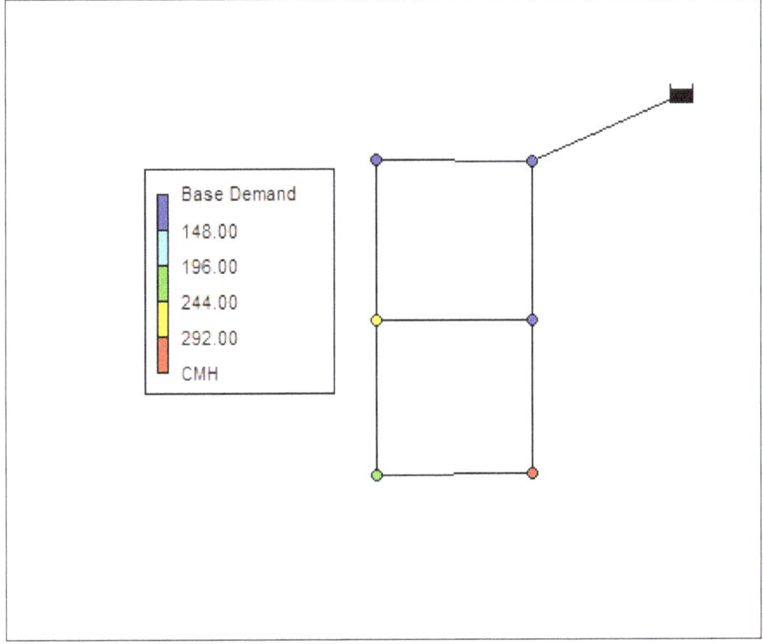

Figure 2. Topology and nodal demand for the 2-loop network.

Considering the size of the network, certainly only one sensor should be installed. Given the simplicity of the problem we only maximise the sensitivity, thereby using just the sensitivity matrix as the input for the fuzzy DEMATEL.

Table 2 details the linguistic evaluations of influence (associated to the TFNs of Table 1) referring to the two-loop network nodes, each one possibly hosting a sensor. Each element has been codified as N_i $(i = 1, \ldots, 6)$. Tables 3 and 4 respectively present the defuzzified DRM and the corresponding

TRM, this last one also presenting the final ranking of elements. For ease of replication, we specify that the graded mean integration approach has been applied to get the crisp values d_{ij} of the DRM matrix:

$$d_{ij} = \frac{a_{ij} + 4b_{ij} + c_{ij}}{6}. \tag{14}$$

The obtained results shown as the nodes occupying the first positions of the ranking (N_6, N_5, N_4) are more suitable to host sensors because they have higher associated sensitivity. By assuming this condition, the monitoring capability can be enhanced in the considered network. From the hydraulic point of view, these nodes can be identified as those presenting lower pressure during the hydraulic simulations.

Figure 3 shows the final chart graphically showing interdependencies. As it is possible to note, the first three nodes of the ranking are in the first quadrant, being characterised by both high prominence and relation.

Table 2. Linguistic evaluations of input.

Elements	N_1	N_2	N_3	N_4	N_5	N_6
N_1	NI	NI	NI	NI	NI	NI
N_2	NI	NI	NI	NI	NI	NI
N_3	NI	NI	NI	LI	LI	LI
N_4	NI	NI	LI	NI	MI	MI
N_5	NI	NI	MI	MI	NI	HI
N_6	NI	LI	MI	MI	HI	NI

Table 3. Defuzzified direct relation matrix (DRM).

Elements	N_1	N_2	N_3	N_4	N_5	N_6
N_1	0.042	0.042	0.042	0.042	0.042	0.042
N_2	0.042	0.042	0.042	0.042	0.042	0.042
N_3	0.042	0.042	0.042	0.250	0.250	0.250
N_4	0.042	0.042	0.250	0.042	0.500	0.500
N_5	0.042	0.042	0.500	0.500	0.042	0.075
N_6	0.042	0.250	0.500	0.500	0.750	0.042

Table 4. Total relation matrix (TRM) and final ranking.

Elements	N_1	N_2	N_3	N_4	N_5	N_6	$A+B$	$A-B$	Ranking
N_1	0.027	0.034	0.065	0.065	0.072	0.072	0.667	0.000	N_6
N_2	0.027	0.034	0.065	0.065	0.072	0.072	1.026	−0.359	N_5
N_3	0.050	0.088	0.250	0.341	0.379	0.379	3.723	−0.751	N_4
N_4	0.068	0.132	0.490	0.400	0.643	0.643	4.612	0.139	N_3
N_5	0.081	0.164	0.681	0.681	0.583	0.837	5.616	0.436	N_2
N_6	0.083	0.241	0.686	0.686	0.842	0.588	5.715	0.535	N_1

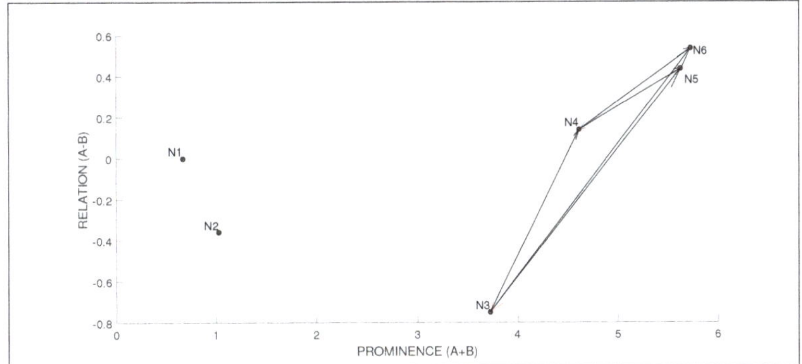

Figure 3. Relational chart.

4. Case Study

In this section, we apply the proposed approach to a moderate-size real-world water network, to check its effectiveness and applicability.

The proposed methodology is applied to the JYN network [40]. The network is composed by 300 nodes and two reservoirs. The mean inlet flow is around 2800 L/s. Figure 4 presents the topology of the network and the mean pressure at the nodes.

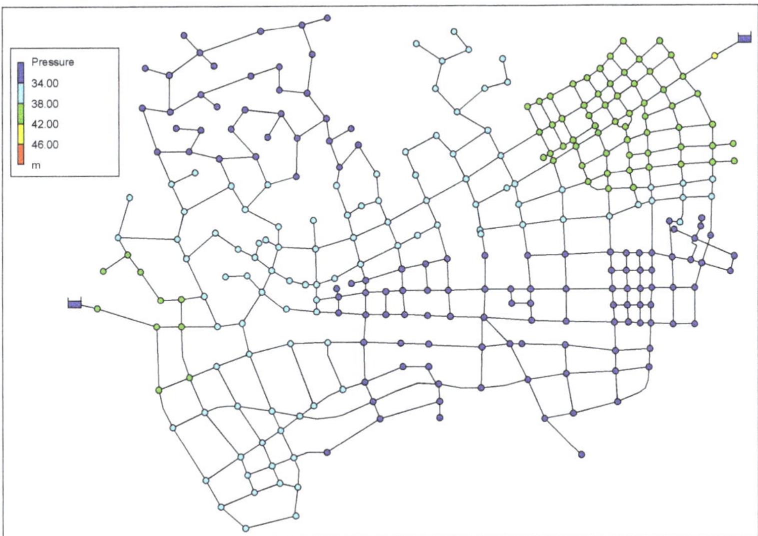

Figure 4. Topology and nodal mean pressure of JYN network.

During the day, the nodal demand varies. This fact makes the calculation of the sensitivity matrix hard. This complexity does not come from the computational effort, but because of the choice of the time step used. To avoid this decision, an extended period simulation is conducted, and the sensitivity time series is represented as a fuzzy number in the format $(\mu - \sigma, \mu, \mu + \sigma)$, where μ is the mean value of $s_{i,j}$ and σ is the standard deviation. For leakage simulations, an emitter coefficient equal to 1 is adopted as discussed on the numerical example.

For a better understanding of the conditional entropy effect, the sensitivity and the conditional entropy matrices are presented in Figure 5a,b. The sensitivity parameter concentrates the nodes with highest scores around the node 300 (north-west side of the network), while the conditional entropy allows concentrations to be more scattered, giving higher scores to other nodes. A comparison between both can be drawn from this joint Figure 5.

Using solely the sensitivity matrix for sensor placement can lead to a concentration of nodes in zones of greater sensitivity. Despite the network sensitivity being maximised in this situation, there is no guarantee of good coverage. In contrast, the conditional entropy matrix, where the probabilities are calculated using the sensitivity matrix, as explained in Section 2.2, distributes the information along other sensitive zones, thereby guaranteeing improved scattering of sensors.

This can be achieved with the use of the fuzzy DEMATEL algorithm applied to the conditional entropy matrix. Using this algorithm enables us to obtain the rankings of nodes and identify those with lower influence on the others. This ranking is used to select a set of nodes to be monitored. The main interest of the methodology comes from combining the selection of high sensitivity nodes, based on the sensitivity matrix, with the lowest influencing nodes; namely, those minimising redundancy.

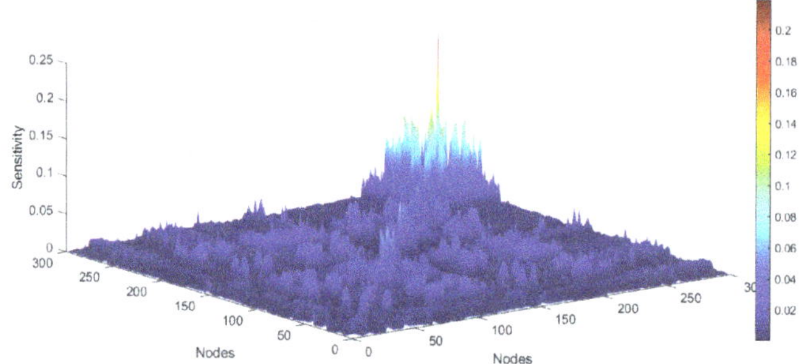

(a) Sensitivity index for JYN's network

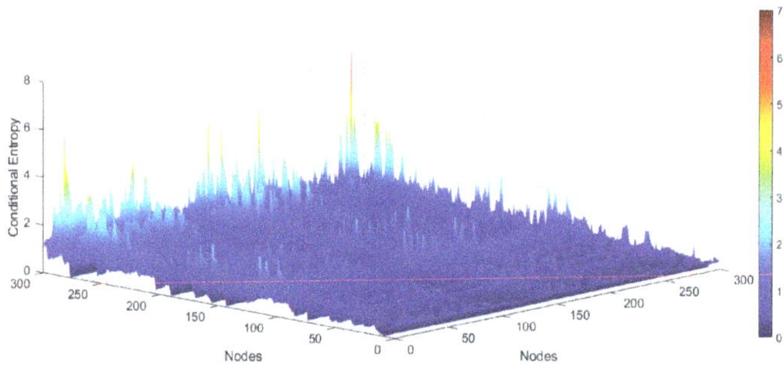

(b) Conditional entropy for JYN's network

Figure 5. Comparison between sensitivity and conditional entropy for JYN's network. (a) Sensitivity index for JYN's network; (b) conditional entropy for JYN's network.

Considering the solution ranking obtained from fuzzy DEMATEL, Figure 6 presents, for the sake of simplicity, a scenario for just four sensors and their location in the water network.

An interesting property related to the hydraulic conditions of the network is that all the sensors are installed in low pressure zones. Usually, low pressure zones are more sensitive to changes in the network, so they have greater sensitivity than other zones. To check the improvement obtained from the use of the conditional entropy with respect to the performance using just the sensitivity matrix, Figure 7 presents the application of the fuzzy DEMATEL method using just the fuzzy sensitivity matrix, for the same scenario with four sensors. The concentration of sensors in the lowest pressure zone of this network can be observed. This zone is also the one identified as the most sensitive by Equation (1). The use of the conditional entropy provides a more widespread distribution of sensors (see Figure 6). Of course, the low pressure zone (excessively) identified by just the sensitivity matrix is not missed when the conditional entropy is used. In addition, this methodology avoids the redundancy of information derived from the concentration of sensors in that low pressure zone, as illustrated in Figure 7, with a more widespread distribution of the four sensors.

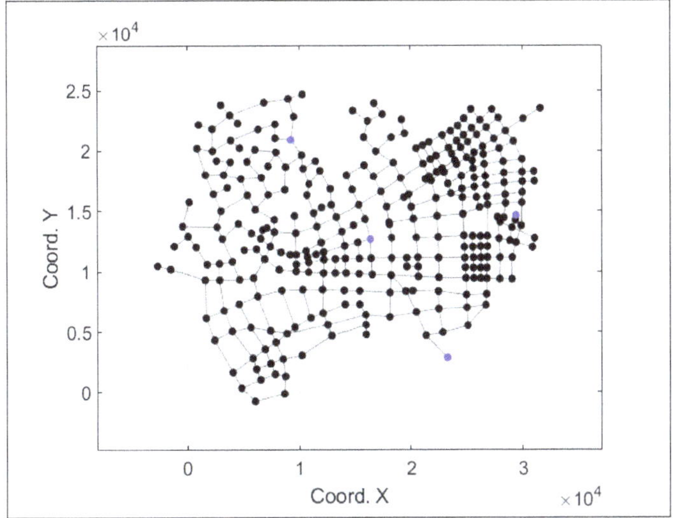

Figure 6. Layout of a sensor network with four sensors using the conditional entropy as input for fuzzy DEMATEL.

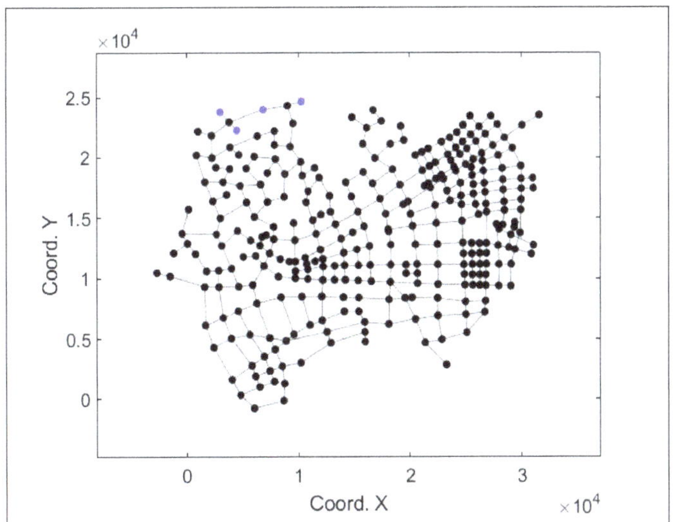

Figure 7. Layout of a sensor network with four sensors using only the sensitivity as input for Fuzzy DEMATEL.

To further evaluate the difference between both solutions, the global sensitivity, which means the sum of the greatest sensitivity of each sensor, and the total entropy of the system, are calculated. Global values for these two metrics are calculated using only the columns of the sensitivity matrix corresponding to the sensors' positions. Using the conditional entropy as input for the fuzzy DEMATEL method, the global sensitivity equals 4.1985, while using as input the sensitivity matrix, the global sensitivity equals 5.4381. In terms of the total entropy, the conditional entropy leads to a value of 5.2990, while just the sensitivity matrix results in 4.7560. In both cases, the conditional entropy-based approach outperforms the results of the sensitivity matrix alone.

To compare the proposed methodology with two classical optimisation approaches, agent swarm optimisation [41] is applied as an optimisation engine.

Among the various published papers presenting optimisation-based approaches for sensor placement, we consider here [7], which applies a bi-objective optimisation, maximising the sensitivity of the sensors' network and the entropy of the system. In [7], the sensitivity is defined as in Equation (1), and the entropy as in Equation (3). Given a solution for the sensor placement problem, $X = (x_1...x_k)$, it is possible to take the corresponding k-columns of the sensitivity matrix S. Using this new sensitivity submatrix S_k, it is possible to identify, for each simulated leak, the most sensitive sensor, as in Equation (5). Then the sensors' network sensitivity and entropy objective functions are calculated as:

$$F_1 = \sum_{x \in X} a(x). \tag{15}$$

$$F_2 = \sum_{x \in X} p(x) \cdot \ln \frac{1}{p(x)}. \tag{16}$$

Finally, to apply single objective optimisation, the authors of [7] combine normalised values of F_1 and F_2.

IngeniousWare®, in a modification of the methodology in [7], created a plugin for WaterIng©, software for optimising pressure sensor placement in water systems. In this case, the entropy based function is modified to guarantee a better spread of sensors in real networks (unpublished results).

The methodology of [7] and the commercial software WaterIng© are used, fixing a number of four sensors, as in the case study solved with the fuzzy DEMATEL approach. With the methodology of [7], the total sensitivity of the sensors' network results in 4.2261 and the total entropy in 5.5912, which are similar results to those obtained with the fuzzy DEMATEL approach. This is because, in general, the sensors are placed in the same region in both cases. WaterIng© produces a sensitivity value of 5.4920 and a total entropy of 5.3195. This approach produces better values for both sensitivity and entropy. Comparing the four approaches, the modified entropy approach proposed by IngeniousWare® manages to get the best values for both objective functions. Nevertheless, the fuzzy DEMATEL approach of this paper gives results which are comparable with [7], but without the need of running any optimisation process. Table 5 summarises the results obtained in this research.

Table 5. Comparison table for the four considered approaches.

Method	Global Sensitivity	Global Entropy
Fuzzy Dematel Conditional Entropy	4.1985	5.4381
Fuzzy Dematel Sensitivity	5.2990	4.7560
Optimisation following [7]	4.2261	5.5912
Optimisation plugin WaterIng©	5.4920	5.3195

Let us finally emphasise that the relation between the number of sensors and the global sensitivity is a good indicator for decision makers about the number of sensors to be installed. Figure 8 presents the increase of sensitivity with the increase of the number of sensors in the network. The bigger increases for small numbers of sensors, and an asymptotic trend for larger numbers of sensors may be verified. The graph in Figure 8 also shows how by increasing the number of sensors beyond a certain point does not entail a corresponding relevant increase in monitoring.

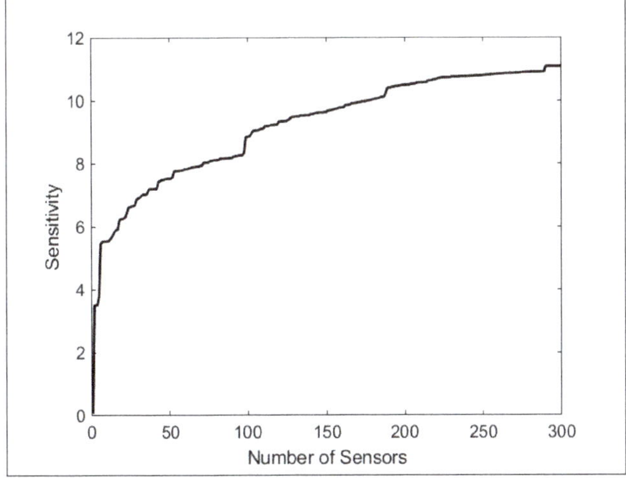

Figure 8. Sensitivity vs. number of sensors for the network under study.

5. Conclusions

Given the fundamental role played by water networks in daily life and business activities, the present research focuses on suitably operating, managing, and maintaining these assets. In this regard, the problem of optimal sensor placement has been faced with the aim of improving the operations of monitoring and the control of the networks.

Most used methods calculate the sensitivity matrix for a specific simulation time, usually the highest consumption time. The fuzzy DEMATEL handles extended period simulations, suitably transformed into fuzzy numbers. The approach enables us to have information for many horizon simulations, making sensor placement more robust. Other possible applications of fuzzy DEMATEL are the simulations of several leakage scenarios with different emitter coefficients. In that case, the sensitivity matrix could be built considering from small to large leaks.

The use of the sensitivity matrix to generate a conditional entropy helps guarantee the spread of sensors in the network. A clear improvement is found, as observed from the comparisons between results from global sensitivity and global entropy of the sensor network, as shown in the case study.

The use of fuzzy DEMATEL for optimal sensor placement helps water companies identify the most suitable monitoring points. Using conditional entropy, the spread of sensors is guaranteed by using the last positions of the DEMATEL ranking. One important positive point of this approach is the absence of optimisation, which usually requires prior knowledge of the number of sensors to be installed. With the presented fuzzy DEMATEL approach, the sensors' network can be implemented in steps, without requiring new simulations.

The fuzzy DEMATEL approach presented in this paper produces similar results to the ones obtained with the optimisation algorithm in [7]. Both methodologies use sensitivity and entropy to place sensors in the network. The main advantage of fuzzy DEMATEL hinges on the final rank obtained: this rank enables placing new sensors without performing new simulations. In optimisation-based approaches, adding new sensors requires new simulations. Incidentally, the approach of WaterIng© finds a solution with better results for both sensitivity and entropy.

Possible future developments of the presented research may regard further investigations about how to choose other quantitative parameters to collect input data. For example, the proposed modified version of the fuzzy DEMATEL may be integrated with other MCDM methodologies to identify a suitable set of parameters, all related to relationships of influence among the considered factors. The selection of the number of sensors should also be further investigated, so as to provide the utilities with a Pareto-like solution enabling them to select the most appropriate number, an aspect not treated in this paper. Such a joint process made of (multi-objective) optimisation plus MCDM methods can be a way of identifying the optimal number of sensors, using the ranking provided for the fuzzy DEMATEL methodology herein developed.

Author Contributions: Conceptualisation, S.C. and J.F.-C.; methodology, B.M.B. and S.C.; software, B.M.B. and I.M.; validation, S.C. and J.F.-C.; formal analysis, J.I. and I.M.; writing—original draft preparation, J.F.-C., B.M.B., and S.C.; writing—review and editing, J.I.; supervision, J.I. and I.M. All authors have read and agreed to the published version of the manuscript.

Funding: This research has been partially supported by the CNPq grant with number 156213/2018-4.

Conflicts of Interest: The authors declare no conflict of interest.

References

1. Li, J.; Wang, C.; Qian, Z.; Lu, C. Optimal sensor placement for leak localization in water distribution networks based on a novel semi-supervised strategy. *J. Process Control* **2019**, *82*, 13–21. [CrossRef]
2. Pérez, R.; Puig, V.; Pascual, J.; Quevedo, J.; Landeros, E.; Peralta, A. Methodology for leakage isolation using pressure sensitivity analysis in water distribution networks. *Control Eng. Pract.* **2011**, *19*, 1157–1167. [CrossRef]
3. Boatwright, S.; Romano, M.; Mounce, S.; Woodward, K.; Boxall, J. Optimal Sensor Placement and leak/burst localisation in a water distribution system using spatially-constrained inverse-distance weighted interpolation. *EPiC Ser. Eng.* **2018**, *3*, 282–289.
4. Blesa, J.; Nejjari, F.; Sarrate, R. Robust sensor placement for leak location: Analysis and design. *J. Hydroinf.* **2015**, *18*, 136–148. [CrossRef]
5. Steffelbauer, D.B.; Fuchs-Hanusch, D. Efficient sensor placement for leak localization considering uncertainties. *Water Resour. Manag.* **2016**, *30*, 5517–5533. [CrossRef]
6. Yoo, D.; Chang, D.; Song, Y.; Lee, J. Optimal Placement of Pressure Gauges for Water Distribution Networks Using Entropy Theory Based on Pressure Dependent Hydraulic Simulation. *Entropy* **2018**, *20*, 576. [CrossRef]
7. De Schaetzen, W.; Walters, G.; Savic, D. Optimal sampling design for model calibration using shortest path, genetic and entropy algorithms. *Urban Water* **2000**, *2*, 141–152. [CrossRef]
8. Cugueró-Escofet, M.À.; Puig, V.; Quevedo, J. Optimal pressure sensor placement and assessment for leak location using a relaxed isolation index: Application to the Barcelona water network. *Control Eng. Pract.* **2017**, *63*, 1–12. [CrossRef]
9. Perelman, L.S.; Abbas, W.; Koutsoukos, X.; Amin, S. Sensor placement for fault location identification in water networks: A minimum test cover approach. *Automatica* **2016**, *72*, 166–176. [CrossRef]
10. Fontela, E.; Gabus, A. *DEMATEL, Innovative Methods, Technical Report No. 2, Structural Analysis of the World Problematique*; Battelle Geneva Research Institute: Geneva, Switzerland, 1974.
11. Fontela, E.; Gabus, A. *The DEMATEL Observe*; Battelle Institute, Geneva Research Center: Geneva, Switzerland, 1976.
12. Carpitella, S.; Carpitella, F.; Certa, A.; Benítez, J.; Izquierdo, J. Managing Human Factors to Reduce Organisational Risk in Industry. *Math. Comput. Appl.* **2018**, *23*, 67. [CrossRef]
13. Addae, B.A.; Zhang, L.; Zhou, P.; Wang, F. Analyzing barriers of Smart Energy City in Accra with two-step fuzzy DEMATEL. *Cities* **2019**, *89*, 218–227. [CrossRef]
14. Dalvi-Esfahani, M.; Niknafs, A.; Kuss, D.J.; Nilashi, M.; Afrough, S. Social media addiction: Applying the DEMATEL approach. *Telemat. Inf.* **2019**, *43*, 101250. [CrossRef]
15. Quezada, L.E.; López-Ospina, H.A.; Palominos, P.I.; Oddershede, A.M. Identifying causal relationships in strategy maps using ANP and DEMATEL. *Comput. Ind. Eng.* **2018**, *118*, 170–179. [CrossRef]

16. Nilashi, M.; Samad, S.; Manaf, A.A.; Ahmadi, H.; Rashid, T.A.; Munshi, A.; Almukadi, W.; Ibrahim, O.; Ahmed, O.H. Factors influencing medical tourism adoption in Malaysia: A DEMATEL-Fuzzy TOPSIS approach. *Comput. Ind. Eng.* **2019**, *137*, 106005. [CrossRef]
17. Zhang, L.; Sun, X.; Xue, H. Identifying critical risks in Sponge City PPP projects using DEMATEL method: A case study of China. *J. Clean. Prod.* **2019**, *226*, 949–958. [CrossRef]
18. Du, Y.W.; Zhou, W. New improved DEMATEL method based on both subjective experience and objective data. *Eng. Appl. Artif. Intell.* **2019**, *83*, 57–71. [CrossRef]
19. Yazdi, M.; Nedjati, A.; Zarei, E.; Abbassi, R. A novel extension of DEMATEL approach for probabilistic safety analysis in process systems. *Saf. Sci.* **2020**, *121*, 119–136. [CrossRef]
20. Chen, Z.; Ming, X.; Zhang, X.; Yin, D.; Sun, Z. A rough-fuzzy DEMATEL-ANP method for evaluating sustainable value requirement of product service system. *J. Clean. Prod.* **2019**, *228*, 485–508. [CrossRef]
21. Wu, W.W.; Lee, Y.T. Developing global managers' competencies using the fuzzy DEMATEL method. *Expert Syst. Appl.* **2007**, *32*, 499–507. [CrossRef]
22. Zadeh, L.A. Fuzzy sets. *Inf. Control* **1965**, *8*, 338–353. [CrossRef]
23. Mahmoudi, S.; Jalali, A.; Ahmadi, M.; Abasi, P.; Salari, N. Identifying critical success factors in Heart Failure Self-Care using fuzzy DEMATEL method. *Appl. Soft Comput.* **2019**, *84*, 105729. [CrossRef]
24. Lin, K.P.; Tseng, M.L.; Pai, P.F. Sustainable supply chain management using approximate fuzzy DEMATEL method. *Resour. Conserv. Recycl.* **2018**, *128*, 134–142. [CrossRef]
25. Vardopoulos, I. Critical sustainable development factors in the adaptive reuse of urban industrial buildings. A fuzzy DEMATEL approach. *Sustain. Cities Soc.* **2019**, *50*, 101684. [CrossRef]
26. Mirmousa, S.; Dehnavi, H.D. Development of criteria of selecting the supplier by using the fuzzy DEMATEL method. *Procedia-Soc. Behav. Sci.* **2016**, *230*, 281–289. [CrossRef]
27. Acuña-Carvajal, F.; Pinto-Tarazona, L.; López-Ospina, H.; Barros-Castro, R.; Quezada, L.; Palacio, K. An integrated method to plan, structure and validate a business strategy using fuzzy DEMATEL and the balanced scorecard. *Expert Syst. Appl.* **2019**, *122*, 351–368. [CrossRef]
28. Chou, J.S.; Ongkowijoyo, C.S. Hybrid decision-making method for assessing interdependency and priority of critical infrastructure. *Int. J. Disaster Risk Reduc.* **2019**, *39*, 101134. [CrossRef]
29. De Winter, C.; Palleti, V.R.; Worm, D.; Kooij, R. Optimal placement of imperfect water quality sensors in water distribution networks. *Comput. Chem. Eng.* **2019**, *121*, 200–211. [CrossRef]
30. Schwaller, J.; Van Zyl, J. Modeling the pressure-leakage response of water distribution systems based on individual leak behavior. *J. Hydraul. Eng.* **2014**, *141*, 04014089. [CrossRef]
31. Giustolisi, O.; Savic, D.; Kapelan, Z. Pressure-driven demand and leakage simulation for water distribution networks. *J. Hydraul. Eng.* **2008**, *134*, 626–635. [CrossRef]
32. Rossman, L.A. EPANET 2: Users Manual. 2000. Available online: https://epanet.es/wp-content/uploads/2012/10/EPANET_User_Guide.pdf (accessed on 10 Fabruary 20120).
33. Christodoulou, S.E.; Gagatsis, A.; Xanthos, S.; Kranioti, S.; Agathokleous, A.; Fragiadakis, M. Entropy-based sensor placement optimization for waterloss detection in water distribution networks. *Water Resour. Manag.* **2013**, *27*, 4443–4468. [CrossRef]
34. Falatoonitoosi, E.E.; Leman, Z.; Sorooshian, S.; Salimi, M. Decision-making trial and evaluation laboratory. *Res. J. Appl. Sci. Eng. Technol.* **2013**, *5*, 3476–3480. [CrossRef]
35. Opricovic, S.; Tzeng, G.H. Defuzzification within a multicriteria decision model. *Int. J. Uncertain. Fuzz. Knowl.-Based Syst.* **2003**, *11*, 635–652. [CrossRef]
36. Meyer, C. *Matrix Analysis and Applied Linear Algebra*; SIAM: Philadelphia, PA, USA, 2000.
37. Sara, J.; Stikkelman, R.M.; Herder, P.M. Assessing relative importance and mutual influence of barriers for CCS deployment of the ROAD project using AHP and DEMATEL methods. *Int. J. Greenh. Gas Control* **2015**, *41*, 336–357. [CrossRef]
38. Alperovits, E.; Shamir, U. Design of optimal water distribution systems. *Water Resour. Res.* **1977**, *13*, 885–900. [CrossRef]
39. Walski, T.; Bezts, W.; Posluszny, E.T.; Weir, M.; Whitman, B.E. Modeling leakage reduction through pressure control. *J.-Am. Water Work. Assoc.* **2006**, *98*, 147–155. [CrossRef]

40. Zheng, F.; Du, J.; Diao, K.; Zhang, T.; Yu, T.; Shao, Y. Investigating effectiveness of sensor placement strategies in contamination detection within water distribution systems. *J. Water Resour. Plan. Manag.* **2018**, *144*, 06018003. [CrossRef]

41. Montalvo, I.; Izquierdo, J.; Pérez-García, R.; Herrera, M. Water distribution system computer-aided design by agent swarm optimization. *Comput.-Aided Civ. Infrastr. Eng.* **2014**, *29*, 433–448. [CrossRef]

Article

Reducing Impacts of Contamination in Water Distribution Networks: A Combined Strategy Based on Network Partitioning and Installation of Water Quality Sensors

Carlo Ciaponi [1], Enrico Creaco [1,2], Armando Di Nardo [2,3,*], Michele Di Natale [2,3], Carlo Giudicianni [2,3], Dino Musmarra [2,3] and Giovanni Francesco Santonastaso [2,3]

[1] Dipartimento di Ingegneria Civile e Architettura, Università di Pavia, Via Ferrata 3, 27100 Pavia, Italy; ciaponi@unipv.it (C.C.); creaco@unipv.it (E.C.)

[2] Action Group CTRL+SWAN of the European Innovation Partnership on Water, EU, Via Roma 29, 81031 Aversa, Italy; michele.dinatale@unicampania.it (M.D.N.); carlo.giudicianni@unicampania.it (C.G.); dino.musmarra@unicampania.it (D.M.); giovannifrancesco.santonastaso@unicampania.it (G.F.S.)

[3] Dipartimento di Ingegneria, Università degli Studi della Campania "Luigi Vanvitelli", Via Roma 29, 81031 Aversa, Italy

* Correspondence: armando.dinardo@unicampania.it; Tel.: +39-081-501-0202

Received: 21 May 2019; Accepted: 21 June 2019; Published: 25 June 2019

Abstract: This paper proposes a combined management strategy for monitoring water distribution networks (WDNs). This strategy is based on the application of water network partitioning (WNP) for the creation of district metered areas (DMAs) and on the installation of sensors for water quality monitoring. The proposed methodology was tested on a real WDN, showing that boundary pipes, at which flowmeters are installed to monitor flow, are good candidate locations for sensor installation, when considered along with few other nodes detected through topological criteria on the partitioned WDN. The option of considering only these potential locations, instead of all WDN nodes, inside a multi-objective optimization process, helps in reducing the search space of possible solutions and, ultimately, the computational burden. The solutions obtained with the optimization are effective in reducing affected population and detection time in contamination scenarios, and in increasing detection likelihood and redundancy of the monitoring system. Last but most importantly, these solutions offer benefits in terms of management and costs. In fact, installing a sensor alongside the flowmeter present between two adjacent DMAs yields managerial advantages associated with the closeness of the two devices. Furthermore, economic benefits due to the possibility of sharing some electronic components for data acquisition, saving, and transmission are derived.

Keywords: water distribution monitoring; optimal sensor placement; water network partitioning; topological centrality

---

1. Introduction

Installing an efficient monitoring and control sensor system gives the possibility to carry out main tasks on Water Distribution Network (WDN) management and protection. Securing these critical infrastructures is a crucial task for ensuring society's welfare and prosperity. In fact, WDNs are strongly vulnerable to malicious and intentional actions [1] since they are made up of thousands of exposed elements. From a practical and economic point of view, securing all the apparatuses is not feasible. Thus, the design of an effective and cost-effective quality monitoring system represents a crucial management strategy for ensuring the delivery of good quality water to users. Optimal sensor placement becomes a necessary step for satisfactory water quality monitoring systems, also allowing identification of the source contamination [2]. These systems should provide a fast and accurate detection, distinguishing between normal variations and contamination events; furthermore, they should be economical, easy to integrate into network, and reliable [3]. This problem has been extensively studied for the past 20 years and several approaches have been proposed to identify optimal locations of sensors (Byoung et al. (1992) [4] defined the concept of maximum coverage to locate sensors formulating the problem as integer programming problem; using the same objective as the maximum coverage, Kumar et al. (1997) [5] employed a mixed-integer programming method; Watson et al.(2004) [6] used a mixed-integer linear programming model, showing that the problem of sensor placement must simultaneously consider multiple design objectives; Berry et al. (2005) [7] pointed out the difficulty of solving sensor placement by means of integer programming optimization; Ostfeld and Salomons (2004) [8] studied the problem in unsteady conditions using a genetic algorithm framework integrated with EPANET; Uber et al. (2004) [9] used a greedy heuristic solution methodology providing a heuristic (non-optimal) solution procedure scalable to large networks, taking into account uncertainty in threat scenario). The problem received lots of attention especially after the events of 11 September 2001. However, although many research works have been carried out in this field, the challenge of the optimal sensor placement is still open in many aspects, such as identification of optimal sensor locations and evaluation of performance and applicability to real-world scenarios. Models and algorithms for solving this arduous problem include deterministic and stochastic optimization techniques, optimizing one (Kessler et al. (1998) [10] defined the total volume of contaminated water consumed ahead of detection; Ostfeld and Salomons (2005) [11] enhanced previous study by taking into account the randomness of flow rate of the intruded pollutant, stochastic demands, and reaction time of the sensors; Berry et al. (2009) [12] incorporated into a mixed-integer programming formulation the probability of sensor failure) or more objectives (McKenna et al. (2007) [13] demonstrated the importance of considering sensor failure rates showing the trade-off between the sensor detection limit and the number of sensors; Dorini et al. (2008) [14] considered four objectives in the model and used a noisy cross-entropy sensor locator algorithm to find the optimal solution; Huang et al.(2008) [15] considered three objectives in their formulation solved by using a competent genetic algorithm while the contamination events were simulated by a development of Monte Carlo method; Propato and Piller (2006) [16] used a mixed-integer linear program methodology including notions of statistical and uncertainty analysis in the design process; Wu and Walski (2008) [17] combined four objectives into a single objective), such as detection likelihood, expected contaminated water volume, affected population, detection time, and the contaminated population. The interested reader can refer to Hart and Murray (2010) [18,19] for a review of this topic. The optimal sensor placement problem was also dealt with at the Battle of the Water Sensor Networks (BWSN) [20]. The main difficulty is that, given WDN complexity, efficient numerical techniques are needed to support optimal monitoring system design and the huge number of all potential contamination events in a WDN makes the problem computationally intractable (as each of these events is characterized by a different injection location, duration, mass rate, and starting time). Indeed, the optimal sensor placement in a network represents a combinatorial optimization problem that has been proven to be NP-hard [21]. For example, Krause et al. (2008) [22] showed that, using 30 parallel processors, it took 8 days to simulate random contamination events that could occur at 5 min intervals over a 24 h period from any of the 12,527 nodes in a medium-sized distribution network. In recent years, new concepts in sensor network design have been studied; Sankary and Ostfeld (2016) [23] investigated the possibility of adopting a mobile wireless sensor network to wirelessly transmit data to fixed transceivers in real time; Rathi et al. 2016 [24] proposed a novel strategy for the selection of contamination events with associated risk to be used in design of sensor network; Zheng et al. 2018 [25] investigated the characteristics of the sensor placement strategy effectiveness using several metrics, and providing guidance for selecting the most appropriate strategy for the preparedness for contamination events.

On the other hand, the "divide and conquer" concept has recently been gaining attention in the field of WDNs, showing to be one of the most efficient management strategies. The option of dividing large-scale networks into smaller and manageable subsystems, called district metered areas (DMAs), offers undisputable advantages for the monitoring and control of consumption, pressure, leakage, and water resource quality. In the scientific literature, numerous works were dedicated to the design of DMAs. Most of them are based on the application of decomposition algorithms [26,27] based on graph [28–32] and spectral theories [33,34], multi-agent method [35], social network theory [36], modularity index [37–39]. Though being significant contributions to the field, the works mentioned above are mostly focused on DMA design. Therefore, they fail to analyze the positive effects brought by the creation of DMAs to WDN management, for reducing the impacts of potential contamination events.

The global aim of this paper is to provide a general management framework for WDN monitoring, while exploring the benefits of water network partitioning (WNP) for:

- reduction in inauspicious consequences of contamination (both accidental and intentional) in terms of limitation of contaminated areas (direct action);
- optimal placement of quality sensors (indirect action).

While the analysis of the former aspect is presented hereinafter as a follow-up of the work of Ciaponi et al. (2018) [40], the analysis of the latter aspect is entirely novel. In this context, the possibility of installing some or all sensors at boundary pipes will be considered, resulting in a two-fold advantage: numerical, due to the reduction in the research space of possible candidate solutions for sensor installation, and managerial, due to easiness of access and to cost savings for the possibility of sharing some electronical components for data acquisition, saving, and transmission.

In the following sections, first the methodology is presented, followed by the applications to a real case study, testing different scenarios and comparing different sensor locations with four water quality-based parameters, in order to validate the results.

2. Materials and Methods

The methodology used in this work is the combination of two main procedures, used for WNP and sensor placement, respectively. These procedures, both derived from the scientific literature, are described in the following Sections 2.1 and 2.2, respectively. Section 2.3 deals with the postprocessing of the sensor placement solutions obtained in Section 2.2.

2.1. Procedure 1—WNP

According to Perelmann et al. (2015) [41], WNP is carried out in two main phases: (a) *clustering*, in which the optimal shape and size of the clusters are defined by minimizing the number of edge cuts (boundary pipes) and by simultaneously balancing the number of nodes of each cluster, and (b) *dividing*, in which clusters are separated from each other by closing isolation valves at some boundary pipes and installing flow meters at the remaining boundary pipes.

In this work, the clustering layout is obtained exploiting the properties of the normalized Laplacian matrix $\mathbf{L} = \mathbf{D} - \mathbf{A}$, in which $\mathbf{D}$ is the diagonal matrix containing the node degree k_i of each node, and $\mathbf{A}$ is the adjacency matrix. In this matrix, the elements $a_{ij} = a_{ji} = 1$ if nodes n_i and n_j are connected by a pipe; otherwise, $a_{ij} = a_{ji} = 0$. Shi and Malik (2000) [42] demonstrated that through the first C smallest eigenvector of the normalized Laplacian matrix, the relaxed version of the min-cut problem can be solved. In fact, it corresponds to the minimization of the Rayleigh quotient. If C is the number of clusters in which the network must be divided, the first C smallest eigenvectors of the Laplacian matrix are considered and used to create a new matrix $\mathbf{U_{nxC}}$. A k-means algorithm is applied to the rows of $\mathbf{U_{nxC}}$ for grouping the nodes of the network in C clusters. The main trick is to change the representation of the nodes in the eigenspace of the first C eigenvectors, which enhances the cluster-properties of the nodes in such a way that they can be trivially detected in the new representation. The spectral clustering algorithm proved to show a superior performance to other clustering procedures, in that the provided clustering layout features both a well-balanced cluster size and a minimum number of edge cuts [43]. The main spectral clustering steps in the case of a WDN are described by Di Nardo et al.

(2018b) [44]. The graph of the WDN can be considered un-weighted (every connection between the nodes has the same importance, $a_{ij} = a_{ji} = 1$) or weighted (the value $a_{ij} = a_{ji}$ can be related to pipe features, such as diameter d and length l). In the applications of this work, a_{ij} and a_{ji} were set at 1. The optimal number of clusters C (from a topological point of view) in which to subdivide the network is chosen as a function of the number n of nodes, according to the relationship $C_{opt} = n^{0.28}$ [45]. The clustering phase provides the optimal cluster layout and, as a result, the edge-cut set, consisting of a group of N_{ec} boundary pipes between clusters. In correspondence to each boundary pipe, the flow transfer between the adjacent clusters must always be known, if it is larger than zero, in order to make the dividing effective. Therefore, the choice must be made whether either a gate valve must be closed, or a flow meter must be installed in the generic boundary pipe. Following this choice, the sum of closed gate valves (as numerous as N_{gv}) and installed flow meters (as numerous as N_{fm}) must be equal to N_{ec}. Closing gate valves has the effect of reducing the service pressure and, therefore, leakage in the WDN. However, if service pressure falls below the desired threshold value h_{des}, this negatively impacts on WDN reliability. In this work, the trade-off between leakage and WDN reliability was explored through the bi-objective optimization, performed through the NSGAII genetic algorithm [46]. In this optimization, several decisional variables equal to N_{ec} was considered, to encode, inside individual genes, gate valve closure (gene value equal to 1) or flow meter installation (gene value equal to 0) at boundary pipes. The first objective function f_1 to minimize was the daily leakage volume V_l (m³):

$$f_1 = V_l \tag{1}$$

where V_l is calculated as the sum of the temporal integral of the nodal leakage outflows, evaluated as a function of nodal pressure heads through the Tucciarelli et al. (1999) [47] formula.

The second objective function f_2 relates to the global resilience failure index *GRF* index proposed by Creaco et al. (2016) [48] to represent the instantaneous power surplus/deficit conditions of the WDN. In fact, *GRF* is dimensionless and is the sum of the resilience (I_r) and failure (I_f) indices evaluated at the generic instant of WDN operation:

$$GRF = I_r + I_f = \frac{max\left(q_{user}^T H - d^T H_{des}, 0\right)}{Q_0^T H_0 - d^T H_{des}} + \frac{min\left(q_{user}^T H - d^T H_{des}, 0\right)}{d^T H_{des}} \tag{2}$$

where d and q_{user} are the vectors of nodal demands d (m³/s) and water discharges q_{user} (m³/s) delivered to users, respectively, at WDN demanding nodes. In this work, q_{user} was evaluated as a function of d and pressure head h (m) at each node though the pressure-driven formula of Tanyimboh and Templeman (2010) [49], with calibration proposed by Ciaponi et al. (2014) [50]. H and H_0 are the vectors of nodal heads (m) at demanding nodes and sources, respectively. H_{des} is the vector of desired nodal heads, which are the sum of nodal elevations and desired pressure heads h_{des} (m). Finally, Q_0 is the vector of the water discharges leaving the sources. The *GRF* index has the advantage of being within range $[-1, 1]$. Higher values of *GRF* indicate higher power delivered to WDN users and, therefore, higher service pressure. With reference to WDN daily operation, the second objective function f_2 to maximize was calculated with the following relationship, as suggested by Creaco et al. (2016) [48]:

$$f_2 = median(GRF) \tag{3}$$

The choice of the median value of *GRF* is because Creaco et al. (2016) [48] proved it to give a suitable and concise representation of a sequence of operation scenarios in the extended period simulation of the WDN. Both f_1 and f_2 can be calculated by applying a pressure-driven WDN solver (e.g., that of Creaco et al. 2016 [48]). They are mutually contrasting objectives: in fact, as the number of closed gates grows, f_1, which has to be minimized, decreases. At the same time, f_2, which has to be maximized, decreases as well due to the decreasing service pressure. This creates a trade-off between the two objectives, which takes the form of a Pareto front of optimal solutions, that is a group of solutions from which to select the final solution for the partitioning. To this end, additional criteria, such as the partitioning cost or demand satisfaction, can be adopted. In fact, the Pareto front of optimal solutions can be re-evaluated in terms of other functions, such as number N_{fm} of flow meters and demand satisfaction rate I_{ds}. In fact, N_{fm} is a surrogate for the partitioning cost [34], whereas I_{ds} represents the effectiveness of the service to WDN users. The latter index can be calculated as:

$$I_{ds} = \frac{w_d}{w_{tot}} \tag{4}$$

where w_d (m^3) and w_{tot} (m^3) are the delivered water volume and the WDN demand, respectively. Variable w_d can be calculated starting from the temporal integral of the water discharge q_{user} delivered to the users at each node.

2.2. Procedure 2—Optimal Sensor Placement

Let a set S of potential contamination events considered in the analysis, each of which featuring a certain location, starting time, duration, and total mass, be defined. In this context, sensor placement was formulated as a bi-objective optimization problem [51], in which the first objective function is $f_3 = N_{sens}$ (number of installed sensors), as a surrogate for the installation cost for WDN monitoring, while the second objective function is:

$$f_4 = pop = \frac{\sum_{r=1}^{S} pop_r}{S} \tag{5}$$

The objective function f_4 is related to the contaminated population pop_r before the first detection of the generic r-th contamination event. This corresponds to the sum of the inhabitants served by the contaminated nodes and can be evaluated using the EPANET quality solver [52], using an unreactive contaminant. The EPANET quality solver can be applied to the flow field obtained following procedure 1. If the r-th event is not detected, pop_r includes all the nodes crossed by the contamination till the whole contaminant mass leaves the WDN. Though numerous objective functions can be used for the optimal installation of sensors, the population exposed to contamination was chosen as the objective function to minimize along with the number of sensors. This choice was made because, compared to other potential objective functions (such as detection likelihood and sensor redundancy), the population exposed to contamination represents more directly the impact of contamination, which is the most meaningful from the viewpoint of risk assessment and mitigation. The time interval Δt_{react} for the activation of emergency operations is set to 0 hr hereinafter for simplifying purposes. This means that contamination is assumed to stop instantaneously after its detection. However, Δt_{react} can be set to other values without loss of validity of the whole methodology. The function f_4 is therefore the average value pop of pop_r. In the bi-objective optimization, functions f_3 and f_4 are minimized simultaneously through the NSGAII genetic algorithm [46]. In fact, the minimization of the former reduces the sensor cost while the minimization of the latter impacts positively on the system security. In the population individuals of NSGAII, the number of genes is equal to the number of network nodes where sensors can be installed. Each gene can take on the two possible values 0 and 1, which stand for absence and presence of the sensor in the node associated with the gene, respectively.

In this paper, four options for sensor locations on the partitioned network were tested:

1. *Option 1*, sensors can be installed at all nodes (typical greedy approach);
2. *Option 2*, sensors can be installed only at the hydraulically upstream nodes of the boundary pipes;
3. *Option 3*, sensors can be installed at the most central nodes of each district, identified through topological considerations;
4. *Option 4*, sensors can be installed at the hydraulically upstream nodes of the boundary pipes and at the most central nodes of each district.

In the last two cases, the idea is to take advantage from the study of WDN topology in order to define which nodes are potential candidates for sensor installation, according to their connectivity centrality. In this paper, the most central nodes were defined using the betweenness centrality [53], defined starting from the shortest paths in a graph. The shortest path $\sigma(s, t)$ between two nodes s and t is the connecting path with the lowest number of links (or the minimum sum of the weights associated with its links in the case of weighted graph). The betweenness centrality of a node i is defined as the sum of the ratios of the number of shortest paths between nodes s and t passing through i to the total number of shortest paths between nodes s and t. It is a measure of the influence of a node i over the flow of information between other nodes. In this paper, for each cluster, the nodes with the highest value of betweenness centrality were selected as possible sensor locations alone (*Option 3*) or in combination with boundary nodes (*Option 4*). *Options 2, 3,* and *4* aim to investigate the possibility of limiting the search for optimal sensor locations to the hydraulically upstream nodes of the flow meter-fitted boundary pipes and to the most central nodes of each district. This choice leads to significant computational simplifications, due to the reduction in the search space. This offers the possibility of better facing the problem of optimal sensor placement also for big-size WDNs (for which the number of all potential scenarios makes the problem computationally intractable). Furthermore, the strategy of locating all or some sensors in the same stations as boundary flowmeters offers easiness and cheapness of inspection and maintenance.

2.3. Procedure 3—Comparison of Sensor Placement Solutions

Sensor placement solutions were evaluated using the following four contamination impact indicators. The first is function f_4 in Equation (5), followed by functions f_5, f_6 and f_7 reported in the following Equations (6)–(8), respectively.

The function f_5 is the detection likelihood (i.e., the probability of detection):

$$f_5 = P_s = \frac{1}{S} \sum_{r=1}^{S} d_r \tag{6}$$

where $d_r = 1$ if contamination scenario r is detected, and zero otherwise; and S denotes the total number of the contamination scenarios considered.

The function f_6 is the detection time. For each detected contamination scenario, the sensor detection time corresponds to the elapsed time from the start of the contamination event, to the first identified presence of a nonzero contaminant concentration. If t_j is the time of the first detection (referred to the j-th sensor location), the detection time (t_d) for the solution for each contamination event, is the minimum among all present sensors $t_d = min(t_j)$; the characteristic detection time of the solution is defined as the mean of all t_d for the contamination scenarios detected by at least one sensor:

$$f_6 = mean(t_d) \tag{7}$$

Finally, the function f_7 is the sensor redundancy. In a generic scenario, the variable *Red* corresponds to the number of sensors (including the first) that detect the contamination within 30 minutes from the first detection; the redundancy *Red* of the solution is defined as the mean of all the values of redundancy *Red* for all the considered contamination scenarios:

$$f_7 = Red = mean(Red) \tag{8}$$

which contributes to the safety of the monitoring systems, especially in the case of sensor failures or false positive/negative detection, conferring a higher reliability.

As for the choice of the objective functions, it must be remarked that theoretically more than two of them could be inserted in the same optimization framework. However, to prevent this framework from becoming overly complex, we preferred to keep only two objective functions (number of sensors and exposed population) in the optimization framework, while other assessment criteria (e.g., detection likelihood, detection time, and sensor redundancy) will be considered in the postprocessing of the optimal solutions.

3. Case Study

The methodology described above was tested on the WDN of Parete [54], which is a small town located in a densely populated area to the south of Caserta (Italy), with population of 11,150 inhabitants. This WDN has 182 demanding nodes (with ground elevations ranging from 53 m a.s.l. to 79 m a.s.l.), 282 pipes and 2 sources with fixed head of 110 m a.s.l. A uniform desired pressure head h_{des} = 19 m was assumed for the demanding nodes, coming from the sum of the maximum building height in the town, which is 9 m in Parete, and 10 m, as prescribed by the Italian guidelines. Reference was made to the day of maximum consumption in the year when the total nodal demand ranges from 7.6 L/s at nighttime to 77.2 L/s in the morning and midday peaks, with an average value of 36.3 L/s. The leakage volume of the networks in the day of maximum consumption adds up to 930 m^3 (about 23% of the total outflow from the sources). The number of users connected to each WDN node was derived based on its average nodal demand.

4. Results and Discussions

In this section, the results of the procedures described in Sections 2.1–2.3 are reported. The first step is the definition of an optimal water network partitioning. In this regard, the clustering phase was applied to produce 5 DMAs. The choice of 5 DMAs was made because the formula $C_{opt} = n^{0.28}$ proposed by Giudicianni et al. (2018) [45] to calculate the optimal number of clusters yields C_{opt} = 4.3 for this WDN. The number of nodes for each DMA are DMA_1 = 20, DMA_2 = 35, DMA_3 = 39, DMA_4 = 41 and DMA_5 = 49, with N_{ec} = 21. For the dividing, the optimization through NSGAII yielded the Pareto front reported in Figure 1a, showing, as expected, growing values of median(GRF) with V_l growing. In fact, both variables are growing functions of the service pressure in the WDN. Figure 1b,c report the number N_{fm} of flow meters and the demand satisfaction rate I_{ds}, respectively, re-evaluated from the Pareto front and plotted against V_l. Globally, Figure 1b highlights that the higher values of N_{fm} tend to be associated with the high values of V_l. This is because V_l tends to grow when fewer gate valves are closed (and then more numerous flow meters are installed) at the boundary pipes. Finally, Figure 1c shows that I_{ds} tend to grow with V_l increasing, since both variables are increasing functions of the service pressure.

From the graphs in Figure 1, the solution with the lowest value of N_{fm} (= 8), highest number of closed valves N_{gv} (= 13), which ensures I_{ds} = 100%, was finally chosen. An important remark to be made is that among the several advantages of the WNP, the adopted partitioning solution enables also reducing leakage, from 930 m^3 (for the un-partitioned layout) to 895 m^3 (partitioned solution with 13 gate valves closed and 8 flow meters installed). This corresponds to a 3.7% leakage reduction without negatively affecting I_{ds} and GRF. In fact, for this solution median(GRF) is equal to 0.32, very close to the value of 0.36 for the un-partitioned network. The layout of the partitioned layout is reported in Figure 2. The optimal sensor placement is then carried out. The following assumptions were made for the construction of the set S of contamination events considered in the optimization:

- all the 182 demanding nodes were considered to be potential locations for contaminant injection;
- 24 possible contamination times in the day (hour 0, 1, 2, ... , 22, 23);
- single value of the mass injection rate equal to 350 g/min;
- single value of the injection duration equal to 60 min.

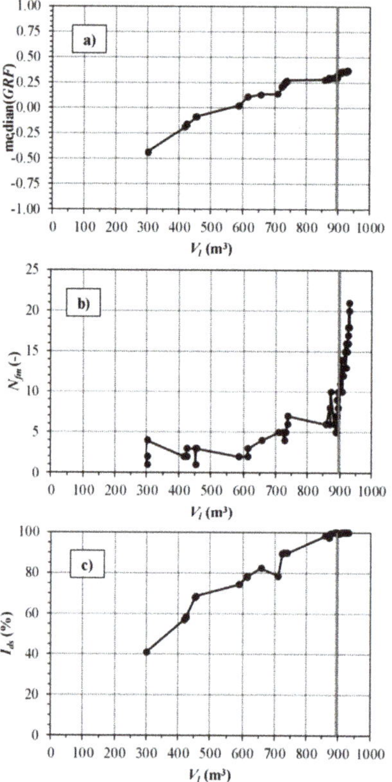

Figure 1. Dividing phase considering the clustered graph of the Parete WDN (Variant 1). Pareto front of optimal solutions in the trade-off between daily median (*GRF*) index and leakage volume V_l (**a**), re-evaluated solutions in terms of number of flow meters N_{fm} (**b**), and of demand satisfaction rate I_{ds} (**c**). In all graphs, the selected solution is highlighted with a grey vertical line.

The values reported above for mass injection and duration were sampled from those proposed by Preis and Ostfeld (2008) [55], using the procedure of Tinelli et al. (2017) [51], with the objective to obtain a representative smaller set of significant contamination events. Due to the previous assumptions, the total number *S* of contamination events was $182 \times 24 \times 1 \times 1 = 4368$.

The water quality simulations were run for 2 days of WDN operation to make sure that even contaminants injected close to the sources at the last instant of the first day had enough time to leave the network. In the optimization for sensor placement, the partitioned WDN layout was indicated as *Var1* to differentiate it from the original layout (*Var0*). Therefore, according to the three optimization options described in Section 2.2, optimizations were organized as follows:

1. *Var1Op1*: Optimal sensor placement on the partitioned WDN allowing sensor installation on all nodes (182 potential locations);
2. *Var1Opt2*: Optimal sensor placement on the partitioned WDN allowing sensor installation only on the nodes hydraulically upstream from the flowmeter fitted boundary pipes (8 potential locations);
3. *Var1Opt3*: Optimal sensor placement on the partitioned WDN allowing sensor installation only on the most central nodes of each district (15 potential locations, i.e., three locations for each district, which feature a much higher betweenness centrality value than the other nodes);

4.   *Var1Opt4*: Optimal sensor placement on the partitioned WDN allowing sensor installation on the nodes hydraulically upstream from the flowmeter fitted boundary pipes and on the most central nodes of each district (23 scenarios).

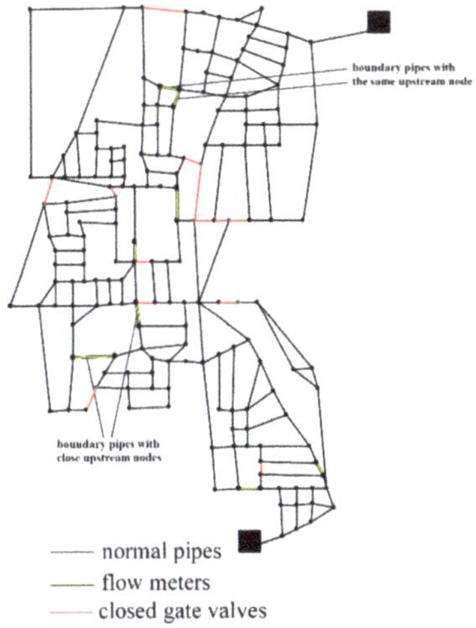

Figure 2. WDN partitioning into 5 DMAs.

Compared to *Var1Opt1*, *Var1Opt2*, *Var1Opt3* and *Var1Opt4* reduce the group of potential sensor locations respectively by 96%, 92% and 87%, resulting in a research space reduction which helps in diminishing the computational burden. The three optimizations were compared with the benchmark *Var0Opt1*, where all the 182 potential sensor locations are explored in the original layout. Table 1 shows the optimization framework, made up of 5 runs. In all of them, NSGAII was applied with a population of 200 individuals and a total number of 200 generations.

Table 1. Framework of optimizations for sensor placement in the Parete WDN.

Option	Variant 0 (Un-Partitioned)	Variant 1 (Partitioned)
1 (all nodes)	*Var0Opt1*	*Var1Opt1*
2 (only boundary nodes)	-	*Var1Opt2*
3 (only central nodes)	-	*Var1Opt3*
4 (boundary nodes + central nodes)	-	*Var1Opt4*

For the optimizations that consider all nodes as potential sensor locations (*Var0Opt1*, *Var1Opt1*), the slow convergence of NSGAII was initially remarked towards interesting solutions for water utilities, which are solutions with a reasonably low number of sensors in comparison with the total number of demanding nodes. This problem was solved by implementing inside NSGAII a heuristic algorithm to correct solutions with numerous sensors, that is $N_{sens} > 20$. In this heuristic algorithm, for each NSGAII solution violating $N_{sens} = 20$, a random integer number within the range (1, 20) is generated, representing the target number of sensors for that solution. Then, starting from the initial value of N_{sens}, the least effective sensors in terms of pop are removed one by one to reach the target. Though increasing the computation time for each NSGAII generation by about 30 times, this algorithm proved to solve the issue of slow convergence. This heuristic algorithm was not applied to the optimizations *Var1Opt2*, *Var1Opt3* and *Var1Opt4*. This made the NSGAII optimizations in the two latter applications much lighter from the computational viewpoint.

Figure 3a reports the Pareto fronts obtained in optimization *Var0Opt1*, on the un-partitioned layout, and in optimizations *Var1Opt1*, *Var1Opt2*, *Var1Opt3* and *Var1Opt4*, on the partitioned layout. As expected, these fronts in Figure 3a show decreasing values of pop as N_{sens} increases up to 20. However, for high values of N_{sens}, the additional benefit of a further sensor installed in the network tends to decrease, as already pointed out by Tinelli et al. (2017) [56]. In the present work, N_{sens} = 6 appears to be the threshold of benefit for the installation of an additional sensor, slightly to right of the knee of the Pareto fronts (which lies around N_{sens} = 3).

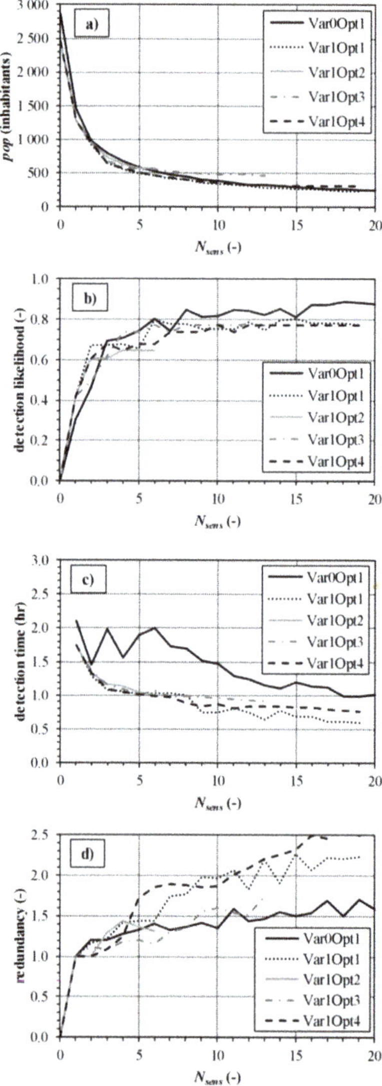

Figure 3. For the original un-partitioned WDN (*Var0Opt1*), reported as benchmark, and for the partitioned WDN (*Var1Opt1*, *Var1Opt2*, *Var1Opt3* and *Var1Opt4*), Pareto front of optimal sensor placement solutions in the trade-off between N_{sens} and contaminated population *pop* (**a**), re-evaluated solutions in terms of N_{sens} and detection likelihood P_s (**b**), N_{sens} and detection time T_{mean} (**c**), and N_{sens} and redundancy *Red* (**d**).

Another point to highlight is that for the partitioned network, the contaminated population corresponding to the case of zero installed sensors (*pop* = 2458) is lower than the corresponding contaminated population for the un-partitioned WDN (*pop* = 2806) as shown in Table 2.

Table 2. Simulation results in terms of exposed population from the four optimizations for sensor placement in the Parete WDN, considering N_{sens} up to 6.

N_{sens}	*Var0Opt1*	*Var1Opt1*	*Var1Opt2*	*Var1Opt3*	*Var1Opt4*
0	2806	2458	2458	2458	2458
1	1438	1274	1274	1274	1274
2	982	919	953	974	953
3	789	648	741	679	653
4	667	559	638	598	569
5	589	500	572	561	515
6	514	462	564	548	472

This points out the first advantage of the partitioning: by reducing the average number of possible paths in the network (due to the closure of some pipes), it produces a reduction in the contaminated population by around 12.4%. This is due to the reduction in the spreading of contamination (direct action). Furthermore, the WNP also enhances the results of optimal sensor placement (indirect action). As is shown in Table 2 for $N_{sens} \leq 6$, *pop* for the un-partitioned WDN (*Var0Opt1*) is always higher than pop for the *Var1Opt1* for all the number N_{sens} of sensors installed in the network. The minimum value of pop = 462 is for *Var1Opt1*. *Var1Opt2* (sensors allowed only upstream from boundary pipes), *Var1Opt3* (sensors allowed only on topologically central nodes in DMAs) and *Var1Opt4* (sensors allowed upstream from boundary pipes and on topologically central nodes in DMAs) give similar results to *Var1Opt1* up to N_{sens} = 2. For N_{sens} > 2, *Var1Opt2* and *Var1Op3* degenerate while the good performance of *Var1Opt4* persists. This is evidence that constraining sensor installation only upstream from boundary pipes or on topologically central nodes may lead to remarkably sub-optimal solutions. However, the combination of locations upstream from the boundary pipes and of topologically central nodes offers a good set of potential locations in the problem of optimal sensor placements. Figure 3b–d report the results of the reprocessing of the optimal solutions in terms of detection likelihood, detection time, and redundancy as a function of N_{sens}. Along with Figure 3a, they give indications on the effectiveness of the solutions obtained in the NSGAII runs. Globally, the *Var1* solutions obtained on the partitioned graph, especially *Var1Opt1*, *Var1Op2*, and *Var1Opt4*, tend to perform better in terms of *pop*, detection time, and sensor redundancy. Conversely, they feature slightly worse values in terms of detection likelihood. This may be because the optimization was carried out considering pop as objective function, which is slightly contrasted with detection likelihood [56]. In fact, the former variable mainly contributes to the system's early warning capacity whereas the latter contributes to the system safety. As for Figure 3, it must be remarked that the curves in Figure 3a are Pareto fronts while those in the other Figure 3b–d are obtained by reprocessing the optimal solutions in terms of other assessment criteria. Since these curves are not Pareto fronts, they are not strictly monotonous. Figure 4 shows the sensor placement solutions obtained for N_{sens} = 6 with three optimizations (*Var0Opt1*, *Var1Opt1*, and *Var1Opt4*). In this context, it must be noted that the *Var1Opt4* solution has three of the six sensors placed close to flowmeters (the other three sensors are in the most central nodes according to the betweenness centrality). This solution yields managerial and economic benefits, due to the closeness of some sensors to installed flow meters and due to the possibility of sharing some electronical components for data acquisition, sharing, and transmission. Summing up, the *Var1Opt4* solution represents a quasi-optimal solution in the explored trade-off between *pop* and N_{sens}, while offering significant potentials for improved management. Another advantage compared to the *Var0Opt1* and *Var1Opt1* solutions with N_{sens} = 6 is that it was obtained at a much lower computation cost (about 1/30), due to the research space reduction mentioned above for Options 2–4. Overall, the advantages in terms of computational lightness during the optimization as well as the possibility of inspecting and maintaining sensors in proximity to flow meters make solutions obtained in *Opt4* preferable from the water utilities' viewpoint. The results highlighted that nodes close to flow meters used for the monitoring of DMAs, which must always be easily accessible sites, represent good sensor locations for WDN monitoring from contaminations, when they are inserted into an optimization framework that also includes topologically central nodes inside DMAs. As for the optimal positions of the sensors in *Var1Opt1* (partitioned network and all nodes as potential candidates) and Var1Opt4 (partitioned network and sensor installations restricted to entry points and central nodes in DMAs), it must be remarked that many locations are similar in the two cases (see Figure 4). This corroborates the fact that entry points and central nodes in DMAs are good candidate locations in the present case study.

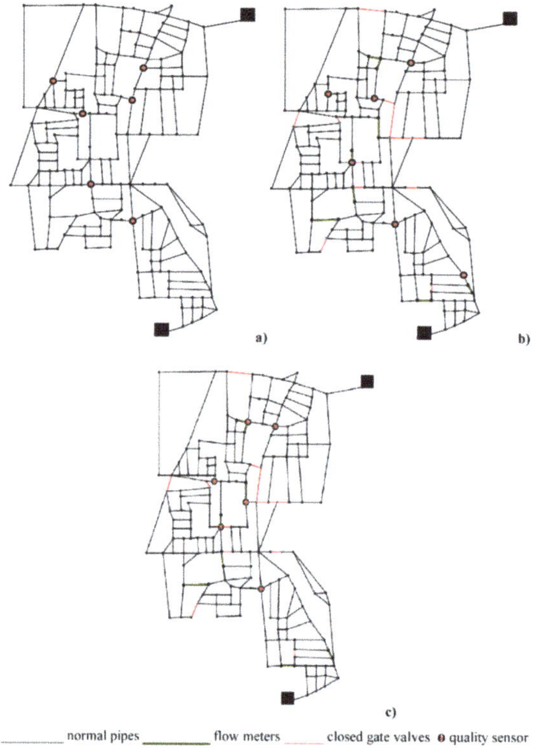

Figure 4. Optimal location of 6 sensors in (**a**) original un-partitioned WDN (*Var0Opt1*), (**b**) partitioned WDN (*Var1Opt1*), and (**c**) partitioned WDN (*Var1Opt4*).

5. Conclusions

In this work, a methodology that combines WNP and optimal sensor installation was proposed, to investigate the benefits of the "divide and conquer" technique for the monitoring of WDNs from contamination events (direct action), and for the effectiveness of optimal sensor placement (indirect action). The applications concerned a real Italian WDN, which was first partitioned into 5 DMAs. Optimal sensor solutions were searched for on the original un-partitioned WDN and on the partitioned layout, in the trade-off between number of installed sensors and affected population for an assigned set of contamination events. Further optimizations were carried out by restricting sensor installation to some pre-selected nodes (nodes hydraulically upstream from the flow meter-fitted boundary pipes and central nodes). The results showed that, for a given number of installed sensors, the monitoring stations installed in the partitioned layouts offer better monitoring performance. On the other hand, the option of considering locations in proximity to flow meters and at most central nodes as the only potential locations in the context of optimal sensor placement has the following advantages:

1. almost identical performance in terms of WDN monitoring, compared to the option of considering all nodes as potential locations;
2. money savings thanks to the possibility of sharing some electronical components for data acquisition, sharing, and transmission;
3. easiness of access to the sensors for maintenance;
4. reduction in the search space and, therefore, in the computational complexity in the optimizations for optimal sensor placement;
5. easier identification of the area from which the contamination starts with the subsequent possibility of isolating the district, assuring a higher resilience of the system to the spreading of the contamination.

With regards to the last issue, it must be noted that the calculations of the present work were carried out on a simple though real WDN. Therefore, the benefits are expected to be much larger in the case of big-size WDNs, for which the problem of optimal sensor placement may become computationally infeasible. Indeed, the topics analyzed in this paper fully match the future research directions identified by Ostfeld et al. (2008) [20] during the Battle of the Water Sensor Networks. In fact, specific reference was made to the problems of aggregation, i.e., the possibility of using a reduced but still significant sample of nodes for investigations into water quality, multi-criteria analysis of sensor performance, choice of optimal number of sensors and multiple use of boundary pipes (for both monitoring flow between DMAs and detecting potential contaminations).

Though topologically central nodes have been considered in this analysis along with DMA entry points, another attractive option is made up of critical sink nodes with lowest head inside DMAs, in which water quality parameters are already monitored. Future works will be dedicated to exploring the solution of critical sink nodes. Future work will be dedicated to investigating how results change when other objective functions from those used in the present work are considered. The methodology presented in this paper will be refined in the future considering also other benchmark networks. Adopting different clustering algorithms and centrality metrics could affect the results; to better investigate the influence on the solutions, new algorithms will be applied. Another aspect that deserves to be further investigated concerns the assumptions made for the definition of the representative set of contamination scenarios. Other prospects could concern the issues of restoration after the generic contamination and of constructing mega-monitoring stations on which to locate all the management devices (chlorine stations, pressure valves, etc.). This will be done with reference to specific real contaminants, while abandoning the simplifying assumption of unreactive and conservative contaminant adopted so far.

Author Contributions: E.C. and A.D.N. wrote the manuscript, C.G. and G.F.S. developed the coding and performed the analysis, C.C., D.M., and M.D.N. revised the paper.

Funding: This research received no external funding.

Acknowledgments: This research is part of the Ph.D. project "Water distribution network management optimization through Complex Network theory" within the Doctoral Course "A.D.I." granted by Università degli Studi della Campania "L. Vanvitelli".

Conflicts of Interest: The authors declare no conflict of interest.

References

1. Kroll, D.J. *Securing Our Water Supply: Protecting a Vulnerable Resource*; PennWell Books: Tulsa, OK, USA, 2006.
2. Adedoja, O.S.; Hamam, Y.; Khalaf, B.; Sadiku, R. Towards Development of an Optimization Model to Identify Contamination Source in a Water Distribution Network. *Water* **2018**, *10*, 579. [CrossRef]
3. Che, H.; Liu, S.; Smith, K. Performance evaluation for a contamination detection method using multiple water quality sensors in an early warning system. *Water* **2015**, *7*, 1422–1436. [CrossRef]

4. Lee, B.H.; Deininger, R.A. Optimal locations of monitoring stations in water distribution system. *J. Environ. Eng.* **1992**, *118*, 4–16. [CrossRef]

5. Kumar, A.; Kansal, M.L.; Arora, G. Identification of monitoring stations in water distribution system. *J. Environ. Eng.* **1997**, *123*, 746–752. [CrossRef]

6. Watson, J.-P.; Greenberg, H.J.; Hart, W.E. A multiple-objective analysis of sensor placement optimization in water networks. In Proceedings of the World Water and Environmental Resources Congress 2004, Salt Lake City, UT, USA, 27 June–1 July 2004; pp. 1–10.

7. Berry, J.W.; Hart, W.E.; Phillips, C.A.; Watson, J.-P. Scalability of integer programming computations for sensor placement in water networks. In Proceedings of the World Water and Environmental Resources Congress 2004, Anchorage, AK, USA, 15–19 May 2005; pp. 1–13.

8. Ostfeld, A.; Salomons, E. Optimal layout of early warning detection stations for water distribution systems security. *J. Water Resour. Plan. Manag.* **2004**, *130*, 377–385. [CrossRef]

9. Uber, J.; Janke, R.; Murray, R.; Meyer, P. Greedy heuristic methods for locating water quality sensors in distribution systems. In Proceedings of the World Water and Environmental Resources Congress 2004, Salt Lake City, UT, USA, 27 June–1 July 2004; pp. 1–9.

10. Kessler, A.; Ostfeld, A.; Sinai, G. Detecting accidental contaminations in municipal water networks. *J. Water Resour. Plan. Manag.* **1998**, *124*, 192–198. [CrossRef]

11. Ostfeld, A.; Salomons, E. Securing water distribution systems using online contamination monitoring. *J. Water Resour. Plan. Manag.* **2005**, *131*, 402–405. [CrossRef]

12. Berry, J.; Carr, R.D.; Hart, W.E.; Leung, V.J.; Phillips, C.A.; Watson, J.-P. Designing contamination warning systems for municipal water networks using imperfect sensors. *J. Water Resour. Plan. Manag.* **2009**, *135*, 253–263. [CrossRef]

13. McKenna, S.A.; Hart, D.B.; Yarrington, L. Impact of sensor detection limits on protecting water distribution systems from contamination events. *J. Water Resour. Plan. Manag.* **2006**, *132*, 305–309. [CrossRef]

14. Dorini, G.; Jonkergouw, P.; Kapelan, Z.; Di Pierro, F.; Khu, S.T.; Savic, D. An efficient algorithm for sensor placement in water distribution systems. In Proceedings of the Eighth Annual Water Distribution Systems Analysis Symposium, Cincinnati, OH, USA, 27–30 August 2006; pp. 1–13.

15. Huang, J.; McBean, E.A.; James, W. Multiobjective optimization for monitoring sensor placement in water distribution systems. In Proceedings of the 8th Annual Symposium on Water Distribution Systems Analysis, Cincinnati, OH, USA, 27–30 August 2006.

16. Propato, M.; Piller, O. Battle of the water sensor networks. In Proceedings of the Annual Symposium on Water Distribution Systems Analysis, Cincinnati, OH, USA, 27–30 August 2006.

17. Wu, Z.Y.; Walski, T. Multiobjective optimization of sensor placement in water distribution systems. In Proceedings of the 8th Annual Water Distribution Systems Analysis Symposium, Cincinnati, OH, USA, 27–30 August 2006.

18. Hart, W.E.; Murray, R. Review of sensor placement strategies for contamination warning systems in drinking water distribution systems. *J. Water Resour. Plan. Manag.* **2010**, *136*, 611–619. [CrossRef]

19. Rathi, S.; Gupta, R.; Ormsbee, L. A review of sensor placement objective metrics for contamination detection in water distribution networks. *Water Supply* **2015**, *15*, 898–917. [CrossRef]

20. Ostfeld, A.; Uber, J.G.; Salomons, E.; Berry, J.W.; Hart, W.E.; Phillips, C.A.; Watson, J.-P.; Dorini, G.; Jonkergouw, P.; Kapelan, Z.; et al. The battle of the water sensor networks (BWSN): A design challenge for engineers and algorithms. *J. Water Resour. Plan. Manag.* **2008**, *134*, 556–568. [CrossRef]

21. Xu, X.; Lu, Y.; Huang, S.; Xiao, Y.; Wang, W. Incremental sensor placement optimization on water network. In Proceedings of the Joint European Conference on Machine Learning and Knowledge Discovery in Databases, Berlin, Germany, 23–27 September 2013; pp. 467–482.

22. Krause, A.; Leskovec, J.; Guestrin, C.; VanBriesen, J.; Faloutsos, C. Efficient sensor placement optimization for securing large water distribution networks. *J. Water Resour. Plan. Manag.* **2008**, *134*, 516–526. [CrossRef]

23. Sankary, N.; Ostfeld, A. Inline mobile sensors for contaminant early warning enhancement in water distribution systems. *J. Water Resour. Plan. Manag.* **2016**, *143*, 04016073. [CrossRef]

24. Rathi, S.; Gupta, R.; Kamble, S.; Sargaonkar, A. Risk-based analysis for contamination event selection and optimal sensor placement for intermittent water distribution network security. *Water Resour. Manag.* **2016**, *30*, 2671–2685. [CrossRef]

25. Zheng, F.; Du, J.; Diao, K.; Zhang, T.; Yu, T.; Shao, Y. Investigating effectiveness of sensor placement strategies in contamination detection within water distribution systems. *J. Water Resour. Plan. Manag.* **2018**, *144*, 06018003. [CrossRef]
26. Deuerlein, J.W. Decomposition model of a general water supply network graph. *J. Hydraul. Eng.* **2008**, *134*, 822–832. [CrossRef]
27. Zheng, F.; Simpson, A.R.; Zecchin, A.C.; Deuerlein, J.W. A graph decomposition-based approach for water distribution network optimization. *Water Resour. Res.* **2013**, *49*, 2093–2109. [CrossRef]
28. Perelman, L.; Ostfeld, A. Topological clustering for water distribution systems analysis. *Environ. Model. Softw.* **2011**, *26*, 969–972. [CrossRef]
29. Perelman, L.S.; Allen, M.; Preisc, A.; Iqbal, M.; Whittle, A.J. Multi-level automated sub-zoning of water distribution systems. In Proceedings of the 7th International Congress on Environmental Modelling and Software, San Diego, CA, USA, 18 June 2014.
30. Galdiero, E.; De Paola, F.; Fontana, N.; Giugni, M.; Savic, D. Decision support system for the optimal design of district metered areas. *J. Hydroinf.* **2016**, *18*, 49–61. [CrossRef]
31. Hajebi, S.; Roshani, E.; Cardozo, N.; Barrett, S.; Clarke, A.; Clarke, S. Water distribution network sectorisation using graph theory and many-objective optimisation. *J. Hydroinf.* **2016**, *18*, 77–95. [CrossRef]
32. Herrera, M.; Abraham, E.; Stoianov, I.; Fernandez, M.H. A graph-theoretic framework for assessing the resilience of sectorised water distribution networks. *Water Resour. Manag.* **2016**, *30*, 1685–1699. [CrossRef]
33. Di Nardo, A.; Di Natale, M.; Giudicianni, C.; Greco, R.; Santonastaso, G.F. Weighted spectral clustering for water distribution network partitioning. *Stud. Comput. Intell. Complex Netw. Their Appl. V* **2016**, *693*, 797–807. [CrossRef] [PubMed]
34. Di Nardo, A.; Di Natale, M.; Giudicianni, C.; Santonastaso, G.F.; Tzatchkov, V.G.; Varela, J.M.R.; Yamanaka, V.H.A. Economic and Energy Criteria for District Meter Areas Design of Water Distribution Networks. *Water* **2017**, *9*, 463. [CrossRef]
35. Herrera, M.; Izquierdo, J.; Pérez-García, R.; Montalvo, I. Multi-agent adaptive boosting on semi-supervised water supply clusters. *Adv. Eng. Softw.* **2012**, *50*, 131–136. [CrossRef]
36. Campbell, E.; Izquierdo, J.; Montalvo, I.; Ilaya-Ayza, A.; Pérez-García, R.; Tavera, M. A flexible methodology to sectorize water supply networks based on social network theory concepts and multiobjective optimization. *J. Hydroinf.* **2015**, *18*, 62–76. [CrossRef]
37. Girvan, M.; Newman, M.E.J. Community structure in social and biological networks. *Proc. Natl. Acad. Sci. USA* **2002**, *99*, 7821–7826. [CrossRef] [PubMed]
38. Ciaponi, C.; Murari, E.; Todeschini, S. Modularity-based procedure for partitioning water distribution systems into independent districts. *Water Resour. Manag.* **2016**, *30*, 2021–2036. [CrossRef]
39. Zhang, Q.; Wu, Z.Y.; Zhao, M.; Qi, J.; Huang, Y.; Zhao, H. Automatic Partitioning of Water Distribution Networks Using Multiscale Community Detection and Multiobjective Optimization. *J. Water Resour. Plan. Manag.* **2017**, *143*, 4017057. [CrossRef]
40. Ciaponi, C.; Creaco, E.; Di Nardo, A.; Di Natale, M.; Giudicianni, C.; Musmarra, D.; Santonastaso, G.F. Optimal Sensor Placement in a Partitioned Water Distribution Network for the Water Protection from Contamination. *Proceedings* **2018**, *2*, 670. [CrossRef]
41. Perelman, L.S.; Allen, M.; Preis, A.; Iqbal, M.; Whittle, A.J. Automated sub-zoning of water distribution systems. *Environ. Model. Softw.* **2015**, *65*, 1–14. [CrossRef]
42. Shi, J.; Malik, J. Normalized Cuts and Image Segmentation. *IEEE Trans. Pattern Anal. Mach. Intell.* **2000**, *22*, 888–905.
43. Di Nardo, A.; Di Natale, M.; Giudicianni, C.; Greco, R.; Santonastaso, G.F. Water Distribution Network Clustering: Graph Partitioning or Spectral Algorithms? *Stud. Comput. Intell. Complex Netw. Their Appl. VI* **2018**, *689*, 1197–1209.
44. Di Nardo, A.; Giudicianni, C.; Greco, R.; Herrera, M.; Santonastaso, G.F. Applications of Graph Spectral Techniques to Water Distribution Network Management. *Water* **2018**, *10*, 45. [CrossRef]
45. Giudicianni, C.; Di Nardo, A.; Di Natale, M.; Greco, R.; Santonastaso, G.F.; Scala, A. Topological taxonomy of water distribution networks. *Water* **2018**, *10*, 444. [CrossRef]
46. Deb, K.; Agrawal, S.; Pratapm, A.; Meyarivan, T. A fast and elitist multi-objective genetic algorithm: NSGA-II. *IEEE Trans. Evol. Comput.* **2002**, *6*, 182–197. [CrossRef]

47. Tucciarelli, T.; Criminisi, A.; Termini, D. Leak Analysis in Pipeline System by Means of Optimal Value Regulation. *J. Hydraul. Eng.* **1999**, *125*, 277–285. [CrossRef]
48. Creaco, E.; Franchini, M.; Todini, E. Generalized resilience and failure indices for use with pressure driven modeling and leakage. *J. Water Resour. Plan. Manag.* **2016**, *142*, 4016019. [CrossRef]
49. Tanyimboh, T.T.; Templeman, A.B. Seamless pressure-deficient water distribution system model. *J. Water Manag.* **2010**, *163*, 389–396. [CrossRef]
50. Ciaponi, C.; Franchioli, L.; Murari, E.; Papiri, S. Procedure for Defining a Pressure-Outflow Relationship Regarding Indoor Demands in Pressure-Driven Analysis of Water Distribution Networks. *Water Resour. Manag.* **2014**, *29*, 817–832. [CrossRef]
51. Tinelli, S.; Creaco, E.; Ciaponi, C. Sampling significant contamination events for optimal sensor placement in water distribution systems. *J. Water Resour. Plan. Manag.* **2017**, *143*, 4017058. [CrossRef]
52. Rossman, L.A. *EPANET2 Users Manual*; United States Environmental Protection Agency: Washington, DC, USA, 2000.
53. Freeman, L.C. A set of measures of centrality based on betweenness. *Sociometry* **1977**, *40*, 35–41. [CrossRef]
54. Di Nardo, A.; Di Natale, M.; Giudicianni, C.; Savic, D. Simplified Approach to Water Distribution System Management via Identification of a Primary Network. *J. Water Resour. Plan. Manag.* **2017**, *144*, 1–9. [CrossRef]
55. Preis, A.; Ostfeld, A. Multiobjective contaminant sensor network design for water distribution systems. *J. Water Resour. Plan. Manag.* **2008**, *134*, 366–377. [CrossRef]
56. Tinelli, S.; Creaco, E.; Ciaponi, C. Impact of objective function selection on optimal placement of sensors in water distribution networks. *Ital. J. Eng. Geol. Environ.* **2018**, 173–178. [CrossRef]

 *water*

Article

An Inverse Transient-Based Optimization Approach to Fault Examination in Water Distribution Networks

Chao-Chih Lin and Hund-Der Yeh *

Institute of Environmental Engineering, National Chiao Tung University, Hsinchu 30010, Taiwan;
tom.r1000000@gmail.com
* Correspondence: hdyeh@mail.nctu.edu.tw; Tel.: +886-3-571-2121 (ext. 31910)

Received: 8 April 2019; Accepted: 24 May 2019; Published: 1 June 2019

Abstract: This research introduces an inverse transient-based optimization approach to automatically detect potential faults, such as leaks, partial blockages, and distributed deteriorations, within pipelines or a water distribution network (WDN). The optimization approach is named the Pipeline Examination Ordinal Symbiotic Organism Search (PEOS). A modified steady hydraulic model considering the effects of pipe aging within a system is used to determine the steady nodal heads and piping flow rates. After applying a transient excitation, the transient behaviors in the system are analyzed using the method of characteristics (MOC). A preliminary screening mechanism is adopted to sift the initial organisms (solutions) to perform better to reduce most of the unnecessary calculations caused by incorrect solutions within the PEOS framework. Further, a symbiotic organism search (SOS) imitates symbiotic relationship strategies to move organisms toward the current optimal organism and eliminate the worst ones. Two experiments on leak and blockage detection in a single pipeline that have been presented in the literature were used to verify the applicability of the proposed approach. Two hypothetical WDNs, including a small-scale and large-scale system, were considered to validate the efficiency, accuracy, and robustness of the proposed approach. The simulation results indicated that the proposed approach obtained more reliable and efficient optimal results than other algorithms did. We believe the proposed fault detection approach is a promising technique in detecting faults in field applications.

Keywords: fault identification; hydraulic transient; inverse transient analysis (ITA); water distribution network; optimization approach

---

1. Introduction

1.1. Background and Problem Statement

Water distribution networks (WDNs) in modern cities are usually large-scale, with complex systems and limited instrumentation. Water may be lost due to system aging, poor maintenance, and improper operations. The effective management of a water supply may be a serious engineering problem faced by cities, and rapid urbanization and infrastructure aging are expected to intensify in the future [1]. Faults in the pipeline system may not only cause problems in water resource management, but may also induce economic problems such as lost revenue or extensive repair times [2]. The faults in a pipeline or a WDN may be divided into three types: Leak, blockage, and deterioration, which may induce various problems. Leaks in pipelines and WDNs may cause large economic losses. Water supply networks leak an average of 20% of their water supply and lose an estimated U.S. $9.6 billion each year [3]. This may also affect environmental health and safety [4–6] and create water quality problems, such as equipment failure, problematic operations management, and errors in pipeline design [4,7–9]. If a pipeline has blockages, this will reduce the pipe carrying capacity of the system and there will be severe safety problems [10]. Pipeline deterioration may not impose imminent threats to the operation of pipeline systems, but it may reduce water transmission efficiency [11] and create water quality problems [12]. Hence, fault detection in WDNs is an important task in the community of water supply engineers.

1.2. Literature Review

Due to different data collection methods, the fault identification problem may be classified into the following two categories: Steady-state methods and transient analysis. Steady-state methods, such as vibration analysis, pulse-echo analysis, and acoustic reflectometry, were developed in previous studies for leak isolation [13–16], blockage detection [17–20], and deterioration determination [21–23]. These methods deliver a large number of results with high precision. However, they are usually developed based on some indispensable customized hardware with a long-term operation, which may lead to high costs [24]. In contrast, the application of the transient-based approach is simple and efficient [25,26]. In transient analysis, a pressure wave with appropriate bandwidth and amplitude is intentionally injected into the system [27]. The faults in the system, such as leakage, blockage, and deterioration can easily affect the head changes in the system when they are compared to those in a steady-state condition. The system responses can be freely obtained through a simple operation. However, this has a big drawback because the pressures created by a transient event may be too high to damage pipelines or even cause catastrophic failure in pipelines.

The heuristic algorithm is capable of searching for global optimal solutions [28]. It is therefore commonly used for detecting leaks in WDNs. Vítkovský et al. [29] combined a genetic algorithm (GA) with inverse transient analysis (ITA) to detect leaks and to calibrate friction factors in water pipelines. A GA was utilized to replace the Levenberg–Marquardt (LM) method used in Reference [30] to minimize the difference between calculated and measured heads. Vítkovský et al. [31] considered the shuffled complex evolution (SCE) algorithm to be an optimization tool in ITA for detecting single and multiple leaks in a pipeline system using laboratory observations with various errors (i.e., data errors, model input errors, and model structure errors). They indicated that a model structure error was the most possible limiting factor in field tests of ITA application. Jung and Karney [32] contrasted the performance of a GA and particle swarm optimization (PSO) in leak detection and friction factor calibration in a developed WDN model. They found that PSO provided faster convergence and produced better results than the GA. Haghighi and Ramos [33] exploited a central force optimization (CFO)-based approach as an inverse problem solver for leak detection in a benchmark leaking pipe network (reported in References [30,31]). The CFO-based approach exhibited excellent accuracy in identifying the friction factor and detecting the leaking node. Covelli et al. [34] highlighted the susceptibility of aged and high-pressure zones in leakage occurrences in WDNs and applied a GA to determine the optimal number, positioning, and setting of pressure reduction valves for reducing background leakages within the network.

Blockage detection is a crucial issue in aged pipelines and pipe networks in energy, chemical, and water industries. A blockage consists of chemical or physical depositions [26] or a valve that has only been partially reopened. It may cause system failures and an increase in water leakage due to the high-pressure redistribution within the system [35]. On the issue of blockage detection development, Wang et al. [10] detected discrete blockages in pipes by analytically using the transient damping of different frequency harmonics. However, detection of the blockage location was not mentioned in their study. Mohapatra et al. [36] developed a technique for detecting partial blockages in a single pipeline using the frequency response method. The patterns and numbers of peaks were used in the pressure frequency response of the system to detect blockage locations and estimate the effective size of two partial blockages. Lee et al. [37] numerically determined the properties of blockage-induced oscillations using the Fourier transform of the inverted peak magnitude in the frequency response diagram. Meniconi et al. [35] investigated two transient-based methods, pressure signal analysis and frequency response analysis, to detect a partial blockage in experimental pipes. The results showed that the former was more accurate in detecting the location of the blockage, while the latter was more reliable in predicting the severity of the blockage. Duan et al. [38] examined wave–blockage interactions under unsteady flow in pressurized pipelines. They revealed that an extensive blockage might change resonant frequencies and amplitudes, but a partial blockage might only affect resonant amplitudes. Lee et al. [27] used analytical, numerical, and experimental methods to investigate the importance of signal bandwidth in fault detection. They suggested that both low and high bandwidth signals should be considered in a transient-state system. A low bandwidth signal was used to identify the regions of suspected damage, while the fault's location and properties were pinpointed by the high bandwidth signal.

The condition of the pipe wall in pressurized pipelines changes with their age or operating condition. Pipe wall deterioration may be due to corrosion, material erosion, and external pressures with system aging. At present, the transient-based approach is recognized as a potential tool for the noninvasive detection of discrete and distributed deterioration in pressurized pipelines [39]. Many previous studies have investigated deterioration detection technologies for water transmission pipelines. Stephens et al. [40,41] applied fluid transients and ITA to detect changes in the thickness of a pipe wall in a field test. They mentioned that the loss of cement mortar lining could lead to wall corrosion and significant changes in wave speed. Hachem and Schleiss [42] presented a transient-based approach to determine the stiffness of a pipe segment and identify the location of a structurally weak segment of a single pipeline. The location and length of the weak segment were identified using two mean wave speed values and the travel time of the reflections from a weak segment. Gong et al. [43] applied time-domain reflectometry (TDR) analysis to detect distributed deterioration in an experimental water transmission pipeline in a laboratory. They found that the size of the pressure wave reflection from a deteriorated section could be affected by any change in the pipeline impedance of the deteriorated section. Recently, Gong et al. [44] developed a new transient pressure wave generator using controlled electrical sparks. They provided high-frequency waves and improved the incident signal bandwidth. The location and length of thinner wall sections in an experimental pipeline system were then determined through a TDR technique.

1.3. Objective

Multiple fault detection in pipeline systems or WDNs using ITA is considered to be a troublesome issue because a large amount of input data and computation time is required. Moreover, the computation time and searching space in the optimization process may be enormous, especially for a complicated WDN with multiple faults. This paper presents a novel and efficient transient-based approach for multiple fault detection, including leak detection, partial blockage identification, and distributed deterioration determination, in a single pipeline or a WDN. An ITA-based hybrid heuristic approach called the Pipeline Examination Ordinal Symbiotic Organism Search (PEOS) was developed based on a combination of an ordinal optimization algorithm (OOA) and a symbiotic organism search (SOS). The proposed approach can simultaneously determine information on various faults via inverse calculation. Two experimental single pipeline cases and two numerical tests with different pipe network configurations were considered to examine the performance and capability of the proposed approach. The performance of PEOS was further compared to different optimization algorithms to demonstrate its accuracy and efficiency in predicting fault information. The reliability and robustness of the proposed approach for fault detection in a complicated WDN (considering data collection issues) was further validated.

2. Methodology

2.1. Pipe Network Simulation

EPANET is a widely used public software package for modeling hydraulic and water quality behavior in pressurized pipe systems. However, it needs an external functionality to model water leakage in a system in simulations [45]. Moreover, it is not easy to simulate the hydraulic behavior of a pressurized pipe system with blockages or deterioration. In order to simulate steady-state water head distribution in a network with various faults, we therefore developed a heuristic optimization algorithm called a pipe network symbiotic organism search (PNSOS) based on the algorithm for pipe network simulated annealing (SA) introduced by Yeh and Lin [46]. The SOS was adopted here to replace the SA in order to deal with a complex network for the sake of computational efficiency. The PNSOS is an efficient tool in estimating the steady-state nodal head and flow rate for a given pipe network system with faults before a transient operation. The Hazen–Williams (H–W) equation is then used to express the relationship between the flow rate and head loss for each pipe [47,48]. The modified loss coefficient ($K_{ij}(t)$) in the H–W equation for a pipe at used year t is defined as

$$K_{ij}(t) = \frac{10.66667 \cdot L_{ij}}{C_{ij}^{HW}(t)^{1.851852} \cdot D_{ij}^{4.870370}}, \tag{1}$$

where ij is defined from node i to node j for the variable, L_{ij} is the length (m) of the pipe, and D_{ij} is the internal pipe diameter (m). The modified H–W coefficient $C_{ij}^{HW}(t)$ (for modeling the effect of pipe aging) is defined as [49]

$$C_{ij}^{HW}(t) = 18 - 37.2 \log\left(\frac{e_{0ij}(t) + t \times a_{ij}(t)}{D_{ij}}\right), \tag{2}$$

where t is the used year of the pipe, $e_{0ij}(t)$ is the initial roughness (mm) of the pipe, and $a_{ij}(t)$ is the roughness growth rate (unique per year) in the pipe at year t. The following equations are used in the proposed approach to calculate the values of e_{0ij} and a_{ij} [49]:

$$\log\left(e_{0ij}(t)\right) = \frac{C_{ij}^{HW}(t-1) - 18}{-37.2} + \log\left(D_{ij}\right), \tag{3}$$

$$a_{ij}(t) = \frac{10^{\left(\frac{0.5C_{ij}^{HW}(t-1)-18}{-37.2}\right)} \times D_{ij} - e_{0ij}(t)}{50}. \tag{4}$$

For a new installed pipe (i.e., $t = 0$), the value of $C_{ij}^{HW}(t-1)$ in Equation (3) is considered to be the initial value of the H–W coefficient at the time of pipe installation (i.e., $C_{ij}^{HW}(0)$). Thus, the modified H–W coefficient for each pipe could be iteratively obtained. On the basis of Equations (1)–(4), the flow rate $Q_{ij}(t)$ (m^3/s) in each pipe at year t could be expressed as

$$Q_{ij}(t) = \left[\frac{\Delta H_{ij}}{K_{ij}(t)}\right]^{0.54}, \tag{5}$$

where ΔH_{ij} is the frictional head loss in a pipe. The equation of mass conservation at node i could be written as

$$MC_i(t) = \sum_{j=1}^{nn} Q_{ij}(t) + QI_i(t), \tag{6}$$

where nn is the number of total neighbor nodes to node i, and $QI_i(t)$ is the demand or the source at node i. The flow rate is positive for flow out of node i and negative for flow into node i, while QI_i is positive for inflow and negative for outflow. The objective function used in the PNSOS is defined as

$$\text{Minimize} \sum_i^{nd} (MC_i(t))^2, \tag{7}$$

where nd is the total number of nodes needed to estimate the nodal heads and flows in a network system.

2.2. Hydraulic Transient Model and Faults in the Pipeline

The unsteady pressurized flow in a pipe network with a known steady-state nodal head and flow rate can be described by a pair of partial differential equations, written as [50]

$$gA\frac{\partial H}{\partial x} + \frac{\partial Q}{\partial t} + \frac{f}{2DA}Q|Q| = 0, \tag{8}$$

$$\frac{\partial H}{\partial t} + \frac{a^2}{gA}\frac{\partial Q}{\partial x} = 0, \tag{9}$$

where g is the acceleration of gravity, A is the pipe cross-sectional area, H is the piezometric head, x is the distance along the pipe, Q is the volume flow rate, t is the time, D is the diameter of the pipe, a is the wave speed, and f is the friction factor, which can be described in steady-, quasi-steady-, or unsteady-state conditions. The friction factor was considered to be steady with a value of 0.02, since this study was numerical verification-oriented. Readers can refer to related studies regarding unsteady friction [51,52]. Equations (8) and (9) are respectively the momentum and continuity equations. By means of the method of characteristics (MOC) and the finite difference method, both equations can be solved with appropriate initial and boundary conditions. Then the hydraulic transient heads and flow rates along the pipelines are solved.

Three kinds of faults (i.e., leaks, partial blockages, and distributed deterioration) are considered and discussed. Both leaks and blockages could be described by the simple orifice equation and implemented as an internal boundary condition in the MOC analysis as [53]

$$Q_O = C_{dO}A_O\sqrt{2g\Delta H_O}, \tag{10}$$

where Q_O is the volumetric flow rate through the orifice, C_{dO} is the discharge coefficient of the orifice, A_O is the orifice area, and ΔH_O is the head loss across the orifice. The leaks represent the flow loss through the offline orifice with no head loss, while the blockages represent the head loss through the inline orifice with no flow loss.

The volumetric flow rate Q_L through leakage is denoted as [53]

$$Q_L = Q^U - Q^D = C_{dL}A_L\sqrt{2g(H_P - H_{Out} - z)} \text{ with } H_P = H_P^U = H_P^D, \tag{11}$$

where Q^U and Q^D are the volumetric flow rates upstream and downstream of the leakage, respectively; $C_{dL}A_L$ is the discharge coefficient of leakage times the leak area of the orifice; H_P and H_{out} are respectively the heads at the leak and outside the leak; z is the pipe elevation at the leak; and H_P^U and H_P^D are respectively the heads upstream and downstream of the leak. The outside head is generally considered to be the atmospheric pressure head and is hence set to zero [53]. The initial value of C_{dL} is set to unity, and the elevation z is assumed to be zero.

Similarly, a discrete (partial) blockage is treated as an inline valve with a constant opening area. The upstream and downstream of the blockage satisfy the continuity conditions of the head and flux. The volumetric flow rate Q_B through the blockage is expressed as [53,54]

$$Q_B|Q_B| = 2g(C_{dB}A_B)^2(H_P^U - H_P^D) \text{ with } Q_B = Q_B^U = Q_B^D, \tag{12}$$

where Q_B^U and Q_B^D are respectively the flow rates upstream and downstream of the blockage; and $C_{dB}A_B$ is the discharge coefficient times the orifice area of the blockage. Note that Equation (12) is a simple model to approximate a blockage of any shape and length [53].

Deterioration (e.g., pipe wall damage or pipeline corrosion) often introduces a decrease in pipe wall thickness, which in turn introduces a change in the pipeline impedance and wave speed, defined as [39,43]

$$B_i^{im} = a_i/(gA_i), \tag{13}$$

where B_i^{im}, a_i, and A_i are respectively the impedance, wave speed, and pipe cross-sectional area of *i*th reach. Their values are known in the MOC analysis.

2.3. Ordinal Optimization Approach (OOA)

Ho et al. [55] introduced the key cogitation of the OOA to reduce the process of searching for global optimal solutions blindly. Ordinal comparison and goal softening procedures are the major processes employed in the OOA. The approach looks for reliable and satisfactory solutions by searching through the relative rankings of each solution instead of directly evaluating the optimal solution in a complex optimization model. Thus, relatively better solutions are selected and used in the optimization process, and the best solution can be obtained without meaningless calculations and iterations of the worst solutions.

2.4. Symbiotic Organism Search (SOS)

The SOS algorithm [56] is an evolutionary metaheuristic algorithm inspired by actual biological interactions in nature, such as mutualism, commensalism, and parasitism. Like other population-based algorithms (e.g., a GA and PSO), the SOS shares the following similar features: (1) Control parameters should be properly settled before operation; (2) it has operators to enhance or improve candidate solutions via the interaction of each solution; (3) it has a selection mechanism to determine the current optimal solution in the solution domain and preserve the current best solution during the process [56,57]. Furthermore, the SOS algorithm requires no algorithm-specific parameters. Only the initial ecosystem (population) size and the maximum number of iterations are needed.

In short, the organisms (solutions) in the ecosystem are guided toward the current best organism in mutualism and commensalism states, while the parasitism state is used to prevent the organisms trapping in a local optimal solution. These three states are repeated until the stopping criterion is achieved. Details about the SOS algorithm are given in the Supplementary Materials.

2.5. Inverse Transient Analysis (ITA)

The ITA introduced by Pudar and Liggett [58] was developed by minimizing the errors between the measured and calculated system state variables (i.e., pressure or flow rates). Various potential faults with unknown parameters (fault information) are tested in a numerical simulator until the measured state variable traces match the calculated ones [4]. A heuristic algorithm is a useful tool for the numerical simulators of ITA because it can explore global or near-global optimum solutions in the search space in an affordable time [28]. However, the ITA method relies on an accurate transient model of the system. A model consisting of transient and boundary conditions with correct system parameters is needed in ITA for obtaining a reliable transient response in the system [5]. The pressure measurements are theoretically more suitable than the volume measurements (i.e., flow rate) because the response of the pressure is more sensitive than that of the flow rate in the ITA [59]. Transient flow is not easy to precisely measure in practice with a very high sampling rate, when only the pressure can be measured. The objective function F in the proposed approach for fault detection is defined as

$$F = \text{Min} \sum_{j=1}^{m} \sum_{i=1}^{n} \left(H_{ij}^{o} - H_{ij}^{s} \right)^2, \tag{14}$$

where m is the total number of observation points in the network; n is the total amount of data at an observation point; and H_{ij}^{o} and H_{ij}^{s} represent the *i*th observed and simulated heads at observation point j, respectively. Thus, an ITA model was set up for a pipe network, in which head specifications were computed as a function of unknown variables (fault information), e.g., L_p, L_L, $C_{dL}A_L$, B_p, B_L, $C_{dB}A_B$, D_p, D_L, L_D, a_D, and A_D (listed and defined in Table 1).

2.6. Development of PEOS

The PEOS is a hybrid heuristic algorithm combining the screening procedure of OOA and the heuristic algorithm SOS to automatically determine fault information in WDNs. The overall operational architecture and steps of PEOS are briefly given below (also in Figure 1):

1. Import the network configurations;
2. Randomly generate candidate solutions (CASes) with different fault information consisting of the unknown variables listed in Table 1;
3. Rearrange the network configurations, since the new fault points (leaks and blockages) and/or new fault pipe reaches (deterioration parts) are added;
4. Use PNSOS to calculate the optimal steady-state nodal heads and piping flow rates within a given WDN for each CAS;
5. Generate hydraulic transient events and apply the MOC to obtain the transient head distribution of each CAS;
6. Utilize Equation (14) to calculate the CASes' objective function values (OFVs) and rank them. The top 5% of CASes with smaller OFVs are selected for the next step;
7. Consider the selected CASes to be initial organisms for the ecosystem of the SOS used in the pipe examination;
8. Execute the fault detection procedure, in which the organisms containing fault information continually move forward to the current best solution (X_{best}), with optimal fault information due to the three states of the SOS;
9. Check whether the optimization process satisfies the stopping criterion. If so, the fault detection procedure is then terminated and moves to the next step. Otherwise, the searching process goes on.

The first stopping criterion is defined as the absolute difference between two successive optimal OFVs (in Equation (14)), which is always less than 10^{-4} within four iterations. The second criterion for fault detection is the iteration reaching the specified maximum limit.

Table 1. Fault information to be determined.

Variable	Description
Leak	
L_p	Leak pipe number
L_L	Leak location
$C_{dL}A_L$	Discharge coefficient times the leak area of the orifice
Blockage	
B_p	Blockage pipe number
B_L	Blockage location
$C_{dB}A_B$	Discharge coefficient times the open orifice area of the blockage
Deterioration	
D_p	Deterioration pipe number
D_L	Deterioration location
L_{Di}	Length of ith distributed deterioration reach
a_{Di}	Wave speed of ith distributed deterioration reach
A_{Di}	Pipe cross-sectional area of ith distributed deterioration reach

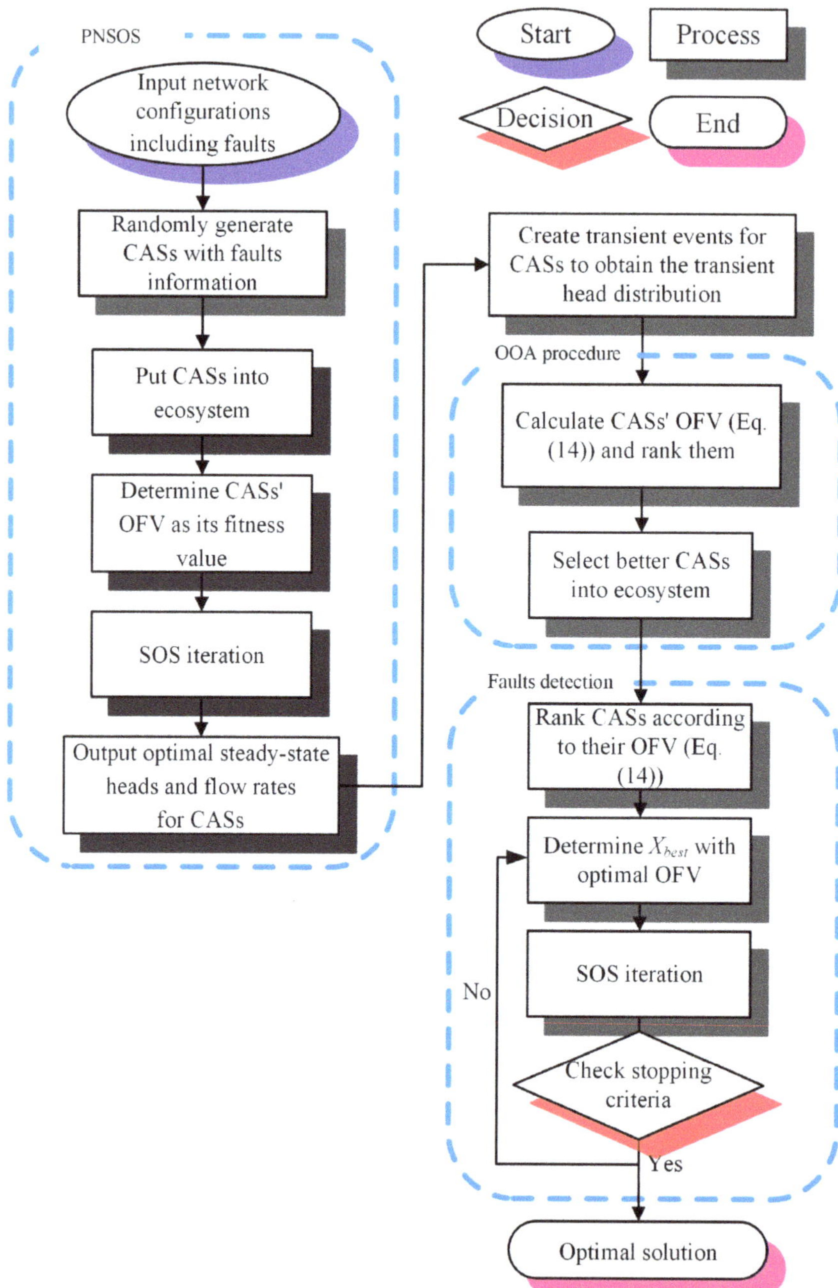

Figure 1. Flowchart of the Pipe Examination Ordinal Symbiotic Organism Search (PEOS).

2.7. Benchmark Evolutionary Algorithms

To validate the ability of the proposed approach in obtaining global optimal fault information, the performance of PEOS will be further compared in a later section to other evolutionary algorithm-based approaches, including a pipe examination genetic algorithm (PEGA), pipe examination particle swarm optimization (PEPSO), and a pipe examination symbiotic organism search (PESOS).

PEGA and PEPSO are benchmark pipe examination techniques that were developed based on the evolutionary algorithms of GA and PSO. A GA and PSO are employed as optimization tools to substitute the algorithm SOS uses in PESOS. PEGA uses a process involving selection, crossover, and mutation to evolve a population of potential solutions toward improved solutions. In PEPSO, the potential solutions, called particles, fly through the problem space by following the current optimum particle. Each particle's movement is influenced by its local best-known position and is also guided toward the global best-known positions in the search space. The readers may refer to References [29,60] for detailed discussions on the use of GAs. In addition, more detail about the application of PSO can be obtained in References [32,61]. The other benchmark approach is PESOS, which is a simplified fault detection approach similar to PEOS but without the preliminary elimination procedure (i.e., the OOA) for the initial organisms. The initial solutions of PEGA, PEPSO, and PESOS are randomly generated from feasible solution domains with corresponding upper and lower bounds. The control and specific parameter settings for each algorithm are listed in Table 2.

Table 2. Specific parameters for each algorithm, with N_P = 10, 20, or 50, and M_{iter} = 10,000 or 20,000.

PEGA	PEPSO	PESOS and PEOS
$m = 0.01$	$w = 0.9\sim0.7$	No specific
$c = 0.8$	$v = X_{min}/10\sim X_{max}/10$	parameters required
$g = 0.9$	-	

Note: N_P = population size/ecosystem size; M_{iter} = maximum iteration; m = mutation rate; c = crossover rate; g = generation gap; w = inertia weight; v = limit of velocity.

3. Laboratory Experiments and PEOS Simulations

3.1. Experiment Configurations

Two cases of experimental reservoir pipe valve (RPV) systems with leaks or blockages that have been reported in the literature were adopted to verify the applicability of PEOS. The first case was carried out in a specially constructed RPV system at Imperial College (IC), London [62]. The system had a pump and tank upstream and a valve at the downstream end. The valve was a transient generation point, and pressure signals were also measured there at the same time. The IC pipe was made of high-density polyethylene (HDPE) with an inner diameter of 50.6 mm and a length of 272 m. Two leaks with different orifice sizes of 1.21×10^{-5} m^2 and 1.50×10^{-5} m^2 occurred at the locations of 65.95 m and 146.32 m, respectively: This was measured from upstream. These two leak orifices were very small, and the discharge coefficient was considered to be one. Thus, the $C_{dL}A_L$s for the two leaks was respectively 1.21×10^{-5} m^2 and 1.50×10^{-5} m^2. The initial flow rate downstream was 1 L/s.

The second case was carried out at the Water Engineering Laboratory (WEL) at the University of Perugia, Italy [63]. A pressurized tank upstream of the system supplied the pipe, and a valve was located at the downstream end for data measurement and transient generation. The WEL pipe was also made of HDPE, with an inner diameter of 93.3 mm and a length of 164.93 m. A partial blockage was located at 88.96 m, measured from upstream. The partial blockage was simulated by an inline valve with a diameter of 38.8 m, and thus the $C_{dB}A_B$ was 1.18×10^{-3} m^2. The initial flow rate downstream was 2.57 L/s.

3.2. PEOS Simulation

In the PEOS application, the IC pipeline system was divided into six series segments with seven nodes. Each segment was assigned a pipe number from 1 to 6 from upstream to downstream. The first five segments had the same length, 50 m, and the last segment had a length 22 m. Two leaks, L1 and L2, which occurred 65.95 m and 146.32 m from the upstream end, were respectively placed in segments

2 and 3. The WEL pipeline system was partitioned into four series segments with five nodes. From upstream to downstream, the segments were given a pipe number from 1 to 4. Segments 1 to 3 had the same length, 50 m, and the last one was 14.93 m. A blockage named B1 was located at segment 2 and was 88.96 m from upstream. A valve was set at the last downstream node for measurement and transient generation for both pipeline systems. The distance interval (Δx) was considered to be 2 m to divide each segment for PEOS to search for leaks. The simulation durations for the IC and WEL pipeline systems were selected to be 15 and 5 s, respectively.

The temporal head distributions predicted by the PEOS for the IC and WEL pipeline systems were respectively displayed in Figure 2a,b. Both predicted temporal head distributions exhibited oscillatory patterns almost identical to the experimental data, indicating that the transient events in the IC and WEL pipeline systems could be precisely simulated by PEOS. Fault information was successfully identified, with the initial values listed in Table 3. In the IC pipeline system, L1 and L2 were respectively detected at 66 m in segment 2, with $C_dAs = 1.23 \times 10^{-5}$ m^2, and at 146 m in segment 3, with $C_dAs = 1.52 \times 10^{-5}$ m^2. Blockage B1 in the WEL pipeline system was identified at 88 m in segment 2, with $C_{dB}A_B = 1.20 \times 10^{-3}$ m^2. The leak and blockage locations in both systems were accurately determined by the proposed approach. The largest relative difference (E) between the actual $C_{dL}A_{LS}/C_{dB}A_B$ and the predicted one was 1.69% for detecting blockage B1 in the WEL pipeline system. The relative differences were insignificant in both systems. The success of PEOS in fault detection indicated that PEOS performs excellently in a pipeline system.

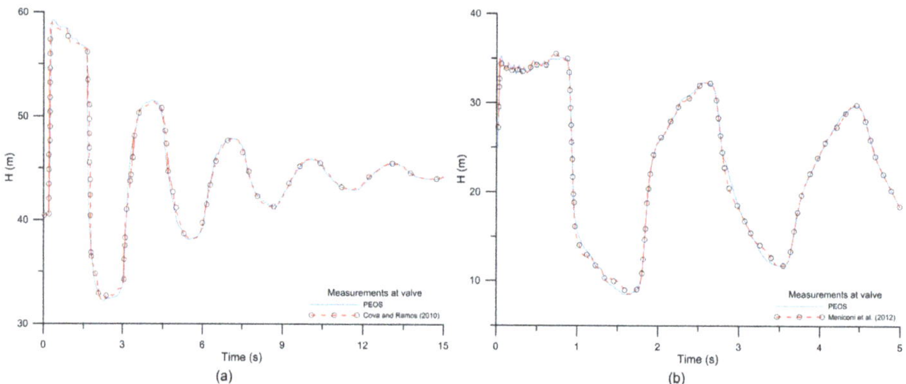

Figure 2. The simulated head distributions at the valve for (**a**) the Imperial College (IC) pipeline and (**b**) the Water Engineering Laboratory (WEL) pipeline.

Table 3. The predicted and actual fault information for the two pipeline systems.

IC pipeline	L1				L2			
	L_p	L_L (m)	$C_{dL}A_L$ (m^2)	E (%)	L_p	L_L (m)	$C_{dL}A_L$ (m^2)	E (%)
Actual	2	15.95	1.21×10^{-5}	-	3	46.32	1.50×10^{-5}	-
PEOS	2	16	1.23×10^{-5}	1.65	3	46	1.52×10^{-5}	1.33

WEL pipeline	B1			
	B_p	B_L (m)	$C_{dB}A_B$ (m^2)	E (%)
Actual	2	38.96	1.18×10^{-3}	-
PEOS	2	38	1.20×10^{-3}	1.69

Note: E = relative difference between the predicted $C_{dL}A_L/C_{dB}A_B$ and the actual one.

4. Fault Detection in a Synthetic Pipe Network

4.1. Simulation Setup and Pipe Network Configuration

A synthetic benchmark WDN (pipe network A) was adopted from Reference [46] to test the applicability of PEOS in fault detection. The associated simulation followed the concept of district metering areas (DMAs), implying that the inflow and outflow of the pipe network system were steady and completely understood. User demands in the pipe network were considered to be constant and could be separated through continuous observation of mass conservations of flow measurements. Pipe network A, shown in Figure 3, is composed of 11 pipes, 9 nodes (with 7 in continuous outflow), 1 potential leak, 1 partial blockage, and 1 distributed deterioration reach. Notice that the characters "N", "P", "L", "B", and "D" represent the node, pipe, leak point, blockage point, and distributed deterioration reach, respectively. The properties of the pipes and nodes of pipe network A are listed in Table 4. The pipe material was considered to be cast iron with an aging effect on the material. Thus, the initial H–W coefficient $C^{HW}(0)$ for each pipe was 130. The H–W coefficient considering the effect of pipe aging ($C^{HW}(t)$) for each pipe was calculated through Equations (2)–(4) and is given in the last column of Table 4. The initial wave speed a_0 of all pipes was postulated as 1000 m/s [25], except for the faulty parts. The impedance of each pipe was calculated by Equation (13) and is given in Table 4. Node N1 was the water supply node, with a constant inflow rate of 400 L/s and a constant head of 120 m. In addition, continuous discharges at N2, N3, N4, N5, N6, N8, and N9 had rates of 80, 40, 35, 35, 40, 80, and 80 L/s, respectively. The leak L1 was located at P11, 300 m away from N3, with $C_{dL}A_L = 2.50 \times 10^{-4}$ m^2 and $Q_L = 3.0$ L/s. A partial blockage B1 was placed at P10, 200 m away from N9. It blocked about 20% of the cross-sectional area of P10, and thus the $C_{dB}A_B$ was 5.6×10^{-2} m^2. In addition, a distributed deterioration reach, D1, occurred at a segment of P1 and was 200 m away from N2. The length and cross-sectional area of D1 were respectively designed to be 80 m and 0.071 m. Its wave speed was assumed to be 800 m/s, and thus the impedance was calculated as 1148.98 s/m^2 from Equation (13). In the simulation, N8 was treated as the transient generation and data measurement point for the simulation of a sudden closure of the valve. The total transient simulation time was considered to be 30 s, with a simulation time interval (Δt) selected as 0.01 s. Thus, the initial Δx was 10 m for the nondeterioration reach and further changed with the wave speed of the deterioration reach. The transient operation was fixed to 5 s for a simulation of the complete closure of the valve.

Table 4. The characteristics of the synthetic water distribution network (WDN) (pipe network A).

Pipe	Node		Diameter (mm)	Length (m)	Impedance (s/m^2)	Year Used (year)	$C^{HW}(t)$
	From	To					
P1	N1	N2	300.0	1000.0	1442.60	10	108.2
P2	N2	N3	300.0	1000.0	1442.60	15	90.2
P3	N3	N4	250.0	1100.0	2077.35	10	105.7
P4	N1	N4	400.0	1250.0	811.47	15	92.2
P5	N4	N5	200.0	500.0	3245.86	5	112.1
P6	N5	N6	400.0	400.0	811.47	5	114.2
P7	N7	N6	200.0	500.0	3245.86	5	112.1
P8	N4	N7	350.0	400.0	1059.87	5	113.6
P9	N7	N8	350.0	600.0	1059.87	5	113.6
P10	N8	N9	300.0	1100.0	1442.60	10	108.2
P11	N3	N9	300.0	1250.0	1442.60	15	90.2

In the following section, the performance of the proposed approach is validated and compared to the other evolutionary algorithm-based approaches mentioned in Section 2.7. The maximum iteration (M_{iter}) was 10,000. Notice that all of the results presented in the following sections were performed on a personal computer with an Intel 2.8 G i5-8400 CPU and 32 GB of RAM.

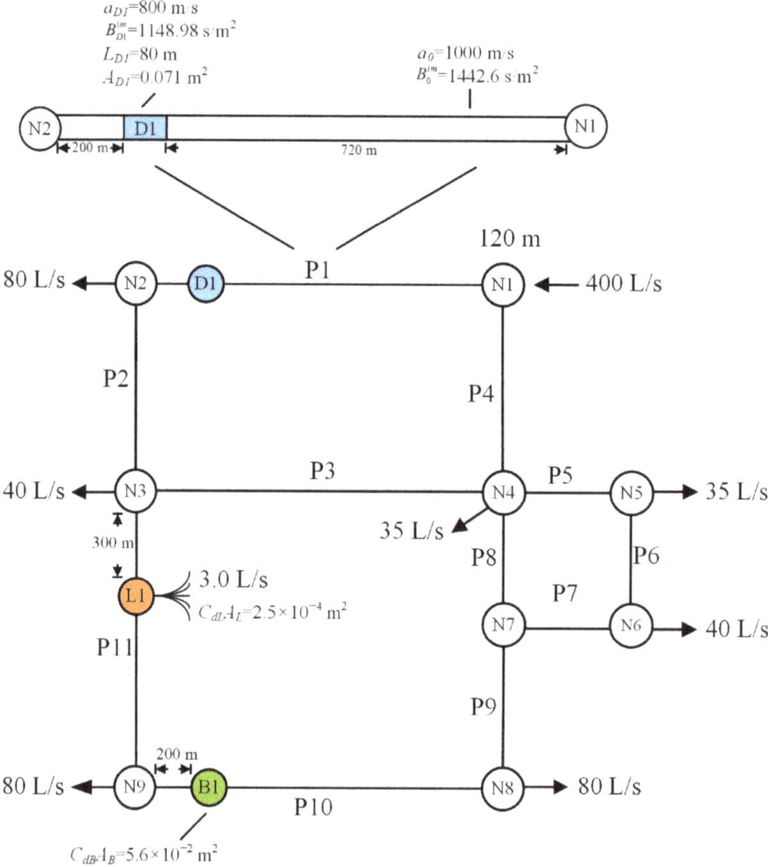

Figure 3. Configuration of pipe network A with a sectional view of P1.

4.2. Validation and Application of PEOS

The steady-state nodal heads and piping flow rates of pipe network A were solved by PNSOS in 52 s. The transient head distributions were further predicted by three benchmark algorithms and the proposed approach. Temporal transient perturbations were observed at N8 by applying different approaches with the various N_P displayed in Figure 4a–d, and the predicted results are given in Table 5. The figures show that the transient perturbations fluctuated between 20 and 140 m with similar oscillatory patterns over 30 s. Figure 4a,b shows that PEGA and PEPSO overestimated the transient perturbations for the case N_P = 10 due to an overestimation of the leakage area size by both algorithms. Such results reflect that the WDN contained a larger total flow rate at the beginning of transient perturbations. Moreover, the blockage at P10 was not detected by either PEGA or PEPSO. Thus, the transmission of water and pressure may not have been affected by the blockage, resulting in the accumulated volumes of water at N8 being higher than other cases when the transient operation point was closed. The predicted head was also overestimated in the case of PEGA for N_P = 20. The calculations in both PEGA and PEPSO were forced to stop because they reached the maximum iteration, M_{iter} = 10,000, in the cases of N_P = 10 and 20. In contrast, the temporal transient perturbations displayed in Figure 4c,d were precisely reconstructed by two SOS-based approaches for all cases of ecosystem size. Deterioration, a blockage, and a leak were detected at P1, P10, and P11, respectively. Table 5 shows that the deterioration, blockage, and leak information was also accurately predicted by two SOS-based approaches. The results prove that those two SOS-based approaches are capable of obtaining optimal fault information even after using fewer initial organisms, reflecting that PESOS and PEOS had great abilities in obtaining the best solution even when using less input data and guessing values. This may have greatly reduced the searching process and computation times. Moreover, Figure 5a,b displays the predicted results of PEOS for impedance and wave speed along P1 and P10, respectively. Both a partial blockage and a deteriorated section can also be identified from the plots of the predicted distributions of the impedance and wave speed in Figure 5. The successful numerical simulation validated the proposed approach to detecting various faults in WDNs.

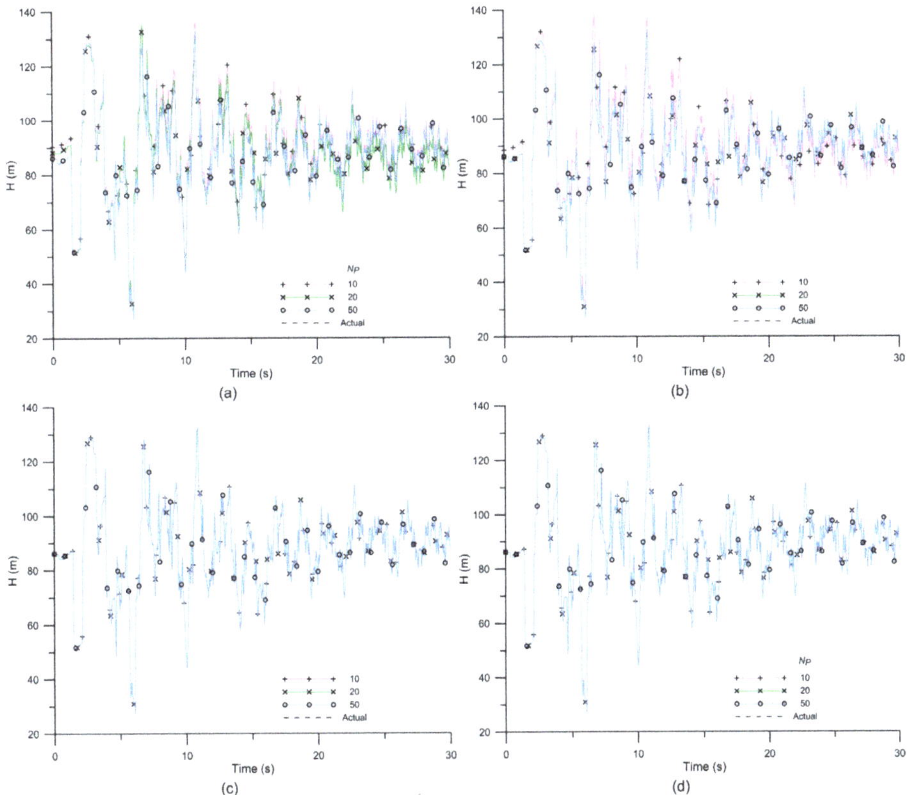

Figure 4. Temporal transient perturbations at N8 of pipe network A predicted by (**a**) PEGA, (**b**) PEPSO, (**c**) PESOS, and (**d**) PEOS with various N_P.

Table 5. Determined fault information of pipe network A.

N_P	Method	L1			B1			D1				
		L_p	L_L (m)	$C_{dL}A_L$ (m^2)	B_P	B_L (m)	$C_{dB}A_B$ (m^2)	D_P	D_L (m)	LD (m)	a_D (m/s)	B_{D1}^{im} (s/m^2)
	Actual	11	300	2.50×10^{-4}	10	200	5.60×10^{-2}	1	200	80	800	1148.98
10	PEGA	2	650	3.27×10^{-4}		Not detected			Not detected			
	PEPSO	11	830	3.19×10^{-4}		Not detected		3	510	100	805	1156.16
	PESOS	11	300	2.49×10^{-4}	10	200	5.58×10^{-2}	1	200	80	800	1148.98
	PEOS	11	300	2.51×10^{-4}	10	200	5.61×10^{-2}	1	200	80	800	1148.98
20	PEGA	11	510	3.34×10^{-4}		Not detected		3	490	70	805	1156.16
	PEPSO	11	300	2.49×10^{-4}	10	200	5.61×10^{-2}	3	700	70	805	1156.16
	PESOS	11	300	2.50×10^{-4}	10	200	5.59×10^{-2}	1	200	80	800	1148.98
	PEOS	11	300	2.50×10^{-4}	10	200	5.60×10^{-2}	1	200	80	800	1148.98
50	PEGA	11	300	2.49×10^{-4}	10	200	5.60×10^{-2}	1	200	80	800	1148.98
	PEPSO	11	300	2.49×10^{-4}	10	200	5.60×10^{-2}	1	200	80	800	1148.98
	PESOS	11	300	2.50×10^{-4}	10	200	5.59×10^{-2}	1	200	80	800	1148.98
	PEOS	11	300	2.50×10^{-4}	10	200	5.60×10^{-2}	1	200	80	800	1148.98

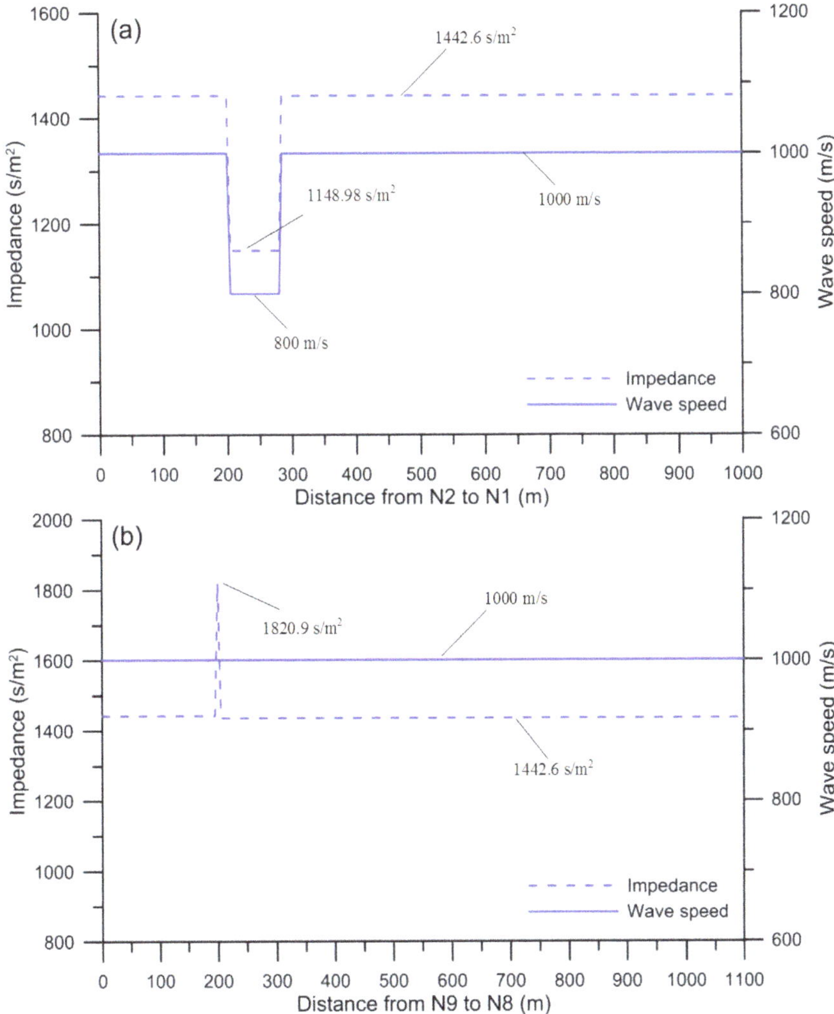

Figure 5. Impedance and wave speed along (**a**) P1 and (**b**) P10, determined by PEOS.

The present techniques were further executed five times to guarantee the reproducibility of the predicted result and to test its efficiency, accuracy, and robustness for obtaining the optimal solution. The N_P was fixed at 50 for all algorithms, and thus all approaches were ensured to deliver accurate predictions, as the results demonstrate above. Table 6 delineates the performance of PEOS and other approaches (five times) in obtaining the optimal fault information of pipe network A. PEGA, PEPSO, and PESOS took about 331.2, 302.2, and 105.4 min and 8072, 7604, and 3882 iterations, respectively, to obtain optimal results over a five-time average. In contrast, PEOS took about 50.6 min and 1382 iterations to complete the searching process and obtain the optimal result. Apparently, PEOS outperformed PEGA and PEPSO, not only in computation time but also in convergence speed. The computational efficiency of PEOS was approximately 84.7% and 83.2% better than PEGA and PEPSO. The computational efficiency of PEOS in fault detection in the WDN significantly improved as a result of using the OOA and SOS. In addition, PEOS saved about 52.8% in computing time and 64% in iterations compared to PESOS, indicating that the OOA could significantly speed up optimization computation by reasonably avoiding blind searches and unnecessary objective function evaluations in the optimization process. PEOS had superiority over the other methods in its fast convergence and effective computation. It also gave more accurate results than the other evolutionary-based algorithms.

Table 6. The performances of the four algorithms.

Method	Round	CPU Time (min)	Average Time (min)	Iterations	Average Iterations
PEGA	1	325		8021	
	2	346		8216	
	3	322	331.2	8124	8072
	4	324		7983	
	5	339		8016	
PEPSO	1	310		7502	
	2	308		7551	
	3	312	302.2	7669	7604
	4	294		7606	
	5	287		7710	
PESOS	1	101		3789	
	2	107		4012	
	3	108	105.4	3883	3882
	4	110		3810	
	5	101		3915	
PEOS	1	56		1415	
	2	49		1371	
	3	46	50.6	1337	1382
	4	52		1396	
	5	50		1391	

Note: CPU time is the computation time.

5. Faults Detection in Large-scale WDN

5.1. Simulation Setup and Large-Scale WDN

PEOS further demonstrated its accuracy and robustness in fault detection on a large-scale drinking WDN by considering different data collection issues. Figure 6 displays the structure of pipe network B with various faults. Pipe network B was modified from Reference [64] with the data of the pipe characteristics listed in Table 7. The pipe network consisted of 74 pipes and 48 nodes, including 11 continual consumption nodes, 2 water supply nodes, and 2 constant-head reservoirs. All pipes were considered to be long-term used cast iron pipes. Hence, the initial H–W coefficient $C^{HW}(0)$ and wave speed a_0 for all pipes in pipe network B were 130 and 1000 m/s, respectively. The $C^{HW}(t)$ for various pipes was also calculated by Equations (2)–(4) and is listed in the last column of Table 7.

Table 7. The characteristics of the large-scale WDN (pipe network B).

Pipe	Node From	Node To	Diameter (mm)	Length (m)	Impedance (s/m^2)	Year Used (year)	$C^{HW}(t)$
P1	N48	N1	950.0	240.0	143.86	5	120.5
P2	N34	N33	900.0	60.0	160.29	10	113.5
P3	N2	N46	1450.0	1830.0	61.75	0	130.0
P4	N43	N2	1150.0	3550.0	98.17	0	130.0
P5	N41	N45	1450.0	1220.0	61.75	0	130.0
P6	N45	N46	1450.0	640.0	61.75	0	130.0
P7	N42	N43	900.0	60.0	160.29	10	113.5
P8	N41	N43	900.0	60.0	160.29	10	113.5
P9	N44	N43	1000.0	50.0	129.83	10	114.6

Table 7. *Cont.*

Pipe	Node		Diameter (mm)	Length (m)	Impedance (s/m^2)	Year Used (year)	C^{HW}(t)
	From	**To**					
P10	N42	N2	900.0	3660.0	160.29	10	113.5
P11	N41	N42	900.0	60.0	160.29	10	113.5
P12	N42	N44	1000.0	60.0	129.83	10	114.6
P13	N40	N42	900.0	800.0	160.29	10	113.5
P14	N37	N41	1450.0	3140.0	61.75	0	130.0
P15	N38	N43	1150.0	3140.0	98.17	0	130.0
P16	N39	N44	1650.0	3140.0	47.69	0	130.0
P17	N38	N36	900.0	60.0	160.29	10	113.5
P18	N38	N39	1000.0	60.0	129.83	10	114.6
P19	N36	N40	800.0	2300.0	202.87	10	112.8
P20	N38	N37	900.0	60.0	160.29	10	113.5
P21	N35	N38	1150.0	4050.0	98.17	0	130.0
P22	N36	N37	900.0	60.0	160.29	10	113.5
P23	N33	N36	800.0	4050.0	202.87	10	112.8
P24	N34	N37	1150.0	4050.0	98.17	0	130.0
P25	N33	N35	900.0	60.0	160.29	10	113.5
P26	N34	N35	900.0	60.0	160.29	10	113.5
P27	N25	N32	800.0	2150.0	202.87	10	112.8
P28	N32	N33	800.0	180.0	202.87	10	112.8
P29	N23	N34	1450.0	2980.0	61.75	0	130.0
P30	N25	N35	1450.0	2980.0	61.75	0	130.0
P31	N31	N30	1650.0	12,000.0	47.69	0	130.0
P32	N22	N24	950.0	670.0	143.86	10	114.0
P33	N29	N28	1000.0	60.0	129.83	10	114.6
P34	N30	N29	1650.0	13400.0	47.69	0	130.0
P35	N13	N11	900.0	80.0	160.29	10	113.5
P36	N11	N15	950.0	4290.0	143.86	5	120.5
P37	N12	N14	900.0	4290.0	160.29	5	115.7
P38	N13	N12	50.0	60.0	51,933.76	10	102.6
P39	N10	N11	970.0	2590.0	137.99	5	120.5
P40	N11	N12	50.0	60.0	51,933.76	10	102.6
P41	N6	N12	900.0	2960.0	160.29	5	115.7
P42	N7	N13	1150.0	2960.0	98.17	0	130.0
P43	N9	N8	1150.0	2280.0	98.17	0	130.0
P44	N8	N10	950.0	370.0	143.86	5	120.5
P45	N8	N7	1000.0	90.0	129.83	0	130.0
P46	N6	N7	50.0	60.0	51,933.76	10	102.6
P47	N5	N6	900.0	1610.0	160.29	5	115.7
P48	N6	N8	50.0	60.0	51,933.76	10	102.6
P49	N3	N5	950.0	1350.0	143.86	5	120.5
P50	N4	N8	50.0	2960.0	51,933.76	10	102.6
P51	N47	N3	950.0	6530.0	143.86	5	120.5
P52	N3	N4	900.0	60.0	160.29	10	113.5
P53	N48	N47	950.0	230.0	143.86	5	120.5
P54	N48	N4	950.0	7200.0	143.86	5	120.5
P55	N27	N26	1000.0	60.0	129.83	10	114.6
P56	N29	N27	1150.0	3200.0	98.17	0	130.0
P57	N26	N25	1450.0	4300.0	61.75	0	130.0
P58	N28	N26	1150.0	3200.0	98.17	0	130.0
P59	N22	N23	800.0	80.0	202.87	10	112.8
P60	N23	N25	750.0	90.0	230.82	0	130.0
P61	N18	N23	950.0	2050.0	143.86	5	120.5
P62	N21	N22	800.0	2380.0	202.87	10	112.8
P63	N20	N23	1150.0	3050.0	98.17	0	130.0
P64	N19	N21	50.0	670.0	51,933.76	5	105.8
P65	N18	N19	50.0	60.0	51,933.76	10	102.6
P66	N19	N20	50.0	60.0	51,933.76	10	102.6
P67	N17	N19	800.0	1830.0	202.87	10	112.8
P68	N18	N20	900.0	60.0	160.29	10	113.5

Table 7. *Cont.*

Pipe	Node		Diameter (mm)	Length (m)	Impedance (s/m²)	Year Used (year)	C^{HW}(t)
	From	**To**					
P69	N14	N17	800.0	1950.0	202.87	10	112.8
P70	N15	N18	950.0	3780.0	143.86	5	120.5
P71	N16	N14	50.0	60.0	51,933.76	5	105.8
P72	N16	N15	900.0	60.0	160.29	10	113.5
P73	N13	N16	1150.0	4290.0	98.17	0	130.0
P74	N14	N15	50.0	60.0	51,933.76	5	105.8

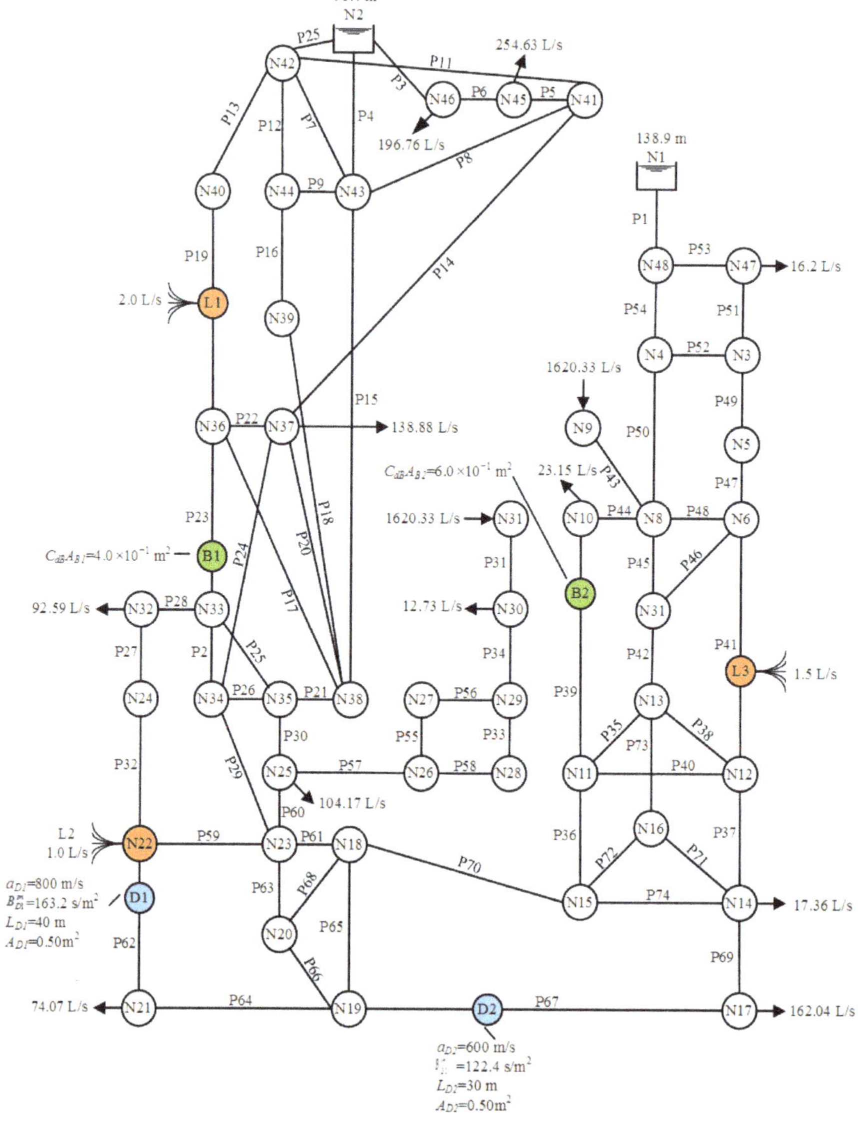

Figure 6. Configuration of the large-scale WDN (pipe network B).

N1 was the first reservoir with a constant-head of 138.9 m, and the second reservoir N2 had a constant-head of 91.4 m. The inflow rates at nodes N9 and N31 were both 1620.33 L/s. The consumption rates at nodes N10, N14, N17, N21, N25, N30, N32, N37, N45, N46, and N47 were respectively 23.15, 17.36, 162.04, 74.07, 104.17, 12.73, 92.59, 138.88, 254.63, 196.76, and 16.2 L/s. Three leaks were separately located at different pipes. Leak L1 was at the middle of P19 and was 1150 m away from N36. Leak L2 was located at P32, 0 m away from N22, implying that leak L2 occurred exactly at N22. Leak L3 was 960 m away from N12 and was located at P41. The $C_{dL}A_L$ values for L1, L2, and L3 were respectively 2.00×10^{-4}, 1.00×10^{-4}, and 1.20×10^{-4} m^2. In addition, Q_{LS} was 2.0, 1.0, and 1.5 L/s for L1, L2, and L3, respectively. Two partial blockages, B1 and B2, were respectively situated at P23 and P39. B1 was 200 m away from N33 and blocked 20% of the cross-sectional area of P23, while B2 was 600 m away from N10 and blocked 15% of the cross-sectional area of P39. Hence, the $C_{dB}A_B$ values of B1 and B2 were 4.0×10^{-1} m^2 and 6.0×10^{-1} m^2, respectively. Moreover, two distributed deterioration reaches, D1 and D2, occurred at P62 and P67, respectively. D1 was located at P62, 400 m away from N22, while D2 was located at P67, 600 m away from N19. The length, wave speed, impedance, and cross-sectional area of D1 were respectively 40 m, 800 m/s, 163.2 s/m^2, and 0.50 m^2, while those of D2 were 30 m, 600 m/s, 122.4 s/m^2, and 0.50 m^2. The properties of the two deterioration reaches are shown in Figure 6 as well. The outflow node N17 was considered to be the transient operation and data collection point for pipe network B. The Δt was also selected to be 0.01 s. Thus, the initial Δx was also considered to be 10 m for the intact pipe reach and was further altered with different wave speeds in the deterioration reach. Because the WDN scale was large and complicated, the transient wave may have taken more time to arrive at the fault points/parts. The total simulation time increased to 60 s. A total of 6001 data points should be collected and used in a complete simulation. Two different cases with different data collection issues were considered to test the reliability of the proposed approach for fault detection in a large-scale WDN. N_P was chosen to be 50, and M_{iter} was updated to 20,000 for possible enormous iterations. The transient excitation period was chosen as 5 or 10 s for the simulation of the complete closure of the valve.

5.2. Case Description and Error Criteria

Three cases were selected to test the capability of PEOS in fault detection in a complex pipe network such as pipe network B, considering the effects of limited observations, measurement errors, and inappropriate transient operation. Case 1 used less data, with a low frequency of 0.1 s (i.e., 10% of the original sampling frequency) to represent instrument limitations in the field survey, and thus 601 data points were collected and used in the simulation of case 1. In case 2, measurement errors were added to the 601 low-frequency data points to depict the uncertainty in data collection. Notice that the white noise ε was normally distributed, with a zero mean and a standard deviation of 0.01 m, which was generated as a random measurement error that was added to each data point in case 2. The observation heads with errors were defined as

$$H^o_{\varepsilon ij} = H^o_{ij} + \varepsilon. \tag{15}$$

Case 3 was designed under the same sampling frequency as case 1, but the transient operation time was extended to 10 s to address the effects of an inappropriate transient operation. There were 601 data points collected after 10 s of transient operation that were used in the simulations of case 3.

In order to evaluate the effects of limited observations and measurement errors on the results predicted by the proposed approach, two error criteria, the standard error of the estimate (SEE) and mean error (ME), were considered. The SEE is a measure of the accuracy of predictions, defined as the square root of the sum of squared errors between the observed and predicted heads divided by the number of degrees of freedom, which equals the number of observed data points minus the number of unknowns. The criterion ME is the average of the sum of errors between the observed and simulated heads.

5.3. Results and Error Analysis

The steady-state hydraulics of pipe network B were predicted by PNSOS in 309 s, and the transient event was then generated by closing the valve at N17. The transient head distributions for cases 1–3 were therefore measured at N17. Table 8 shows the results of fault detection for cases 1–3. In case 1, the information about deterioration reaches D1 and D2 was accurately determined with its corresponding parameters. The locations of three leaks and two blockages were also precisely detected by PEOS. It is noteworthy that leak L2 at node N22 was isolated by the proposed approach, indicating that PEOS was capable of handling the case of pipe junction leakage. In case 1, the E between the actual $C_{dL}A_{LS}/C_{dB}A_B$ values and the predicted ones was insignificant. Table 9 shows the values of the ME and SEE, which for case 1 were 3.41×10^{-6} m and 1.27×10^{-4} m, respectively. The results denote that the predicted heads were not affected by the use of limited observations. The results for case 1 and the small ME and SEE values indicate that PEOS had the potential to deliver moderately good results in a field survey even when only a few observations were available. The success of using fewer measurements indicates that PEOS may not be restricted by instrument limitations. In addition, the data measurement period can therefore be reduced, and the system impact due to a transient event may be slight while using PEOS.

Table 8 shows that PEOS provided relatively good results for deterioration detection in case 2. The locations of the deterioration segments, determined at 390 m for P62 and 610 m for P67, deviated slightly from the actual ones, which were instead located at 400 m for P62 and 600 m for P67. The lengths of D1 and D2 were accurately determined. The impedances for D1 and D2 were respectively estimated as 162.0 and 121.5 s/m^2, with corresponding wave speeds of 794.3 and 595.8 m/s. For leak and blockage detection in case 2, the predicted locations of three leaks and two blockages were close to the real locations, implying that the measurement errors may not have affected location detection. There were errors in the predictions of $C_{dL}A_L$ and $C_{dB}A_B$ in case 2. The relative differences between the predicted $C_{dL}A_L$ values and the actual ones were about 6%, 2%, and 5.83% for L1, L2, and L3, respectively. The relative differences between the determined $C_{dB}A_B$ values and the real ones were about 5.25% for B1 and 4.17% for B2. The results showed that the predicted $C_{dL}A_L$ values and $C_{dB}A_B$ may have been more sensitive than location to measurement errors. This was due to the fact that the OFVs used in PEOS for fault detection were directly related to the head difference (i.e., Equation (14)), which may have been directly influenced by the change in leak area and blockage area. The MEs and SEEs for case 2 are listed in Table 9 and were respectively 1.73×10^{-4} m and 6.35×10^{-2} m, which were both two orders larger than those of case 1. Such a result indicates that measurement errors may have affected accuracy in determining the leak area and blockage area. Thus, data uncertainty should be of concern as an important issue in fault detection in a large-scale pipe network or in future field applications.

In case 3, leaks, blockages, and deterioration segments were also accurately determined by PEOS, with its associated parameters listed in Table 8. The locations of various faults were precisely detected by PEOS. The sizes of leaks and blockages were slightly overestimated compared to case 1, with the largest relative difference, 2.5%, for L3. The values for the ME and SEE for case 3 were respectively 3.29×10^{-6} m and 1.12×10^{-4} m, as shown in Table 9. The results indicate that the predicted heads were not affected, while the transient operation was inadequate. Note that the concept of ITA is to minimize errors between the measured and calculated system state variables. Measurements with an unsuitable transient operation still work well based on the objective function of ITA. The results of case 3 reveal that PEOS can provide good predictions when using different transient operation durations. However, a rapid transient operation is recommended, because it produces large system response data, thus improving the performance of the ITA [31].

Table 8. The optimal fault information of pipe network B predicted by PEOS for three cases.

Case			Leak					Blockage						Deterioration				
	No.	L_p	L_L (m)	$C_{dL}A_L$ (m^2)	E (%)		No.	B_P	B_L (m)	$C_{dB}A_B$ (m^2)	E (%)		No.	D_P	D_L (m)	L_D (m)	a_D (m/s)	B_D^{im} (s/m^2)
Actual	L1	19	1150	2.00×10^{-4}	-		B1	23	200	4.00×10^{-1}	-		D1	62	400	40	800	163.2
	L2	32	0	1.00×10^{-4}	-		B2	39	600	6.00×10^{-1}	-		D2	67	600	30	600	122.4
	L3	41	960	1.20×10^{-4}	-					-	-					-		
Case 1	L1	19	1150	1.98×10^{-4}	1.00		B1	23	190	3.98×10^{-1}	0.50		D1	62	400	40	799.2	163.0
	L2	32	0	1.01×10^{-4}	1.00		B2	39	600	6.04×10^{-1}	0.67		D2	67	600	30	603.1	123.0
	L3	41	950	1.18×10^{-4}	1.67					-	-					-		
case 2	L1	19	1160	1.88×10^{-4}	6.00		B1	23	200	3.79×10^{-1}	5.25		D1	62	390	40	794.3	162.0
	L2	32	0	0.98×10^{-4}	2.00		B2	39	610	5.75×10^{-1}	4.17		D2	67	610	30	595.8	121.5
	L3	41	950	1.11×10^{-4}	5.83					-						-		
Case 3	L1	19	1150	1.96×10^{-4}	2.00		B1	23	190	3.94×10^{-4}	1.50		D1	62	400	40	798.5	162.9
	L2	32	0	0.99×10^{-4}	1.00		B2	39	600	6.07×10^{-4}	1.16		D2	67	600	30	598.2	122.1
	L3	41	950	1.17×10^{-4}	2.50					-	-					-		

Note: E = relative difference between the predicted $C_{dL}A_L/C_{dB}A_B$ and the actual one.

Table 9. The prediction errors for three cases. ME: mean error; SEE: standard error of the estimate.

Case	Prediction Errors	
	ME (m)	SEE (m)
1	3.41×10^{-6}	1.27×10^{-4}
2	1.73×10^{-4}	6.35×10^{-2}
3	3.29×10^{-6}	1.12×10^{-4}

6. Conclusions

This paper demonstrates an inverse transient-based heuristic optimization approach called PEOS for pipe examination in a pipeline or pipe network system. The application of PEOS was verified by two experimental RPV systems in the literature, and PEOS was further applied to identify fault information in synthetic pipe networks. PEOS was used to detect faults in an experimental pipeline (carried out at Imperial College London) and in a pipeline at the Water Engineering Laboratory at the University of Perugia. The head distributions predicted by PEOS agreed well with those from the experiments reported in the literature. The leak and blockage information in both systems was accurately determined by the proposed approach. The results indicated that PEOS provided good predictions in fault detection in a real pipeline system.

The proposed approach was further compared to three evolutionary-based algorithms in fault detection in a synthetic benchmark pipe network. Temporal head distribution and fault information were accurately predicted by PEOS and agreed well with the actual ones, even when using only 10 initial input organisms. PEOS on average took about 50.6 min and 1382 iterations to obtain the optimal results, which is significantly faster than other algorithms. The results indicated that the OOA made the proposed approach avoid most unnecessary calculations of incorrect solutions and quickly converge to the optimal result via three states of SOS. In other words, PEOS not only provided predictions with better accuracy and robustness, but also performed better at computational efficiency. The proposed approach with these two advantages obviously outperformed other algorithms.

To illustrate the applicability of PEOS in fault detection in real-world problems, a large-scale WDN with three data collection statuses was considered as a field study to represent practical issues. The results indicated that PEOS performed well in solving the fault detection problem, considering the effects of limited observations and measurement errors in a complicated WDN. The effect of limited observations on the estimated result was not significant, but the measurement errors induced some inaccuracy. When the observations contained measurement errors, the predicted $C_{dL}A_L$ and $C_{dB}A_B$ had slight deviations compared to the actual ones, indicating that PEOS could achieve good results if the measurements were well collected. Moreover, the results revealed that inappropriate transient operation may not have affected the performance of PEOS in predicting head distribution and fault information.

In summary, we demonstrated via the simulations that PEOS has the ability to simultaneously detect various faults in a pipeline and pipe networks and can outperform other existing evolutionary-based algorithms. Another superiority of PEOS over competing algorithms is the small number of parameters that must be tuned. Fault information can be precisely predicted even when observations are collected with issues. The cases presented in this study were for relatively simple pipe system configurations and operations. Extending the current work from numerical simulations to solving the problems of real-world complicated WDNs would be an interesting direction for further research.

Supplementary Materials: The details of the SOS algorithm are available online at http://www.mdpi.com/2073-4441/11/6/1154/s1.

Author Contributions: C.-C.L. designed the numerical experiment, analyzed the data, and wrote the paper. H.-D.Y. is the supervisor of the proposed research.

Acknowledgments: The authors would like to thank the editor and three anonymous reviewers for their valuable and constructive comments, which greatly improved the manuscript.

Conflicts of Interest: The authors declare no conflict of interest.

References

1. Bartos, M.; Wong, B.; Kerkez, B. Open storm: A complete framework for sensing and control of urban watersheds. *Environ. Sci. Water Res. Technol.* **2018**, *4*, 346–358. [CrossRef]

2. Scola, I.R.; Besançon, G.; Georges, D. Blockage and leak detection and location in pipelines using frequency response optimization. *J. Hydraul. Eng. ASCE* **2017**, *143*, 04016074. [CrossRef]

3. Moser, G.; Paal, S.G.; Smith, I.F.C. Leak detection of water supply networks using error-domain model falsification. *J. Comput. Civil Eng.* **2018**, *32*, 04017077. [CrossRef]

4. Colombo, A.F.; Lee, P.; Karney, B.W. A selective literature review of transient-based leak detection methods. *J. Hydro Environ. Res.* **2009**, *2*, 212–227. [CrossRef]

5. Puust, R.; Kapelan, Z.; Savić, D.A.; Koppel, T. A review of methods for leakage management in pipe networks. *Urban Water J.* **2010**, *7*, 25–45. [CrossRef]

6. Xin, K.; Li, F.; Tao, T.; Xiang, N.; Yin, Z. Water losses investigation and evaluation in water distribution system—The case of sa city in china. *Urban Water J.* **2015**, *12*, 430–439. [CrossRef]

7. Al-Khomairi, A. Leak detection in long pipelines using the least squares method. *J. Hydraul. Res.* **2008**, *46*, 392–401. [CrossRef]

8. Shamloo, H.; Haghighi, A. Optimum leak detection and calibration of pipe networks by inverse transient analysis. *J. Hydraul. Res.* **2010**, *48*, 371–376. [CrossRef]

9. Huang, Y.C.; Lin, C.C.; Yeh, H.D. An optimization approach to leak detection in pipe networks using simulated annealing. *Water Resour. Manag.* **2015**, *29*, 4185–4201. [CrossRef]

10. Wang, X.J.; Lambert, M.F.; Simpson, A.R. Detection and location of a partial blockage in a pipeline using damping of fluid transients. *J. Water Resour. Plan. Manag. ASCE* **2005**, *131*, 244–249. [CrossRef]

11. Tran, D.H.; Perera, B.J.C.; Ng, A.W.M. Hydraulic deterioration models for storm-water drainage pipes: Ordered probit versus probabilistic neural network. *J. Comput. Civil Eng.* **2010**, *24*, 140–150. [CrossRef]

12. Vreeburg, J.H.G.; Boxall, J.B. Discolouration in potable water distribution systems: A review. *Water Res.* **2007**, *41*, 519–529. [CrossRef] [PubMed]

13. Juliano, T.M.; Meegoda, J.N.; Watts, D.J. Acoustic emission leak detection on a metal pipeline buried in sandy soil. *J. Pipeline Syst. Eng. Pract.* **2013**, *4*, 149–155. [CrossRef]

14. Martini, A.; Troncossi, M.; Rivola, A. Vibroacoustic measurements for detecting water leaks in buried small-diameter plastic pipes. *J. Pipeline Syst. Eng. Pract.* **2017**, *8*, 04017022. [CrossRef]

15. Martini, A.; Rivola, A.; Troncossi, M. Autocorrelation Analysis of Vibro-Acoustic Signals Measured in a Test Field for Water Leak Detection. *Appl. Sci.* **2018**, *8*, 2450. [CrossRef]

16. Yazdekhasti, S.; Piratla, K.R.; Atamturktur, S.; Khan, A. Experimental evaluation of a vibration-based leak detection technique for water pipelines. *Struct. Infrastruct. Eng.* **2017**, *14*, 46–55. [CrossRef]

17. Wang, X.; Lennox, B.; Turner, J.; Lewis, K.; Ding, Z.; Short, G.; Dawson, K. Blockage detection in long lengths of pipeline using a new acoustic method. In Proceedings of the 16th International Congress on Sound and Vibration, Krakow, Poland, 5–9 July 2009.

18. Lile, N.L.T.; Jaafar, M.H.M.; Roslan, M.R.; Azmi Muhamad, M.S. Blockage detection in circular pipe using vibration analysis. *Int. J. Adv. Sci. Eng. Inf. Technol.* **2012**, *2*, 54–57. [CrossRef]

19. Lile, N.L.T.; Hadi, H.; Roslan, M.R. Vibration Analysis of Blocked Circular Pipe Flow. *Appl. Mech. Mater.* **2012**, *165*, 197–201. [CrossRef]

20. Duan, W.; Kirby, R.; Prisutov, J.; Horoshenkov, K.V. On the use of power reflection ratio and phase change to determine the geometry of a blockage in a pipe. *Appl. Acoust.* **2015**, *87*, 190–197. [CrossRef]

21. Holley, M.; Diaz, R.; Giovanniello, M. Acoustic Monitoring of Prestressed Concrete Cylinder Pipe: A Case History. In Proceedings of the Pipeline Division Specialty conference 2001, San Diego, CA, USA, 15–18 July 2001.

22. Delgadillo, H.H.; Loendersloot, R.; Akkerman, R.; Yntema, D. Development of an inline water mains inspection technology. In Proceedings of the IEEE International Ultrasonics Symposium (IUS), Tours, France, 18–21 September 2016.

23. Wang, X.H.; Jiao, Y.L.; Yang, J.; Niu, Y.C. The acoustic emission detection and localisation technology of the pipeline crack. *Int. J. Sen. Net.* **2016**, *20*, 111–118. [CrossRef]

24. Li, R.; Huang, H.; Xin, K.; Tao, T. A review of methods for burst/leakage detection and location in water distribution systems. *Water Sci. Technol. Water Supply* **2015**, *15*, 429–441. [CrossRef]

25. Chaudhry, M.H. *Applied Hydraulic Transients*, 3th ed.; Springer: New York, NY, USA, 2014.

26. Datta, S.; Sarkar, S. A review on different pipeline fault detection methods. *J. Loss Prev. Process Ind.* **2016**, *41*, 97–106. [CrossRef]

27. Lee, P.J.; Duan, H.F.; Tuck, J.; Ghidaoui, M. Numerical and experimental study on the effect of signal bandwidth on pipe assessment using fluid transients. *J. Hydraul. Eng. ASCE* **2015**, *141*, 04014074. [CrossRef]

28. Sheikholeslami, R.; Talatahari, S. Developed swarm optimizer: A new method for sizing optimization of water distribution systems. *J. Comput. Civil Eng.* **2016**, *30*, 04016005. [CrossRef]

29. Vítkovský, J.P.; Simpson, A.R.; Lambert, M.F. Leak detection and calibration using transients and genetic algorithms. *J. Water Resour. Plan. Manag. ASCE* **2000**, *126*, 262–265. [CrossRef]

30. Liggett, J.A.; Chen, L.C. Inverse transient analysis in pipe networks. *J. Hydraul. Eng. ASCE* **1994**, *120*, 934–955. [CrossRef]

31. Vítkovský, J.P.; Lambert, M.F.; Simpson, A.R.; Liggett, J.A. Experimental observation and analysis of inverse transients for pipeline leak detection. *J. Water Resour. Plan. Manag. ASCE* **2007**, *133*, 519–530. [CrossRef]

32. Jung, B.S.; Karney, B.W. Systematic exploration of pipeline network calibration using transients. *J. Hydraul. Res.* **2008**, *46*, 129–137. [CrossRef]

33. Haghighi, A.; Ramos, H.M. Detection of leakage freshwater and friction factor calibration in drinking networks using central force optimization. *Water Resour. Manag.* **2012**, *26*, 2347–2363. [CrossRef]

34. Covelli, C.; Cozzolino, L.; Cimorelli, L.; Della Morte, R.; Pianese, D. Optimal location and setting of prvs in wds for leakage minimization. *Water Resour. Manag.* **2016**, *30*, 1803–1817. [CrossRef]

35. Meniconi, S.; Duan, H.F.; Lee, P.J.; Brunone, B.; Ghidaoui, M.S.; Ferrante, M. Experimental investigation of coupled frequency and time-domain transient test-based techniques for partial blockage detection in pipelines. *J. Hydraul. Eng. ASCE* **2013**, *139*, 1033–1040. [CrossRef]

36. Mohapatra, P.K.; Chaudhry, M.H.; Kassem, A.A.; Moloo, J. Detection of partial blockage in single pipelines. *J. Hydraul. Eng. ASCE* **2006**, *132*, 200–206. [CrossRef]

37. Lee, P.J.; Vítkovský, J.P.; Lambert, M.F.; Simpson, A.R.; Liggett, J.A. Discrete blockage detection in pipelines using the frequency response diagram: Numerical study. *J. Hydraul. Eng. ASCE* **2008**, *134*, 658–663. [CrossRef]

38. Duan, H.F.; Lee, P.J.; Ghidaoui, M. Transient wave-blockage interaction in pressurized water pipelines. *Procedia Eng.* **2014**, *70*, 573–582. [CrossRef]

39. Gong, J.; Lambert, M.F.; Simpson, A.R.; Zecchin, A.C. Detection of localized deterioration distributed along single pipelines by reconstructive moc analysis. *J. Hydraul. Eng. ASCE* **2014**, *140*, 190–198. [CrossRef]

40. Stephens, M.L.; Simpson, A.R.; Lambert, M.F. Internal wall condition assessment for water pipelines using inverse transient analysis. In Proceedings of the 10th Annual Symposium on Water Distribution Systems Analysis, American Society of Civil Engineers, Kruger National Park, South Africa, 17–20 August 2008.

41. Stephens, M.L.; Lambert, M.F.; Simpson, A.R. Determining the internal wall condition of a water pipeline in the field using an inverse transient. *J. Hydraul. Eng. ASCE* **2013**, *139*, 310–324. [CrossRef]

42. Hachem, F.E.; Schleiss, A.J. Detection of local wall stiffness drop in steel-lined pressure tunnels and shafts of hydroelectric power plants using steep pressure wave excitation and wavelet decomposition. *J. Hydraul. Eng. ASCE* **2012**, *138*, 35–45. [CrossRef]

43. Gong, J.; Simpson, A.R.; Lambert, M.F.; Zecchin, A.C.; Kim, Y.i.; Tijsseling, A.S. Detection of distributed deterioration in single pipes using transient reflections. *J. Pipeline Syst. Eng. Pract.* **2013**, *4*, 32–40. [CrossRef]

44. Gong, J.; Lambert, M.F.; Nguyen, S.T.N.; Zecchin, A.C.; Simpson, A.R. Detecting thinner-walled pipe sections using a spark transient pressure wave generator. *J. Hydraul. Eng. ASCE* **2018**, *144*, 06017027. [CrossRef]

45. Cobacho, R.; Arregui, F.; Soriano, J.; Cabrera, E. Including leakage in network models: An application to calibrate leak valves in EPANET. *J. Water Supply Res. Technol. Aqua* **2015**, *64*, 130–138. [CrossRef]

46. Yeh, H.D.; Lin, Y.C. Pipe network system analysis using simulated annealing. *J. Water Supply Res. Technol. Aqua* **2008**, *57*, 317–327. [CrossRef]

47. Mays, L.W. *Water Supply Systems Security*; McGraw-Hill: New York, NY, USA, 2004.

48. Savić, D.A.; Banyard, J.K. *Water Distribution Systems*, 2nd ed.; ICE: London, UK, 2011.

49. Seifollahi-Aghmiuni, S.; Bozorg Haddad, O.; Omid, M.H.; Mariño, M.A. Effects of pipe roughness uncertainty on water distribution network performance during its operational period. *Water Resour. Manag.* **2013**, *27*, 1581–1599. [CrossRef]

50. Larock, B.E.; Jeppson, R.W.; Watters, G.Z. *Hydraulics of Pipeline Systems*, 1st ed.; CRC Press: Boca Raton, FL, USA, 2000.

51. Duan, H.F.; Ghidaoui, M.; Lee, P.J.; Tung, Y.K. Unsteady friction and visco-elasticity in pipe fluid transients. *J. Hydraul. Res.* **2010**, *48*, 354–362. [CrossRef]

52. Reddy, H.P.; Silva-Araya, W.F.; Chaudhry, M.H. Estimation of decay coefficients for unsteady friction for instantaneous, acceleration-based models. *J. Hydraul. Eng. ASCE* **2012**, *138*, 260–271. [CrossRef]

53. Bergant, A.; Tijsseling, A.S.; Vítkovský, J.P.; Covas, D.I.C.; Simpson, A.R.; Lambert, M.F. Parameters affecting water-hammer wave attenuation, shape and timing—Part 1: Mathematical tools. *J. Hydraul. Res.* **2008**, *46*, 373–381. [CrossRef]

54. Stephens, M.L.; Simpson, A.R.; Lambert, M.F. Hydraulic transient analysis and discrete blockage detection on distribution pipelines: Field tests, model calibration, and inverse modeling. In Proceedings of the World Environmental and Water Resources Congress 2007, Tampa, FL, USA, 15–19 May 2007.

55. Ho, Y.C.; Cassandras, C.G.; Chen, C.H.; Dai, L. Ordinal optimisation and simulation. *J. Oper. Res. Soc.* **2000**, *51*, 490–500. [CrossRef]

56. Cheng, M.Y.; Prayogo, D. Symbiotic organisms search: A new metaheuristic optimization algorithm. *Comput. Struct.* **2014**, *139*, 98–112. [CrossRef]

57. Cheng, M.Y.; Prayogo, D.; Tran, D.H. Optimizing multiple-resources leveling in multiple projects using discrete symbiotic organisms search. *J. Comput. Civil Eng.* **2016**, *30*, 04015036. [CrossRef]

58. Pudar, R.S.; Liggett, J.A. Leaks in pipe networks. *J. Hydraul. Eng. ASCE* **1992**, *118*, 1031–1046. [CrossRef]

59. Abhulimen, K.E.; Susu, A.A. Liquid pipeline leak detection system: Model development and numerical simulation. *Chem. Eng. J.* **2004**, *97*, 47–67. [CrossRef]

60. Simpson, R.A.; Dandy, C.G.; Murphy, J.L. Genetic algorithms compared to other techniques for pipe optimization. *J. Water Resour. Plan. Manag. ASCE* **1994**, *120*, 423–443. [CrossRef]

61. Jung, B.S.; Karney, B. Fluid transients and pipeline optimization using ga and pso: The diameter connection. *Urban Water J.* **2004**, *1*, 167–176. [CrossRef]

62. Covas, D.; Ramos, H. Case Studies of Leak Detection and Location in Water Pipe Systems by Inverse Transient Analysis. *J. Water Resour. Plan. Manag. ASCE* **2010**, *136*, 248–257. [CrossRef]

63. Meniconi, S.; Brunone, B.; Ferrante, M. Water-hammer pressure waves interaction at cross-section changes in series in viscoelastic pipes. *J. Fluids Struct.* **2012**, *33*, 44–58. [CrossRef]

64. Chin, K.K.; Gay, R.K.L.; Chua, S.H.; Chan, C.H.; Ho, S.Y. Solution of water networks by sparse matrix methods. *Int. J. Numer. Methods Eng.* **1978**, *12*, 1261–1277. [CrossRef]

Article

Comparison of Flow-Dependent Controllers for Remote Real-Time Pressure Control in a Water Distribution System with Stochastic Consumption

Philip R. Page [1] and Enrico Creaco [2,*]

[1] Built Environment, Council for Scientific and Industrial Research (CSIR), Pretoria 0184, South Africa; pagepr7@gmail.com

[2] Dipartimento di Ingegneria Civile e Architettura, University of Pavia, Via Ferrata 3, 27100 Pavia, Italy

* Correspondence: creaco@unipv.it

Received: 14 January 2019; Accepted: 20 February 2019; Published: 27 February 2019

Abstract: The control of pressure at a remote critical node using a pressure control valve is a highly effective way to attain pressure management. To perform real-time control, various kinds of controllers can be used, including flow-dependent controllers. These controllers calculate valve setting adjustment based both on the deviation of the pressure from the set-point and on the flow rate at the valve site. After putting all the flow-dependent controllers present in the scientific literature within the same framework, this paper presents a numerical comparison of their performance under realistic conditions of stochastic demand. Two controllers were selected for the comparison, namely the simple LCF (parameter-less proportional controller with known constant pressure control valve flow); and LVF (parameter-less controller with known variable pressure control valve flow), which uses a flow rate forecast. Indeed, this study considered an upgrade of LVF, in which the flow rate forecast was tailored to the conditions of stochastic demand. The application in a specific example network proved the performance of these controllers to be quite similar, although LCF was preferable due to its simple structure. For LCF, the average pressure at the critical node had a clear relationship to the consumption pattern. LVF outperformed when the hourly variation dominates the fluctuations in the flow. The conditions under which this out-performance occurred are comprehensively discussed.

Keywords: hydraulic modelling; pressure control valve; pressure management; remote real-time control; stochastic consumption; water distribution system

---

1. Introduction

Decreased pressure reduces water leakage from pipes, lowers pipe burst frequency and may reduce water consumption [1]. The device that is by far the most widely used to reduce pressure in a water distribution system (WDS) is a pressure control valve (PCV): commonly a pressure reducing valve (PRV) [2,3].

A closed-loop technique uses measurements in the WDS, while an open-loop technique does not. Earlier advanced pressure management techniques, with the first being the simplest, include: (1) time modulation (open-loop) [1]; (2) flow modulation (closed-loop) [4,5]; and (3) remote node modulation, which is not real-time (closed-loop) [6]. These earlier techniques reduce the pressure better than a classical PCV with no controller, but do not reduce the pressure as low as possible.

A WDS node where the pressure is sensitive to PCV adjustment, and whenever possible, also has the lowest pressure, is called a critical node (CN) [7]. To keep the pressure at the CN continually constant [8], the PCV setting must be changed in real-time, i.e., not manually, intermittently or only at specific times. This is usually accomplished by adjusting the setting every time-step, where the time-step is typically of the order of minutes [7].

The remote real-time control (RRTC) technique strives to make the pressure throughout the WDS as low as possible [9,10], by attempting to set the pressure at a remote CN in real-time to a low and constant target set-point value [11,12]. This is made possible by recent advances in the availability of wireless technology [13–15].

A laboratory experiment demonstrated how RRTC with a PRV can be attained by use of a controller [16]. An example of a field demonstration is the one in the district of Benevento, Italy [10].

The controllers in [7,17,18] only use the pressure measurement at the CN. Controllers that also use the flow rate through the PCV, taking changes in WDS conditions into account, were subsequently developed [19–23]. The water flow rate in a pipe equipped with the PCV needs to be measured.

Consumption in a real WDS is stochastic in nature. Recently, several numerical RRTC studies take this into account: either approximating it as random fluctuations [18], or using a comprehensive bottom-up approach [22,23]. This paper reports numerical results on two closely related flow-dependent PCV controllers in the latter approach. One of these was formulated and studied for the first time with stochastic consumption, because this may critically affect the controller's viability.

2. Head-Loss Controller

In this and the next section, a derivation of various controllers, outlining assumptions made, is presented. This is done in an effort to bring them all together and to put them in the same rigorous framework. The aim is to emphasise that the controllers are important from the viewpoint of hydraulic theory. Particularly, the derivation is within the context of a WDS where there is not significant time-dependence on a time-scale shorter than the control time-step T_c, or on a time-scale of a few T_c (see Appendix A).

Let $\tilde{H}$ be the head-loss across the PCV, and H the head at the CN. It can be argued from the Newton–Raphson numerical method (see Appendix A) that an appropriate controller, called the "head-loss" controller [21], would calculate the adjusted head-loss

$$\tilde{H}_{i+1} = \tilde{H}_i - S_i \left(H_i - H_{sp} \right) \tag{1}$$

from the current head-loss $\tilde{H}_i$. Here, H_{sp} is the target set-point head of the CN; and the notation and sensitivity S_i are defined in Appendix A (see also [24]). The information at iteration i determines the next iteration $i + 1$. The iterations are separated by T_c. The value of S_i varies for different iterations, and is impractical to determine for a real-world WDS without a hydraulic model [21]. When the CN head depends very sensitively on the PCV head-loss, $S_i = -1$ for a PRV. Using this value in Equation (1) yields the controller employed in [9,11,20].

Equation (1) represents the choice of controller, from the viewpoint of theory [12,19,21,23]. However, the controller evaluates hydraulic quantities at a specific time, and hence is not sensible for quantities that exhibit significant time-dependence.

3. Controllers Based on Known PCV Flow Rate

A PCV is conventionally modelled by [17,19,20]:

$$\tilde{H} = \frac{\zeta}{2g} v^2 \qquad v = \frac{Q}{A} \tag{2}$$

where ζ is the (dimension-less) PCV head-loss coefficient, v is the water velocity, Q is the flow rate through the PCV, A is the area of the port opening within the PCV, and g is the acceleration due to gravity. Substituting Equation (2) into Equation (1) implies that the adjusted head-loss coefficient can be calculated as

$$\zeta_{i+1} = \zeta_i \left(\frac{v_i}{v_{i+1}} \right)^2 - \frac{2g S_i}{v_{i+1}^2} \left(H_i - H_{sp} \right) \tag{3}$$

from the current head-loss coefficient ξ_i. Equation (3) can be used as the very general form of a controller, as conceived in [23] for $S_i = -1$. It is called the "general parameter-less controller with known variable PCV flow" (GVF). It is parameter-less, because it contains no tunable parameter. Specifically, S_i is not tunable.

The right hand side of Equation (3) is separated into two parts. The first part does not involve the future (does not depend on v_{i+1}), and is important because it does not require modelling the future. The second part is the remainder (denoted by Φ and Ψ). Separating Equation (3) into these parts yields

$$\xi_{i+1} = \xi_i - \frac{2gS_i}{v_i^2}\left(H_i - H_{sp}\right) + \Phi_i + \Psi_i \tag{4}$$

where

$$\Phi_i = -\xi_i f_i \qquad \Psi_i = \frac{2gS_i}{v_i^2} f_i \left(H_i - H_{sp}\right) \qquad f_i = 1 - \frac{1}{\left(1 + \frac{\Delta v_i}{v_i}\right)^2} \tag{5}$$

with $\Delta v_i \equiv v_{i+1} - v_i$. All dependence on the future in the remainder part is through the dimensionless f_i, and hence through Δv_i, which needs to be modelled.

Neglecting Δv_i leads to a controller with

$$\Phi_i \approx 0 \qquad \Psi_i \approx 0 \tag{6}$$

Equations (4) and (6) define the "parameter-less proportional controller with known constant PCV flow" (LCF) [21]. With $S_i = -1$, it is first derived in [19]; and is also called "valve resistance" (RES) control [11,20].

Another controller can be obtained by only keeping the dominant terms in Φ and Ψ, which are linear and up to first order in the difference terms. f_i is proportional to Δv_i. Hence, Ψ_i is the only term in Equation (4) that is proportional to *two* difference terms (Δv_i and $H_i - H_{sp}$) and can accordingly be neglected. f_i can be evaluated to lowest order in the difference term Δv_i, leading to a controller with

$$\Phi_i \approx -\frac{2\xi_i}{v_i}\Delta v_i \qquad \Psi_i \approx 0 \tag{7}$$

Equations (4) and (7) define the "parameter-less controller with known variable PCV flow" (LVF), first derived in [21].

4. Modification of Controllers: Stochastic Consumption

In a real WDS, there is usually significant time-dependent behaviour due to stochastic water consumption. For this situation, the controllers need to be modified, so that a hydraulic quantity evaluated at a specific time is postulated to be replaced by an average. Let $Z(t_{i-n}, t_i)$ be the average of Z in the time interval from $t_{i-n} = t_i - nT_c$ to t_i. Then, it is natural to define the controllers in Equations (3)–(7) to be used with H_i and v_i replaced by $H(t_{i-1}, t_i)$ and $v(t_{i-1}, t_i)$, respectively, as done in [11,22] for the LCF controller with $S_i = -1$. Similarly, S_i is replaced by $S(t_{i-1}, t_i)$.

The future change Δv_i can be modelled by estimating it from the past. Since Δv_i is a velocity change in an interval T_c, it is proposed that it should be replaced in Equation (7) by

$$\frac{v(t_{i-n}, t_i) - v(t_{i-2n}, t_{i-n})}{n} \tag{8}$$

Hence, n is an indicator of how far back the past mean values are used to predict the future. For each n, the corresponding LVF controller is denoted LVFn.

5. Numerical Study in the Example WDS

Controllers LCF and LVF were applied to the RRTC of a network in northern Italy (see skeletonised layout in Figure 1), which caters to about 30,000 inhabitants. This network has already been used for investigations in the area of pressure management [19,22,25].

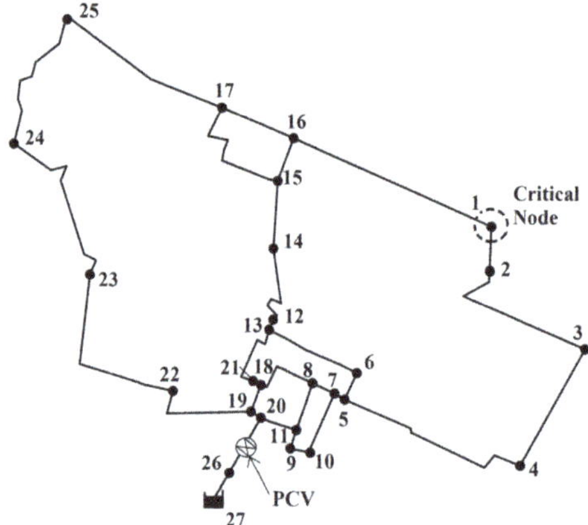

Figure 1. Example water distribution system.

The network consists of a single source node (node 27), although networks with more sources could also be considered. There are 26 demanding nodes with ground elevation of 0 m a.s.l. and 32 pipes. The network can be considered as a single pressure zone, because of its size and the uniform ground elevation.

The references above give further details about the characteristics of the network. The source node has a head varying around 40 m a.s.l. [22]. A single DN300 plunger valve is the PRV, located at the end of pipe 26-11. The PRV has the head-loss coefficient $\zeta(\alpha)$ given by

$$\zeta = 10^{c_1 - c_2 \log_{10}(1-\alpha)} \tag{9}$$

where data from the valve manufacturer allow the coefficients $c_1 = 1.5$ and $c_2 = 2.8$ to be calculated. The setting α is adjusted by the controller; and is constrained to range from 0 (completely open) to 0.95 (nearly completely closed). The maximum value $\alpha = 0.95$ was chosen consistent with the real use of control valves, the objective of which is to modulate flow, rather than to interrupt it. However, it must be noted that this upper boundary does not affect the results of the simulations, as shown below.

The lowest pressure values during the day are found at node 1, which was hence selected as the CN where pressure control was applied. The RRTC of the PCV was performed to enable the pressure head at the CN to be near the target set-point head of $H_{sp} = 25$ m.

The bottom-up approach detailed in [22,25] was used to obtain the consumption for each node. This approach is based on consumption pulse generation through the Poisson model [26], considering pulse duration and intensity to be dependent random variables, both of which are expressed through the beta distribution. The pulse arrival rate at each node was calculated to obtain the expected average nodal demand, while considering the pattern of total demand observed in the WDS in a single day. More details of the bottom-up approach for demand generation are presented in [22], along with the parameter values used for demand generation.

To describe the hydraulics of the WDS, the model described in [22] was used. This model enables the unsteady flow modelling of the WDS and the accurate reproduction of the hydraulic behaviour of the valve. Compared to other software available in the market, this proprietary model has the advantage of considering unsteady flow pipe resistances, thus yielding more realistic results.

6. Results

Since preliminary investigations proved that $S_i = -1$ gave good results, this value was kept throughout all the calculations. As an example of the results, Figure 2 reports calculations over a day for LCF with $T_c = 3$ min. The instantaneous and averaged flow-rate Q_{av} through the PCV, valve setting and instantaneous pressure at the CN are shown. As for the flow-rates, these values include the WDS pulsed demand [22] (Figure 4) and leakage, which added up on average to approximately 20% of the total output from the source. As for the valve setting, it must be noted that it always stayed far from the lower and upper boundaries, attesting to the proper regulation behaviour of the valve.

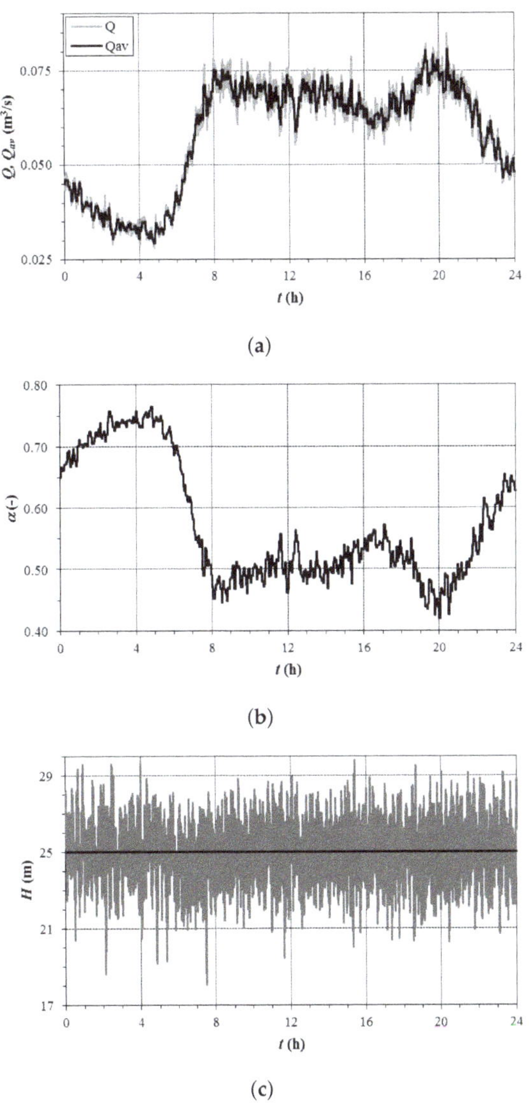

Figure 2. (**a**) Flow-rate Q every second and its value Q_{av} averaged over 3 min; (**b**) valve setting α (evaluated every 3 min); and (**c**) ressure head H every second.

The LVF controller models the future from the past. All LVFn controllers that use velocities for a period $2nT_c \leq 42$ min into the past were investigated. A different method from Equation (8), which uses a regression fit of past Q_{av} values to predict the future flow, has also been proposed [23].

Let $|e|_{mean}$ and e_{mean} denote the average of $|H - H_{sp}|$ and $H - H_{sp}$, respectively, where H is evaluated every second. These two performance measures determine the deviation of the pressure at the CN from the target set-point. In accordance with [22], the primary measure was $|e|_{mean}$. In addition, let the performance measure $\Sigma|\Delta\alpha|$ be the sum of the actuator setting absolute corrections evaluated at each iteration. This is a measurement of the wear and tear on the PCV due to setting changes. Performance is the best when the performance measures are as low as possible.

Undesirable behaviour due to PRV self-interactions was observed for $T_c \leq 1$ min [22], thus $T_c = 3, 5$ and 10 min were considered. Of the time-steps studied, $T_c = 3$ min gave the lowest $|e|_{mean}$ for both LCF and LVF. The results for the full 24-h period with this time-step are now discussed.

For LVFn, $|e|_{mean}$ and $\Sigma|\Delta\alpha|$ decreased monotonically as n increased (Figure 3). Evidently, decreases became insignificant nearing $n = 7$. It was also found that this monotonic decrease happened in each individual hourly period. The best performing LVFn was hence LVF7, which uses velocities for a period 42 min into the past. However, $|e|_{mean}$ and $\Sigma|\Delta\alpha|$ were perfectly reasonable for LVF3, if a controller looking less into the past were desired.

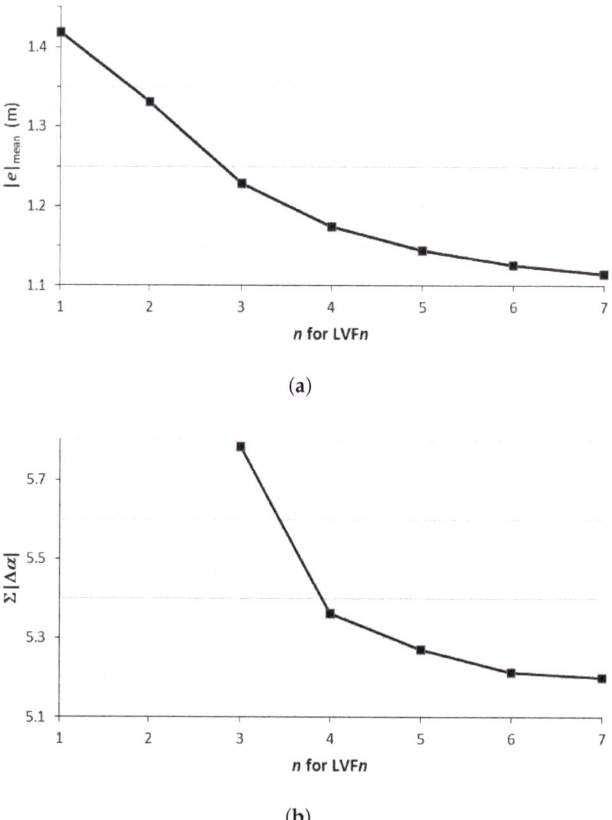

Figure 3. Results for LVFn with $T_c = 3$ min over one day: (**a**) $|e|_{mean}$; and (**b**) $\Sigma|\Delta\alpha|$. For $n = 1, 2$, the values are out of range at 11.1 and 7.4, respectively.

For the day, $|e|_{mean}(LVF)-|e|_{mean}(LCF) = 0.016$ m, thus LCF outperformed insignificantly. Evaluating a similar difference for $\Sigma|\Delta\alpha|$, it was found that LCF outperformed LVF7 by a tiny 0.85%.

It is interesting to determine the variation of the pressure deviation during the day. Comparison with the consumption pattern can more easily be done by evaluating hourly averages, in order to reduce the effect of stochastic fluctuation in consumption. In Figure 4, $|e|_{mean}$ is shown as a function of time. It is noticeable that the results for LCF and LVF7 are very similar at a certain time, and that there is apparently random variation from hour to hour. This suggests that $|e|_{mean}$ has significant stochastic fluctuation. On the other hand, e_{mean} as a function of time shows clear patterns (especially for LCF), and appears to be less dependent on stochastic fluctuation (Figure 5a).

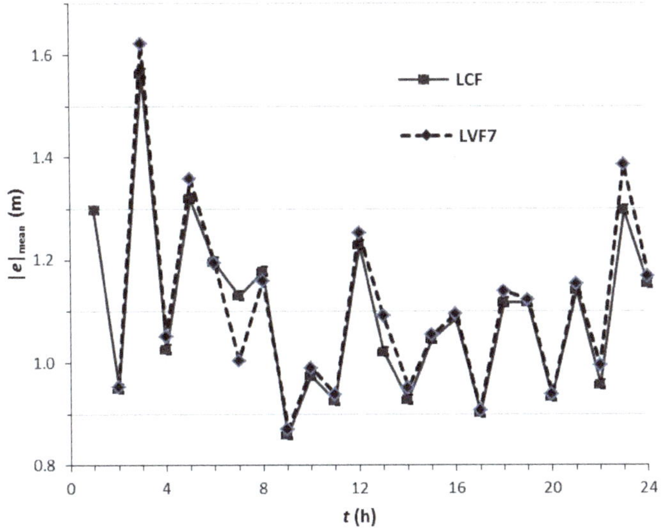

Figure 4. $|e|_{mean}$ evaluated during an hour period preceding the time of the datum shown $T_c = 3$ min.

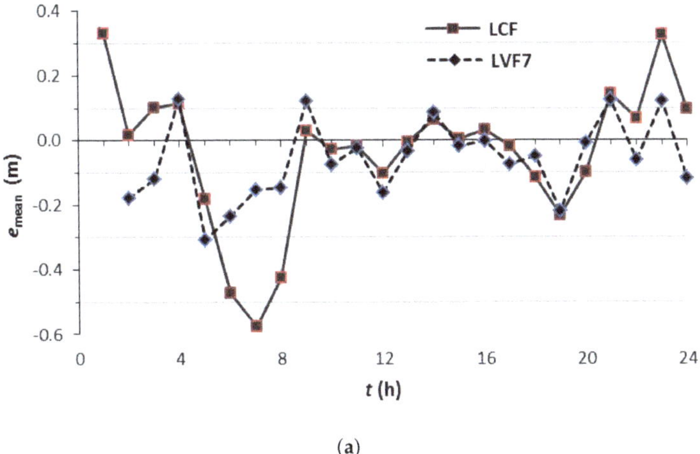

(a)

Figure 5. *Cont.*

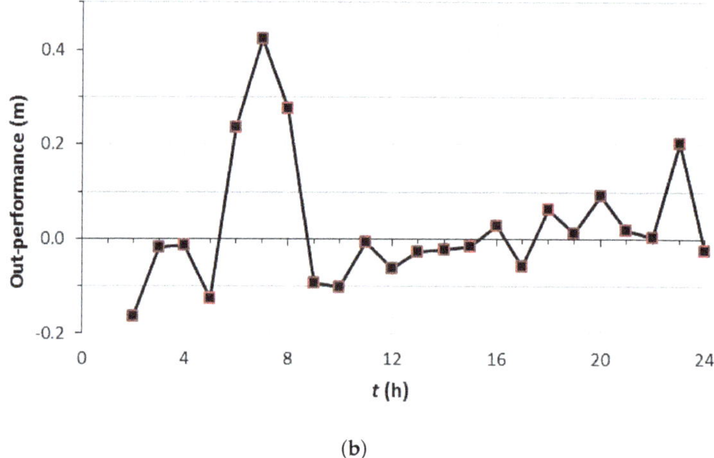

(b)

Figure 5. e_{mean} evaluated during an hour period preceding the time of the datum shown ($T_c = 3$ min): (**a**) e_{mean}; and (**b**) out-performance of LVF7 over LCF, defined using e_{mean}. Out-performance is positive if the e_{mean} of LVF7 is nearer to 0 than the e_{mean} of LCF.

Assume that the flow rate through the PCV in Figure 2a is fitted by a smoothly varying Q_{trend}, which indicates the trend in the flow rate (the hourly variation). The results for LCF in Figure 5a have an interesting pattern. Q_{trend} increased noticeably during 5–8 h and 17–20 h. Accordingly, e_{mean} was negative for the points in Figure 5a representing these hours. In addition, the deviation of e_{mean} from zero was largest for the period 6–7 h when the rate of change of flow was the largest of the entire day. The flow decreased noticeably during 0–1 h and 20–24 h. Fittingly, e_{mean} was positive for the points in Figure 5a representing these hours.

All of these observations for LCF were consistent with a related study predicting that the deviation of the pressure is approximately proportional to $-Q'/Q$ in the context of non-stochastic consumption [27], where Q' is the rate of change of Q.

For non-stochastic consumption, numerical results for LCF were previously pointed out to suggest that the deviation is driven by $-Q'$ [21]. This can be confirmed by Figures 7, 10, 12 and 13 of [11], and in the pressure shown in Figure 7 (7-RES) [12]. For stochastic consumption, this behaviour can also be seen in Figure 7b of [22].

Figure 5a shows that LVF7 was less prone than LCF to deviate significantly from the target set-point pressure. Figure 5b indicates that LVF7 substantially outperformed LCF during Hours 5–8. This coincided with the hours when Q_{trend} changed the most quickly. In addition, during the second fastest flow change during 20–24 h, LVF7 outperformed LCF.

The sum of the out-performance amounts for each hour in Figure 5b was 0.028 m. Hence, with the performance measure e_{mean}, LVF7 outperformed LCF insignificantly over the day. Taking into account the results from all three performance measures, it is fair to say that the performance of LVF7 was the same as LCF over the entire day.

7. Discussion

Water consumption shows stochastic fluctuation in a real WDS. There are also unsteady flow processes that can cause sudden variations in flow and pressure [4,5,28]. For RRTC, the effect on a proportional-integral controller of adding random consumption fluctuation at each time-step T_c to smooth water consumption is initially studied in [18]. The bottom-up approach used in this work incorporates both fully stochastic consumption fluctuation and unsteady flow processes.

From the viewpoint of the derivation of the controllers, LVF should at first glance be an improvement on LCF. However, LVF depends on a future change Δv_i, which can only be modelled by estimating it from the past [11]. The way to decrease the effect of stochastic fluctuation in consumption is to estimate Δv_i by looking far into the past (Figure 3). However, relying on the far past is undesirable, as shown by the case of no fluctuation. (Assuming Q is a smooth function of time, estimating Δv_i from the most recent past velocities should be the most accurate). Hence, the larger is the fluctuation, the greater is the performance of LVF weakened.

The performance of LVF relative to LCF depends on a large number of factors. For a given WDS, these were argued to include $|Q'_{trend}|$ at a certain time. It is postulated that this is to be compared to the magnitude of the average fluctuation at a certain time. The following study indicates what happens if $|Q'_{trend}|$ dominates the fluctuations. For non-stochastic consumption, LVF1 was found to strongly outperform LCF at almost all times in two WDSs (Figures 3, 5 and 6 of [27]). In another study, a controller that uses future flow forecasting (LCb [23]) shows a clear advantage above the case when the future forecasting is neglected (LCa), when the fluctuations in consumption are small compared to its hourly variation. The advantage is obtained for a flow rate that has smaller fluctuations and much larger hourly variation than in Figure 2a.

In the current study for $T_c = 3$ min, it was found that LVF7 outperformed LCF when $|Q'_{trend}|$ was large at a certain time. On the other hand, LVF7 and LCF performed the same during the entire day, for the assumed consumption pattern. However, it is expected that LVFn can outperform over the entire day (for some n) when there are more hours when $|Q'_{trend}|$ dominates the fluctuations, or there are hours when $|Q'_{trend}|$ strongly dominates the fluctuations.

8. Conclusions

Of the three time-steps considered, 3 min was used because it gave the best performance for the example network and consumption. Extensive care was taken to construct this realistic example, for which the flow-dependent LCF and LVF7 controllers were found to have the same overall performance. However, since LVF7 was more complicated, LCF Was preferable, and should be considered as the controller of choice.

The performance of the LVF controller relative to LCF is expected to be better when:

1. The magnitude of the average stochastic fluctuation in consumption decreases.
2. There are many hours with a sizeable magnitude of the rate of change of the flow rate through the valve $|Q'_{trend}|$.
3. There are hours with a large $|Q'_{trend}|$.

Unless the factors listed above are particularly favourable, the flow-dependent controller, which does not require modelling the future (LCF here), is an adequate choice with stochastic consumption, even though it may not perform as well as the controller that requires modelling the future (LVF here). This remains true even in the case where it is preferable or necessary to replace the sensitivity with a dimension-less tunable parameter [23].

Further research can address ways to improve controllers that model the future.

Author Contributions: Conceptualization, P.P. and E.C.; methodology, E.C.; software, E.C.; validation, E.C.; formal analysis, P.P.; investigation, P.P. and E.C.; resources, E.C.; data curation, E.C.; writing—original draft preparation, P.P.; writing—review and editing, P.P. and E.C.; visualization, P.P. and E.C.; supervision, P.P.; project administration, P.P. and E.C.; funding acquisition, P.P. and E.C.

Funding: This research received no external funding.

Conflicts of Interest: The authors declare no conflict of interest.

Appendix A. Notation and Derivation of Head-Loss Controller

Let t_{ci} be a time period which differs from iteration to iteration; and $t_{ci} < T_c$. At time t_i, the PCV head-loss, velocity, flow rate and head-loss coefficient are, respectively, $\tilde{H}_i$, v_i, Q_i and ξ_i; and the head at the CN is H_i. For all quantities X listed here, except ξ, X_i is defined as the quantity $X(t)$ *evaluated* at $t = t_i$. The PCV adjustment process commences soon after time t_i, and continues until time $t_i + t_{ci}$, when the PCV is completely adjusted to the new coefficient ξ_{i+1}. At time $t_{i+1} \equiv t_i + T_c$, the coefficient is still ξ_{i+1}; and the head-loss, velocity and flow are denoted by $\tilde{H}_{i+1}$, v_{i+1} and Q_{i+1}, respectively.

The Newton–Raphson numerical method has as its goal to find z such that $f(z) = 0$, i.e., find the root of a function of one variable. z is found by the iteration [29]

$$z_{i+1} = z_i - \frac{f(z_i)}{f'(z_i)} \tag{A1}$$

For the sake of argument, assume a WDS *with no* time-dependence, with only changes in $\tilde{H}$ allowed. Identifying z with $\tilde{H}$ and defining $f(\tilde{H}) = H(\tilde{H}) - H_{sp}$ means the goal is to find $\tilde{H}$ such that $H(\tilde{H}) = H_{sp}$, as required [30]. Applying Equation (A1) leads to Equation (1). At this point, i is simply an iteration variable, with no notion of time attached to it. For the method to be applicable, f, and hence H, must be a continuous and differentiable function of $\tilde{H}$. The iteration i can be chosen to refer to time t_i, because the WDS has no time-dependence. Particularly, the sensitivity

$$\frac{1}{S_i} \equiv \frac{dH}{d\tilde{H}} \tag{A2}$$

is evaluated at time t_i.

In a general WDS *with* time-dependence, a head-loss controller can then be postulated (not derived) by applying Equation (1) even when there is time-dependence. The more there is time-dependence from one iteration to the next, over a few iterations, the less reliable the head-loss controller is expected to be. Such situations are when there is significant time-dependence on a time-scale shorter than T_c, or on a time-scale of a few T_c.

References

1. Vicente, D.J.; Garrote, L.; Sánchez, R.; Santillán, D. Pressure management in water distribution systems: Current status, proposals, and future trends. *J. Water Resour. Plan. Manag.* **2015**, *142*, 04015061. [CrossRef]
2. Ates, S. Hydraulic modelling of control devices in loop equations of water distribution networks. *Flow Meas. Instrum.* **2017**, *53*, 243–260, doi:10.1016/j.flowmeasinst.2016.12.002. [CrossRef]
3. Adedeji, K.B.; Hamam, Y.; Abe, B.T.; Abu-Mahfouz, A.M. Pressure management strategies for water loss reduction in large-scale water piping networks: A review. In *Advances in Hydroinformatics*; Gourbesville, P., Cungem, J., Caignaert, G., Eds.; Springer Water; Springer: Singapore, 2018; pp. 465–480.
4. Prescott, S.L.; Ulanicki, B. Improved control of pressure reducing valves in water distribution networks. *J. Hydraul. Eng.* **2008**, *134*, 56–65. [CrossRef]
5. Ulanicki, B.; Skworcow, P. Why PRVs tends to oscillate at low flows. *Procedia Eng.* **2014**, *89*, 378–385. [CrossRef]
6. Bakker, M.; Rajewicz, T.; Kien, H.; Vreeburg, J.H.G.; Rietveld, L.C. Advanced control of a water supply system: A case study. *Water Pract. Technol.* **2014**, *9*, 264–276. [CrossRef]
7. Campisano, A.; Creaco, E.; Modica, C. RTC of valves for leakage reduction in water supply networks. *J. Water Resour. Plan. Manag.* **2010**, *136*, 138–141. [CrossRef]
8. Nicolini, M.; Zovatto, L. Optimal location and control of pressure reducing valves in water networks. *J. Water Resour. Plan. Manag.* **2009**, *135*, 178–187. [CrossRef]
9. Sanz, E.; Pérez, R.; Sánchez, R. Pressure control of a large scale water network using integral action. In Proceedings of the 2nd IFAC Conference on Advances in PID Control, Brescia, Italy, 28–30 March 2012; pp. 270–275.

10. Fontana, N.; Giugni, M.; Glielmo, L.; Marini, G. Real-time control of pressure for leakage reduction in water distribution network: Field experiments. *J. Water Resour. Plan. Manag.* **2018**. [CrossRef]

11. Giustolisi, O.; Ugarelli, R.M.; Berardi, L.; Laucelli, D.B.; Simone, A. Strategies for the electric regulation of pressure control valves. *J. Hydroinform.* **2017**, *19*, 621–639. [CrossRef]

12. Berardi, L.; Simone, A.; Laucelli, D.B.; Ugarelli, R.M.; Giustolisi, O. Relevance of hydraulic modelling in planning and operating real-time pressure control: Case of Oppegård municipality. *J. Hydroinform.* **2017**, *20*, 535–550. [CrossRef]

13. Campisano, A.; Cabot Ple, J.; Muschalla, D.; Pleau, M.; Vanrolleghem, P.A. Potential and limitations of modern equipment for real time control of urban wastewater systems. *Urban Water J.* **2013**, *10*, 300–311. [CrossRef]

14. Kruger, C.P.; Abu-Mahfouz, A.M.; Hancke, G.P. Rapid prototyping of a wireless sensor network gateway for the Internet of Things using off-the-shelf components. In Proceedings of the 2015 IEEE International Conference on Industrial Technology (ICIT), Seville, Spain, 17–19 March 2015; pp. 1926–1931.

15. Abu-Mahfouz, A.M.; Hamam, Y.; Page, P.R.; Djouani, K.; Kurien, A. Real-time dynamic hydraulic model for potable water loss reduction. *Procedia Eng.* **2016**, *154*, 99–106. [CrossRef]

16. Fontana, N.; Giugni, M.; Glielmo, L.; Marini, G.; Verrilli, F. Real-time control of a PRV in water distribution networks for pressure regulation: Theoretical framework and laboratory experiments. *J. Water Resour. Plan. Manag.* **2018**, *144*, 04017075. [CrossRef]

17. Campisano, A.; Modica, C.; Vetrano, L. Calibration of proportional controllers for the RTC of pressures to reduce leakage in water distribution networks. *J. Water Resour. Plan. Manag.* **2012**, *138*, 377–384. [CrossRef]

18. Campisano, A.; Modica, C.; Reitano, S.; Ugarelli, R.; Bagherian, S. Field-oriented methodology for real-time pressure control to reduce leakage in water distribution networks. *J. Water Resour. Plan. Manag.* **2016**, *142*, 04016057. [CrossRef]

19. Creaco, E.; Franchini, M. A new algorithm for real-time pressure control in water distribution networks. *Water Sci. Technol. Water Supply* **2013**, *13*, 875–882. [CrossRef]

20. Giustolisi, O.; Campisano, A.; Ugarelli, R.; Laucelli, D.; Berardi, L. Leakage management: WDNetXL pressure control module. *Procedia Eng.* **2015**, *119*, 82–90. [CrossRef]

21. Page, P.R.; Abu-Mahfouz, A.M.; Yoyo, S. Parameter-less remote real-time control for the adjustment of pressure in water distribution systems. *J. Water Resour. Plan. Manag.* **2017**, *143*, 04017050. [CrossRef]

22. Creaco, E.; Campisano, A.; Franchini, M.; Modica, C. Unsteady flow modeling of pressure real-time control in water distribution networks. *J. Water Resour. Plan. Manag.* **2017**, *143*, 04017056. [CrossRef]

23. Creaco, E. Exploring numerically the benefits of water discharge prediction for the remote RTC of WDNs. *Water* **2017**, *9*, 961. [CrossRef]

24. Piller, O.; Elhay, S.; Deuerlein, J.; Simpson, A. Local sensitivity of pressure-driven modeling and demand-driven modeling steady-state solutions to variations in parameters. *J. Water Resour. Plan. Manag.* **2017**. [CrossRef]

25. Creaco, E.; Pezzinga, G.; Savic, D. On the choice of the demand and hydraulic modeling approach to WDN real-time simulation. *Water Resour. Res.* **2017**, *53*, 6159–6177. [CrossRef]

26. Creaco, E.; Farmani, R.; Kapelan, Z.; Vamvakeridou-Lyroudia, L.; Savic, D. Considering the mutual dependence of pulse duration and intensity in models for generating residential water demand. *J. Water Resour. Plan. Manag.* **2015**. [CrossRef]

27. Page, P.R.; Abu-Mahfouz, A.M.; Piller, O.; Mothetha, M.L.; Osman, M.S. Robustness of parameter-less remote real-time pressure control in water distribution systems. In *Advances in Hydroinformatics*; Gourbesville, P., Cungem, J., Caignaert, G., Eds.; Springer Water; Springer: Singapore, 2018; pp. 449–463.

28. Piller, O.; van Zyl, J.E. Modeling control valves in water distribution systems using a continuous state formulation. *J. Hydraul. Eng.* **2014**, *140*, 04014052. [CrossRef]

29. Press, W.H.; Teukolsky, S.A.; Vetterling, W.T.; Flannery, B.P. *Numerical Recipes in FORTRAN*; Cambridge University Press: Cambridge, UK, 1992; p. 355.
30. Page, P.R.; Abu-Mahfouz, A.M.; Mothetha, M.L. Pressure management of water distribution systems via the remote real-time control of variable speed pumps. *J. Water Resour. Plan. Manag.* **2017**. [CrossRef]

Review

A Systematic Review of the State of Cyber-Security in Water Systems

Nilufer Tuptuk [1,*], Peter Hazell [2], Jeremy Watson [3] and Stephen Hailes [1]

[1] Department of Computer Science, University College London, London WC1E 6EA, UK; s.hailes@ucl.ac.uk

[2] Department of Technology, Enterprise, Change and Data Science—Information & Cyber-Physical Security, Yorkshire Water, Bradford BD6 2SZ, UK; peter.hazell@yorkshirewater.co.uk

[3] Department of Science, Technology, Engineering and Public Policy, University College London, London WC1E 6BT, UK; jeremy.watson@ucl.ac.uk

* Correspondence: n.tuptuk@ucl.ac.uk

Abstract: Critical infrastructure systems are evolving from isolated bespoke systems to those that use general-purpose computing hosts, IoT sensors, edge computing, wireless networks and artificial intelligence. Although this move improves sensing and control capacity and gives better integration with business requirements, it also increases the scope for attack from malicious entities that intend to conduct industrial espionage and sabotage against these systems. In this paper, we review the state of the cyber-security research that is focused on improving the security of the water supply and wastewater collection and treatment systems that form part of the critical national infrastructure. We cover the publication statistics of the research in this area, the aspects of security being addressed, and future work required to achieve better cyber-security for water systems.

Keywords: smart water systems; cyber–physical security; cyber-security; cyber–physical attacks

Citation: Tuptuk, N.; Hazell, P.; Watson, J.; Hailes, S. A Systematic Review of the State of Cyber-Security in Water Systems. *Water* **2021**, *13*, 81. https://dx.doi.org/10.3390/w13010081

Received: 30 November 2020
Accepted: 25 December 2020
Published: 1 January 2021

Publisher's Note: MDPI stays neutral with regard to jurisdictional claims in published maps and institutional affiliations.

1. Introduction

Water is becoming scarcer. According to the United Nations World Water Development Report published in 2018 [1], nearly half the world's population, around 3.6 billion people, face water-scarcity for at least one month per year, and it is expected that over 5 billion people will suffer some water shortage by 2050. The World Bank estimates that around 45 million cubic meters of water are lost each day in developing countries, costing over US$3 billion per year [2]. This loss is mainly due to inefficient infrastructure, ageing infrastructure that leaks, and non-revenue water due to lack of billing or inaccuracies in costing such as metering issues [2]. It affects both developed and developing countries. In England and Wales 2954 million litres of water are leaked each day from distribution networks and supply pipes [3].

Climate change, water pollution, increasing urbanisation and population growth, ageing and inefficient infrastructure, compliance with tighter regulation and water quality standards are some of the challenges faced by water sector in seeking to maintain their services. To resolve these challenges, water and wastewater providers are moving towards smart water systems [4–6] that are reliable, efficient and that support real-time decision-making. This is particularly true in the UK, where the UK government has established strategic priorities for the period from 2020 to 2025 aimed at securing long-term resilience in the water industry; these are supported by major investments by water companies and providers [7,8].

Water systems are a type of cyber–physical system (CPS) that integrate computational and physical capabilities to control and monitor physical processes. In the past, water system security was achieved largely through isolation, limiting access to control components. However, with the emergence of IoT, water systems, as with other critical infrastructure services, are increasingly using a smart systems philosophy. This promotes the incorporation of IoT and analytics into industrial control systems (ICS) to improve the sensing and control capacity and ensure better integration with business processes. Collectively, this is known as the Industrial Internet of Things (IIoT), often labelled Industry 4.0, in which IoT is applied to industrial applications. It relies on connecting multiple layers of cyber–physical systems to facilitate autonomous decentralised decision-making and to improve the use of real-time data and predictive analytics to promote reliability, efficiency and productivity. With these technological advances, water systems that collect, treat, transport and distribute water to customers are undergoing a similar transformation, becoming highly connected and facing new technological challenges in the drive to provide safe water reliably.

ICS deployment often follows a hierarchical architectural approach that is sometimes characterised using the Purdue reference model [9], as shown in Figure 1. This spans multiple layers, encompassing the variety of equipment and communication protocols and the range of goals and complexity that are likely to be found in these environments [9].

Level 5, the enterprise network, is the level at which business decisions are made, and in which the regular corporate systems (enterprise desktops and servers) operate. At Level 4, the site business planning and logistics applications and systems are found. At Level 3, the operations network, operations management systems such as domain controllers, data collection servers (historians) and application servers are found. Level 2, supervisory control, consists of devices that monitor and control the process at the lower levels. Typically, these consist of supervisory interfaces for the operators, engineering workstations, and distributed control servers that monitor and control various parts of production. At Level 1, controllers monitor and control a set of devices autonomously and/or based on decisions that come from the supervisory system. They receive inputs from instrumentation equipment (e.g., field devices) such as sensors, and send output signals to other devices (actuators). Level 0 is where the actual process takes place, containing the sensors and actuators connected via a fieldbus network.

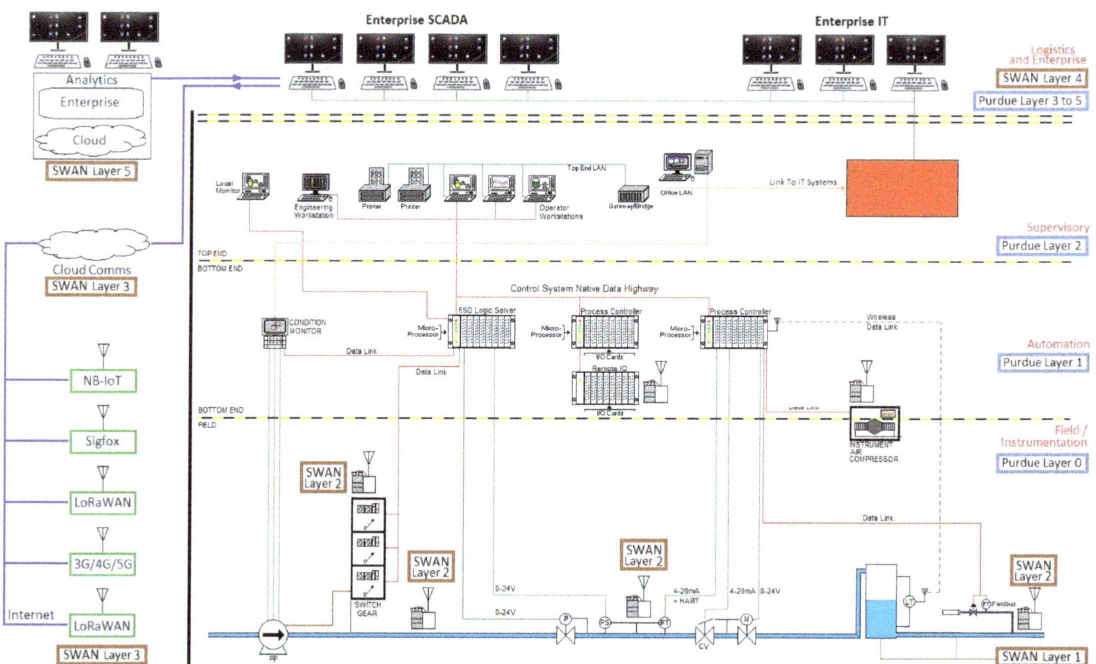

Figure 1. Purdue reference model with SWAN layers.

According to the Smart Water Networks Forum (SWAN) [10], a global non-profit hub consisting of international water companies, academics, regulators, and other water experts, smart water networks are the "entire system of data technologies connected to or serving the water distribution network [and] it is informative to separate its components into layers." These layers [10] are similar to those found in Purdue reference model, as indicated in Figure 1:

- Level 1: Physical layer is composed of physical devices that provide the distribution and delivery of water services. This includes pipes, pumps, valves, reservoirs and endpoints for delivering water.
- Level 2: Sensing and control layer is composed of equipment and sensors responsible for gathering measurements for monitoring and controlling water delivery and distribution; and remote-controlled actuators to remotely operate water networks.
- Level 3: Collection and communications layer provides the data collection, transmission, and storage between layer 2 and level 4 where the instructions for sensors and actuators are computed. All network protocols used for data transfer are found in this layer.
- Level 4: Data management and display layer is responsible for gathering and managing data from different sources. Supervisory control and data acquisition (SCADA) systems, control systems, visualisation systems and tools such as human-machine interface (HMI), data storage repositories and control systems are found in this layer. This is where decisions taken by upper layers are interpreted into control and other commands such as settings for devices at lower layers.
- Level 5: The data fusion and analysis layer is where raw data is processed into information and where the "smart" emerging technologies are deployed. These include modelling and optimisation systems, network infrastructure monitoring, and other supporting and decision support systems for managing water networks.

The adoption of network communication, the increasing use of commercial-off-the-shelf (COTS) components and the deployment of wireless systems in Purdue and SWAN architecture layers bring new security challenges as they have the potential to expose water systems to a wide variety of adversaries. The number of reported attacks targeting cyber–physical systems that are critical for national infrastructure services has been on the increase and, as the evidence from successful attacks such as Stuxnet [11], DuQu [12], BlackEnergy [13] and Havex [14] shows, such attacks can have catastrophic consequences. The criticality of water to human life and the ecosystem means that water systems are an obvious target for political, military and terrorist actors [15,16].

Table 1 reports some of the incidents against water infrastructure services that have been made public. These indicate the potential for successful attacks to exploit a wide variety of vulnerabilities and so cause both direct disruption of services and damage to control equipment and communication networks that, in turn, may affect essential services. The broader impacts of such attacks lie in the health of both the public and the ecosystem, as well as in financial and reputational losses for the companies affected. Hassanzadeh et al. [17] report a review of 15 water incidents, including some of the attacks summarised in Table 1. A widely referenced source for cyber-security incidents in the water sector is the work carried out by Industrial Control Systems Cyber Emergency Response Team (ICS-CERT) in the United States. This tells us that, in 2015, the US Department of Homeland Security (DHS) recorded 25 cyber-security incidents from the water sector [18].

Table 1. Past attacks on water systems.

Reference	Year	Target	Attribution	Infection Vector	Details	Impact
Israel's water system [19]	2020	OP	Hacktivist/ Nation state	Unknown	Israeli government reported cyber-attacks against water supply and treatment facilities and urged these facilities to change passwords.	Unknown.
Northern Colorado [20]	2019	OP	Cybercrime	Ransomware	Locked access to technical and engineering data.	Disruption, took about three weeks to unlock data.
Cryptojacking [21]	2018	OP	Cybercrime	Cryptocurrency mining	Cryptocurrency malware installed on HMI on the SCADA network.	Unknown.
Kemuri water [22]	2016	OP	Hacktivist	Remote access	Accessed PLC responsible for controlling water treatment chemicals.	Engineers were able to identify and reverse the changes made to process control parameters.
Bowman Avenue Dam [23,24]	2016	OP	Hackers/ Nation state	Remote access	According to US authorities, hackers linked to Iranian Armed Forces infiltrated ICS of Bowman Avenue Dam and accessed the SCADA for the dam.	Data exfiltration and over $30k on remediation costs. Physical damage was not possible due to disconnected sluice gates.
Florida Wastewater [25]	2012	IT	Ex-Employee	Remote access	Stolen login credentials were used to access district's computer system.	Deleting and modifying information. Ex-employee was arrested on account of computer crime.
Tehama-Colusa Canal [26]	2007	OP	Ex-employee	Physical access	Installed malware on SCADA system responsible for controlling agricultural irrigation [26].	Damage to equipment, and additional unknown amount of monetary loss due to replacing production.
Harrisburg water plant [27]	2006	IT	Hackers	Remote Access	Compromised and installed malware on an employee's laptop which could have been used as an entry point to reach water treatment system.	Unknown.
Maroochy Shire [28,29]	2000	OP	Ex-employee of a contractor	Physical access	Masqueraded as a controller using stolen equipment and sent fake commands to the pumping station.	Approximately 800,000 litres of sewage was released into the environment, harming local parks and rivers, impacting public health, killing marine life, and caused large monetary loss.

Cyber-attacks against infrastructure services are often not made public and attribution of these incidents can be a complex and uncertain process, requiring well-developed skills and capabilities [30] to identify the actors. Nevertheless, publicly reported incidents show that the sources of cyber-attacks against water systems appear to include a wide variety of actors. These include hacktivists who perform cyber-attacks often based on a political ideology; disgruntled former employees seeking revenge; cybercriminal networks motivated by monetary gain; and hacker hobbyists who attack for fun, curiosity, or the desire for recognition [31]. Other potential adversaries include nation-state-sponsored attacks for political gain and industrial espionage; rival organisations or companies seeking business advantage; terrorist groups attacking national security; and insiders motivated by problems at work, political or monetary gain, fear/coercion or just for the thrill or fun.

The current history of incidents suggests that the design and performance of advanced targeted attacks against operational processes (OP) require actors with more than just IT skills [32]. Until recently, most of the cyber-attacks against cyber–physical processes were carried out by insiders, with most of the remainder conducted by nation states. In other words, most attacks have been conducted by those with the knowledge, skills and resources needed to cause a real physical impact. More recently, however, there has been an increasing incidence of cyber-criminals targeting industrial processes, with the aim of installing ransomware [33].

In this paper, we present a systematic literature review and evaluate the current state of cyber-security of cyber–physical systems within the water sector, focusing on process control layers, as the corporate IT layers are primarily affected by security problems covered by traditional information security. Our aim is to identify what is being done, by whom, where, how and what aspects of cyber-security are being covered.

The remainder of this paper is structured as follows. Section 2 provides brief overview of cyber–physical system security. Section 3 describes the research questions and methodology used for carrying out the systematic review. Key research findings are reported and discussed in Section 4. Section 5 highlights the limitations of existing studies and discusses some direction for future research. Finally, Section 6 concludes the paper.

2. Cyber–Physical Systems

The term "cyber–physical system" (CPS) was first coined by Helen Gill at the National Science Foundation (NSF) in 2006 to describe "physical, biological and engineered systems whose operations are integrated, monitored, and/or controlled by a computational core" [34]. Since then, CPS have attracted significant research effort, including initiatives in Industry 4.0, the Internet of Things and the Industrial Internet of Things. As computer scientist Edward A. Lee points out [35], terms such as the Internet of Things (IoT), Industry 4.0, the Industrial Internet (II), Machine to Machine (M2M), the Industrial Internet of Things (IIoT) and other similar terms have been strongly connected with CPS, and sometimes used interchangeably and sometimes for specific sectors (e.g., Industry 4.0 for manufacturing). However, these terms cover "implementation approaches (e.g., the "Internet" in IoT) or particular applications (e.g., Industry 4.0)" [35]. CPS are found in a broad range of sectors including health care and medicine, materials, manufacturing, automotive, aerospace, utilities, chemical, civil infrastructure and transportation [34]. Despite the differences in interpretation, many industry sectors share common technologies and, by extension, share similar concerns relating to their security. A common concern for all these sectors in adopting new enabling technologies for CPS is to ensure security in the face of cyber-attacks.

2.1. Securing Cyber–Physical Systems

The National Institute of Standards and Technology (NIST) defines cyber-security as "the process of protecting information by preventing, detecting and responding to attacks" [36]. The prevention of attacks against information technology systems is defined in terms of three security goals: confidentiality, integrity and availability, known as the CIA triad. These goals are also applied to CPS to maintain security.

Confidentiality ensures data or system resources "are not disclosed to unauthorised individuals, processes, or devices" [37]. The operation of CPS requires, *inter alia*, data from instrumentation devices, controllers, supervisory control systems, monitoring and safety systems. Unauthorised access to this data is potentially useful for preparing and implementing attacks and for industrial espionage. Integrity deals with "guarding against improper information modification or destruction, and includes ensuring information non-repudiation and authenticity" [38]. Violating integrity could interfere with the operation of CPS and undermine the reliability and safety of the CPS process. Availability deals with "timely, reliable access to data and information services for authorised entities" [39]. Many CPS are continuous systems and loss of availability can cause systems to shut down and interrupt the production process. Usually, integrity and availability are the most important concern for critical cyber–physical systems [40], but the priority given to each of these security goals depends on the risks associated with loss of these properties in the context of a particular system.

Cyber–physical systems have control properties that need to be maintained. These include stability, observability, controllability, safety and efficiency [41], as well as accuracy, responsiveness, rapid disturbance rejection and low control effort. Security attacks aimed at sabotaging CPS involve the manipulation of these properties; thus, the maintenance of these properties, even when the system is under attack, is an essential component of ensuring the security of CPS.

2.2. Attacks against Cyber–Physical Systems

Figure 2 shows the typical components of a networked CPS. The controller is given a process reference (Setpoint-SP) as the desired process output to maintain. The sensor measures the output of the physical process (Measured Process Value-PV) and sends this over a network to the controller. The controller (for example a PLC) receives these values, compares the PV against the desired SP reference value, calculates a control command (Manipulated Variable-MV) and sends this, through the network, to the actuator. The actuator acts on this command and outputs a physical control action that modifies the process. Attacks against CPS involve attacking components of CPS to achieve either data exfiltration, which involves gathering sensitive information about the CPS, or sabotage, which involves disrupting the process.

Adversaries use a range of tools to carry out attacks against elements of Figure 2. These include attacks that compromise sensors, actuators and controllers to modify their settings or configurations so that incorrect signals are sent to relevant components; for example, incorrect control commands from controller to actuator or incorrect PVs from sensor to controller. Attacks can be carried out against the network: modifying the data in transit (replaying old data, dropping data, injecting false data); denying or delaying the flow of data (e.g., DoS, jamming attacks); or impersonating another actor (for example IP and ARP spoofing and communication hijacking). Eavesdropping attacks against networks can be carried out to gather information related to the operation of CPS, such as identifying communication protocols, open ports, hosts and applications, and sniffing network traffic. Physical attacks can be carried out against CPS components, e.g., to modify the location of devices; change device calibration; install rogue devices on the network; install malware via portable devices (e.g., USB sticks); cause changes in sensor values by manipulating the physical environment of the devices; and cause physical damage to devices.

The success of an attack depends on the resources and skills available to adversaries as well as system vulnerabilities and the absence of appropriate independent layers of protection designed to prevent mal-operation due to operator error, random equipment failure or cyber-attack. Vulnerabilities are typically introduced into CPS due to: poor security design; insecure network communication protocols; insecure backdoors and holes in the virtual or physical network perimeter; insecure software and hardware; poor management of security or ineffective policies and inappropriate physical access [40]. To exploit a CPS, a highly motivated adversary with high skills and resources can purchase zero-day vulnerabilities that are, by definition, not yet public, as seen in the past (e.g., Stuxnet [11]).

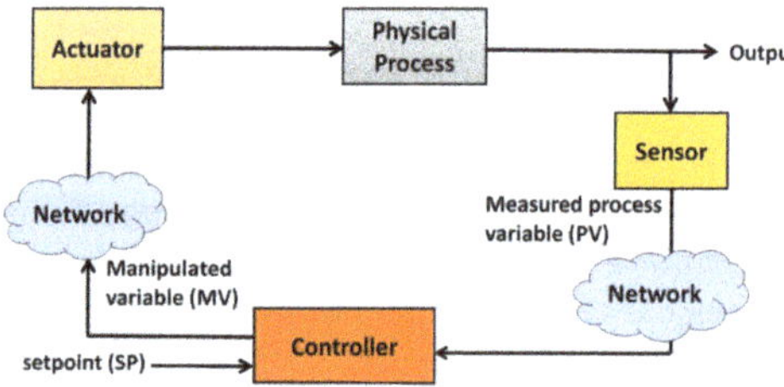

Figure 2. Typical cyber–physical system.

Adversaries have a wide variety of motivations, and impact goals depend on these motivations. Potential impacts include process disruption; damage to production, equipment, safety and the environment; data disclosure; data loss; disruption to assets; injuries and loss of life; damage to reputation; and financial damage.

2.3. Security Measures for Cyber–Physical Systems

Security mechanisms to protect systems against malicious behaviour can be divided into three main categories: *preventive, reactive* and *responsive* measures. *Preventive* measures are security controls implemented to prevent attacks such as authentication; access control; network segmentation; maintaining confidentiality and integrity of transmitted data and in storage using cryptographic techniques; patching software vulnerabilities; deploying usable and effective security management policies that defines roles and procedures for managing and maintaining security; personnel awareness and training programs to understand threats; and measures for protecting the supply chain [40]. *Reactive* or *detection-based* measures are security controls implemented to identify attacks and anomalous behaviour such as intrusion/anomaly-based monitoring and detection for process and host; antivirus and other malware monitoring tools; and safety management systems. After an attack is detected, *response* strategies include measures to reduce damage; for example, reconfiguring the network; restricting access to network; systems or devices; deploying designed-in redundancies; and shutting down the system.

3. Methodology for Systematic Review

Our aim in this paper is to review and gain an understanding of cyber-security research targeted at protecting cyber–physical systems in the water sector, thence to identify areas that require future research. The Preferred Reporting Items for Systematic Reviews (PRISMA) [42] guidelines were followed, as illustrated in Figure 3. A set of question research questions were devised to analyse and evaluate the relevant publications. A set of electronic databases and a search strategy was designed to identify the publications. Inclusion and exclusion criteria were used to assess the eligibility of each publication. The eligible publications were then manually inspected to extract relevant evidence for analysis.

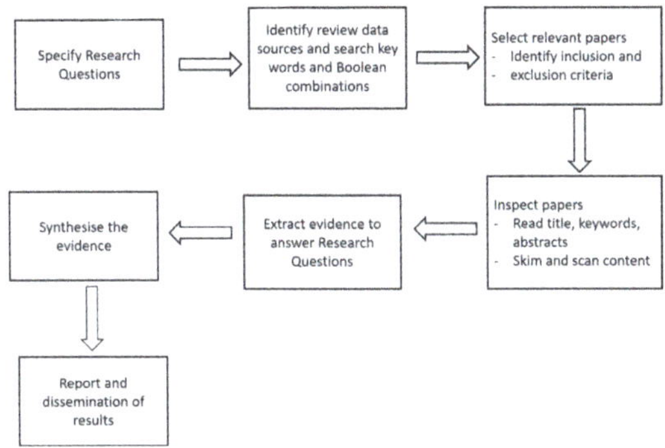

Figure 3. Systematic literature review process, adapted from [42].

3.1. Research Questions

To identify, classify and evaluate the existing cyber-security work within water sector, a set of research questions were identified.

- *RQ1 How did the number of publications change over the years?* To understand the publication trends over the years, and to understand if the topic is gaining more research focus with moves towards IIoT and Industry 4.0. Answering this question might also enable us to see any trends that might have motivated more work from the research community.
- *RQ2 What is the geographic distribution of these studies?* To understand by whom and from where these studies are being conducted. Answering RQ2 will help to determine

countries investing the least and most in research in these areas, and why this could be the case. Security of national infrastructure services such as water often require a joint effort from academia, governmental bodies and industry.

- *RQ3 What is the distribution of academic, governmental and industry studies?* To identify the level of involvement, and the support of government and industry in research studies. Answering this question will enable assessment of whether relevant government and industry bodies are participating in these studies. Their involvement is crucial for these studies, as they are essentially the clients that will deploy and implement security solutions.
- *RQ4 What are the target venues for publishing these studies?* To identify publication venues targeted by these studies. Answering this question will help to identify the top target venues for publication, and gain some understanding of the maturity and quality of publications by analysing the rating of the journals and conferences.
- *RQ5 Which security aspects are covered in these studies?* To understand the security themes of interest, proposed solutions and focus of these studies. Answering this question will inform the security problems that are being solved.
- *RQ6 Can one classify security aspects in RQ5 further?* To see if there are popular areas of research that can be classified further. If there are popular research aspects, answering this question could help to compare different approaches.

3.2. Identification of Sources and Search Term

The search strategy for identifying publications was primarily through online databases: Springer Link, IEEE Xplore, ACM, Science Direct and ASCE library. These are the most common libraries for publishing conference proceedings and journal publications within the field of cyber-security in cyber–physical systems. Google Scholar returned articles that were covered in these databases; however, we also used it to identify relevant publications that appeared in other databases or venues. The search strings used for the databases were "water and cyber-security" or "cyber-security". Table 2 shows the search string for each database. When a basic search on databases returned many papers, advanced searching was used to filter irrelevant papers. For example, searching Google Scholar using combinatorial search keywords such as "water" AND "cyber-security" resulted in a high number of papers (over 17,900) that were not relevant to this systematic review. Instead, the search was limited to terms appearing in the title: "water" and "cyber" to identify studies that primarily focused on cyber-security of water systems. A list of security keywords was also used in conjunction, to search the databases for relevant publications. These qualifiers included: water, integrity, confidentiality, availability, integrity, authentication, authorisation, access control, threat, vulnerabilities, attacks, and detection. However, these failed to capture any new publications. Searching was limited to publications that had been published from 2000 to 2020.

Table 2. Search string used for each data source.

Source	Search String
Springer	where the title contains: Water AND with at least one of the words: cyber-security OR cybersecurity
ACM Digital Library	[Document Title: water] AND [[Abstract: cyber-security] OR [Abstract: cybersecurity]]
IEEE Xplore	"All Metadata": water cyber-security
ScienceDirect	Find articles with these terms: cyber-security OR cybersecurity, title, abstract, keywords: water
ASCE Library	water AND (cyber-security OR cybersecurity)
Google Scholar	allintitle: water cyber

Figure 4 shows the number of publications retrieved from online databases. Duplicates were removed from this pool of publications and the remaining publications were included for further review.

To complement online database searching, a manual review of reference lists of eligible papers and any notable journals (e.g., Water and Environment Journal), conferences (e.g., World Environmental and Water Resources Congress) and workshops (e.g., International Workshop on Cyber–Physical Systems for Smart Water Networks) was carried out to identify any relevant publications that might have been missed in the database search.

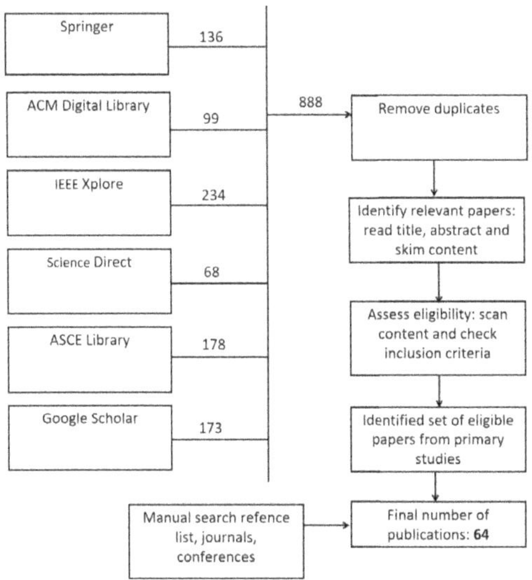

Figure 4. Publication selection process.

3.3. Criteria for Selection of Papers

Selection criteria for identifying publications for systematic review were as follows:

- Must address cyber–physical systems in water.
- Must have a technical content and address cyber-security.
- Must be peer-reviewed and must have appeared in an international journal, conference or workshop.

Books, book chapters, theses, editorials, feature or opinion pieces, essays, governmental and industry guidelines, other non-peer-reviewed or non-research publications, non-English publications, and publications appearing in local conferences, workshops or journals were excluded from the search. Review papers were not included in the analysis, but their content was analysed in the manual reference search and, where relevant, they are mentioned.

3.4. Paper Inspection

Online database searching resulted in 888 publications, and details of these were exported into a CSV file for further processing. After removing any duplicates, the remaining peer-reviewed publications published in internationally recognised conferences, workshops or journals were selected for further inspection. Selection of the eligible list of publications for analysis was based on inclusion and exclusion criteria by inspecting title and abstract, and text skimming. As a result, a set of 64 publications was finalised for analysis to answer the research questions.

3.5. Extraction of Appropriate Information

To analyse the content of the publications, the reviewed publications were classified into categories according to application domains, date of publication, number of citations, publication type, publication venue, affiliation, authors' countries of affiliation, and security aspects covered by the publication. Citation numbers for publications retrieved through online databases were not always accurate, so Google Scholar was used as a cross reference to retrieve the citation numbers. The data extracted was recorded in an Excel spreadsheet to facilitate analysis.

4. Analysis of Results

4.1. Publication Trends

Figure 5 shows the application domains of the security studies. The majority of studies were carried out on drinking water systems: 39 studies focused on security of water distribution systems (WDS) including water distribution networks; 3 studies included water supply and distribution systems; and 2 studies focused on water supply systems. Another 16 studies investigated security of drinking water treatment systems. Only four studies focused on non-drinking water systems: 3 studies focused on canal automation systems used for irrigation; and one study covered wastewater systems. There is a clear imbalance between studies covering water systems designed to provide drinking water versus those designed for other forms of water.

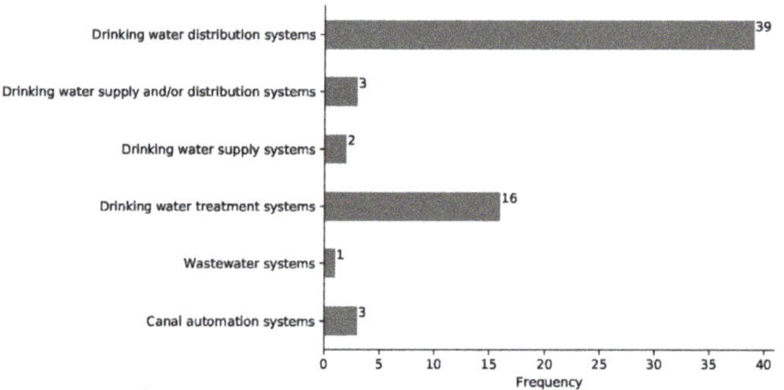

Figure 5. Application domains.

Figure 6 shows the timeline of publication. The earliest publication found dates from 2004, but most of the research effort (56 papers) was published after 2015. Answering **RQ1**, there has been increasing interest in the security of water systems over the years, likely as a result of the emergence of new resources and corresponding effort that made use of them.

These resources include the deployment of two important testbeds: the Secure Water Treatment (SWaT) testbed [43] and water distribution testbed (WADI) [44], and associated datasets [45] at the iTrust Centre for research in cyber-security at Singapore University of Technology and Design [46], and the BATADAL (BATtle of the Attack Detection Algorithms) competition organised by iTrust center and their international collaborators [47] to detect cyber-attacks against water distribution systems (WSD). This corresponds to a period (post 2016) in which associated open-source attack detection has become more available and European Commission (EC) projects such as FACIES (online identification of Failure and Attack on interdependent Critical InfrastructurES) [48] and STOP-IT [49] have been investigating physical and cyber-security of critical water infrastructures. This trend is supported by the number of publications per country involved in these projects.

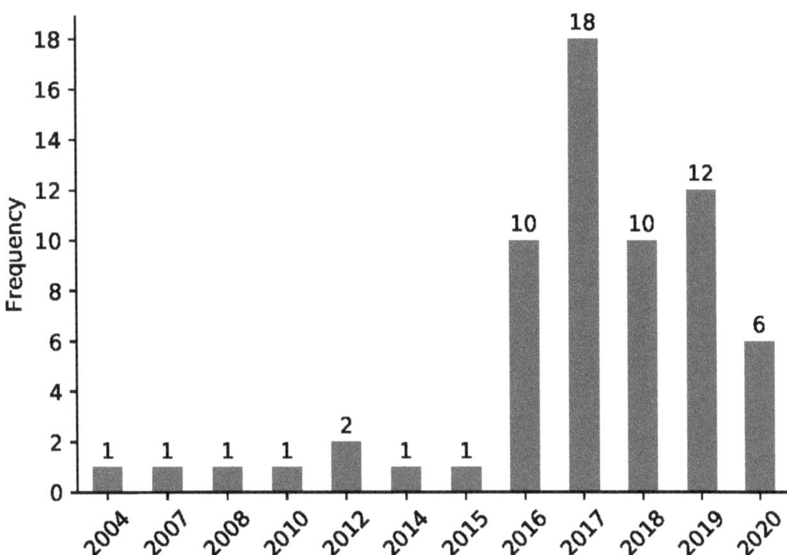

Figure 6. Number of publications over the years.

Figure 7 shows the distribution of studies per country based on the location of the authors. If the authors of the publication were located in multiple countries, for example several authors from Singapore and one author from Israel, both countries were added to the statistics. Figure 7 provides an answer to **RQ2** indicating that most of the existing research has been carried out by authors in countries that have made investments in this area: Singapore and their collaborators (Israel, USA) and countries involved in projects funded by the EC.

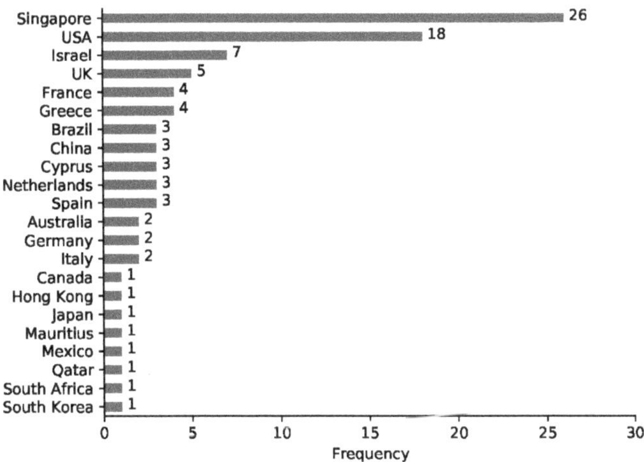

Figure 7. Country of publication based on location of authors.

Figure 8 shows the results to answer **RQ3**. Most of the research has been carried out by academia (85.1%); 6.8% was based in private organisations that provide security consulting services; 6.8% is provided by independent or public funded research organisations; and one paper (1.4%) was supported by a government agency. Interestingly, we failed to identify any research papers that were co-written with authors from water companies.

Figure 9 illustrates the distribution of publications based on venue type. Most publications (54.7%) were published in conferences, 31.2% were published in journals and the remaining 14.1% were published in workshops. Table 3 shows the publication venues for these papers. To answer **RQ4**, the most targeted conference is the World Environmental and Water Resources Congress with 11 papers published; the remaining conference papers were published in a wide range of conferences. The International Workshop on Cyber–Physical Systems for Smart Water Networks, which was established in 2015 and brings together researchers and engineers working on smart water systems, is the most targeted workshop. The most popular journal targeted for publishing security-related papers for water systems is the Journal of Water Resources Planning and Management, published by the American Society of Civil Engineers since the early 1990s. There was not enough data to reliably investigate the role of the conference and journal influencing the publication citations.

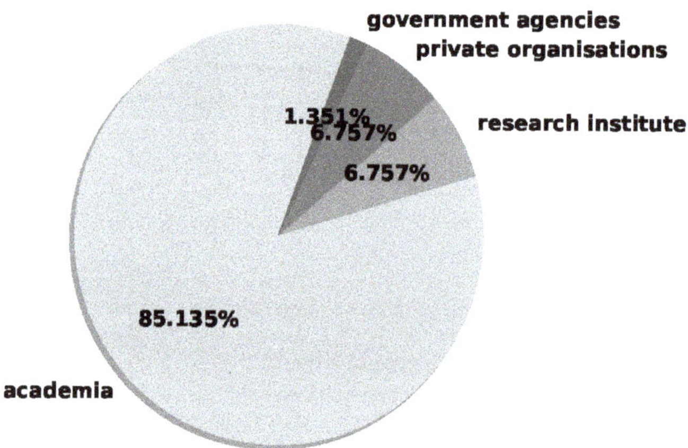

Figure 8. Affiliation of authors.

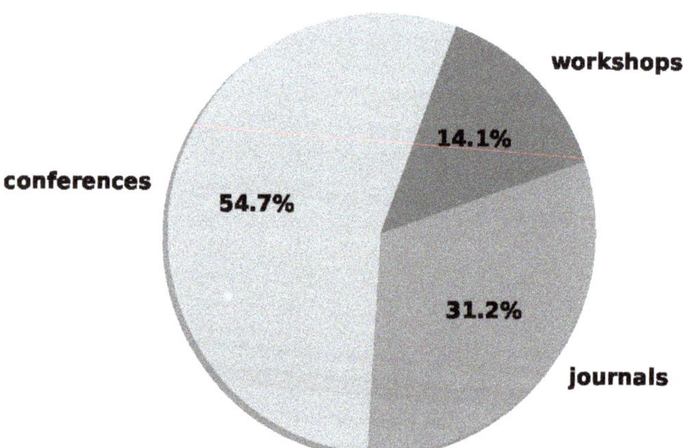

Figure 9. Venues for publication.

Figure 10 shows the results for **RQ5**, the security aspects covered by the publications. Most of the existing work focuses on detection mechanisms. The availability of datasets such as SWaT and WADI [45] has encouraged more research in this area. 31 papers investigated detection models; 10 papers investigating attacks against water systems and determining their impact; 9 papers on simulation or testbeds; 5 papers used modelling approaches for security analysis; 3 papers developed approaches for risk and resilience management; 2 papers were on datasets; 2 papers covered case studies; 2 papers examined benchmarking; a single paper addressed the development of a security framework; and another paper looked at improving security monitoring capabilities for water systems. In the following sections, we introduce the security aspects covered by the publications and provide a review.

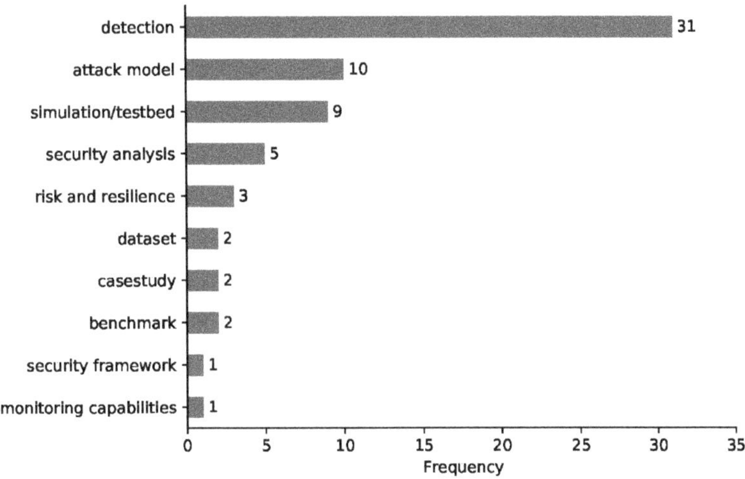

Figure 10. Security aspects covered by publication.

Table 3. Publication venues.

Type	Name	Count
conference	World Environmental and Water Resources Congress	11
workshop	International Workshop on Cyber-Physical Systems for Smart Water Networks	6
journal	Journal of Water Resources Planning and Management	5
journal	Journal of Environmental Engineering	3
conference	IEEE International Conference on Software Quality, Reliability and Security	3
conference	International Conference on Critical Information Infrastructures Security	2
conference	ACM on Asia Conference on Computer and Communications Security	2
journal	IEEE Transactions on Control Systems Technology	2
workshop	International Workshop on the Security of Industrial Control Systems and CPS	1
workshop	International Workshop on Critical Information Infrastructures Security	1
workshop	IEEE/ACM International Workshop on Software Engineering for Smart CPS	1
workshop	ACM Workshop on Cyber–Physical Systems Security and Privacy	1
journal	Water Resources Management	1
journal	Water Research	1
journal	Journal of Systems Science and Systems Engineering	1
journal	International Journal of Critical Infrastructure Protection	1
journal	IEEE Transactions on Dependable and Secure Computing	1
journal	IEEE Signal Processing Magazine	1
journal	IEEE Design and Test	1
journal	Human-centric Computing and Information Services	1
journal	Future Internet	1
journal	Environmental Modelling and Software	1
conference	Pipeline Division Specialty Congress	1
conference	International Symposium on Computer Science and Intelligent Control	1
conference	International Conference on Technology Trends	1
conference	International Conference on Harmony Search Algorithm	1

Table 3. *Cont.*

Type	Name	Count
conference	International Conference on Critical Infrastructure Protection	1
conference	International Conference on Auditory Display	1
conference	IFIP TC 11 International Conference on ICT Systems Security and Privacy Protection	1
conference	IFAC Conference on Cyber–Physical and Human Systems	1
conference	IEEE/ACM Int'l Conference on Cyber, Physical and Social Computing	1
conference	IEEE Pacific Rim International Symposium on Dependable Computing	1
conference	IEEE International Symposium on High Assurance Systems Engineering	1
conference	IEEE International Conference on Machine Learning and Applications	1
conference	IEEE International Conference on Data-Mining Workshops	1
conference	IEEE International Conference on Big Data	1
conference	ACM international conference on Hybrid systems: Computation and Control	1
conference	Annual Computer Security Applications Conference	1

4.2. Classification of Studies

Existing studies were categorised into the following areas: testbeds, simulation and datasets; cyber-attack models; cyber-attacks detection models; model-based security analysis; risk and resilience management; security frameworks; and security benchmarks and case studies. These categories help to answer **RQ6**, showing type of research contributions.

4.2.1. Testbeds, Simulation and Datasets

As it is typically neither possible nor safe to carry out cyber-security research studies that include attacks on operational cyber–physical systems, researchers have been using testbeds and simulation to reproduce the operation and characteristics of real-world systems. A number of testbed and simulation platforms have been proposed for the security of water systems. Table 4 outlines reported tools that have been used to support security research for water systems, including developing datasets for testing intrusion detection and validating mitigation techniques. The most widely known and reputable of these are the Secure Water Treatment (SWaT) testbed [43] and water distribution testbed (WADI) [44], both of which were implemented and deployed at iTrust Centre for research in cyber-security at Singapore University of Technology and Design [46]. SWaT consists of a six-stage water treatment process: raw water processing, chemical dosing, ultrafiltration, water purification (reverse osmosis) and backwashing [46]. The testbed also includes a real layered communication network consisting of layer 0 (sensors, actuators, PLCs) and layer 1 (SCADA, HMI, workstation and historians) of the Purdue model, using both wired and wireless network protocols. The WADI testbed is composed of set of tanks (e.g., reservoir tanks, consumer tanks, raw and returned water tanks), chemical dosing systems, and supporting equipment for water storage and distribution. WADI was designed as an extension to the SWaT [46] testbed and, by combining the capabilities of both testbeds, researchers were able to form a complete and fully functional water treatment, storage and distribution testbed for security research. Both testbeds were designed with international collaborators and engineers from the water sector and the combination has facilitated investigations that include the cascading effects of cyber-attacks between different components of the two testbeds. Researchers have also provided the cyber-security research community with datasets [45] containing normal operation and attack scenarios to allow detection methods to be evaluated. These datasets are multivariate time-series collected from real-time data sources such as sensors and actuators. One of the widely studied datasets in cyber-security research is the SWaT dataset [50] containing normal data streams collected from 51 sensors and actuators collected over 7 days, and attack data consisting of 41 attacks carried out over a period of 4 days. The WADI dataset [45] contains data from 123 sensors and actuators collected over a period of 14 days, and two days with attacks. Given the care in their design and their uniqueness, it is no surprise that a significant amount of research has been carried out using these testbeds and datasets. The iTrust Centre also runs schemes for other local and international researchers to request access to testbeds, subject to availability and an hourly charge.

Table 4. Testbeds and simulation tools used for cyber-security studies.

Publication	Details	Dataset
WaterBox (2015) [51]	A small-scale cyber–physical testbed designed for an in-lab environment to simulate smart water networks using components designed from acrylic, Arduino boards, small-scale sensors (pressure sensor, flow meter) and a motorised valve (using a small stepper motor).	-
SWaT (2016) [43,46]	An operational small-scale water treatment testbed with real cyber and physical equipment to investigate cyber-security research in 2015 by Singapore University of Technology and Design. It consists of a six-stage water treatment process with the modern-day components.	Available [45,50]
WADI (2016) [44,46]	A testbed launched by Singapore University of Technology and Design funded in 2016 as an extension of SWaT testbed to form a complete water treatment, storage and distribution system.	Available [45]
epanetCPA (2016) [52,53]	EPANET-based toolbox that is designed to assess the impact of cyber–physical attacks.	-
FACIES (2017) [54]	A water distribution system prototype funded by EU project FACIES based on a small fictitious city distributing water to different residential areas with a reservoir represented as tanks of different sizes.	-
RISKNOUGHT (2018) [55–57]	A cyber–physical stress testing platform leveraging EPANET software library to simulate the physical process and a custom network model for SCADA system.	-
Water storage control (2018) [58]	A SCADA testbed simulating water storage control consisting of water tank, PLC, historian, HMI, water level sensors and actuators (pumps and valve). The testbed was used to evaluate machine learning detection models against reconnaissance, command injection, and DoS attacks.	-

Other identified testbeds include WaterBox [51], a small-scale cyber-physical testbed designed as an in-lab facility built using Arduino boards, pressure sensors, flow meters, motorised valves, and acrylic structure to simulate smart water networks to carry out experiments related to water systems research including cyber-security and control optimisation. Teixeira et al. [58] developed a SCADA testbed system designed for controlling a water storage tank, simulating the process of water treatment and distribution, to test developed solutions such as machine learning based cyber attack detection models. This testbed includes a PLC (Schneider model M241CE40), HMI, water tanks, water pumps, valves, and sensors for water levels, and uses Modbus communication protocol. Miciolino et al. [54] reports FACIES testbed, emulating a water supply and distribution system for a fictional city to study security of water systems as part of EU project FACIES. The testbed consists of acrylic water tanks, sensors and actuators that are connected to PLCs (Modicon M340, Schneider), a SCADA system and a HMI. The communication protocol used by SCADA and PLC is Modbus over TCP protocol.

Simulation tools developed to study security of water systems include EPANET [59] based tools: epanetCPA [52,53], a simulation toolbox designed for simulating water distribution networks; and RISKNOUGHT (2018) [55–57] developed by STOP-IT project as a "cyber-physical stress testing platform for water distribution networks" including functionalities to simulate the flow of information between physical (hydraulic model) and cyber layers (SCADA networks).

4.2.2. Cyber-Attack Models

The modelling of attacks is an important part of cyber-security research, because it helps in understanding: the vulnerabilities of cyber–physical systems; the resources required to carry out successful attacks; the impact of attacks; and the resilience of countermeasures. Over the past decade, attacks against cyber–physical systems have attracted increased interest from the security research community to understand the resources required for attackers to carry out effective attacks.

We identified several papers that developed attack models to examine the behaviour of water systems and the impact of attacks. In [60], researchers investigated stealthy attacks that could cause damage while evading detection. They assumed an attacker with advanced skills and developed resources such as system dynamics, system diagnostic schemes, and the ability to manipulate PV (sensor) data. Attacks were carried out on the Gignac (Southern France) canal network's SCADA system. Researchers were able to design attacks that evaded the diagnostic scheme, which was based on unknown input observers for fault detection and isolation.

Adepu and Mathur [61] investigated single-point cyber-attacks against SWaT testbed and proposed attack detection based on system response to the attacks. Adepu et al. [62] and Tomic et al. [63] investigated jamming attacks against wireless communications in water systems. In [62], researchers carried out attacks against different parts of the SWaT testbed and, in [63], researchers used the Waterbox testbed [51] to investigate the robustness of process control schemes against jamming attacks using different attack strategies. Such attacks have the potential to halt or slow down a process and cause components to fail [62].

Robles-Durazno et al. [64] investigated memory corruption attacks against a PLC used in a water supply process, demonstrating their research using a Festo MPA rig. Researchers investigated memory corruption attacks in three location: attacking PLC inputs by overwriting memory allocated to connected sensors; attacking PLC outputs by overwriting memory for actuators; and attacking PLC working memory, targeting runtime code that contained setpoint variables. Researchers proposed a detection model based on monitoring energy consumption and voltage signals of sensors and actuators. Amin et al. [65] demonstrate stealthy deception attacks against SCADA systems used within water infrastructures.

RISKNOUGHT [55–57] simulation platform developed interaction between physical processes, and the computational and networking layers to simulate a range of cyber–physical threats including cyber-attacks targeting sensors, actuators, PLCs, SCADA and historians, causing physical damage to hydraulic components such as pumps, valves and pipes. Similarly, Taormina et al. [66] included a range of attack scenarios with the epanetCPA [53] toolkit to simulate cyber and physical attacks that target sensors, actuators, PLCs and SCADA, and communication between these components.

Erba et al. [67] investigated adversarial machine learning against ICS used in water distribution systems using WADI and BATADAL datasets. They present two models for concealment attacks to evade detectors that were trained using deep neural networks: (i) a white box attacker that has knowledge of the system and detection model and uses optimisation to generate adversarial samples that are close to the normal operating values of sensors; and (ii) a black box attacker, where the attacker has no knowledge of the detection and uses deep neural networks to learn the behaviour of expected ICS behaviour and produce adversarial sensor readings that resemble real data.

4.2.3. Cyber-Attack Detection Models

Designing effective detection techniques for cyber–physical systems is an important and dynamic area of research. A general list of cyber–physical systems detection models is reported in [68]. In this section, we review models proposed for detecting cyber-attacks in water systems.

A wide variety of approaches have been used to detect abnormal behaviour in water systems. These approaches are illustrated in Table 5. These can be divided into: model-based detection, which tries to model the physical evolution of systems; machine learning models, which learn representative characteristics of a system using data; and statistical models, which use statistical analysis to detect attacks.

Table 5. Papers related to the cyber-attackattack detection.

Publication	Attacks	Application Environment	Dataset	Detection Model
Amin et al. [69]	deception attacks against PVs	a simplified canal hydrodynamic model	-	model-based
Adepu and Mathur [70–73]	bias attacks [74]	SWaT testbed	-	model-based: invariants
Yoong and Heng [75]	-	SWaT testbed	-	machine learning invariants
Miciolino et al. [54]	DoS, replay	FACIES	-	standard deviation
Zohrevand et al. [76]	attacking water flow	water supply system	operational water supply system in Canada	hidden Markov model
Ahmed et al. [77]	false data injection and zero-alarm attacks against PVs and MVs	simulation: EPANET	-	model-based
Moazeni and Khazaei [78]	-	simulation: MATLAB OPTi toolbox	-	model-based: MINLP
Inoue et al. [79]	deception attacks against PVs and MVs	-	SWaT	LSTM and one-class SVM
Hindy et al. [80]	DoS, spoofing	physical testbed	-	classic machine learning methods
Studies using BATADAL dataset [47]	deception attacks, replay against PVs and MVs	-	BATADAL	autoencoders [81,82], MLP and PCA [83,84], data-mining [85,86], NARX [87], rule-based and deep learning [88], model-based (MILP) [89,90], model-based(feature extraction and random forest) [91], PCA, EWMA and RBC [92], ensemble (SOD, LOF and QDA) [93],
Kadosh et al. [94]	deception attacks, replay	C-Town, E-Town WDSs	BATADAL and generated dataset	SVDD
Bakalos et al. [95]	deception attacks against PVs, physical intrusions	water infrastructure SCADA systems	STOP-IT	TDL-CNN
Min et al. [96]	deception attacks against PVs and MVs	simulation: EPANET	-	ANN
Macas et al. [97]	deception attacks against PVs and MVs	-	SWaT	deep autoencoders
Zou et al. [98]	-	WDS in US	-	data-driven estimation (ANNs) and one-class SVM
Ghaeini and Tippenhauer [99]	network attacks	SWaT testbed	-	deep packet inspection

Amin et al. [69] propose a theoretical model-based detection scheme based on hydrodynamic models to detect cyber-attacks against sensor measurements and other anomalous behaviour in canal systems. Adepu and Mathur [70] used the SWaT testbed to detect cyber-attacks using invariants, the physical conditions that must be true for a process at a given state. Researchers test their approaches using a selection of bias attacks, in which attackers modified sensor outputs and actuator commands by adding a small constant each time [74]. Researchers extended their work in [71,72] to detect bias attacks [74] against sensors and actuators using physics-based invariants for each state of the process, derived from process design for both single-point attacks happening at a single stage, and multiple point attacks that affect multiple sensors and actuators at a single stage [72], and proposed a distributed attack detection method in [73] to detect coordinated cyber-attacks. Yoong and Heng [75] proposed a security framework to develop and evaluate machine learning invariants to detect anomalies, and tested their framework using the SWaT testbed. They used an autoregressive model with exogenous inputs (ARX) combined with group searching to construct machine learning invariants to detect anomalies. The proposed framework is capable of being tested in real-life water treatment plants without causing any disturbances.

Miciolino et al. (2017) [54] proposed a fault detection and network anomaly-based detection models for FACIES testbed by monitoring data generated by sensors and network traffic between PLCs and SCADA which uses Modbus over TCP protocol. Detection uses standard deviation between the normal behaviour and actual observations. Normal behaviour of sensors and network traffic is determined by using statistical averages calculated using data from normal runs.

Zohrevand et al. (2016) [76] used a hidden Markov model (HMM) to design an anomaly-based detection model for a water supply system. Training data was collected from a SCADA-based water supply system in the City of Surrey in British Columbia (Canada) between 2011 and 2014. Working with domain experts, researchers generated anomalous cases and inserted these into the normal data as potential attack data. Four anomalies were constructed by targeting the flow capacity of water: maximum flow, minimum flow, continuous overflow and frequent overflow. Ahmed et al. (2017) [77] used EPANET to simulate a water distribution network to demonstrate a model-based attack detection technique. Detection involves determining the input-output dynamical model of the water distribution network as a set of Linear Time Invariant (LTI) equations. A Kalman Filter is then used to estimate the state of the physical process. The difference between actual measurements and estimations are used to obtain residuals which are then fed into a change detection procedure, CUSUM (cumulative sum control chart) to identify abnormal behaviour. Generated attacks include false data injection (sending modified PVs to controller; and sending false signals to actuators); and controller zero-alarm attack where the attacker changes sensor measurements in such a way that residuals do not cause any alarms. Moazeni and Khazaei [78] proposed a mixed integer nonlinear programming (MINLP) approach to estimate state variables, and tested this on a simulated 6-node water distribution system modelled using the MATLAB OPTi toolbox.

Many machine learning techniques, both supervised and unsupervised, have been used to detect anomalous behaviour. Inoue et al. [79] used a SWaT dataset [50], which consists of 41 cyber and physical attacks [45] against sensors, actuators and controllers including modifying PVs and MVs. Researchers used unsupervised learning approaches from deep learning (long short-term memory neural networks) and one-class support vector machines to detect anomalies.

Hindy et al. [80] built a water system testbed composed of two water tanks, a PLC, a Modicon M238 logic controller, pumps and five sensors that measures various water levels and the presence of water in the tanks. The testbed has two mode of operation, simulating water distribution, and storage. Sensor measurements are sent to the control and monitoring units using the Modbus protocol. Anomalous behaviour is generated as a result of cyber-attacks (DoS, spoofing), system faults and physical attacks (e.g., humans hitting tanks). Classic machine learning algorithms are used to classify anomalous behaviour and affected components using the data gathered and reported by the PLCs. These algorithms are logistic regression, Gaussian naive Bayes, k-nearest neighbors (K-NN), support vector machine (SVM), decision trees and random forests [80]. They report that the K-NN model achieved the highest accuracy.

Several teams participated in the BATADAL challenge competition [47], developing attack detection for the fictitious C-Town water distribution network (WDN) benchmark [100]. This was built using the epanetCPA water distribution modelling toolkit, and presented at the 2017 World Environmental and Water Resources Congress organized by the Environmental and Water Resources Institute of the American Society of Civil Engineers (EWRI/ASCE). Three datasets [45], one with normal operational data, and two datasets (one for training, one for testing) containing cyber-attacks, were given to each competing team. Generated cyber-attacks were deception attacks (against PVs and MVs and SCADA data) and replay attacks. Taormina and Galelli [81,82] used autoencoders (deep neural networks) in detecting attacks. Abokifa et al. [83,84] proposed a detection approach composed of three layers to detect anomalies in the BATADAL datasets; first removing outliers using statistical analysis then, using a feed forward artificial neural network (ANN), a multilayer perceptron (MLP) to identify anomalies and, finally, principal component analysis (PCA) to identify multiple affected sensors. Giacomoni et al. [85] developed two detection approaches based on data-mining. The first of these is a method using actuator rules to ensure readings from the SCADA are within defined normal ranges. The second method uses an optimization routine that extracts low-dimensionality components of the data, and thereby separates normal operation data from attack data. Pasha et al. [86,101] also used a data-mining approach on BATADAL datasets based on extracting control rules, pattern recognition, PCA, and relationship between hydraulic and system parameters. Brentan et al. [87] applied autoregressive networks with exogenous inputs (NARX), a recurrent neural network. Housh and Ohar [89,90] used physical simulation to model the system to detect cyber-attacks. Their model-based approach uses mixed integer linear programming (MILP) to estimate the hydraulic processes of the water distribution systems under normal operating conditions to produce expected errors between the actual measurements and estimated model. The difference between the expected and actual value is used to detect attacks. Chandy et al. [88] developed an ensemble model comprising two models to detect attacks for the BATADAL detection challenge competition. The first uses physical and operational rules and violations to generate events. The second uses these events along with raw data to train a deep learning model, a convolutional variational autoencoder, to detect attacks. Aghashahi et al. [91] first extracted features related to the characteristics of the attack and no-attack data by using a covariance matrix and distance measure of every data point. Then, a random forest classifier was used to classify these characteristics as attack and normal operation. A detailed description of the competition and a discussion of results can be found in [47]. MarcosQuiñones-Grueiro [92] combined widely used signal processing techniques, PCA, the adaptive exponential weighted moving average chart (EWMA) and the reconstruction-based contribution (RBC) method to detect attacks and to diagnose the area of the network that was under attack using the BATADAL dataset. Ramotsoela et al. [93] used the BATADAL dataset to evaluate some of the traditional anomaly detection approaches to detect attacks in WDS, and proposed an ensemble technique. The proposed ensemble technique combines the subspace outlier degree (SOD) algorithm, a distance-based shared nearest neighbors approach designed to detect outliers in high-dimensional data [102] and a local outlier factor (LOF) algorithm [103] to detect outliers in low-dimensional data. Both algorithms are run in parallel for each predicted datapoint and feed their outputs to a quadratic discriminant analysis (QDA) process to classify datapoints into anomalous or normal. Kadosh et al. [94] used a support vector data description (SVDD) classifier to propose a one-class cyber-attack detection model to detect attacks in WSD using both the BATADAL dataset and epanetCPA.

Bakalos et al. [95] developed a cyber-attack detection approach for water systems using multimodal data fusion and adaptive deep learning. Multimodal data fusion involves combining different channels of information, including visual data from thermal camera streams, Wi-Fi reflection, and ICS data. The weight attached to each of these streams of data is determined through a deep learning model process. The proposed adaptive deep learning approach uses a tapped delay line (TDL) convolutional neural network (CNN) with autoregressive moving average [95]. The data used to evaluate the approach is from STOP-IT project.

Min et al. [96] used an artificial neural network to detect attacks against a water distribution network using the EPANET simulator [84]. Macas et al. [97] used an "unsupervised attention-based spatio-temporal autoencoder for anomaly detection (STAE-AD)" model to detect attacks against water infrastructures using the SWaT dataset. Zou et al. [98] proposed a hybrid model making use of an MLP and a one-class SVM. MLP was used to forecast measurement parameters, and prediction errors were used to train a one-class SVM to classify outliers; finally, Bayesian sequence analysis was used to detect contamination attacks against water distribution systems.

Majority of cyber-attack detection models reviewed focus on detecting anomalous behaviour by monitoring and analyzing physical process variables, and failed to monitor industrial control network traffic and use this knowledge to detect cyber-attacks. Ghaeini and Tippenhauer [99] proposed a hierarchical monitoring intrusion detection system (HAMIDS) for ICS to collect network events in different layers of industrial networks. HAMIDS extends the Bro, an open-source tool for monitoring and analyzing network traffic. IDS sensors are installed on different layers of industrial networks to monitor network events. These events are then aggregated and processed in a central cluster to detect malicious behaviour. HAMIDS was validated using a range of network attacks (e.g., ARP poising, network flooding and man in the middle attacks) against SWaT testbed.

Proposed detection approaches are evaluated for effectiveness using (i) operational data from real-world systems; (ii) testbeds; and (iii) simulation. Existing studies show a wide variety of techniques that were applied to detect cyber-attacks against water systems; however, making a reliable comparison among detection approaches is not feasible due to a lack of common performance metrics and/or missing reported performance data, different datasets and sizes.

4.3. Model-Based Security Analysis

Several research studies focused on using modelling approaches to analyse the security of water systems and to identify vulnerabilities.

Kang et al. [104] proposed a model-based security analysis for a water treatment system. Testing their approach on SWaT, they modelled the interaction between the physical plant and controller using approximate, discrete models to discover and explore potential attacks. The model is constructed using a first-order modelling language Alloy to capture, as state transition rules, connections among various components and the behaviour of the plant.

Motivated by malware techniques that hide critical information from operators while executing an attack (e.g., Stuxnet), Patloll et al. [105] proposed a multiple security domain non-deducibility (MSDND) model [106] using belief, information transfer and trust (BIT) logic [107] to identify critical information that attackers may hide. BIT logic is used to reason about the reliability of data moving between entities, defined as the belief and trust one entity has in information received from another entity. A system is decomposed into components, and each component that could change the state of the state is treated as a separate domain. Requiring development of invariants, an information execution flow across these domains starting from source to destination is monitored to identify when vulnerabilities that have been exploited have resulted in invariant violation. Mishra et al. [108] proposed an agent-based modelling framework to model critical CPS and their interdependencies, to understand the impact of attacks on interconnected critical infrastructures; they evaluated the application of the model to a water distribution system and used invariant-based method [70] to generate rules to detect attacks.

Taormina et al. [66] and Hunter et al. [109] proposed a modelling approach to quantify the hydraulic behaviour of the system (such as tank overflow, variation in pumps) under cyber–physical attacks by defining components of a system, and specifying attack variables (starting time, duration). They give simulation results using the epanetCPA toolbox and the C-Town network [100].

4.4. Risk and Resilience Management

A small number of studies worked on methods to support risk and resilience management.

Moraitis et al. [110] describes a methodology to quantify the impact of cyber–physical attacks on water distribution networks. The methodology is based on quantifying failures described under categories (magnitude, propagation, severity, crest factor, rapidity) against user-defined service levels. A proposed model is demonstrated using the C-Town WDN.

Jeong [111] discusses the development of a risk management framework for water infrastructure against intentional attacks, including cyber-attacks based on vulnerability assessment and consequence assessment of attacks. The proposed vulnerability assessment involves the development of a hierarchical structure of the system to identify all water infrastructure components, using expert knowledge and fuzzy hierarchical analysis. The recommended consequence assessment is based on the time to restore the system to its normal operation, and the areas affected by the attack, and the expected damage is based on attacker's and defender's capabilities.

Shin et al. [112] investigated resilience strategies against water CPS. Resilience is characterized in terms of four capabilities [112]: (i) ability to withstand disruption; (ii) absorptive capability (if disruption is unavoidable then minimize undesirable consequences; (iii) adaptive capability (adjusting to disrupted and undesirable conditions); (iv) restorative capability (recover quickly to completely normal operation). A resilience metric is proposed to measure the resilience of water systems against cyber-attack, and the C-town benchmark water distribution system is used as a case study to demonstrate the proposed metric.

4.5. Security Frameworks

Modern water treatment infrastructures consist of interconnected systems layered in a hierarchy, such as a supervisory layer consisting of SCADA systems, and a control layer composed of PLCs, sensors and actuators. Data flows occur between these layers via multiple communication networks. Mathur [113] proposes a multilayer security framework composed of seven layers of countermeasures applied to different network layers to secure water treatment systems. Proposed countermeasures include attack prevention mechanisms (firewalls), attack detection mechanisms (intrusion detection systems, process anomaly detection), and post-attack mechanisms that could bring the process back to a normal or manageable state. A partial implementation of the proposed framework was tested on the SWaT testbed.

4.6. Security Benchmarks and Case Studies

TNO (Netherlands Organisation for Applied Scientific Research—an independent research organisation) and the NICC (the Netherlands Infrastructure Cybercrime unit), carried out a study [114] to understand the current state of cyber-security of process control in the drinking water sector in the Netherlands. Researchers report that a large variance of security posture was found among organisations; the data collected exposed serious weaknesses in each company. As the study contained sensitive national data, confidentiality of the organisations was maintained and the reported analyses were based on artificially aggregated data. The study was effective and resulted in the development of good practices for SCADA security for drinking water organisations, which are available both in Dutch and English [115]. Building on this work, Burghouwt et al. [116] measured the cybersecurity state of the 19 water management organisations in the Netherlands through an improved questionnaire. Researchers identified a lack of uniformity on security postures between organisations, partly due to ineffective management of security responsibilities. They designed and built DESI [116], a simulator to demonstrate cyber–physical attack scenarios and improve cyber-attack knowledge.

A case study paper was presented in [117] investigating access control mechanisms in industrial control systems conducted on the WADI testbed, to show how the lack of effective access control could lead to malicious behaviour. Researchers revealed that a lack of access control in network protocols, systems and field devices used in ICS is making these systems vulnerable to attacks.

A critical case study for security of water systems is the Marooch water breach incident. Slay and Miller [29] discusses this incident and reports the lessons learned from the incident emphasising the need for effective, reliable and economically viable security countermeasures including intrusion detection systems for SCADA networks, better management of security policies and procedures, investment in security training for staff, and a wider and sustainable collaboration between academia, industry, vendors and government agencies to tackle existing and future security threats.

4.7. Security Monitoring Capabilities

One of the papers identified dealt with improving security monitoring capabilities for water distribution systems. In [118], researchers propose sonification, data in audio, to help system operators avoid cognitive overloading with visual information to raise alarms for cyber-attacks on water distribution systems. Motivated by prior work on sonification, designed to improve monitoring capabilities, researchers designed a sonification system to reduce the overload of human operators faced with visual channels, to support better decision-making for a water facility by sonifying the outputs of an anomaly detection model. Current anomaly detection models are represented as visual diagrams showing anomalous data points at a given time and often an alarm is raised when a threshold is reached.

5. Open Issues and Future Research Areas

Results obtained from the systematic review show that research has made a significant contribution to the security of water systems. In the following sections we discuss limitations of existing studies and highlight some areas for future research.

5.1. Building Testbeds for Water Systems

Much of the existing research in this area involves a pool of resources (SWaT and WADI testbeds, epanetCPA toolbox, and datasets) provided by the iTrust Centre for research in cyber-security. Researchers associated with the iTrust Centre demonstrate the importance of developing and providing access to a real physical testbed for carrying out security research. Most of the existing studies have focused on drinking water systems, primarily those responsible for water distribution. Given the diversity of water and wastewater systems, more work in this area would provide obvious benefits, especially through testbeds involving water systems such as sewer and wastewater systems, and irrigation systems; these could be used to further validate the applicability of existing research. Although of immense value, building and maintaining realistic operational testbeds is not an easy task and requires significant and ongoing access to resources, skills and people.

5.2. Threat and Attack Models

Existing attack models primarily make use of manual and single-point attacks targeting single measurement variables (sensor readings) or control commands. However, stealthy attacks, those trying to cause damage and at the same time remain undetected, may necessitate multi-point attacks if they are to evade detection mechanisms and operators. This area is starting to receive increased attention from researchers investigating the security of CPS [67]; however, more effort is required to understand how these attacks can be performed and what the limits on their effectiveness might be. Consequently, few studies have verified the effectiveness of existing detection models against these attacks.

5.3. Attack Detection Models

Many studies designed to detect attacks against water CPS use machine learning-based anomaly detection models, in which normal operational data is the primary (or sole) resource as there is often insufficient anomalous data to create models using supervised approaches. It is not readily possible to compare the performance of existing detection models, or to determine their generality or the reproducibility of their results. This is due both to a lack of datasets, leading to poor diversity in assumptions and plant models, and to a lack of common performance metrics. Where common datasets and performance metrics have been used, as in the case of, say, the SWaT and WADI datasets, reported results suggest that deep learning-based anomaly detection models perform better than conventional anomaly detection models. However, further studies are required to build confidence that such performance improvement is real.

As is usually the case with intrusion detection studies for CPS, the effectiveness of the proposed solutions were measured using conventional performance metrics, including accuracy, precision, recall, F-score, false positives and false negatives. These performance metrics were not designed for multivariate time-series datasets of CPS, in which anomalies usually occur in bursts [119]. Even when using these conventional performance metrics, some fail to report false positives and none of the studies reported detection latency, which is an important metric for critical systems [68] as early detection is critical for CPS.

Over the last decade, there has been an increase in number of CPS applying deep learning models to detect anomalous behaviour and datasets such as BATADAL, SWaT and WADi have contributed to some of these studies. However, studies from other fields have shown that machine learning-based approaches are rather vulnerable to accidental or intentional corruption of training data sets; thus, say, adversarial attacks can influence detection outcomes [120]. At the same time, there is a significant number of research studies that focus on improving the robustness of such models [121]. At present, however, such work is invariably targeted at other fields of study, most notably computer vision, and we are yet to understand the possible risks in the application of learning models to CPS.

The generation of attack or anomalous behaviour for testing detection models is often done manually. Typically, measurement values or control signals are modified, and performance data is collected both with and without these variations. However, such an approach assumes that the modifications are representative of those that will be experienced in reality, and this assumption is tenuous at best. Furthermore, over time, CPS actuators and sensors degrade as a result of ageing and become more prone to noise. As a result, normal behaviour is itself non-stationary and it will be necessary either to use richer training sets and models that capture temporal change, or to use online learning. The latter is again vulnerable to changes induced by an adversary that are intended to pervert the detection mechanism. There is therefore a pressing need to increase the attention paid to the practicalities associated with actual deployment, including the usability and maintainability of proposed detection models.

Identifying anomalous behaviour should ideally be followed by the raising of an alert that identifies the potential cause and so determines a strategy to be followed for mitigation. However, existing studies often stop at detection. Future work is therefore required to investigate approaches that identify the root cause of anomalous behaviour, locate compromised devices and respond and mitigate further damage in a timely manner.

5.4. Collaboration with Industry

Although several studies have consulted with engineers who have experience in dealing with water systems, we failed to identify any publications that were written by industry. There is currently a lack of collaborative work between industry and academia in this area. Securing water systems requires a multidisciplinary effort that involves both the designers and operators of these systems and academics working at the leading edge of technology to ensure that security research pushes the boundaries of the possible while remaining applicable and usable.

6. Conclusions

In this paper, we have systematically reviewed the existing peer-reviewed research efforts to secure water systems, and have identified limitations in those research efforts and possible future directions for securing next generation of smart water CPS. This study provides guidance for understanding the existing security research for developing secure smart water systems.

In comparison to other utilities such as electricity, the security of water systems has not received much research attention in the past, but this is changing, and there has been an increase in the number of studies since 2016 supported by EC research and innovation funding programs and international funding opportunities. The studies reviewed in this paper are encouraging, but they require further work for validation and deployment on real water systems. Most of the existing studies, including testbeds, simulation tools and datasets, have focused on drinking water treatment, supply and distribution. Further studies are required to build testbeds, simulation and datasets that investigate security of non-drinking water sectors such as wastewater treatment systems, stormwater management and systems for agriculture and irrigation.

Finally, development of a comprehensive usable security framework that covers different aspects of security, from prevention to detection, response and mitigation requires a multidisciplinary approach involving academia-industry-government cooperation.

Author Contributions: Conceptualization, N.T., S.H. and J.W.; methodology, N.T. and S.H.; data curation, N.T.; writing—original draft preparation, N.T.; writing—review and editing, N.T., P.H., S.H. and J.W.; visualization, N.T. and P.H.; supervision, S.H. and J.W. All authors have read and agreed to the published version of the manuscript.

Funding: This research was funded by the UK Engineering and Physical Sciences Research Council (EPSRC), under The PETRAS National Centre of Excellence for IoT Systems Cybersecurity, grant number EP/S035362/1.

Institutional Review Board Statement: Not applicable.

Informed Consent Statement: Not applicable.

Data Availability Statement: Not applicable.

Conflicts of Interest: The authors declare no conflict of interest.

References

1. WWAP (United Nations World Water Assessment Programme)/UN-Water The United Nations World Water Development Report 2018: Nature-based Solution for Water. Paris, UNESCO. 2018. Available online: www.unwater.org/publications/world-water-development-report-2018/ (accessed on 6 November 2020).
2. Bank, W. The World Bank and the International Water Association to Establish a Partnership to Reduce Water Losses. 2016. Available online: https://www.worldbank.org/en/news/press-release/2016/09/01/the-world-bank-and-the-international-water-association-to-establish-a-partnership-to-reduce-water-losses (accessed on 6 November 2020).
3. Discoverwater. Leaking Pipes. 2020. Available online: https://discoverwater.co.uk/leaking-pipes (accessed on 14 November 2020).
4. Li, J.; Yang, X.; Sitzenfrei, R. Rethinking the Framework of Smart Water System: A Review. *Water* **2020**, *12*, 412. [CrossRef]
5. Giudicianni, C.; Herrera, M.; Nardo, A.D.; Adeyeye, K.; Ramos, H.M. Overview of Energy Management and Leakage Control Systems for Smart Water Grids and Digital Water. *Modelling* **2020**, *1*, 134–155. [CrossRef]
6. Adedeji, K.B.; Hamam, Y. Cyber-Physical Systems for Water Supply Network Management: Basics, Challenges, and Roadmap. *Sustainability* **2020**, *12*, 9555. [CrossRef]
7. Ofwat. *PR19 Draft Determinations: UK Government Priorities 2019 Price Review Draft Determinations*; Technical Report; Ofwat: Birmingham, UK, 2019.
8. Ofwat. *Time to Act, Together: Ofwat's Strategy*; Technical Report; Ofwat: Birmingham, UK, 2019.
9. Schickhuber, G.; McCarthy, O. Distributed fieldbus and control network systems. *Comput. Control Eng. J.* **1997**, *8*, 21–32. [CrossRef]
10. SWAN Forum. A Layered View of Smart Water Networks. Available online: https://www.swan-forum.com/swan-tools/a-layered-view (accessed on 1 November 2020).

11. Falliere, N.; Murchu, L.O.; Chien, E. *W32.Stuxnet Dossier (Version 1.4)*; White Paper, Symantec Security Response; Symantec: Mountain View, CA, USA, 2011.
12. Symantec. *W32.Duqu: The Precursor to the Next Stuxnet (Version 1.4)*; White Paper, Symantec Security Response; Symantec: Mountain View, CA, USA, 2011.
13. Kaspersky. BlackEnergy APT Attacks in Ukraine. Available online: https://www.kaspersky.co.uk/resource-center/threats/blackenergy (accessed on 30 November 2020).
14. Havex Hunts For ICS/SCADA Systems. 2014. Available online: https://www.f-secure.com/weblog/archives/00002718.html (accessed on 30 October 2020).
15. Gleick, P.H. Water and terrorism. *Water Policy* **2006**, *8*, 481–503. [CrossRef]
16. Interpol. The Protection of Critical Infrastructure against Terrorist Attacks: Compendium of Good Practices. Compiled by CTED and UNOCT in 2018. 2018. Available online: https://www.un.org/sc/ctc/wp-content/uploads/2019/01/Compendium_of_Good_Practices_Compressed.pdf (accessed on 1 August 2020).
17. Hassanzadeh, A.; Rasekh, A.; Galelli, S.; Aghashahi, M.; Taormina, R.; Ostfeld, A.; Banks, M.K. A Review of Cybersecurity Incidents in the Water Sector. *J. Environ. Eng.* **2020**, *146*, 03120003. [CrossRef]
18. Clark, R.M.; Panguluri, S.; Nelson, T.D.; Wyman, R.P. Protecting Drinking Water Utilities from Cyberthreats. *J. AWWA* **2017**, *109*, 50–58. [CrossRef]
19. ZDNet. Israel Government Tells Water Treatment Companies to Change Passwords. 2020. Available online: https://www.zdnet.com/article/israel-says-hackers-are-targeting-its-water-supply-and-treatment-utilities/ (accessed on 6 November 2020).
20. The Coloradoan. Cyberattacker Demands Ransom from Northern Colorado Utility. 2019. Available online: https://eu.coloradoan.com/story/money/2019/03/14/cyberattacker-demands-ransom-colorado-utility/3148951002/ (accessed on 11 September 2020).
21. Eweek. Water Utility in Europe Hit by Cryptocurrency Malware Mining Attack. 2018. Available online: https://www.eweek.com/security/water-utility-in-europe-hit-by-cryptocurrency-malware-mining-attack (accessed on 11 September 2020).
22. The Registry. Water Treatment Plant Hacked, Chemical Mix Changed for Tap Supplies. 2016. Available online: https://www.theregister.com/2016/03/24/water_utility_hacked (accessed on 14 November 2020).
23. The New York Times. A Dam, Small and Unsung, Is Caught Up in an Iranian Hacking Case. 2016. Available online: https://www.nytimes.com/2016/03/26/nyregion/rye-brook-dam-caught-in-computer-hacking-case.html (accessed on 11 September 2020).
24. The United States Department of Justice. United States District Court Southern District of New York: Sealed Indictment. 2016. Available online: https://www.justice.gov/opa/file/834996/download (accessed on 31 December 2020).
25. Govtech. Report: Hacking Lands Florida Wastewater Official in Hot Water. 2012. Available online: https://www.govtech.com/public-safety/Report-Hacking-Lands-Florida-Wastewater-Official-in-Hot-Water.html (accessed on 31 December 2020).
26. Computer World. Insider charged with hacking California canal system. 2007. Available online: https://www.computerworld.com/article/2540235/insider-charged-with-hacking-california-canal-system.html (accessed on 10 October 2020).
27. TechRepublic. Pennsylvania Water System Hack Demonstrates Lax Security. 2006. Available online: https://www.techrepublic.com/blog/it-security/pennsylvania-water-system-hack-demonstrates-lax-security/ (accessed on 11 September 2020).
28. The MITRE Corporation. Malicious Control System Cyber Security Attack Case Study–Maroochy Water Services, Australia. 2008. Available online: http://www.mitre.org/sites/default/files/pdf/08_1145.pdf (accessed on 11 September 2020).
29. Jill, J.S.; Miller, M. Lessons Learned from the Maroochy Water Breach. In *Critical Infrastructure Protection*; Goetz, E., Shenoi, S., Eds.; Springer: New York, NY, USA, 2008; Volume 253, pp. 73–82.
30. Rid, T.; Buchanan, B. Attributing Cyber Attacks. *J. Strateg. Stud.* **2015**, *38*, 4–37. [CrossRef]
31. Rogers, M.K. A two-dimensional circumplex approach to the development of a hacker taxonomy. *Digit. Investig.* **2006**, *3*, 97–102. [CrossRef]
32. Green, B.; Krotofil, M.; Abbasi, A. On the Significance of Process Comprehension for Conducting Targeted ICS Attacks. In Proceedings of the 2017 Workshop on Cyber-Physical Systems Security and PrivaCy, CPS '17, Dallas, TX, USA, 3 November 2017; Association for Computing Machinery: New York, NY, USA, 2017; pp. 57–67. [CrossRef]
33. Dragos. Cyber Threat Perspective Manufacturing Sector. 2020. Available online: https://www.dragos.com/resource/manufacturing-threat-perspective/ (accessed on 5 December 2020).
34. Gill, H. From Vision to Reality: Cyber-Physical Systems. In Proceedings of the HCSS National Workshop on New Research Directions for High Confidence Transportation CPS: Automotive, Aviation, and Rail, Washington, DC, USA, 18–20 November 2008.
35. Lee, E.A. The Past, Present and Future of Cyber-Physical Systems: A Focus on Models. *Sensors* **2015**, *15*, 4837–4869. [CrossRef] [PubMed]
36. Stouffer, K.; Zimmerman, S.; Timothy, T.C.; Lubell, J.; Cichonski, J.; McCarthy, J. *NISTIR 8183: Cybersecurity Framework Manufacturing Profile*; Technical Report; National Institute of Standards and Technology: Gaithersburg, MD, USA, 2017.
37. Hu, V.; Ferraiolo, D.; Kuhn, R. *Assessment of Access Control Systems*; Technical Report; National Institute of Standards and Technology: Gaithersburg, MD, USA, 2006.
38. Initiative, J.T.F.T. *Security and Privacy Controls for Federal Information Systems and Organizations, NIST Special Publication 800-53 Revision 4*; Technical Report; National Institute of Standards and Technology: Gaithersburg, MD, USA, 2015.

39. Ross, R.; McEvilley, M.; Oren, C.J. *Systems Security Engineering Considerations for a Multidisciplinary Approach in the Engineering of Trustworthy Secure Systems, NIST Special Publication 800-160*; Technical Report; National Institute of Standards and Technology: Gaithersburg, MD, USA, 2016.

40. Stouffer, K.; Lightman, S.; Pillitteri, V.; Abrams, M.; Hahn, A. *NIST Special Publication 800-82: Guide to Industrial Control Systems (ICS) Security*; Technical Report; National Institute of Standards and Technology: Gaithersburg, MD, USA, 2014.

41. Hahn, A.; Thomas, R.K.; Lozano, I.; Cardenas, A. A multi-layered and kill-chain based security analysis framework for cyber-physical systems. *Int. J. Crit. Infrastruct. Prot.* **2015**, *11*, 39–50. [CrossRef]

42. Moher, D.; Liberati, A.; Tetzlaff, J.; Altman, D.; The PRISMA Group. Preferred Reporting Items for Systematic Reviews and Meta-Analyses: The PRISMA Statement. *PLoS Med.* **2009**, *6*, 1–6. [CrossRef] [PubMed]

43. Mathur, A.P.; Tippenhauer, N.O. SWaT: A water treatment testbed for research and training on ICS security. In Proceedings of the 2016 International Workshop on Cyber-physical Systems for Smart Water Networks (CySWater), Vienna, Austria, 11 April 2016; pp. 31–36.

44. Ahmed, C.M.; Palleti, V.R.; Mathur, A.P. WADI: A Water Distribution Testbed for Research in the Design of Secure Cyber Physical Systems. In Proceedings of the 3rd International Workshop on Cyber-Physical Systems for Smart Water Networks, Pittsburgh, PA, USA, 18–21 April 2017; Association for Computing Machinery: New York, NY, USA, 2017; pp. 25–28. [CrossRef]

45. ITrust. Dataset Characteristics: SWaT, WADI and BATADAL. Available online: https://itrust.sutd.edu.sg/itrust-labs_datasets/dataset_info/ (accessed on 8 November 2020).

46. iTrust—Singapore University of Technology and Design (SUTD). Testbeds. Available online: https://itrust.sutd.edu.sg/testbeds (accessed on 30 November 2020).

47. Taormina, R.; Galelli, S.; Tippenhauer, N.O.; Salomons, E.; Ostfeld, A.; Eliades, D.G.; Aghashahi, M.; Sundararajan, R.; Pourahmadi, M.; Banks, M.K.; et al. Battle of the Attack Detection Algorithms: Disclosing Cyber Attacks on Water Distribution Networks. *J. Water Resour. Plan. Manag.* **2018**, *144*, 04018048. [CrossRef]

48. Facies Project. Available online: http://facies.dia.uniroma3.it/ (accessed on 30 November 2020).

49. The STOP-IT Project. Available online: https://stop-it-project.eu/ (accessed on 30 November 2020).

50. Goh, J.; Adepu, S.; Junejo, K.N.; Mathur, A. A Dataset to Support Research in the Design of Secure Water Treatment Systems. In Proceedings of the International Conference on Critical Information Infrastructures Security, Paris, France, 10–12 October 2017; Havarneanu, G., Setola, R., Nassopoulos, H., Wolthusen, S., Eds.; Springer International Publishing: Cham, Switzerland, 2017; pp. 88–99.

51. Kartakis, S.; Abraham, E.; McCann, J.A. WaterBox: A Testbed for Monitoring and Controlling Smart Water Networks. In Proceedings of the 1st ACM International Workshop on Cyber-Physical Systems for Smart Water Networks, CySWater'15, Seattle, WA, USA, 14–16 April 2015; Association for Computing Machinery: New York, NY, USA, 2015. [CrossRef]

52. Taormina, R.; Galelli, S.; Tippenhauer, N.; Ostfeld, A.; Salomons, E. Assessing the Effect of Cyber-Physical Attacks on Water Distribution Systems. In Proceedings of the World Environmental and Water Resources Congress 2016, Palm Beach, FL, USA, 22–26 May 2016; pp. 436–442. [CrossRef]

53. Taormina, R.; Galelli, S.; Douglas, H.; Tippenhauer, N.; Salomons, E.; Ostfeld, A. A toolbox for assessing the impacts of cyber-physical attacks on water distribution systems. *Environ. Model. Softw.* **2019**, *112*, 46–51. [CrossRef]

54. Etchevés Miciolino, E.; Setola, R.; Bernieri, G.; Panzieri, S.; Pascucci, F.; Polycarpou, M.M. Fault Diagnosis and Network Anomaly Detection in Water Infrastructures. *IEEE Des. Test* **2017**, *34*, 44–51. [CrossRef]

55. Nikolopoulos, D.; Makropoulos, C.; Kalogeras, D.; Monokrousou, K.; Tsoukalas, I. Developing a Stress-Testing Platform for Cyber-Physical Water Infrastructure. In Proceedings of the 2018 International Workshop on Cyber-physical Systems for Smart Water Networks (CySWater), Porto, Portugal, 10–13 April 2018; pp. 9–11. [CrossRef]

56. Nikolopoulos, D.; Moraitis, G.; Bouziotas, D.; Lykou, A.; Karavokiros, G.; Makropoulos, C. RISKNOUGHT: A cyber-physical stress-testing platform for water distribution networks. In Proceedings of the 11th World Congress on Water Resources and Environment (EWRA 2019) Managing Water Resources for a Sustainable Future, Madrid, Spain, 25–29 June 2019.

57. Nikolopoulos, D.; Moraitis, G.; Bouziotas, D.; Lykou, A.; Karavokiros, G.; Makropoulos, C. Cyber-Physical Stress-Testing Platform for Water Distribution Networks. *J. Environ. Eng.* **2020**, *146*, 04020061. [CrossRef]

58. Teixeira, M.; Salman, T.; Zolanvari, M.; Jain, R.; Meskin, N.; Samaka, M. SCADA System Testbed for Cybersecurity Research Using Machine Learning Approach. *Future Internet* **2018**, *10*, 76. [CrossRef]

59. EPANET Application for Modeling Drinking Water Distribution Systems. United States Environmental Protection Agency. Available online: https://www.epa.gov/water-research/epanet (accessed on 31 December 2020).

60. Amin, S.; Litrico, X.; Sastry, S.; Bayen, A.M. Cyber Security of Water SCADA Systems—Part I: Analysis and Experimentation of Stealthy Deception Attacks. *IEEE Trans. Control Syst. Technol.* **2013**, *21*, 1963–1970. [CrossRef]

61. Adepu, S.; Mathur, A. An Investigation into the Response of a Water Treatment System to Cyber Attacks. In Proceedings of the 2016 IEEE 17th International Symposium on High Assurance Systems Engineering (HASE), Orlando, FL, USA, 7–9 January 2016; pp. 141–148.

62. Adepu, S.; Prakash, J.; Mathur, A. WaterJam: An Experimental Case Study of Jamming Attacks on a Water Treatment System. In Proceedings of the 2017 IEEE International Conference on Software Quality, Reliability and Security Companion (QRS-C), Prague, Czech Republic, 25–29 July 2017; pp. 341–347.

63. Tomić, I.; Breza, M.J.; Jackson, G.; Bhatia, L.; McCann, J.A. Design and Evaluation of Jamming Resilient Cyber-Physical Systems. In Proceedings of the 2018 IEEE International Conference on Internet of Things (iThings) and IEEE Green Computing and Communications (GreenCom) and IEEE Cyber, Physical and Social Computing (CPSCom) and IEEE Smart Data (SmartData), Halifax, NS, Canada, 30 July–3 August 2018; pp. 687–694.

64. Robles-Durazno, A.; Moradpoor, N.; McWhinnie, J.; Russell, G.; Maneru-Marin, I. *Implementation and Detection of Novel Attacks to the PLC Memory of a Clean Water Supply System*; Botto-Tobar, M., Pizarro, G., Zúñiga-Prieto, M., D'Armas, M., Zúñiga Sánchez, M., Eds.; Technology Trends; Springer International Publishing: Cham, Switzerland, 2019; pp. 91–103.

65. Amin, S.; Litrico, X.; Sastry, S.S.; Bayen, A.M. Stealthy Deception Attacks on Water SCADA Systems. In Proceedings of the 13th ACM International Conference on Hybrid Systems: Computation and Control, HSCC '10, Stockholm, Sweden, 12–15 April 2010; Association for Computing Machinery: New York, NY, USA, 2010; pp. 161–170. [CrossRef]

66. Taormina, R.; Galelli, S.; Tippenhauer, N.O.; Salomons, E.; Ostfeld, A. Characterizing Cyber-Physical Attacks on Water Distribution Systems. *J. Water Resour. Plan. Manag.* **2017**, *143*, 04017009. [CrossRef]

67. Erba, A.; Taormina, R.; Galelli, S.; Pogliani, M.; Carminati, M.; Zanero, S.; Tippenhauer, N.O. Constrained Concealment Attacks against Reconstruction-Based Anomaly Detectors in Industrial Control Systems. In Proceedings of the Annual Computer Security Applications Conference, ACSAC '20, Austin, TX, USA, 7–10 December 2020; Association for Computing Machinery: New York, NY, USA, 2020; pp. 480–495. [CrossRef]

68. Mitchell, R.; Chen, I.R. A Survey of Intrusion Detection Techniques for Cyber-Physical Systems. *ACM Comput. Surv.* **2014**, *46*. [CrossRef]

69. Amin, S.; Litrico, X.; Sastry, S.S.; Bayen, A.M. Cyber Security of Water SCADA Systems—Part II: Attack Detection Using Enhanced Hydrodynamic Models. *IEEE Trans. Control Syst. Technol.* **2013**, *21*, 1679–1693. [CrossRef]

70. Adepu, S.; Mathur, A. Using Process Invariants to Detect Cyber Attacks on a Water Treatment System. In Proceedings of the ICT Systems Security and Privacy Protection, Ghent, Belgium, 30 May–1 June 2016; Hoepman, J.H., Katzenbeisser, S., Eds.; Springer International Publishing: Cham, Switzerland, 2016; pp. 91–104.

71. Adepu, S.; Mathur, A. Distributed Detection of Single-Stage Multipoint Cyber Attacks in a Water Treatment Plant. In Proceedings of the 11th ACM on Asia Conference on Computer and Communications Security, ASIA CCS '16, Xi'an, China, 30 May 2016; Association for Computing Machinery: New York, NY, USA, 2016; pp. 449–460. [CrossRef]

72. Adepu, S.; Mathur, A. From Design to Invariants: Detecting Attacks on Cyber Physical Systems. In Proceedings of the 2017 IEEE International Conference on Software Quality, Reliability and Security Companion (QRS-C), Prague, Czech Republic, 17 July 2017; pp. 533–540.

73. Adepu, S.; Mathur, A. Distributed Attack Detection in a Water Treatment Plant: Method and Case Study. *IEEE Trans. Dependable Secur. Comput.* **2018**. [CrossRef]

74. Cárdenas, A.A.; Amin, S.; Lin, Z.S.; Huang, Y.L.; Huang, C.Y.; Sastry, S. Attacks against Process Control Systems: Risk Assessment, Detection, and Response. In Proceedings of the 6th ACM Symposium on Information, Computer and Communications Security, ASIACCS '11, Hong Kong, China, March, 20–21 March 2011; Association for Computing Machinery: New York, NY, USA, 2011; pp. 355–366. [CrossRef]

75. Yoong, C.H.; Heng, J. Framework for Continuous System Security Protection in SWaT. In Proceedings of the 2019 3rd International Symposium on Computer Science and Intelligent Control, ISCSIC 2019, Amsterdam, The Netherlands, 25–27 September 2019; Association for Computing Machinery: New York, NY, USA, 2019. [CrossRef]

76. Zohrevand, Z.; Glasser, U.; Shahir, H.; Tayebi, M.A.; Costanzo, R. Hidden Markov based anomaly detection for water supply systems. In Proceedings of the 2016 IEEE International Conference on Big Data (Big Data), Washington, DC, USA, 5–8 December 2016; IEEE Computer Society: Los Alamitos, CA, USA, 2016; pp. 1551–1560. [CrossRef]

77. Ahmed, C.M.; Murguia, C.; Ruths, J. Model-Based Attack Detection Scheme for Smart Water Distribution Networks. In Proceedings of the 2017 ACM on Asia Conference on Computer and Communications Security, ASIA CCS '17, New York, NY, USA, 2–6 April 2017; Association for Computing Machinery: New York, NY, USA, 2017; pp. 101–113. [CrossRef]

78. Moazeni, F.; Khazaei, J. MINLP Modeling for Detection of SCADA Cyberattacks in Water Distribution Systems. In Proceedings of the World Environmental and Water Resources Congress 2020, Henderson, NV, USA, 17–21 May 2020; pp. 340–350. [CrossRef]

79. Inoue, J.; Yamagata, Y.; Chen, Y.; Poskitt, C.M.; Sun, J. Anomaly Detection for a Water Treatment System Using Unsupervised Machine Learning. In Proceedings of the 2017 IEEE International Conference on Data Mining Workshops (ICDMW), Atlantic City, NY, USA, 14–17 November 2017; pp. 1058–1065.

80. Hindy, H.; Brosset, D.; Bayne, E.; Seeam, A.; Bellekens, X. *Improving SIEM for Critical SCADA Water Infrastructures Using Machine Learning*; Katsikas, S.K., Cuppens, F., Cuppens, N., Lambrinoudakis, C., Antón, A., Gritzalis, S., Mylopoulos, J., Kalloniatis, C., Eds.; Computer Security; Springer International Publishing: Cham, Switzerland, 2019; pp. 3–19.

81. Taormina, R.; Galelli, S. Real-Time Detection of Cyber-Physical Attacks on Water Distribution Systems Using Deep Learning. In Proceedings of the World Environmental and Water Resources Congress 2017, Sacramento, CA, USA, 21–25 May 2017; pp. 469–479. [CrossRef]

82. Taormina, R.; Galelli, S. Deep-Learning Approach to the Detection and Localization of Cyber-Physical Attacks on Water Distribution Systems. *J. Water Resour. Plan. Manag.* **2018**, *144*, 04018065. [CrossRef]

83. Abokifa, A.A.; Haddad, K.; Lo, C.S.; Biswas, P. Detection of Cyber Physical Attacks on Water Distribution Systems via Principal Component Analysis and Artificial Neural Networks. In Proceedings of the World Environmental and Water Resources Congress 2017, Sacramento, CA, USA, 21–25 May 2017; pp. 676–691. [CrossRef]

84. Abokifa, A.A.; Haddad, K.; Lo, C.; Biswas, P. Real-Time Identification of Cyber-Physical Attacks on Water Distribution Systems via Machine Learning Based Anomaly Detection Techniques. *J. Water Resour. Plan. Manag.* **2019**, *145*, 04018089. [CrossRef]

85. Giacomoni, M.; Gatsis, N.; Taha, A. Identification of Cyber Attacks on Water Distribution Systems by Unveiling Low-Dimensionality in the Sensory Data. In Proceedings of the World Environmental and Water Resources Congress 2017, Sacramento, CA, USA, 21–25 May 2017; pp. 660–675. [CrossRef]

86. Pasha, M.F.K.; Kc, B.; Somasundaram, S.L. An Approach to Detect the Cyber-Physical Attack on Water Distribution System. In Proceedings of the World Environmental and Water Resources Congress 2017, Sacramento, CA, USA, 21–25 May 2017; pp. 703–711. [CrossRef]

87. Brentan, B.M.; Campbell, E.; Lima, G.; Manzi, D.; Ayala-Cabrera, D.; Herrera, M.; Montalvo, I.; Izquierdo, J.; Luvizotto, E. On-Line Cyber Attack Detection in Water Networks through State Forecasting and Control by Pattern Recognition. In Proceedings of the World Environmental and Water Resources Congress 2017, Sacramento, CA, USA, 21–25 May 2017; pp. 583–592. [CrossRef]

88. Chandy, S.E.; Rasekh, A.; Barker, Z.A.; Campbell, B.; Shafiee, M.E. Detection of Cyber-Attacks to Water Systems through Machine-Learning-Based Anomaly Detection in SCADA Data. In Proceedings of the World Environmental and Water Resources Congress 2017, Sacramento, CA, USA, 21–25 May 2017; pp. 611–616. [CrossRef]

89. Housh, M.; Ohar, Z. Model Based Approach for Cyber-Physical Attacks Detection in Water Distribution Systems. In Proceedings of the World Environmental and Water Resources Congress 2017, Sacramento, CA, USA, 21–25 May 2017; pp. 727–736. [CrossRef]

90. Housh, M.; Ohar, Z. Model-based approach for cyber-physical attack detection in water distribution systems. *Water Res.* **2018**, *139*, 132–143. [CrossRef] [PubMed]

91. Aghashahi, M.; Sundararajan, R.; Pourahmadi, M.; Banks, M.K. Water Distribution Systems Analysis Symposium: Battle of the Attack Detection Algorithms (BATADAL). In Proceedings of the World Environmental and Water Resources Congress 2017, Sacramento, CA, USA, 21–25 May 2017; pp. 101–108. [CrossRef]

92. Quiñones-Grueiro, M.; Prieto-Moreno, A.; Verde, C.; Llanes-Santiago, O. Decision Support System for Cyber Attack Diagnosis in Smart Water Networks. *IFAC-PapersOnLine* **2019**, *51*, 329–334, Part of special issue: 2nd IFAC Conference on Cyber-Physical and Human Systems CPHS 2018, Miami, Florida. [CrossRef]

93. Ramotsoela, D.; Hancke, G.; Abu-Mahfouz, A. Attack detection in water distribution systems using machine learning. *Hum. Centric Comput. Inf. Sci.* **2019**, *9*, 13. [CrossRef]

94. Kadosh, N.; Frid, A.; Housh, M. Detecting Cyber-Physical Attacks in Water Distribution Systems: One-Class Classifier Approach. *J. Water Resour. Plan. Manag.* **2020**, *146*, 04020060. [CrossRef]

95. Bakalos, N.; Voulodimos, A.; Doulamis, N.; Doulamis, A.; Ostfeld, A.; Salomons, E.; Caubet, J.; Jimenez, V.; Li, P. Protecting Water Infrastructure From Cyber and Physical Threats: Using Multimodal Data Fusion and Adaptive Deep Learning to Monitor Critical Systems. *IEEE Signal Process. Mag.* **2019**, *36*, 36–48. [CrossRef]

96. Min, K.W.; Choi, Y.H.; Al-Shamiri, A.K.; Kim, J.H. Application of Artificial Neural Network for Cyber-Attack Detection in Water Distribution Systems as Cyber Physical Systems. In *Advances in Harmony Search, Soft Computing and Applications*; Kim, J.H., Geem, Z.W., Jung, D., Yoo, D.G.,Yadav, A., Eds.; Springer International Publishing: Cham, Switzerland, 2020; pp. 82–88.

97. Macas, M.; Wu, C. An Unsupervised Framework for Anomaly Detection in a Water Treatment System. In Proceedings of the 2019 18th IEEE International Conference On Machine Learning And Applications (ICMLA), Boca Raton, FL, USA, 16–19 December 2019; pp. 1298–1305.

98. Zou, X.Y.; Lin, Y.L.; Xu, B.; Guo, Z.B.; Xia, S.J.; Zhang, T.Y.; Gao, N.Y. A Novel Event Detection Model for Water Distribution Systems Based on Data-Driven Estimation and Support Vector Machine Classification. *Water Resour. Manag.* **2019**, *33*, 4569–4581. [CrossRef]

99. Ghaeini, H.R.; Tippenhauer, N.O. HAMIDS: Hierarchical Monitoring Intrusion Detection System for Industrial Control Systems. In Proceedings of the 2nd ACM Workshop on Cyber-Physical Systems Security and Privacy, CPS-SPC 2016, Vienna, Austria, 28 October 2016; Association for Computing Machinery: New York, NY, USA, 2016; pp. 103–111. [CrossRef]

100. Ostfeld, A.; Salomons, E.; Ormsbee, L.; Uber, J.G.; Bros, C.M.; Kalungi, P.; Burd, R.; Zazula-Coetzee, B.; Belrain, T.; Kang, D.; et al. Battle of the Water Calibration Networks. *J. Water Resour. Plan. Manag.* **2012**, *138*, 523–532. [CrossRef]

101. Pasha, M.F.K. Development of an Effective Hybrid Method to Detect Cyber-Physical Attack on Water Distribution Systems. In Proceedings of the World Environmental and Water Resources Congress 2018, Minneapolis, MI, USA, 3–7 June 2018; pp. 410–421. [CrossRef]

102. Aggarwal, C.C. High-Dimensional Outlier Detection: The Subspace Method. In *Outlier Analysis*; Springer New York: New York, NY, USA, 2013; pp. 135–167. [CrossRef]

103. Breunig, M.M.; Kriegel, H.P.; Ng, R.T.; Sander, J. LOF: Identifying Density-Based Local Outliers. In Proceedings of the 2000 ACM SIGMOD International Conference on Management of Data, SIGMOD '00, Dallas, TX, USA, 16–18 May 2000; Association for Computing Machinery: New York, NY, USA, 2000; pp. 93–104. [CrossRef]

104. Kang, E.; Adepu, S.; Jackson, D.; Mathur, A.P. Model-Based Security Analysis of a Water Treatment System. In Proceedings of the 2016 IEEE/ACM 2nd International Workshop on Software Engineering for Smart Cyber-Physical Systems (SEsCPS), Austin, TX, USA, 16 May 2016; pp. 22–28.

105. Patlolla, S.S.; McMillin, B.; Adepu, S.; Mathur, A. An Approach for Formal Analysis of the Security of a Water Treatment Testbed. In Proceedings of the 2018 IEEE 23rd Pacific Rim International Symposium on Dependable Computing (PRDC), Taipei, Taiwan, 4–8 December 2018; pp. 115–124.
106. Howser, G.; McMillin, B. A Modal Model of Stuxnet Attacks on Cyber-physical Systems: A Matter of Trust. In Proceedings of the 2014 Eighth International Conference on Software Security and Reliability (SERE), San Francisco, CA, USA, 30 June–2 July 2014; pp. 225–234. [CrossRef]
107. Liau, C.-J. Belief, information acquisition, and trust in multi-agent systems—A modal logic formulation. *Artif. Intell.* **2003**, *149*, 31–60. [CrossRef]
108. Mishra, V.K.; Palleti, V.R.; Mathur, A. A modeling framework for critical infrastructure and its application in detecting cyber-attacks on a water distribution system. *Int. J. Crit. Infrastruct. Prot.* **2019**, *26*, 100298. [CrossRef]
109. Douglas, H.C.; Taormina, R.; Galelli, S. Pressure-Driven Modeling of Cyber-Physical Attacks on Water Distribution Systems. *J. Water Resour. Plan. Manag.* **2019**, *145*, 06019001. [CrossRef]
110. Moraitis, G.; Nikolopoulos, D.; Bouziotas, D.; Lykou, A.; Karavokiros, G.; Makropoulos, C. Quantifying Failure for Critical Water Infrastructures under Cyber-Physical Threats. *J. Environ. Eng.* **2020**, *146*, 04020108. [CrossRef]
111. Jeong, H.S.; Abraham, D.M.; Qiao, J.; Lawley, M.A.; Richard, J.P.P.; Yih, Y. Issues in Risk Management of Water Networks Against Intentional Attacks. In Proceedings of the ASCE Pipeline Division Specialty Congress—Pipeline Engineering and Construction, San Diego, CA, USA, 1–4 August 2004; pp. 1–10. [CrossRef]
112. Shin, S.; Lee, S.; Burian, S.J.; Judi, D.R.; McPherson, T. Evaluating Resilience of Water Distribution Networks to Operational Failures from Cyber-Physical Attacks. *J. Environ. Eng.* **2020**, *146*, 04020003. [CrossRef]
113. Mathur, A. SecWater: A Multi-Layer Security Framework for Water Treatment Plants. In Proceedings of the 3rd International Workshop on Cyber-Physical Systems for Smart Water Networks, CySWATER '17, Pittsburgh, PA, USA, 21 April 2017; Association for Computing Machinery: New York, NY, USA, 2017; pp. 29–32. [CrossRef]
114. Luiijf, E.; Ali, M.; Zielstra, A. Assessing and Improving SCADA Security in the Dutch Drinking Water Sector. In *Critical Information Infrastructure Security*; Setola, R., Geretshuber, S., Eds.; Springer: Berlin/Heidelberg, Germany, 2009; pp. 190–199.
115. Falliere, N.; Murchu, L.O.; Chien, E. *SCADA Security Good Practices for the Drinking Water Sector*; TNO Defence, Security and Safety; Report: TNO-DV 2008 C096; TNO: Den Haag, The Netherlands, 2008.
116. Burghouwt, P.; Maris, M.; van Peski, S.; Luiijf, E.; van de Voorde, I.; Spruit, M. Cyber Targets Water Management. In *Critical Information Infrastructures Security*; Havarneanu, G., Setola, R., Nassopoulos, H., Wolthusen, S., Eds.; Springer International Publishing: Cham, Switzerland, 2017; pp. 38–49.
117. Adepu, S.; Mishra, G.; Mathur, A. Access Control in Water Distribution Networks: A Case Study. In Proceedings of the 2017 IEEE International Conference on Software Quality, Reliability and Security (QRS), Prague, Czech Republic, 25–29 July 2017; pp. 184–191.
118. Lenzi, S.; Terenghi, G.; Taormina, R.; Galelli, S.; Ciuccarelli, P. Disclosing cyber attacks on water distribution systems: An experimental approach to the sonification of threats and anomalous data. In Proceedings of the International Conference on Auditory Display, Tyne, UK, 23–27 June 2019.
119. Tatbul, N.; Lee, T.J.; Zdonik, S.; Alam, M.; Gottschlich, J. Precision and Recall for Time Series. In Proceedings of the 32nd International Conference on Neural Information Processing Systems, NIPS 2018, Denver, CO, USA, 3–8 December 2018; Curran Associates Inc.: Red Hook, NY, USA, 2018; pp. 1924–1934.
120. Kurakin, A.; Goodfellow, I.; Bengio, S. Adversarial Machine Learning at Scale. *arXiv* **2016**, arXiv:1611.01236.
121. Madry, A.; Makelov, A.; Schmidt, L.; Tsipras, D.; Vladu, A. Towards Deep Learning Models Resistant to Adversarial Attacks. *arXiv* **2019**, arXiv:1706.06083.

water

MDPI

Review

Urban Water Consumption at Multiple Spatial and Temporal Scales. A Review of Existing Datasets

Anna Di Mauro [1], Andrea Cominola [2,3,*], Andrea Castelletti [4] and Armando Di Nardo [1,5]

1   Department of Engineering, Università degli studi della Campania Luigi Vanvitelli, Via Roma 29, 81031 Aversa, Italy; anna.dimauro@unicampania.it (A.D.M.); armando.dinardo@unicampania.it (A.D.N.)
2   Chair of Smart Water Networks, Technische Universität Berlin, Straße des 17. Juni 135, 10623 Berlin, Germany
3   Einstein Center Digital Future, Wilhelmstraße 67, 10117 Berlin, Germany
4   Department of Electronics, Information and Bioengineering, Politecnico di Milano, Piazza Leonardo da Vinci, 32, 20133 Milano, Italy; andrea.castelletti@polimi.it
5   Institute of Complex Networks (Italian National Research Council), Via dei Taurini, 19, 00185 Roma, Italy
*   Correspondence: andrea.cominola@tu-berlin.de

Abstract: Over the last three decades, the increasing development of smart water meter trials and the rise of demand management has fostered the collection of water demand data at increasingly higher spatial and temporal resolutions. Counting these new datasets and more traditional aggregate water demand data, the literature is rich with heterogeneous urban water demand datasets. They are characterized by heterogeneous spatial scales—from urban districts, to households or individual water fixtures—and temporal sampling frequencies—from seasonal/monthly up to sub-daily (minutes or seconds). Motivated by the need of tracking the existing datasets in this rapidly evolving field of investigation, this manuscript is the first comprehensive review effort of the state-of-the-art urban water demand datasets. This paper contributes a review of 92 water demand datasets and 120 related peer-review publications compiled in the last 45 years. The reviewed datasets are classified and analyzed according to the following criteria: spatial scale, temporal scale, and dataset accessibility. This research effort builds an updated catalog of the existing water demand datasets to facilitate future research efforts end encourage the publication of open-access datasets in water demand modelling and management research.

Keywords: urban water consumption; water demand data; water data accessibility; data resolution; smart meter

Citation: Di Mauro, A.; Cominola, A.; Di Nardo, A.; Castelletti, A. Urban Water Consumption at Multiple Spatial and Temporal Scales. A Review of Existing Datasets. *Water* **2021**, *13*, 36. https://dx.doi.org/10.3390/w13010036

Received: 31 October 2020
Accepted: 14 December 2020
Published: 28 December 2020

Publisher's Note: MDPI stays neutral with regard to jurisdictional claims in published maps and institutional affiliations.

1. Introduction

Population growth, urbanization, and climate change are expected to increase the stress on freshwater resources and the burden over urban water systems [1–3]. Adaptive planning and management strategies are thus needed to address seasonal or prolonged water scarcity in drought-prone areas and meet water demands with reduced operational expenditure, overall increasing the resilience of critical urban water network infrastructure systems [4].

In the last decades, demand-side management has increasingly emerged as a key approach to complement traditional water supply operations [5]. Different water demand management strategies (WDMS) have been proposed in the literature to foster water conservation and more efficient water demands [6,7]. These include technological, financial, legislative, maintenance, and educational interventions [8]. The rise of demand-side water management has motivated the development of more and more sophisticated technologies and mathematical models to monitor, characterize, and predict water demands at different spatial and temporal scales, and capture the existing relationships between water demand and its potential climatic and socio-demographic determinants [9–11].

At the coarser urban and suburban scales, the state-of-the-art literature is rich with studies focused on improving the efficiency of water distribution network (WDN) opera-

259

tions (e.g., [12–14]). In these studies, water demands are often considered as a stationary or seasonal input to the hydraulic model of the WDN, with a spatial level of aggregation referred to the city or the district scale. Such spatial scales are typically relevant for infrastructure planning, WDN design, and WDN partitioning. More recently, various techniques for water demand forecasting have also been proposed in the literature. They include regression analysis, time series analysis, and techniques based on black box models, including different Artificial Neural Network architectures (e.g., [15]). Demand prediction models have been developed at different spatial and temporal scales, with the majority of the studies focusing on urban and suburban scales, and temporal resolutions spanning from hourly to monthly intervals (e.g., [16–18]). A disruptive phase in the development of water demand studies is represented by the advent of smart metering technologies [8,19]. The development of smart meters allowed gathering water demand data with an unprecedented level of spatiotemporal detail. Water demand data became potentially available at the spatial scale of individual households and data logging intervals of a few seconds [20]. While understanding the full range of potential benefits of smart meters for water utilities and customers is still a topic for active discussion [21], the variety of studies in the literature based upon smart meter data demonstrates the diversity of data-driven opportunities that high-resolution smart meter data opened up in the context of water demand modelling and management. These include, e.g., water demand profiling and customer segmentation [22], post meter leak detection and water loss management [23], end use studies for fixture-level water demand breakdown and detailed demand forecasting [24], and behavioral studies [25].

The continuously increasing amount of smart meter trials and demand modelling and management studies since the middle of the 1990s [8] suggests that several high-resolution water demand datasets have been recently compiled. The availability of high-resolution datasets opens up several opportunities for advanced applications, including the development of water end use disaggregation algorithms and machine learning techniques for user profiling. Such applications could benefit from open datasets to enhance comparative applications, benchmarking, and facilitate the development of general algorithms trained on combined datasets with water consumption data from different sources and locations. High-resolution datasets, considered in combination with the more traditional water demand datasets gathered at coarser spatial and temporal resolutions would represent a valuable resource for researchers and scientific efforts targeting the development and validation of mathematical models of water demand at different spatial and temporal scales, or the development of advanced smart metering analytics.

Yet, information and metadata on individual water demand datasets are scattered in the literature, and to the authors' knowledge, a comprehensive review of the existing datasets is still missing. Existing data are frequently difficult to access or use, and existing literature reviews on urban water consumption focus on demand modelling or other data-driven applications, rather than on analyzing the heterogeneity of existing datasets, their spatial and temporal scales, and accessibility. Motivated by the recent development and availability of datasets gathered with increasingly high spatial and temporal resolution, the aim of this paper is to gather information on the datasets to identify current trends and gaps and help future data-driven research, along with research benchmarking and reproducibility.

This review contributes the first effort of classification and analysis of 92 water demand datasets and 120 related peer-review publications that have been compiled in the last 45 years to monitor urban water consumption data at different spatial and temporal scales and provide data for water demand modelling and management studies. We characterize the reviewed datasets according to their heterogeneous spatial and temporal scales, and investigate their accessibility. Moreover, since digital disruption has transformed the electricity industry earlier and some lessons learned may apply also in the water or multi-utility sectors [26], we additionally explore similarities and differences between the reviewed subset of high-resolution water demand datasets and 57 comparable high-resolution electricity demand data.

We thus analyze the reviewed datasets and publications to address these five research questions (see Figure 1):

Q1. How are the existing urban water demand datasets distributed across different spatial scales?

Q2. How are the existing urban water demand datasets distributed across different temporal scales?

Q3. What are the main domains of application of the reviewed datasets, within water demand modelling and management studies?

Q4. What is the access policy for the reviewed datasets?

Q5. Is there any synergy with comparable datasets in the electricity sector?

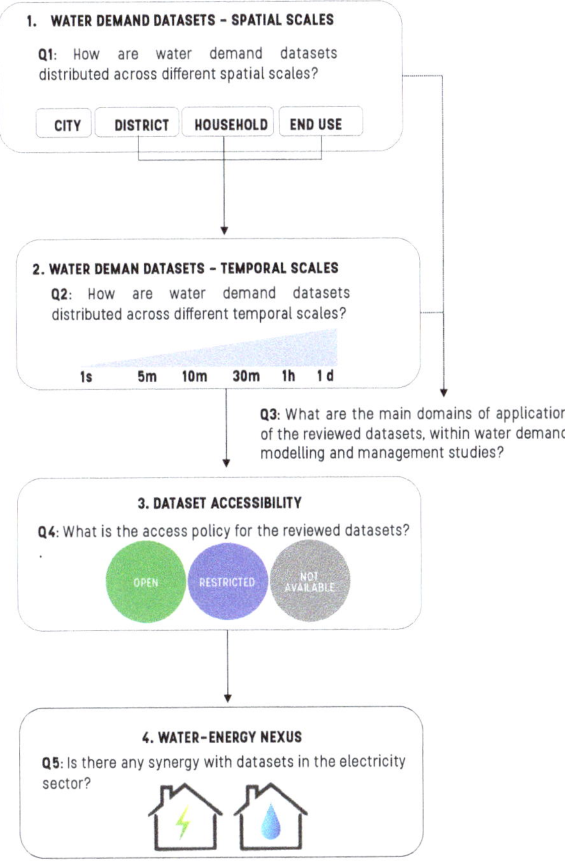

Figure 1. State-of-the-art water demand datasets review: summary of the research questions and multi-stage analysis.

The ultimate goal of this review is to compile an updated catalog of the existing water demand datasets and facilitate future research efforts in this rapidly evolving field of investigation. Researchers performing water demand studies could refer to this review to identify data readily available in formats, spatial scales, and temporal scales that suit their research needs. This review will finally also help identifying water demand datasets that are accessible free of charge, in the attempt to promote further publication of open-access datasets to foster reproducible research, benchmarking, and the development/validation of existing software tools to generate reliable and realistic synthetic data [27–29].

The paper is structured as follows. The dataset review methods and the considered spatial and temporal scales are presented in Section 2; an overview of the dataset search outcomes is presented in Section 3; Sections 4–6 analyzes the reviewed datasets in terms of (i) spatial scales, (ii) temporal scales, and (iii) accessibility; Section 7 analyzes similarities and synergies between some of the reviewed water demand datasets and alike electricity demand datasets; finally, Section 8 draws some final remarks and directions for follow-up research.

2. Datasets Review Methods

To address the research questions formulated in Figure 1, we searched for water demand datasets collected at different spatial and temporal scales and referenced in the peer-reviewed scientific literature on water demand modelling and management. We searched on different web search engines and scientific databases, namely, Google Scholar (https://scholar.google.com/), Mendeley (https://mendeley.com/), Mendeley Data (https://data.mendeley.com/), and data.world (https://data.world/datasets/). We followed the following 3-step procedure:

1. We searched for the following combinations of keywords on Google Scholar and Mendeley: Water demand/Water consumption/Household water demand/Residential end use water/Residential water consumption/Residential water demand/Water demand data/Water demand dataset/Water demand data set/Water demand forecasting/Water demand city/Water demand district/Water end-use/Water consumption patterns/Domestic water use/Urban water demand/Water use behavior/District water demand.
2. We searched for the following combinations of keywords on Mendeley Data and filtered the obtained results to include only two data types, i.e., "Dataset" and "Data Repositories": Water demand/Water consumption/Household water consumption/End use water consumption/Urban water consumption/Urban water demand/District water demand/Water supply demand.
3. We searched for open datasets in data.world, an online catalog for data and analysis. We restricted our research to datasets included in the data topic "water" and selected only datasets mentioned in peer-reviewed articles. More specifically, we searched for the following combinations of keywords: Water demand/Water consumption/Residential water consumption/Domestic water demand/Demand management.

In addition to the datasets retrieved with the above search, we included in this review other high-resolution datasets retrived from two articles strongly focused on residential water demand, i.e., [30,31].

After compiling an inventory with the datasets and related publications retrieved with the above search methods, we reviewed, classified, and critically analyzed the inventory according to three main criteria: (i) spatial scale (Section 4), (ii) temporal scale (Section 5), and dataset accessibility/access policy (Section 6).

Spatial and Temporal Scales of Interest

Depending on the spatial scale of interest, different metering and monitoring tools for water consumption data gathering can be adopted. For instance, end use metering usually requires ad hoc, customized, solutions [20,32], while household or district water consumption can be monitored with commercial flow meters [33]. Datasets collected at different spatial scales will thus represent different levels of aggregation of water demand and will possibly have implications on data privacy and ownership (e.g., water utilities vs individual water consumers). Numerous benefits can derive from high-resolution data, both for water utilities and water consumers [21]. Such data enable, for instance, accurate modelling of water demand patterns, peaks, and anomalies (e.g., leaks) [28]. However, large and high-resolution data implies also several potential drawbacks, e.g., privacy concerns, need for cloud resources for data storage and new skills for data analytics [34]. We identified four scales of interest for urban water consumption monitoring and analysis, from the coarser to the finer:

- *City*. We refer to a city as an urban centre with its own government and administration. The city scale can be composed of multiple districts and it includes the whole water distribution network.
- *District*. A district is a component of an urban center. The district spatial scale refers to a group of residential buildings in one or more municipalities. In many cases, districts coincide with the water network district meter areas (DMAs), i.e., sub-regions of a water network delimited by closing boundary valves. In the case of small cities or villages, the district and city scale can coincide.
- *Household*. The household scale implies a single dwelling, or a single-family residential building connected to an individual water meter. In this category we also include multi-family homes, when connected to one water meter. Depending on the type of household, its water consumption can be attributed to indoor usage only or both indoor and outdoor usage.
- *End use*. The end use scale refers to an individual water fixture within a single apartment/household. End uses can refer to indoor (e.g., shower, dishwasher, toilet, etc.) or outdoor uses (e.g., garden, swimming pool, etc.).

In this review, we keep into account the spatial scale dependencies of the reviewed datasets and classify them according to the three suburban scales included in the *city* level: District, Household, and End Use. In the literature, the spatial scale of interest is related to the type of application that requires water demand datasets (WDDs). WDDs at the district scale, for instance, are mainly used to investigate water network partitioning [35,36], compute water balances [37], assess the hydraulic performance of the network system [38], and perform leakage identification and localization [39,40]. The level of aggregation of these WDDs depends on the network configuration and/or DMA design, and often refers to water demands at network nodes [41,42]. At the household scale, WDDs represent domestic water demands and are primarily used to build descriptive and predictive models of water demand, estimate demand peak timing and magnitude to inform water network operations, and inform conservation campaigns and demand management interventions [43,44]. Finally, at end use scale, WDDs are used to improve our understanding of residential water consumption behaviors, develop disaggregation models to estimate the share of household water consumption of individual fixtures, develop customized water demand management strategies and billing reports, and overall increase customer engagement and help water utilities and customers promote efficient water usage [45,46]. In keeping with the different spatial and temporal scales considered in this study, this review includes both water consumption data retrieved with digital water meters and data measured with low resolution meters or retrieved from water bills [47–49]. Furthermore, when a dataset or publication considers multiple spatial scales, we classify it according to the finest level of data granularity.

Beside the spatial dimension, we also explore how datasets differ in terms of temporal scale (or time sampling frequency). Previous literature has shown that water demand data gathered at monthly or quarterly resolution is mainly used to inform strategic regional planning and to calculate water bills [11], while a number of additional applications, including post-meter leak detection and water end use disaggregation can be enabled by sub-daily data (e.g., recorded with a time sampling frequency of 1 h or a few minutes/seconds) [28]. Here, we characterize the datasets collected at the district, household, and end use scales according to their time sampling resolution, with primary focus on daily and sub-daily frequencies. We consider datasets to have a *low resolution* when they include data with a daily or lower time sampling frequencies (e.g., monthly). In turn, we consider as *high resolution* datasets those gathered with a sub-daily frequency (e.g., hourly, 1 min, 10 s).

3. Overview of Dataset Search Outcome

As an outcome of the dataset search explained above, we retrieved information on 92 unique datasets referenced in 120 scientific works, which in the last 45 years contributed to the literature on water demand modelling and management. The complete catalogue of the datasets and publications reviewed in this study is publicly available at [50]. We have also stored it in a public GitHub repository where pull requests can be submitted, so that our dataset collection can be collaboratively updated as more datasets become available (the repository is accessible at https://github.com/AnnaDiMauro/WDDreview).

A general overview of the reviewed datasets (Figure 2) suggests that, first, the majority of the reviewed datasets contain water consumption data at high spatial resolutions (i.e., end use and household). Second, the temporal distribution of the reviewed publications (Figure 3) is skewed to the right, with a major increase of household and end use studies after 2010. This is likely due to the increasing development of smart meter technologies during the period 2011–2015 [8], following the pioneering studies and prototypes that first appeared in the 1990s (the first end use study reviewed dates back to the 1991–1995 interval in Figure 3).

Finally the worldwide geographical distribution of the reviewed publications (Figure 4) shows an uneven spatial distribution, with more than 50% of the reviewed studies located either in the USA or Europe: 28% USA, 25% EU, 17% Australia and New Zealand, 13% United Kingdom, 9% Asia, 6% Canada, 2% Africa.

A more detailed analysis on the distribution of the reviewed datasets across spatial and temporal scales, along with a critical analysis on their accessibility, are presented in the next sections.

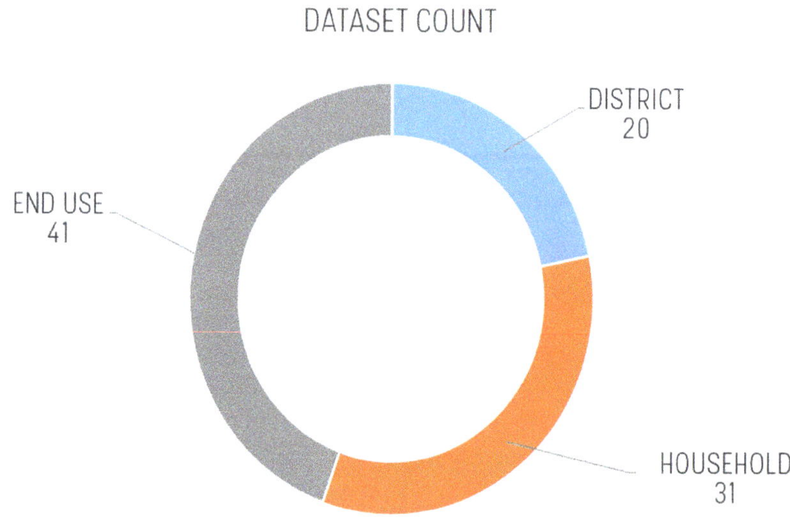

Figure 2. Distribution of the 92 reviewed datasets across three spatial scales, i.e., district, household, and end use.

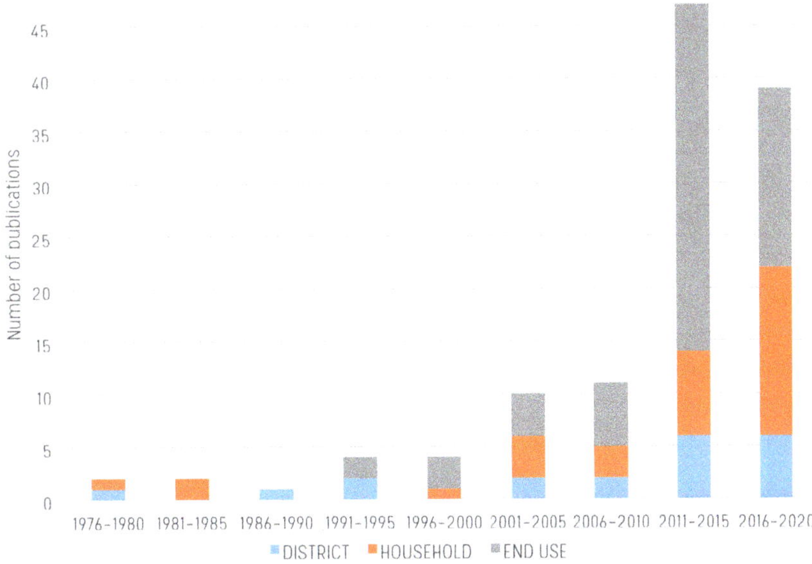

Figure 3. Five–year count of the 120 scientific publications reviewed in this study and referencing the 92 reviewed datasets.

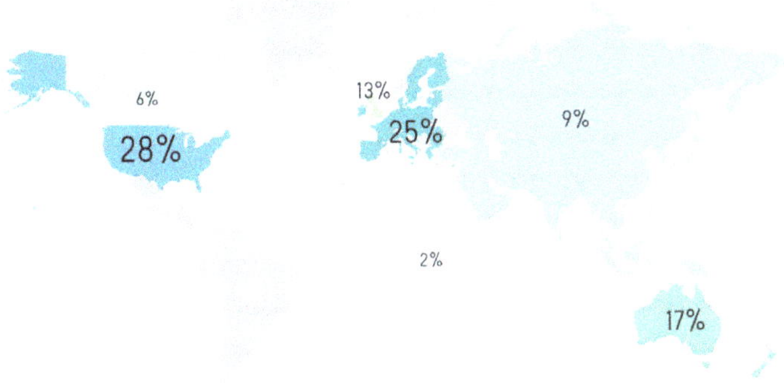

Figure 4. Geographical distribution of the 120 publications reviewed in this study.

4. Dataset Spatial Scales

To answer the first research question reported in Figure 1, we here investigate the distribution of the 92 reviewed datasets across different spatial resolutions, along with their implications for demand modeling and management.

As already reported in Figure 2, we identify only 20 datasets at the district scale. Water demand data collected at this scale relate to specific areas of a water distribution network. They are primarily used to monitor aggregate water demand patterns in the network, or to provide input information to simulation models of water distribution systems. Among these datasets, it is worth highlighting the presence of comprehensive, multi-network datasets, such as the WDSRD database for research applications [51]. This dataset includes data for over 40 different distribution networks, collected by the ASCE Task Committee on Research Databases for Water Distribution Systems for the water distribution system community to develop and test new algorithms for network design, analysis, and operations. A typical problem that requires such type of data is the optimal sensor placement in a partitioned water distribution network [52]. This problem, consisting of finding the optimal sensor location that minimizes the economic costs, while maximizing the amount of information required for network operations and diagnosis, still represents an open challenge for utilities and researchers [53,54]. The datasets classified in the district spatial scale are generally gathered by water utilities for ad hoc analysis on specific case studies within their controlled water network facilities. As the data ownership belongs to water utilities, such data is generally not released to the public, but only released to researchers under non-disclosing agreements. If demand data come from individual household-scale water meters, privacy-protection schemes, e.g., data anonymization, are usually required before data are actually shared.

The majority of the reviewed datasets was collected at the household (31 datasets) or end use (41 datasets) scale. Datasets as such high spatial resolutions have been emerging in the literature in the last 20–30 years, driven by the increasing scientific interest towards smart water metering technology. Smart meters can be defined as digital sensors able to measure, store, and transmit water use data at the household level and with a sub-daily temporal sampling resolution, down to a few seconds [28,55]. Mining smart meter information with advanced data analytics is enabling new opportunities also for developing automatic tools to estimate the water consumption of individual fixtures in a household [56,57], quantify the impact of individual and collective human behaviors on residential water consumption and water conservation [58], and acquire a better understanding on which socio-demographic determinants primarily drive residential water consumption in different geographical contexts [59,60]. Water data at the household/end use scale are of great interest for behavioral studies and provide key information for fostering water conservation, designing water tariffs, promoting more sustainable uses of resource, characterizing water demand during peak hours, and improving demand forecasting and management capabilities [61]. These topics have been already extensively reviewed in the literature, and several comprehensive reviews analyzed the usage and benefits of smart metering for data collection and detailed water demand modelling and management [8,21,62,63].

We report a detailed summary of the metadata of the datasets identified at the district, household and end use scales in Tables 1 (district), 2 (household) and 3 (end use), sorted in chronological order. These metadata include the year when the dataset first appeared in the literature, its size (number of districts/households), time series length, time sampling resolution, access policy (classified in Open (O), Restricted (R), Not Available (NA)), and main goals and dataset applications in the related publications. When a dataset is found to be open access, we include the link to the repository where it is stored at the time of this review.

Some common features and trends can be identified from the information reported in the three tables. First, there is an inverse correlation between the dataset size (or the time series length) and the time sampling resolution. Datasets comprising hundreds or thousands of homes (e.g., [48,49,64–66]) generally include data collected with a monthly or daily time sampling resolution, while datasets with a sub-daily time sampling resolution only include a few units or tens of homes (e.g., [67–69]). This may be attributed to the experimental extent of most high-resolution studies, their usually short-term duration, and the costs of deploying large-scale smart metering systems. Second, while datasets collected at the district scale have been primarily used for WDN optimization, WDN design, understanding the effects of socio-economic determinants on aggregate water demand, and leak detection, we identify four categories of state-of-the-art studies that have used, so far, datasets at the household scale listed in Table 2. These four categories, defined based on the scope of the listed studies, are: water demand forecasting, water demand pattern recognition, water conservation and customer awareness, and water end use disaggregation. The problem of water demand forecasting has been investigated for decades with different modelling techniques. Several recent applications exploit Artificial Neural Networks and other machine learning techniques to predict future water demands [44,66,70] and use this information to optimize water network operations or design water use efficiency programs [49,71–73]. Eight studies can be included in this category, among those listed in Table 2. A second category of studies (e.g., [31,74–76]) exploited household-scale water demand data combined with pattern recognition techniques to inform effective water allocation and reduce water demand to enhance urban water service infrastructure. Other 9 studies from those in Table 2 can be included in this category. Third, 11 datasets among those in Table 2 were gathered as part of water conservation and customer awareness research efforts and projects, including [65,77–79]. These studies investigate the potential of smart meter technologies, often coupled with data analytics and digital platforms, for data communication to water consumers, to increase users' awareness on water consumption and sustainable water usage behaviors. Finally, 3 household-scale datasets were primarily used for water demand disaggregation to estimate water use at individual fixture levels with a non-intrusive approach, i.e., coupling the data from a single-point smart meter with a disaggregation algorithm and avoiding the installation of several intrusive sensors to directly monitor the water consumption of each end use [64,80,81].

Water end use disaggregation can be identified as the link between WDDs at the household and the end use level. Since intrusive smart meter installations at the end use level turn out to be costly and unlikely acceptable and/or accepted by water consumers, thus non-viable for large-scale deployments, non-intrusive techniques represent a valid solution. Yet, non-intrusive end use disaggregation algorithms require ground truth data collected at the fixture level, at least for a limited time span, for algorithm training, validation, and performance assessment. For this reason, the majority of the reviewed WDDs classified in the end use spatial scale (see Table 3) has been used to develop and train different end use disaggregation algorithms, including machine learning-based algorithms (see, for instance, [67,68,82,83]). Differently from the WDDs at the household scales, end use datasets feature a short time series duration (a few days or weeks) and a high time sampling resolution, with data collected primarily with a sampling frequency of 5–10 s. These datasets, mainly collected in the last 10 years, usually include samples collected in two heterogeneous periods (e.g., summer and winter) to account for the seasonal variations of some end uses, e.g., outdoor water demand for irrigation. Whereas developing and testing end use disaggregation methods remains the main purpose of collecting water demand data at the end use level, some of the WDDs listed in Table 3 have been also used to evaluate water consumer behaviors and attitudes toward individual residential water uses (e.g., [84,85]), or test the effectiveness of water conservation strategies based on appliance retrofit and efficiency upgrades [86,87], customized tariffs [88], and awareness campaigns [89,90].

Table 1. Metadata of the 20 reviewed datasets at the district scale. Different goals and applications are considered (see last column): WDNO = Water Distribution Network Optimization; SD = socio-economic studies; DMAD = District Meter Areas Design; LD = Leak detection.

Dataset Name	Authors	Year	Location	Dataset Size	Time Series Lenght	Time Sampling Resolution	Access Policy	Goal and Applications
/	Cassuto, A, et al. [91]	1979	United States	1 districts	5 years (January 1970–December 1975)	1 month	R	SD
/	Billings, R.B., et al. [47]	1989	United States	3 districts in Tucson Arizona	1974 and 1980	1 month	R	WDNO
/	Russac, D.A.V., et al.[92]	1991	United Kingdom	1 district in Potters Bar	3 months (1 April–30 June 1989)	day	R	WDNO
/	Molino, B., et al. [93]	1991	Italy	2 districts in Naples	3 year	1 s	R	WDNO
/	Alvisi, S., et al. [94]	2003	Italy	8 districts in Castelfranco Emilia	1 year (2000)	1 min	R	WDNO
/	Gato, S., et al. [95]	2005	Australia	1 district in East Doncaster	10 years (April 1991–April 2001)	1 day	R	WDNO
/	Worthington, A.C., et al. [96]	2009	Australia	1 district in Queensland local govermenets	10 years (1994 to 2004)	1 month	R	WDNO
/	Mounce, S.R., et al. [41]	2010	United Kingdom	146 DMAs	1 year (2008)	30 min	R	DMAD
/	Gato-Trinidad, S., et al .[97]	2011	Australia	5 districts in Greater Melbourne	1 year	5 min	R	WDNO
/	Bakker, M., et al. [98]	2013	Netherland	6 DMAs	5 years (2006 to 2011)	15 min	R	WDNO
/	Jolly, M.D., et al. [99] and Hernandez, E., et al. [51]	2014; 2016	United States	First release: 12-district model Network Second release: more than 40-district model Network	/	/	O [51]	DMAD and WDNO
/	Boracchi, G., et al. [100]	2014	Spain	10 Districts	82 days	10 min	R	LD
/	Ji, G., et al. [101]	2014	China	1 DMA	1 years	1 h	R	WDN optimization
/	Avni, N., et al. [102]	2015	Israel	/	19 year (1994–2012)	1 month	R	WDNO
/	Vries, D., et al. [103]	2016	Netherlands	6 DMAs	1 year (2013–2014)	5 min	R	LD
/	Gargano, R., et al. [29]	2016	Italy	4 DMA in Piedimonte San Germano	50 days	1 min	R	WDNO
/	Leyli-Abadi, M., et al. [104]	2017	France	1 district in Paris	3 months (January–March 2014)	1 h	R	WDNO
/	Quesnel, K.J., et al. [105]	2017	United States	20 districts	10 years (2005-2015)	2 months	R	SD
/	Di Nardo, A., et al. [106]	2019	Italy	DMAs in Castellammare	1 year (May 2016–2017)	/	R	DMAD
/	Smolak, K., et al. [107]	2020	Poland	28 DMAs	51 days (21 January–12 March 2018)	10 min	R	WDNO

Table 2. Metadata of the 31 reviewed datasets at the household scale. Different goals and applications are considered (see last column): WEUD = Water End Use Disaggregation; WCCA = Water Conservation and Customer Awareness; WDF = Water Demand Forecasting; WDPR = Water Deman Pattern Recognition.

Dataset Name	Authors	Year	Location	Dataset Size	Time Series Lenght	Time Sampling Resolution	Access Policy	Goal and Applications
/	Danielson, L.E. [108]	1979	United States	261 houses	5 year (May 1969–December 1974)	1 day	NA	WDF
Concord, New Hampshire	Hamilton, L.C. [109]	1982	United States	431 houses	6 years (1975–1981)	1 month	NA	WCCA
/	Buchberger, S.G., and Wells, G.J. [110]	1996	United States	4 houses	1 year (July 1993–June 1994)	1 s	R	WDPR
Ohio	Guercio, R., et al. [111]	2001	Italy	85 houses	2 weeks in January 2001; 2 weeks in April 2001	1 m	R	WDPR
/	Silva-Araya, W.F., et al. [112]	2002	Porto Rico	4 houses	1 week	10 s	R	WDPR
DWUS	Loh, M., and Coghlan, P. [71]	2003	Australia	1 phase: 120 houses; 2 phase: 124 houses	1 phase: 20 months (November 1998–June 2000); 2 phase: 14 months (September 2000–November 2001)	/	R	WDF
/	Buchberger, S.G., et al. [113]	2003	United States	21 houses	7 months (April–October 1998)	1 s	R	WDPR
/	Moughton, L.J., et al. [114]	2007	United States	21 houses	36 week (March–November 1997)	1 s	R	WDPR
/	Kenney, D.S., et al. [48]	2008	United States	10,000 houses	5 years (2000–2005)	1 month	R	WCCA
/	Magini, R., et al. [115]	2008	Italy	82 houses	2 years	1 s	R	WDPR
/	Umapathi, S., et al. [116]	2013	Australia	20 houses	12 months (between April 2010–November 2011)	1 m	R	WDPR
/	Cole, G., and Stewart, R.A. [64]	2013	Australia	2884 houses	1 year (1 July 2008–30 June 2009)	1 day	R	WEUD
/	Tanverakul, S.A., and Lee, J. [65]	2013	United States	1000 houses	3 years (October 2008–December 2011)	1 month	R	WCCA
/	Cardell-Oliver, R. [80]	2013	Australia	11,000 houses	35,000 days	1 h	R	WCCA
SmartH2O project	Rizzoli et al. [77]	2014	Switzerland and Spain	/	/	1 h	O [117]	WCCA

Table 2. *Cont.*

Dataset Name	Authors	Year	Location	Dataset Size	Time Series Lenght	Time Sampling Resolution	Access Policy	Goal and Applications
/	Joo, J.C., et al. [74]	2015	Korea	80 houses	1 year (January–December 2011)	30 m	R	WCCA
/	Loureiro, D., et al. [75]	2015	Portugal	311 houses	4 months (January–April 2009)	1 day	R	WCCA
/	Shan, Y., et al. [118]	2015	Greece and Poland	77 houses from Greece; 41 from Poland	2 months (November–December 2014)	/	R	WCCA
/	Liu, A., et al. [78]	2016	Australia	68 houses	November 2012; January 2012	1 day	R	WEUD
/	Makwiza, C., and Jacobs, H. E. [119]	2016	Africa	6 houses	January 2009; December 2014	1 Month	R	WDF
/	Lee, J. [76]	2016	United States	1000 houses	10 years (2002–2011)	1 month	R	WDPR
/	March, H., et al. [120]	2017	Spain	98,228 houses	5 years (2011–2016)	15 m	R	WCCA
/	Leyli-Abadi, M., et al. [104]	2017	France	1000 houses	3 months (1 January–31 March 2014)	1 h	R	WDF
/	Cominola, A., et al. [22]	2018	United States	1107 houses	191 days (28 June–8 December 2015)	1 h	R	WCCA/WDPR
iWIDGET dataset	Kossieris,P. et al. [121]	2018	Greece	11 houses	1 to 2 year (2014–2016)	15 min	R	WDPR
/	Duerr, I., et al. [66]	2018	Florida	973 houses	137 months	1 month	R	WDF
/	Xenochristou, M., et al. [44]	2018	United Kingdom	2000 houses	3 years (October 2014–September 2017)	15 min and 30 min	R	WDF
/	Chen, Y.J., et al. [122]	2019	Nepal	1500 houses	1 year (2014–2015)	30 m	R	WEUD
/	Randall, T., and Koech, R. [123]	2019	Australia	158 houses	6 months (1 August 2015–31 January 2016)	1 h	R	WCCA
/	Rees, P., et al. [49]	2020	United Kingdom	19,238 houses	9 years (2006–2015)	1 day	O [124]	WDF
/	Pesantez, J.E., et al. [70]	2020	Unites States	100 houses	12 months since January 2017	1 h	R	WDF

Table 3. Metadata of the 41 reviewed datasets at the end use scale. Different goals and applications are considered (see last column): WEUD = Water End Use Disaggregation; WCCA = Water Conservation and Customer Awareness.

Dataset Name	Authors	Year	Location	Dataset Size	Time Series Lenght	Time Sampling Resolution	Access Policy	Goal and Applications
/	Butler, D. [125]	1993	United Kingdom	300 homes	7 days (13 December–20 December 1987)	/	NA	WEUD
/	Edwards, K., and Martin, L. [126]	1995	United Kingdom	100 houses	1 year (October 1993–September 1994)	15 m	NA	WEUD
/	DeOreo, W. B. et al. [127]	1996	United Kingdom	16 houses	3 weeks (between June–September 1994)	10 s	NA	WEUD
REUWS	Mayer P.W., et al. [55]	1999	United States	1188 houses	1 month (2 weeks in summer and winter)	10 s	R	WEUD
SHWCS	Mayer, P.W., et al. [128]	2000	United States	37 houses	2 weeks	10 s	R	WCCA
EBMUD	Mayer, P.W., et al. [129]	2003	United States	33 houses	2 weeks	10 s	R	WCCA
/	Mayer P, et al. [130]	2004	United States	26 houses	2 weeks	/	R	WCCA
REUMS	Roberts, P. [131]	2005	Australia	100 houses	2 weeks in February 2004; 2 weeks in August 2004	5 s	R	WCCA
Weep	Heinrich, M. [86]	2007	New Zeland	12 houses	8 months	10 s	R	WEUD
/	Kim, S.H., et al. [132]	2007	Korea	145 houses	3 year (December 2002–February 2006)	1 h	R	WEUD
Gold Coast	Willis, R., et al. [84]	2009	Australia	151 houses	14 days (Winter 2007–2008)	10 s	R	WEUD
/	Froehlich, J.E.,et al. [133]	2009	United States	10 houses	/	/	R	WEUD
AWUS	Heinrich, M., and Roberti, H. [134]	2010	New Zeland	51 houses	4 weeks (between February–March); 5 weeks (between Jun–Jul)	10 s	R	WEUD
SEQ First read	Beal, C.D., et al. [135]	2011	Australia	1500 houses	First read 2 weeks (14 June–28 June 2010)	5 s	R	WEUD
SEQ End-use dataset	Beal, C., et al. [85]	2011	Australia	252 houses	First read 2 weeks (14 June–28 June 2010); Second read 2 weeks (1 December 2010-21 February 2011); Third read (1 June–June 15)	5 s	O [136]	WEUD

Table 3. *Cont.*

Dataset Name	Authors	Year	Location	Dataset Size	Time Series Lenght	Time Sampling Resolution	Access Policy	Goal and Applications
/	Gato-Trinidad, S., et al. [88]	2011	Australia	13 houses	3 weeks in February 2004 and in August 2004	5 s	R	WCCA
/	Otaki, Y., et al. [137]	2011	Thailand	63 houses in Chiang Mai and 59 in Khon Kaen	1 month	/	R	WEUD
/	Srinivasan, V., et al. [67]	2011	United States	2 houses	7 days	7 s	R	WEUD
/	Suero, F. J., et al. [87]	2012	United States	96 houses	3 year (2000–2003)	10 s	R	WEUD
MCW	MidCoast Water. [138]	2012	Australia	141 houses	2 to 5 weeks between December/January and June/August	1 m	R	WCCA
/	Lee, D.J., et al. [139]	2012	Korea	146 households	4 years (2002–2006)	10 min	R	WEUD
/	Borg, M., et al. [140]	2013	United States	3 houses	1 week	/	R	WCCA
/	Neunteufel, R., et al. [141]	2014	Austria	4 houses	2 year (2010–2012)	10 s	R	WEUD
/	Gurung, T. R.,et al. [142]	2015	Australia	130 households	7 different periods of 2 weeks between 2010–2013	5 s	R	WEUD
/	Rathnayaka, K., et al. [143]	2015	Australia	337 houses	2 weeks in Winter 2010 and Summer 2012	5 s	R	WEUD
/	Nguyen, K.A., et al. [144]	2015	Australia	500 homes	3 years (2010–2012)	5 s	R	WEUD
/	Makonin, S., et al. [145]	2016	Canada	1 house	2 years (2012–2014)	1 m	O [146]	WEUD
REU II 2016	William B DeOreo, et al. [147]	2016	United States	762 houses	3 years (2010–2013)	10 s	R	WEUD
/	Kozlovskiy, I., et al. [68]	2016	United States	1 house	21 days (17 April–8 May 2016)	1 s	R	WEUD
2 Data sets: HWU study and MHOW study	Liu, A., et al. [90]	2017	Australia	HWU study: 68 households; MHOW study: 120 households	HWU study: May–September 2013 MHOW study: January–December 2014	1 m	R	WCCA
/	Carranza J.C.I., et al. [148]	2017	Spain	300 houses	9 years since 2008	/	R	WEUD
/	Vitter, J.S., and Webber, M. [69]	2018	United States	1 house	3 week	7 s	O [149]	WEUD

Table 3. *Cont.*

Dataset Name	Authors	Year	Location	Dataset Size	Time Series Lenght	Time Sampling Resolution	Access Policy	Goal and Applications
/	Kofinas, D.T. et al. [150]	2018	Greece	16 house	13 months since February 2015	30 s	O [151]	WEUD
/	Clifford, E., et al. [152]	2018	Ireland	Dataset 1: 745 houses Dataset 2: 1200 houses	/	Dataset 1: 1s Datset 2: 15 m	R	WEUD
/	Nguyen, K.A., et al. [79]	2018	Australia	1000 houses	Winter 2010 (14–28 June). Summer 2010-2011 (1 December 2010–21 February 2011). Winter 2011(1–15 June)	5 s	R	WEUD
/	Omaghomi, T., et al. [153]	2020	United States	1038 houses	14 days	10 s	R	WEUD
/	Meyer, B.E., et al. [81]	2020	Africa	63 houses	247 days	15 s	R	WEUD
/	Pacheco, C.J.B., [154]	2020	Unites States	5 houses	1 month	4 s	R	WEUD
/	Di Mauro, A., et al. [155]	2020	Italy	1 house	8 months (March–October 2019)	1 s	O (website under construction)	WEUD
/	Meyer, B.E., et al. [81]	2020	Africa	63 houses	247 days	15 s	R	WEUD

5. Dataset Temporal Scales

In this section, we address Q2 (see Figure 1) by analyzing the temporal scale of the 92 reviewed WDDs, i.e., we investigate which time sampling resolutions characterize the datasets spatially gathered at the district, household, and end use scales.

As defined in Section 2, water demand data can be recorded with a low resolution characterized by daily or monthly time sampling frequency, or with high resolution, when sub-daily measurements are recorded. The sampling represents a limiting factor for the type of analysis that can be performed [28,115]. Considering the 92 WDDs included in this review, the datasets gathered at the district scale mainly include data collected with a low temporal resolution. These data, recorded with a daily, and more often, monthly, or coarser temporal resolution, consist of measures obtained from billing reports, or periodic meter observations. This is consistent with the main needs of the studies using such datasets for, e.g., the estimation of aggregate water demand for water network design, the resolution of optimal sensor placement problems, and the optimization of water network operations. Only some exceptions include data with a time sampling resolution of 15 min (e.g., [94,100,107]). In turn, the household and end use datasets include data gathered with higher time sampling resolution. The classification of these datasets based on their time sampling resolution (Figure 5) reveals that the majority of the end use-scale datasets contain data gathered with a sub-minute resolution, while most of the household-scale datasets contain data recorded with a time frequency of 15 min to 1 day.

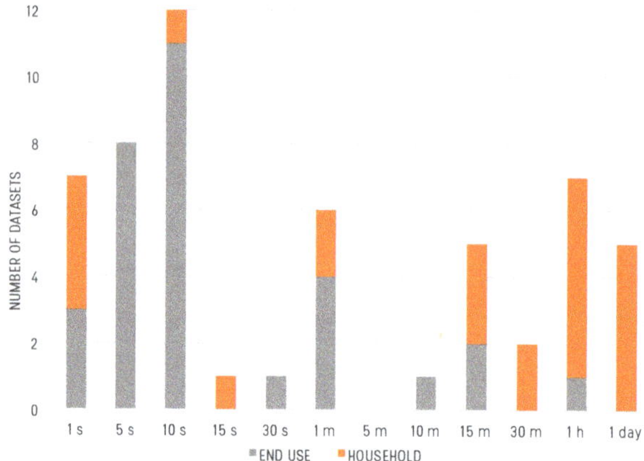

Figure 5. Dataset count for different time sampling frequencies. Only the reviewed datasets gathered at the household (gray) and end use scale (orange) are included.

The distribution of the end use datasets in Figure 5 is an empirical validation of the findings of a previous study by Cominola et al. [28], which demonstrated that only data gathered with time sampling resolutions of a few seconds or, at most, 1 min, can be used to accurately estimate the contribution, peak, and time of use of individual water fixtures, especially when multiple end uses are active. Besides facilitating accurate end use disaggregation [67–69,156–158], such high resolution data also allow a detailed characterization of consumer behaviors [77,155,159,160], and the design of customized water demand strategies [88,123,142,161,162].

Conversely, the distribution of the household-scale datasets in Figure 5 confirms that data sampled with lower frequency suffice for water demand pattern analysis at the household level, i.e., with no detailed end-use analysis. Sub-daily resolution still allow extracting water use patterns and recurring routines [28,66,76], identify anomalies [163], and forecast water demand [49,104].

Cross-correlating information on the time sampling resolution with the metadata previously described in Tables 2 and 3, a trade-off between the time sampling resolution and the size of a dataset emerges.

6. Data Accessibility

Open and free access to scientific datasets can provide valuable support to more reproducible and reusable research [164]. The availability of benchmark datasets accessible by different researchers worldwide would, for instance, help minimize redundant experiments, facilitate benchmarked numerical results on common datasets, and foster reproducibility and incremental research—which in turn drive innovation [165,166]. Yet, data accessibility presents significant challenges in many research fields, due to data ownership, sharing limitations, privacy concerns, technical data management, and security risks [167]. Furthermore, currently available data often lack a standardized format or organized database structure [167,168], or they might not be explicitly referenced in scientific publications, and thus, can be hard to track. Considering the literature on urban water demand modelling and management, WDDs are usually collected as part of large-scale scientific projects carried out by research groups or water utilities at the national and international level [77,86,99,169], or from spatially-constrained experimental settings deployed with the main purpose of creating open-access datasets to be shared for research activities [24,135,145,170].

Here, we aim to answer to Q4 (see Figure 1) and distinguish three main categories of data accessibility to categorize the revised water consumption datasets, namely open, restricted, and not available:

- *Open* WDDs are those available in the literature and downloadable from the web free of charge (when available, the link to each dataset classified as open is reported in Tables 1–3).
- *Restricted* datasets are those WDDs that are available online either only for purchase, or by privately contacting authors/water utilities that own/have direct access to the data.
- *Not available* WDDs are those used and/or cited in the literature (primarily in papers published in the 1970s/80s/90s), but with no information on how to access them.

For the datasets reviewed in this paper, a trade-off emerges between dataset creation and data availability. While there is an increasing amount of water demand data collected at different spatial and temporal scales and related publications (see Figure 3), we found that data sets accessibility is mostly restricted. The datasets we reviewed at the district scale are usually provided by water utilities for specific projects or case studies. As they are owned by water utilities and only released to scientists with non-disclosure agreements for the duration of the relative project, their accessibility is usually restricted or not available. Conversely, the datasets reviewed at the household and end use scales include at least some open and many accessible, but restricted, datasets. Data anonymization, access restriction, or access control filters are usually implemented to protect water consumers privacy [171]. While for many years synthetic household and end use data generation methods have been developed because of limited data availability (e.g., [27,172]), there is an increasing trend of open and restricted household/end use datasets, visible from the number of datasets and access type over time in Figures 6 and 7. The sample of datasets and studies suggests that digital technologies and experimental research are two factors that can foster data availability. Indeed, the majority of the datasets that we classified with *Restricted* or *Open* access, have been collected as part of experimental smart meter trials. In such a context, data are often collected from a sample of volunteer households and are made available by design as part of the research, thus they are not prevented from further usage by utility regulations or ownership rights. Figures 6 and 7 are discussed in detail in the following sections.

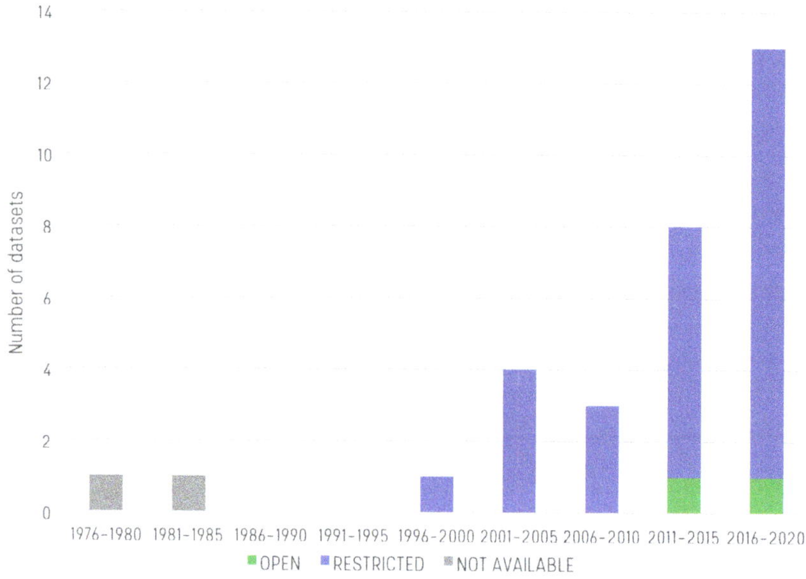

Figure 6. Household scale dataset count and accessibility over time.

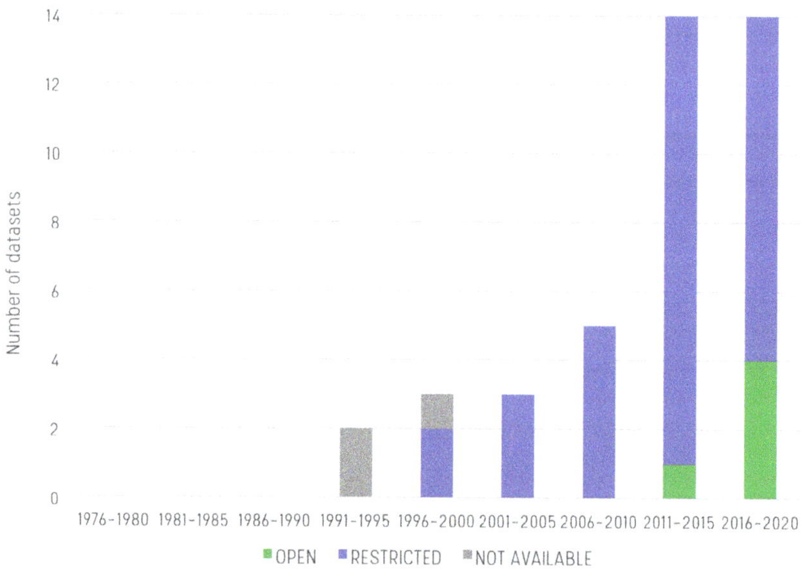

Figure 7. End use scale dataset count and accessibility over time.

6.1. Household-Scale Datasets Accessibility

At the household scale (see Figure 6), there is a more than linear increase in dataset creation. While the few datasets gathered between 1975 and 1995 are not available, almost all those created between 1996 and the time of this review are accessible with restrictions. This may be motivated by the utilities' and researchers' need to protect sensitive customer data, even if they are usually anonymized, or by the interest to control the access to a potentially high-value asset constituted by a limited resource (household/smart meter data, in this case). Only a few datasets gathered in the last 10 years are openly accessible to the scientific community and the public. We found that this limited set of data is usually composed of datasets delivered as outputs of specific research projects in the European area, e.g., the EU-funded SmartH2O project [77] and the studies in London and the Thames Valley [49,173].

6.2. End Use-Scale Dataset Accessibility

Consistently with the household-scale datasets, the majority of end use-scale datasets has restricted access. Yet, some open end use datasets exist since the end of the 1990s. As reported in Figure 7, it also seems that the last 5 years have witnessed an increase of open-access datasets, compared to the total amount of end use datasets. While datasets collected at the household scale are usually owned by utilities, end use datasets are usually collected by researchers as part of experimental research efforts and smart meter/end use studies. This is one of the reasons why more end use-scale datasets are open access, compared with household-scale datasets. According to the experience of the authors, even those datasets declared open are not often easy to access (e.g., download link is broken, website is not updated), but some encouraging preliminary publications, e.g., ([24,170]) suggest that further detailed high-resolution open datasets, collected in controlled environments and provided with groud truth end use labels, will be soon available for research.

All the 41 end use-scale datasets reviewed in this paper have been referenced in at least one peer-reviewed publication on water demand analysis or end use disaggregation. However, a detailed analysis of the usage frequency of the different end use datasets (see Figure 8) reveals that, after excluding those datasets with no identification name and used only for ad hoc individual case studies and trial applications ("no name " datasets in Figure 8), only two datasets were used in more than 5 publications, namely the SEQ and the GOLD COAST datasets. The SEQ dataset has been dominating the scientific scene of the last years and contains the largest collection of sub-minute resolution data estimated for different water end uses. It is the output of a residential end-use study carried out in Australia, i.e., the South East Queensland Residential End Use Study (SEQREUS) [135]. The SEQREUS project aimed to quantify and characterise the main water end uses in a sample of 250 single homes. The SEQ dataset contains water demand with a resolution of 5 sec obtained through the installation of smart meters at the household level. Moreover, end use water demand estimations were achieved using a mixed disaggregation method combining information on the smart metering equipment, household stock inventory surveys, and flow trace analysis [127,144]. Three separate water end use analysis occurred during the SEQREUS project. The first reading campaigns were conducted in the winter (14–28 June 2010); the second one was carried out in the summer (1 December 2010–21 February 2011); the third one in winter 2011 (1–15 June). The SEQ dataset has been so far used in the scientific community to investigate pattern recognition of water usage [174], assess the impact of user awarness on water conservation [89], develop end use disaggregation algorithms [175], and develop demand side management programs [83]. Similarly, the GOLD COAST dataset includes data from the Gold Coast Watersaver End Use Project that was conducted in winter 2008 [84]. It includes data for 151 homes located in the Gold Coast, Australia. The project aimed to explore the degree of influence of household socioeconomic features on end uses. The GOLD COAST dataset contains water demand with a time sampling resolution of 10 seconds, obtained with high-resolution water meters and data loggers to enable the identification of heterogeneous water end uses.

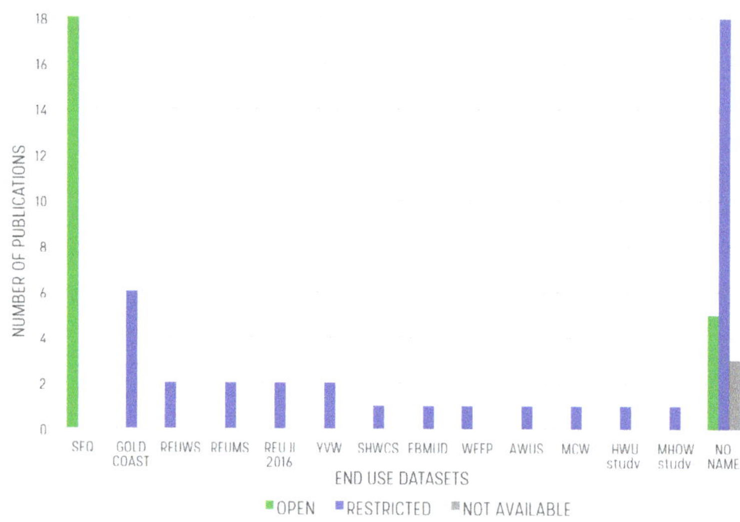

Figure 8. Usage frequency of different reviewed end use datasets. Each dataset is labelled with its name. The "no name" category includes datasets with no identification name and used only for ad hoc individual case studies and trial applications.

7. Nexus Considerations: Outlook and Comparison with Datasets in the Electricity Sector

Motivated by the strong link between water and energy flows in the urban metabolism [176], as well as by the digital transformation of both the water and the energy industry, coordinated actions that account for the water-energy nexus are receiving increasing attention to archive sustainable resource management [177,178] and foster the development of integrated multi-utility services driven by digital transformation [26]. An increasing number of research studies investigated water and electricity correlations to perform customer segmentation analysis and end use classification of residential water-electricity demand data [22,69,145,179]. Most of these studies and other research efforts on water end use disaggregation and water demand profiling were inspired by previous advances in the electricity sector. With a more advanced and consolidated development of smart metering and Internet of Things (IoT) technologies in the electricity sector, high-resolution household and end use electricity datasets became available earlier than similar datasets in the water sector. Indeed, smart meter developments in the water and electricity sectors followed so far two different timelines and speeds of deployment. They also present some technological differences affecting data gathering. The dependence of smart water meters on their battery, for instance, limits their operating life and their data streaming frequency, while electricity meters are fed by a power source by design.

Yet, we recognize some similarities, e.g., also in the electricity sector the availability of end use datasets was pushed by research efforts on building, training, and testing different end use disaggregation algorithms [180,181]. Moreover, while traditional energy system modelling focuses on the national/international scale to assist utilities and authorities in managing the electricity grid, smart electricity metering at the building level is aimed at improving users' awareness and promoting sustainable behaviours and energy savings possibilities [182,183], similarly to water conservation and demand management in the water sector. Also, similarly to the water sector, the temporal scale for electricity demand data gathering is strictly related to the spatial scale. Daily or monthly electricity data are usually required for demand modelling at national scale, while sub-daily resolution is usually adopted for smart metering at building scale. At this fine scale, both water and electricity data are used to enhance the efficiency of consumer behaviors, improve demand forecasting, foster money/resources-saving opportunities, investigate different customer segments, and potentially design customized billing schemes [184,185].

Acknowledging that water and electricity demand modelling and management present both differences and synergies, here we address the research question Q5 listed in Figure 1. We cross-compare the accessibility of water and electricity datasets to assess differences and similarities in data availability, while we do not aim to compare tools for water/electricity modelling. Adopting similar research criteria to those explained in the dataset review methods (Section 2), we retrieved 57 electricity datasets gathered at the household or end use scale. Complete information on these datasets is reported in Supplementary Tables S1 and S2.

We then compared them with the water datasets discussed in the previous section on data accessibility. The outcome of this comparison is represented in Figure 9. The figure reveals that, first, there is a slight majority of electricity datasets gathered at the end use level. This is consistent with what emerges from the reviewed water datasets. Second, the bar plot in Figure 9 shows that most of the electricity end use datasets we retrieved are mainly open. It is worth noting that this might have been facilitated by the availability of low cost and easy-to-install devices, such as smart sockets and Wi-Fi smart plugs, which allow direct end use data gathering [186]. Moreover, the community of researchers working on electricity Non-Intrusive Load Monitoring (NILM) has been very active and open in the last years. The availability of many open end use datasets has been pushed by the need of benchmarking the increasing amount of NILM algorithms on common datasets [187–189], as well as by individual initiatives of some researchers making available data retrieved from their household, or an experimental site equipped with appliance-level sensors, e.g., [145]. Overall, we consider the research efforts in household and end use electricity data collection and analysis as precursors of the trend that is developing in the water sector during the last years. We expect that further developments in the water sector will help fill the gap between available open electricity and water data at the household and end use scales. Similar research will also foster the portability of algorithms and data analytics originally developed for electricity application to water or combined water-energy applications [190,191].

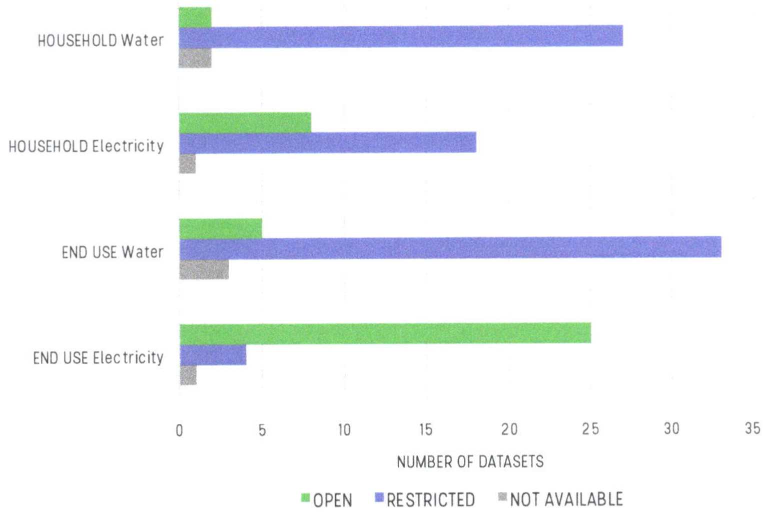

Figure 9. Comparison between water and electricity dataset accessibility at the household and end use scales.

8. Discussion and Conclusions

In the last decades, demand-side water management emerged as a key strategy to pursue efficient water demands and complement supply-side interventions to enhance the overall resilience of urban water systems. The rise of demand-side water management, coupled with the development of digital water metering technologies, has fostered the collection of water demand data at increasingly higher spatial and temporal resolutions. The availability of water demand data at the spatial scale of individual households or end uses, and with a time sampling resolution of a few seconds or minutes, opened up unprecedented opportunities to improve our understanding of water consumer behaviors and modelling water demand. As a consequence of this transformative process, the literature is now rich with urban water demand datasets collected over time with different spatial and temporal resolutions, and archived with different levels of accessibility.

In this paper, we reviewed 92 water demand datasets and 120 related peer-review publications compiled over the last 45 years. We analyzed the datasets and classified them according to their spatial scale, temporal scale, and level of accessibility. Moreover, we analyzed their domains of application within water demand modelling and management studies, and compared them with similar datasets in the electricity sector. As a result of this review and classification effort, we can summarize the following takeaways and address the research questions introduced in Figure 1.

Q1. How are the existing urban water demand datasets distributed across different spatial scales? We found that the majority of the reviewed datasets was collected at the household (31 datasets) or end use scale (41 datasets). Only 20 datasets were identified at the district scale. This is likely due to the increasing number of water demand studies that developed after the advent of digital water meters. Moreover, the datasets gathered at the district scale are usually owned by water utilities, which make them available to researchers usually only temporarily and for ad hoc case study analyses.

Q2. How are the existing urban water demand datasets distributed across different temporal scales? Focusing on the finest spatial scales analyzed, i.e., the household and end use scales, we found that most of the analyzed datasets contain data sampled with a time frequency in the range of 1 s to 1 day. Yet, differences exist: most of the end use-scale datasets contain data gathered with a sub-minute resolution, while household-scale data are characterized by time sampling resolutions of 15 min to 1 day. This is primarily due to the high temporal resolution required by residential water end use disaggregation models.

Q3. What are the main domains of application of the reviewed studies, within water demand modelling and management studies? Our review reveals that the datasets reviewed at district level are mainly used to estimate aggregate demand patterns used in water distribution networks models to investigate water network partitioning, hydraulic performance, network anomalies, and leakage detection. Household-scale datasets have been primarily used to develop data-driven models for water demand forecasting, as well as for explorative analysis to identify water demand determinants. Consistently with our findings for Q2, end use datasets are primarily gathered to develop, train, and validate end use disaggregation algorithms. Both household and end use datasets have also been used to inform water conservation/demand management programs and monitor their effectiveness to change water demand patterns.

Q4. What is the access policy for the reviewed data sets? Most of the reviewed datasets are not open access. Usually, they have a restricted access, i.e., are available for purchase, or can only be obtained by contacting the researchers or water utilities that compiled and own the dataset. However, some households- and end use-scale datasets became openly available, primarily in the last 5 years. This is an encouraging signal for future data sharing and research reproducibility.

Q5. Is there any synergy with comparable datasets in the electricity sector? Similarities exist in the spatial and temporal scales of interest for both the water and the electricity sector, and the amount of reviewed datasets is comparable. Yet, the datasets in these two domains are still very different for what regard their accessibility. Open access datasets are more easily available in the electricity sector, primarily because of the extensive research efforts developed in the last three decades on the problem of electricity end-use disaggregation.

Overall, this paper can provide researchers in the water demand modelling and management sector with useful information to identify data readily available in formats and spatial and temporal scales that suit their research needs. We also identify a roadmap of priorities to enable a complete disclosure of the information value of urban water demand datasets. First, the scientific community would benefit from increased accessibility to open data. We acknowledge that water demand data are sensitive and anonymization and privacy-protection measures need to be undertaken before they can be made openly available. Sharing high-resolution data, consumer data, and sensitive digital data imply potential risks for the privacy and security of private or personal information. Sensitive datasets could potentially be used by third parties for profit and intimidation, or to intrusively track private activities [168]. In response to privacy and security concerns, data protection regulations such as the General Data Protection Regulation (GDPR) implemented by the EU in 2018 and other policies initiated after it in other countries worldwide should be established at the regulatory level [192]. When guaranteed in compliance with privacy protection and data security frameworks, an increasing availability of open access datasets would guarantee better reproducible research, create opportunities for research benchmarking, and foster more transparent and possibly collaborative development and validation of analytic tools.

Second, this review is focused solely on water demand datasets, with primary focus on the household and the end use scales, and only a general overview of possible applications at different temporal and spatial resolutions is provided. Future work could look at systematically reviewing the different goals of existing urban water demand studies at different suburban and urban scales, including those focused on outdoor water use [193], urban landscape water conservation [194], economics and price influences [91], socioeconomic factors and drivers of water demand [195], and metropolitan water planning [196]. Especially these last categories of studies and applications entail cross-domain analysis which combine water consumption data with data from other sources (e.g., socio-economics, climate, behavioral data). Beside requiring proper analytic tools for data analysis, proper data management and sharing frameworks and protocols should be designed to facilitate data fusion among private/public water utilities and the other stakeholders involved in these inter-sectoral studies.

Third, the reviewed datasets are unevenly geographically spread worldwide (some geographical hot spots in USA, Europe, and Australia were identified) and come with different spatial and temporal resolutions. Research efforts aimed at quantitatively comparing water demand data (water consumption volumes, peaks, patterns) gathered across different scales and geographical contexts would advance the generalization of water demand models and contribute to upscale the findings from currently localized water demand studies. In addition, important aspects related to the use of water consumption data from different meters include data standardization and meter accuracy. Data from various sources need a standardized format to facilitate and improve the use of WDDs and increase data portability, interoperability, and overall data quality [197,198]. Moreover, future research could focus on assessing and comparing datasets in the catalogue we have built in this work in terms of measurement precision and accuracy.

Finally, we expect that the current challenges posed to the resilience of interconnected critical infrastructure will foster efforts aimed at overcoming data silos and encourage the development and transfer of multi-sectoral analytic tools to inform resilience planning across sectors (e.g., smart electricity grids, green infrastructure), and scales [26].

Supplementary Materials: The complete catalog with the 92 state-of-the-art water demand datasets and 120 publications reviewed in this paper is available on Zenodo (https://doi.org/10.5281/zenodo. 4390460 [50]) and in this public GitHub repository: https://github.com/AnnaDiMauro/WDDreview. The complete list and metadata of the additional 57 electricity datasets at the end use and household scales that we reviewed in this paper is reported in Supplementary Tables S1 (end use scale) and S2 (household scale). The following are available online at https://www.mdpi.com/2073-4441/13/1/36/s1.

Author Contributions: All authors designed the research. A.D.M. compiled the catalog of the reviewed datasets and peer-reviewed publications, and performed the review. A.D.M. and A.C. (Andrea Cominola) analyzed the outcomes of the review. A.C. (Andrea Cominola), A.C. (Andrea Castelletti), and A.D.N. supervised the research. All authors reviewed the manuscript. All authors have read and agree to the published version of the manuscript.

Funding: The research was conducted as part of the activities financed with the awarding of the V:ALERE: 2019 project of the University of Campania Luigi Vanvitelli.

Conflicts of Interest: The authors declare no conflict of interest.

References

1. Huang, D.; Vairavamoorthy, K.; Tsegaye, S. Flexible design of urban water distribution networks. In Proceedings of the World Environmental and Water Resources Congress 2010: Challenges of Change—Proceedings of the World Environmental and Water Resources Congress 2010, Providence, RI, USA, 16–20 May 2010.
2. McDonald, R.I.; Green, P.; Balk, D.; Fekete, B.M.; Revenga, C.; Todd, M.; Montgomery, M. Urban growth, climate change, and freshwater availability. *Proc. Natl. Acad. Sci. USA* **2011**, *108*, 6312–6317. [CrossRef] [PubMed]
3. Wu, W.; Maier, H.R.; Dandy, G.C.; Arora, M.; Castelletti, A. The changing nature of the water—Energy nexus in urban water supply systems: A critical review of changes and responses. *J. Water Clim. Chang.* **2020**, *11*, 1095–1122. [CrossRef]
4. Cosgrove, W.J.; Loucks, D.P. Water management: Current and future challenges and research directions. *Water Resour. Res.* **2015**, *51*, 4823–4839. [CrossRef]
5. Eggimann, S.; Mutzner, L.; Wani, O.; Schneider, M.Y.; Spuhler, D.; Moy De Vitry, M.; Beutler, P.; Maurer, M. The Potential of Knowing More: A Review of Data-Driven Urban Water Management. *Environ. Sci. Technol.* **2017**, *51*, 2538–2553. [CrossRef] [PubMed]
6. Maggioni, E. Water demand management in times of drought: What matters for water conservation. *Water Resour. Res.* **2015**, *51*, 125–139. [CrossRef]
7. Stavenhagen, M.; Buurman, J.; Tortajada, C. Saving water in cities: Assessing policies for residential water demand management in four cities in Europe. *Cities* **2018**, *79*, 187–195. [CrossRef]
8. Cominola, A.; Giuliani, M.; Piga, D.; Castelletti, A.; Rizzoli, A.E. Benefits and challenges of using smart meters for advancing residential water demand modeling and management: A review. *Environ. Model. Softw.* **2015**, *72*, 198–214. [CrossRef]
9. Giurco, D.; Carrard, N.; McFallan, S.; Nalbantoglu, M.; Inman, M.; Thornton, N.; White, S. *Residential End-Use Measurement Guidebook: A Guide to Study Design, Sampling and Technology*; Technical Report; Institute for Sustainable Futures, UTS: Ultimo, Australia, 2008.
10. Sharma, S.K.; Vairavamoorthy, K. Urban water demand management: Prospects and challenges for the developing countries. *Water Environ. J.* **2009**, *23*, 210–218. [CrossRef]
11. House-Peters, L.A.; Chang, H. Urban water demand modeling: Review of concepts, methods, and organizing principles. *Water Resour. Res.* **2011**, *47*. doi:10.1029/2010WR009624. [CrossRef]
12. Evans, R.G.; Sadler, E.J. Methods and technologies to improve efficiency of water use. *Water Resour. Res.* **2008**, *44*. doi:10.1029/2007WR006200. [CrossRef]
13. Seifollahi-Aghmiuni, S.; Haddad, O.B.; Omid, M.H.; Mariño, M.A. Long-term efficiency of water networks with demand uncertainty. *Proc. Inst. Civ. Eng. Water Manag.* **2011**, *164*, 147–159. [CrossRef]
14. De Souza, E.V.; Costa Da Silva, M.A. Management system for improving the efficiency of use water systems water supply. *Procedia Eng.* **2014**, *70*, 458–466. [CrossRef]
15. Mouatadid, S.; Adamowski, J. Using extreme learning machines for short-term urban water demand forecasting. *Urban Water J.* **2017**, *14*, 630–638. [CrossRef]
16. Firat, M.; Turan, M.E.; Yurdusev, M.A. Comparative analysis of neural network techniques for predicting water consumption time series. *J. Hydrol.* **2010**, *384*, 46–51. [CrossRef]
17. Herrera, M.; Torgo, L.; Izquierdo, J.; Pérez-García, R. Predictive models for forecasting hourly urban water demand. *J. Hydrol.* **2010**, *387*, 141–150. [CrossRef]
18. Adamowski, J.; Fung Chan, H.; Prasher, S.O.; Ozga-Zielinski, B.; Sliusarieva, A. Comparison of multiple linear and nonlinear regression, autoregressive integrated moving average, artificial neural network, and wavelet artificial neural network methods for urban water demand forecasting in Montreal, Canada. *Water Resour. Res.* **2012**, *48*. doi:10.1029/2010WR009945. [CrossRef]
19. Gurung, T.R.; Stewart, R.A.; Sharma, A.K.; Beal, C.D. Smart meters for enhanced water supply network modelling and infrastructure planning. *Resour. Conserv. Recycl.* **2014**, *90*, 34–50. [CrossRef]

20. Giurco, D.; Carrard, N.; Wang, X.; Inman, M.; Nguyen, M. *Innovative Smart Metering Technology and Its Role in End-Use Measurement: Assessing Policy Measures on Addressing Urban Water Scarcity in Can Tho City in the Context of Climate Uncertainty and Urbanization View Project Developing a Truly Intelligent Water*; Technical Report; Australian Water Association: Surfers Paradise, Australia, 2008.

21. Monks, I.; Stewart, R.A.; Sahin, O.; Keller, R. Revealing unreported benefits of digital water metering: Literature review and expert opinions. *Water* **2019**, *11*, 838. [CrossRef]

22. Cominola, A.; Spang, E.S.; Giuliani, M.; Castelletti, A.; Lund, J.R.; Loge, F.J. Segmentation analysis of residential water-electricity demand for customized demand-side management programs. *J. Clean. Prod.* **2018**, *172*, 1607–1619. [CrossRef]

23. Britton, T.C.; Stewart, R.A.; O'Halloran, K.R. Smart metering: Enabler for rapid and effective post meter leakage identification and water loss management. *J. Clean. Prod.* **2013**, *54*, 166–176. [CrossRef]

24. Di Mauro, A.; Di Nardo, A.; Santonastaso, G.F.; Venticinque, S. An IoT system for monitoring and data collection of residential water end-use consumption. In Proceedings of the International Conference on Computer Communications and Networks, ICCCN, Valencia, Spain, 29 July–1 August 2019.

25. Sønderlund, A.L.; Smith, J.R.; Hutton, C.J.; Kapelan, Z.; Savic, D. Effectiveness of smart meter-based consumption feedback in curbing household water use: Knowns and unknowns. *J. Water Resour. Plan. Manag.* **2016**, *142*, 04016060. [CrossRef]

26. Stewart, R.A.; Nguyen, K.; Beal, C.; Zhang, H.; Sahin, O.; Bertone, E.; Vieira, A.S.; Castelletti, A.; Cominola, A.; Giuliani, M.; et al. Integrated intelligent water-energy metering systems and informatics: Visioning a digital multi-utility service provider. *Environ. Model. Softw.* **2018**, *105*, 94–117. [CrossRef]

27. Blokker, E.J.; Vreeburg, J.H.; van Dijk, J.C. Simulating residential water demand with a stochastic end-use model. *J. Water Resour. Plan. Manag.* **2010**, *136*, 19–26. [CrossRef]

28. Cominola, A.; Giuliani, M.; Castelletti, A.; Rosenberg, D.E.; Abdallah, A.M. Implications of data sampling resolution on water use simulation, end-use disaggregation, and demand management. *Environ. Model. Softw.* **2018**, *102*, 199–212. [CrossRef]

29. Gargano, R.; Di Palma, F.; de Marinis, G.; Granata, F.; Greco, R. A stochastic approach for the water demand of residential end users. *Urban Water J.* **2016**, *13*, 569–582. [CrossRef]

30. Scheepers, H.M.; Jacobs, H.E. Simulating residential indoor water demand by means of a probability based end-use model. *J. Water Supply Res. Technol.—AQUA* **2014**, *63*, 476–488. [CrossRef]

31. Carragher, B.J.; Stewart, R.A.; Beal, C.D. Quantifying the influence of residential water appliance efficiency on average day diurnal demand patterns at an end use level: A precursor to optimised water service infrastructure planning. *Resour. Conserv. Recycl.* **2012**, *62*, 81–90. [CrossRef]

32. Sønderlund, A.L.; Smith, J.R.; Hutton, C.; Kapelan, Z. Using smart meters for household water consumption feedback: Knowns and unknowns. *Procedia Eng.* **2014**, *89*, 990–997. [CrossRef]

33. Pericli, A.; Jenkins, J.O. *Review of Current Knowledge Smart Meters and Domestic Water Usage Cover Page Image © Thames Water Utilities Ltd Review of Current Knowledge*; Technical Report; Foundation for Water Research: Marlow, NH, USA, 2015.

34. Zuiderwijk, A.; Janssen, M. *The Negative Effects of Open Government Data—Investigating the Dark Side of Open Data*; ACM International Conference Proceeding Series; ACM: New York, NY, USA, 2014.

35. Di Nardo, A.; Di Natale, M.; Santonastaso, G.F.; Tzatchkov, V.; Alcocer Yamanaka, V.H. Divide and conquer partitioning techniques for smart water networks. *Procedia Eng.* **2014**, *89*, 1176–1183. [CrossRef]

36. Di Nardo, A.; Di Natale, M.; Di Mauro, A.; Martínez Díaz, E.; Blázquez Garcia, J.A.; Santonastaso, G.F.; Tuccinardi, F.P. An Advanced Software to Manage a Smart Water Network with Innovative Metrics and Tools Based on Social Network Theory. In Proceedings of the HIC 2018. 13th International Conference on Hydroinformatics, Palermo, Italy, 1–5 July 2018; Volume 3, pp. 582–592. [CrossRef]

37. Knobloch, A.; Guth, N.; Klingel, P. Automated water balance calculation for water distribution systems. *Procedia Eng.* **2014**, *89*, 428–436. [CrossRef]

38. Massoud, T.; Zia, A. Dynamic management of water distribution networks based on hydraulic performance analysis of the system. *Water Sci. Technol. Water Supply* **2003**, *3*, 95–102. [CrossRef]

39. Farah, E.; Shahrour, I. Leakage Detection Using Smart Water System: Combination of Water Balance and Automated Minimum Night Flow. *Water Resour. Manag.* **2017**, *31*. doi:10.1007/s11269-017-1780-9. [CrossRef]

40. Bragalli, C.; Neri, M.; Toth, E. Effectiveness of smart meter-based urban water loss assessment in a real network with synchronous and incomplete readings. *Environ. Model. Softw.* **2019**, *112*, 128–142. [CrossRef]

41. Mounce, S.R.; Boxall, J.B. Implementation of an on-line artificial intelligence district meter area flow meter data analysis system for abnormality detection: A case study. *Water Sci. Technol. Water Supply* **2010**, *10*, 437–444. [CrossRef]

42. Di Nardo, A.; Di Natale, M.; Gargano, R.; Giudicianni, C.; Greco, R.; Santonastaso, G.F. Performance of partitioned water distribution networks under spatial-temporal variability of water demand. *Environ. Model. Softw.* **2018**, *101*, 128–136. [CrossRef]

43. Pullinger, M.; Anderson, B.; Browne, A.L.; Medd, W. New directions in understanding household water demand: A practices perspective. *J. Water Supply Res. Technol.—AQUA* **2013**, *62*, 496–506. [CrossRef]

44. Xenochristou, M.; Kapelan, Z.; Hutton, C.; Hofman, J. Smart Water Demand Forecasting: Learning from the Data. EasyChair. In Proceedings of the HIC 2018 13th International Conference on Hydroinformatics Palermo, Italy, 1–5 July 2018; Volume 3, pp. 2351–2358.

45. Salleh, N.S.M.; Rasmani, K.A.; Jamil, N.I. The Effect of Variations in Micro-components of Domestic Water Consumption Data on the Classification of Excessive Water Usage. *Procedia—Soc. Behav. Sci.* **2015**, *195*, 1865–1871. [CrossRef]

46. Rathnayaka, K.; Malano, H.; Arora, M.; George, B.; Maheepala, S.; Nawarathna, B. Prediction of urban residential end-use water demands by integrating known and unknown water demand drivers at multiple scales I: Model development. *Resour. Conserv. Recycl.* **2017**, *117*, 85–92. [CrossRef]

47. Billings, R.B.; Day, W.M. Demand management factors in residential water use: The southern Arizona experience. *J. Am. Water Work. Assoc.* **1989**, *81*, 58–64. [CrossRef]

48. Kenney, D.S.; Goemans, C.; Klein, R.; Lowrey, J.; Reidy, K. Residential water demand management: Lessons from Aurora, Colorado. *J. Am. Water Resour. Assoc.* **2008**, *44*, 192–207. [CrossRef]

49. Rees, P.; Clark, S.; Nawaz, R. Household Forecasts for the Planning of Long-Term Domestic Water Demand: Application to London and the Thames Valley. *Popul. Space Place* **2020**, *26*, e2288. [CrossRef]

50. Di Mauro, A.; Cominola, A.; Castelletti, A.; Di Nardo, A. [Database] Urban Water Consumption at Multiple Spatial and Temporal Scales. A Review of Existing Datasets. 2020. Available online: https://doi.org/10.5281/zenodo.4390460. doi:10.5281/zenodo.4390460. (accessed on 23 December 2020).

51. Hernandez, E.; Hoagland, S.; Ormsbee, L.E. WDSRD: A Database for Research Applications. University of Kentucky Libraries. 2016. Available online: https://uknowledge.uky.edu/wdsrd/ (accessed on 10 December 2020).

52. Di Nardo, A.; Di Natale, M.; Di Mauro, A.; Martínez Díaz, E.; Blázquez Garcia, J.A.; Santonastaso, G.F.; Tuccinardi, F.P. An advanced software to design automatically permanent partitioning of a water distribution network. *Urban Water J.* **2020**, *17*, 259–265. doi:10.1080/1573062X.2020.1760322. [CrossRef]

53. Aral, M.M.; Guan, J.; Maslia, M.L.; Grayman, W.M. Optimization model and algorithms for design of water sensor placement in water distribution systems. In Proceedings of the 8th Annual Water Distribution Systems Analysis Symposium 2006, Cincinnati, OH, USA, 27–30 August 2006.

54. Casillas, M.V.; Puig, V.; Garza-Castañón, L.E.; Rosich, A. Optimal sensor placement for leak location in water distribution networks using genetic algorithms. *Sensors* **2013**, *13*, 14984–15005. [CrossRef] [PubMed]

55. Mayer, P.W.; Deoreo, W.B.; Opitz, E.M.; Kiefer, J.C.; Davis, W.Y.; Dziegielewski, B.; Nelson, J.O. *Residential End Uses of Water*; Aquacraft, Inc. Water Engineering and Management: Boulder, CO, USA, 1999.

56. Gilbertson, M.; Hurlimann, A.; Dolnicar, S. Does water context influence behaviour and attitudes to water conservation? *Australas. J. Environ. Manag.* **2011**, *18*. doi:10.1080/14486563.2011.566160. [CrossRef]

57. Parker, J.M.; Wilby, R.L. Quantifying Household Water Demand: A Review of Theory and Practice in the UK. *Water Resour. Manag.* **2013**, *27*, 981–1011. [CrossRef]

58. Jorgensen, B.; Graymore, M.; O'Toole, K. Household water use behavior: An integrated model. *J. Environ. Manag.* **2009**, *91*, 227–236. [CrossRef] [PubMed]

59. Fielding, K.S.; Russell, S.; Spinks, A.; Mankad, A. Determinants of household water conservation: The role of demographic, infrastructure, behavior, and psychosocial variables. *Water Resour. Res.* **2012**, *48*. doi:10.1029/2012WR012398. [CrossRef]

60. Russell, S.V.; Knoeri, C. Exploring the psychosocial and behavioural determinants of household water conservation and intention. *Int. J. Water Resour. Dev.* **2019**, *36*, 940–955. [CrossRef]

61. Willis, R.M.; Stewart, R.A.; Giurco, D.P.; Talebpour, M.R.; Mousavinejad, A. End use water consumption in households: Impact of socio-demographic factors and efficient devices. *J. Clean. Prod.* **2013**, *60*, 107–115. [CrossRef]

62. Liu, A.; Mukheibir, P. Digital metering feedback and changes in water consumption—A review. *Resour. Conserv. Recycl.* **2018**, *134*, 136–148. [CrossRef]

63. Rahim, M.S.; Nguyen, K.A.; Stewart, R.A.; Giurco, D.; Blumenstein, M. Machine learning and data analytic techniques in digitalwater metering: A review. *Water* **2020**, *12*, 294. [CrossRef]

64. Cole, G.; Stewart, R.A. Smart meter enabled disaggregation of urban peak water demand: Precursor to effective urban water planning. *Urban Water J.* **2013**, *10*, 174–194. [CrossRef]

65. Tanverakul, S.A.; Lee, J. Residential water demand analysis due to water meter installation in California. In *World Environmental and Water Resources Congress 2013: Showcasing the Future—Proceedings of the 2013 Congress*; American Society of Civil Engineers (ASCE): Reston, VA, USA, 2013; pp. 936–945.

66. Duerr, I.; Merrill, H.R.; Wang, C.; Bai, R.; Boyer, M.; Dukes, M.D.; Bliznyuk, N. Forecasting urban household water demand with statistical and machine learning methods using large space-time data: A Comparative study. *Environ. Model. Softw.* **2018**, *102*, 29–38. [CrossRef]

67. Srinivasan, V.; Stankovic, J.; Whitehouse, K. WaterSense: Water Flow Disaggregation Using Motion Sensors. In Proceedings of the Third ACM Workshop on Embedded Sensing Systems for Energy-Efficiency in Buildings, Seattle, WA, USA, 1–4 November 2011.

68. Kozlovskiy, I.; Schöb, S.; Sodenkamp, M. Non-intrusive disaggregation of water consumption data in a residential household. In *Lecture Notes in Informatics (LNI), Proceedings—Series of the Gesellschaft fur Informatik (GI)*; Gesellschaft fur Informatik (GI): Bremen, Germany, 2016; Volume P-259, pp. 1381–1387.

69. Vitter, J.S.; Webber, M. Water Event Categorization Using Sub-Metered Water and Coincident Electricity Data. *Water* **2018**, *10*, 714. [CrossRef]

70. Pesantez, J.E.; Berglund, E.Z.; Kaza, N. Smart meters data for modeling and forecasting water demand at the user-level. *Environ. Model. Softw.* **2020**, *125*, 104633. [CrossRef]

71. Loh, M.; Coghlan, P. Domestic Water Use Study. In Perth, Western Australia 1998–2001. *Hydro 2000: Interactive Hydrology*; Institution of Engineers: Barton, Australia, 2003; p. 36.

72. Nawaz, R.; Rees, P.; Clark, S.; Mitchell, G.; McDonald, A.; Kalamandeen, M.; Lambert, C.; Henderson, R. Long-Term Projections of Domestic Water Demand: A Case Study of London and the Thames Valley. *J. Water Resour. Plan. Manag.* **2019**, *145*, 05019017. [CrossRef]

73. Zounemat-Kermani, M.; Matta, E.; Cominola, A.; Xia, X.; Zhang, Q.; Liang, Q.; Hinkelmann, R. Neurocomputing in surface water hydrology and hydraulics: A review of two decades retrospective, current status and future prospects. *J. Hydrol.* **2020**. doi:10.1016/j.jhydrol.2020.125085. [CrossRef]

74. Joo, J.C.; Oh, H.J.; Ahn, H.; Ahn, C.H.; Lee, S.; Ko, K.R. Field application of waterworks automated meter reading systems and analysis of household water consumption. *Desalin. Water Treat.* **2015**, *54*, 1401–1409. [CrossRef]

75. Loureiro, D.; Rebelo, M.; Mamade, A.; Vieira, P.; Ribeiro, R. Linking water consumption smart metering with census data to improve demand management. *Water Sci. Technol. Water Supply* **2015**, *15*, 1396–1404. [CrossRef]

76. Lee, J. Residential water demand analysis of a Low-Income Rate Assistance Program in California, United States. *Water Environ. J.* **2016**, *30*, 49–61. [CrossRef]

77. Rizzoli, A.E.; Castelletti, A.; Cominola, A.; Fraternali, P.; Diniz Dos Santos, A.; Storni, B.; Wissmann-Alves, R.; Bertocchi, M.; Novak, J.; Micheel, I. The SmartH2O project and the role of social computing in promoting efficient residential water use: A first analysis. In Proceedings of the 7th International Congress on Environmental Modelling and Software: Bold Visions for Environmental Modeling, iEMSs 2014, San Diego, CA, USA, 15–19 June 2014; Volume 3, pp. 1559–1567.

78. Liu, A.; Giurco, D.; Mukheibir, P. Urban water conservation through customised water and end-use information. *J. Clean. Prod.* **2016**, *112*, 3164–3175. [CrossRef]

79. Nguyen, K.A.; Stewart, R.A.; Zhang, H.; Sahin, O.; Siriwardene, N. Re-engineering traditional urban water management practices with smart metering and informatics. *Environ. Model. Softw.* **2018**, *101*, 256–267. [CrossRef]

80. Cardell-Oliver, R. Discovering water use activities for smart metering. In Proceedings of the 2013 IEEE 8th International Conference on Intelligent Sensors, Sensor Networks and Information Processing: Sensing the Future, ISSNIP 2013, Melbourne, Australia 2–5 April 2013; Volume 1, pp. 171–176.

81. Meyer, B.E.; Jacobs, H.E.; Ilemobade, A. Extracting household water use event characteristics from rudimentary data. *J. Water Supply Res. Technol.—AQUA* **2020**, *69*, 387–397. [CrossRef]

82. Nguyen, K.A.; Zhang, H.; Stewart, R.A. Application of dynamic time warping algorithm in prototype selection for the disaggregation of domestic water flow data into end use events. In Proceedings of the 34th IAHR Congress 2011—Balance and Uncertainty: Water in a Changing World, Incorporating the 33rd Hydrology and Water Resources Symposium and the 10th Conference on Hydraulics in Water Engineering, Brisbane, Australia, 26 June–1 July 2011.

83. Bennett, C.; Stewart, R.A.; Beal, C.D. ANN-based residential water end-use demand forecasting model. *Expert Syst. Appl.* **2013**, *40*, 1014–1023. [CrossRef]

84. Willis, R.; Stewart, R.A.; Panuwatwanich, K.; Capati, B.; Giurco, D. Gold coast domestic water end use study. *Water* **2009**, *36*, 84–90.

85. Beal, C.; Makki, A.; Stewart, R. Identifying the Drivers of Water Consumption: A Summary of Results from the South East Queensland Residential End Use Study. *Sci. Forum Stakehold. Engag. Build. Linkages, Collab. Sci. Qual.* **2011**, 126–132.

86. Heinrich, M. Water End-Use and Efficiency Project (Weep)—A Case Study. In Proceedings of the SB07 Transforming Our Built Environment, Auckland, New Zealand, 14–16 November 2007; Volume 136, p. 8.

87. Suero, F.J.; Mayer, P.W.; Rosenberg, D.E. Estimating and Verifying United States Households' Potential to Conserve Water. *J. Water Resour. Plan. Manag.* **2012**, *138*, 299–306. [CrossRef]

88. Gato-Trinidad, S.; Jayasuriya, N.; Roberts, P. Understanding urban residential end uses of water. *Water Sci. Technol.* **2011**, *64*, 36–42. [CrossRef]

89. Walton, A.; Hume, M. Creating positive habits in water conservation: The case of the Queensland Water Commission and the Target 140 campaign. *Int. J. Nonprofit Volunt. Sect. Mark.* **2011**, *16*, 215–224. [CrossRef]

90. Liu, A.; Giurco, D.; Mukheibir, P. Advancing household water-use feedback to inform customer behaviour for sustainable urban water. *Water Sci. Technol. Water Supply* **2017**, *17*, 198–205. [CrossRef]

91. Cassuto, A.E.; Ryan, S. Effect of price on the residential demand for water within an agency. *JAWRA J. Am. Water Resour. Assoc.* **1979**, *15*, 345–353. [CrossRef]

92. Russac, D.A.; Rushton, K.R.; Simpson, R.J. Insights into Domestic Demand from a Metering Trial. *Water Environ. J.* **1991**, *5*, 342–351. [CrossRef]

93. Molino, B.; Rasulo, G.; Taglialatela, L. Peak coefficients of household potable water supply. *Water Resour. Manag.* **1991**, *4*, 283–291. [CrossRef]

94. Alvisi, S.; Franchini, M.; Marinelli, A. A stochastic model for representing drinking water demand at residential level. *Water Resour. Manag.* **2003**, *17*, 197–222. [CrossRef]

95. Gato, S.; Jayasuriya, N.; Hadgraft, R.; Roberts, P. A simple time series approach to modelling urban water demand. *Australas. J. Water Resour.* **2005**, *8*, 153–164. [CrossRef]

96. Worthington, A.C.; Higgs, H.; Hoffmann, M. Residential water demand modeling in Queensland, Australia: A comparative panel data approach. *Water Policy* **2009**, *11*, 427–441. [CrossRef]

97. Gato-Trinidad, S.; Gan, K. Characterizing maximum residential water demand. *WIT Trans. Built Environ.* **2011**, *122*, 15–24.

98. Bakker, M.; Vreeburg, J.H.; van Schagen, K.M.; Rietveld, L.C. A fully adaptive forecasting model for short-term drinking water demand. *Environ. Model. Softw.* **2013**, *48*, 141–151. [CrossRef]

99. Jolly, M.D.; Lothes, A.D.; Bryson, L.S.; Ormsbee, L. Research database of water distribution system models. *J. Water Resour. Plan. Manag.* **2014**, *140*, 410–416. [CrossRef]

100. Boracchi, G.; Roveri, M. Exploiting self-similarity for change detection. In Proceedings of the International Joint Conference on Neural Networks, Beijing, China, 6–11 July 2014; Institute of Electrical and Electronics Engineers Inc.: Piscataway, NJ, USA, 2014; pp. 3339–3346.

101. Ji, G.; Wang, J.; Ge, Y.; Liu, H. Urban water demand forecasting by LS-SVM with tuning based on elitist teaching-learning-based optimization. In Proceedings of the 26th Chinese Control and Decision Conference, CCDC 2014, Changsha, China, 31 May–2 June 2014; pp. 3997–4002.

102. Avni, N.; Fishbain, B.; Shamir, U. Water consumption patterns as a basis for water demand modeling. *Water Resour. Res.* **2015**, *51*, 8165–8181. [CrossRef]

103. Vries, D.; Van Den Akker, B.; Vonk, E.; De Jong, W.; Van Summeren, J. Application of machine learning techniques to predict anomalies in water supply networks. *Water Sci. Technol. Water Supply* **2016**, *16*, 1528–1535. [CrossRef]

104. Leyli-Abadi, M.; Samé, A.; Oukhellou, L.; Cheifetz, N.; Mandel, P.; Féliers, C.; Chesneau, O. Predictive classification of water consumption time series using non-homogeneous markov models. In Proceedings of the 2017 International Conference on Data Science and Advanced Analytics, DSAA 2017, Tokyo, Japan, 19–21 October 2017; Institute of Electrical and Electronics Engineers Inc.: Piscataway, NJ, USA, 2017; Volume 2018-January, pp. 323–331.

105. Quesnel, K.J.; Ajami, N.K. Changes in water consumption linked to heavy news media coverage of extreme climatic events. *Sci. Adv.* **2017**, *3*, e1700784. [CrossRef] [PubMed]

106. Di Nardo, A.; Di Natale, M.; Di Mauro, A.; Santonastaso, G.F.; Palomba, A.; Locoratolo, S. Calibration of a water distribution network with limited field measures: The case study of Castellammare di Stabia (Naples, Italy). In *Lecture Notes in Computer Science (Including Subseries Lecture Notes in Artificial Intelligence and Lecture Notes in Bioinformatics)*; Springer: Berlin/Heidelberg, Germany, 2019; Volume 11353 LNCS, pp. 433–436.

107. Smolak, K.; Kasieczka, B.; Fialkiewicz, W.; Rohm, W.; Siła-Nowicka, K.; Kopańczyk, K. Applying human mobility and water consumption data for short-term water demand forecasting using classical and machine learning models. *Urban Water J.* **2020**, *17*, 32–42. [CrossRef]

108. Danielson, L.E. An analysis of residential demand for water using micro time-series data. *Water Resour. Res.* **1979**, *15*, 763–767. [CrossRef]

109. Hamilton, L.C. Response to Water Conservation Campaigns: An Exploratory Look. *Eval. Rev.* **1982**, *6*, 673–688. [CrossRef]

110. Buchberger, S.G.; Wells, G.J. Intensity, duration, and frequency of residential water demands. *J. Water Resour. Plan. Manag.* **1996**, *122*, 11–19. [CrossRef]

111. Guercio, R.; Magini, R.; Pallavicini, I. Instantaneous residential water demand as stochastic point process. *Prog. Water Resour.* **2001**, 129–138. doi:10.2495/WRM010121. [CrossRef]

112. Silva-Araya, W.F.; Artiles-Léon, N.; Romero-Ramírez, M. Dynamic simulation of water distribution systems with instantaneous demands. In *Water Studies*; WITPress: Ashurst, UK, 2002; Volume 10, pp. 379–387.

113. Buchberger, S.G.; Carter, J.; Lee, Y.; Schade, T. *Random Demands, Travel Times and Water Quality in Dead-Ends, Prepared for American Water Works Association Research Foundation*; Technical Report; Water Research Foundation: Denver, CO, USA, 2003.

114. Moughton, L.J.; Buchberger, S.G.; Boccelli, D.L.; Filion, Y.R.; Karney, B.W. Effect of time step and data aggregation on cross correlation of residential demands. In Proceedings of the 8th Annual Water Distribution Systems Analysis Symposium 2006, Cincinnati, OH, USA, 27–30 August 2007; p. 42.

115. Magini, R.; Pallavicini, I.; Guercio, R. Spatial and temporal scaling properties of water demand. *J. Water Resour. Plan. Manag.* **2008**, *134*, 276–284. [CrossRef]

116. Umapathi, S.; Chong, M.N.; Sharma, A.K. Evaluation of plumbed rainwater tanks in households for sustainable water resource management: A real-time monitoring study. *J. Clean. Prod.* **2013**, *42*, 204–214. [CrossRef]

117. The-SmartH2O-Project. The SmartH2O Anonymized Datasets of Terre di Pedemonte and Valencia. 2017. Available online: https://zenodo.org/record/556152 (accessed on 10 December 2020).

118. Shan, Y.; Yang, L.; Perren, K.; Zhang, Y. Household water consumption: Insight from a survey in Greece and Poland. In *Procedia Engineering*; Elsevier Ltd.: Amsterdam, The Netherlands, 2015; Volume 119, pp. 1409–1418.

119. Makwiza, C.; Jacobs, H.E. Assessing the impact of property size on residential water use for selected neighbourhoods in Lilongwe, Malawi. *J. Water Sanit. Hyg. Dev.* **2016**, *6*, 242–251. [CrossRef]

120. March, H.; Morote, Á.F.; Rico, A.M.; Saurí, D. Household smart water metering in Spain: Insights from the experience of remote meter reading in alicante. *Sustainability* **2017**, *9*, 582. [CrossRef]

121. Kossieris, P.; Makropoulos, C. Exploring the statistical and distributional properties of residential water demand at fine time scales. *Water* **2018**, *10*, 1481. [CrossRef]

122. Chen, Y.J.; Chindarkar, N.; Zhao, J. Water and time use: Evidence from Kathmandu, Nepal. *Water Policy* **2019**, *21*, 76–100. [CrossRef]

123. Randall, T.; Koech, R. Smart water metering technology for water management in urban areas. *Water e-J.* **2019**, *4*, 1–14. [CrossRef]

124. Rees, P.; Clark, S.; Nawaz, R. Household Forecasts for the Planning of Long-Term Domestic Water Demand: Application to London and the Thames Valley—Dataset. University of Leeds. 2020. Available online: http://archive.researchdata.leeds.ac.uk/639/ (accessed on 10 December 2020).
125. Butler, D. The influence of dwelling occupancy and day of the week on domestic appliance wastewater discharges. *Build. Environ.* **1993**, *28*, 73–79. [CrossRef]
126. Edwards, K.; Martin, L. A Methodology for Surveying Domestic Water Consumption. *Water Environ. J.* **1995**, *9*, 477–488. [CrossRef]
127. DeOreo, W.B.; Heaney, J.P.; Mayer, P.W. Flow trace analysis to access water use. *J.—Am. Water Work. Assoc.* **1996**, *88*, 79–90. [CrossRef]
128. Mayer, P.W.; DeOreo, W.B.; Lewis, D.M. Seattle Home Water Conservation Study the Impacts of High Efficiency Plumbing Fixture Retrofits in Single Family Homes. In *Environmental Protection*; Aquacraft Inc.: Boulder, CO, USA, 2000; p. 170.
129. Mayer, P.W.; DeOreo, W.B.; Lewis, D.M. *Residential Indoor Water Conservation Study: Evaluation of High Efficiency Indoor Plumbing Fixture Retrofits in Single Family Homes in the East Bay Municipal Utility District Service Area*; Technical Report; Aquacraft, Inc. Water Engineering and Management: Boulder, CO, USA, 2003.
130. Mayer, P.; DeOreo, W.; Towler, E.; Martien, L.; Lewis, D. *Tampa Water Department rEsidential Water Conservation Study: The Impacts of High Efficiency Plumbing Fixture Retrofits in Single-Family Homes*; Aquacraft, Inc. Water Engineering and Management: Boulder, CO, USA, 2004.
131. Roberts, P. *Yarra Valley Water 2004 Residential End Use Measurement Study*; Yarra Valley Water: Melbourne, Australia, 2005.
132. Kim, S.H.; Choi, S.H.; Koo, J.Y.; Choi, S.I.; Hyun, I.H. Trend analysis of domestic water consumption depending upon social, cultural, economic parameters. *Water Sci. Technol. Water Supply* **2007**, *7*, 61–68. [CrossRef]
133. Froehlich, J.E.; Larson, E.; Campbell, T.; Haggerty, C.; Fogarty, J.; Patel, S.N. HydroSense: Infrastructure-mediated single-point sensing of whole-home water activity. In Proceedings of the 11th International Conference on Ubiquitous Computing, Orlando, FL, USA, 30 September–3 October 2009.
134. Heinrich, M.; Roberti, H. Auckland Water Use Study: Monitoring of Residential Water End Uses. In Proceedings of the CIB SB10 New Zealand, Nelson, New Zealand, 26–28 May 2010.
135. Beal, C.D.; Stewart, R.A.; Huang, T.; Rey, E. SEQ residential end use study. In *Smart Water Systems & Metering*; Urban Water Security Research Alliance: Brisbane, Australia, 2011.
136. South East Queensland Residential End Use Study. Dataset. South East Queensland Residential End Use Study. 2011. Available online: https://code.google.com/archive/p/smart-meter-information-portal/downloads (accessed on 10 December 2020).
137. Otaki, Y.; Otaki, M.; Sugihara, H.; Mathurasa, L.; Pengchai, P.; Aramaki, T. Comparison of residential indoor water consumption patterns in Chiang Mai and Khon Kaen, Thailand. *J.—Am. Water Work. Assoc.* **2011**, *103*, 104–110. [CrossRef]
138. MidCoast-Water. *Smart Meter Water Usage Study Report for Pressure Management Group: Tea. Gardens and Hawks Nest*; Technical Report; MidCoast Water: Taree, Australia, 2012.
139. Lee, D.J.; Park, N.S.; Jeong, W. End-use analysis of household water by metering: The case study in Korea. *Water Environ. J.* **2012**, *26*, 455–464. [CrossRef]
140. Borg, M.; Edwards, O.; Kimpel, S. A Study of Individual Household Water Consumption. *Univ. Calif. Davis Water Manag. Res. Inst.* **2013**, *7*, 2017.
141. Neunteufel, R.; Richard, L.; Perfler, R. Water demand: The Austrian end-use study and conclusions for the future. *Water Sci. Technol. Water Supply* **2014**, *14*, 205–211. [CrossRef]
142. Gurung, T.R.; Stewart, R.A.; Beal, C.D.; Sharma, A.K. Smart meter enabled water end-use demand data: Platform for the enhanced infrastructure planning of contemporary urban water supply networks. *J. Clean. Prod.* **2015**, *87*, 642–654. [CrossRef]
143. Rathnayaka, K.; Malano, H.; Maheepala, S.; George, B.; Nawarathna, B.; Arora, M.; Roberts, P. Seasonal demand dynamics of residential water end-uses. *Water* **2015**, *7*, 202–216. [CrossRef]
144. Nguyen, K.A.; Stewart, R.A.; Zhang, H.; Jones, C. Intelligent autonomous system for residential water end use classification: Autoflow. *Appl. Soft Comput. J.* **2015**, *31*, 118–131. [CrossRef]
145. Makonin, S.; Ellert, B.; Bajić, I.V.; Popowich, F. Electricity, water, and natural gas consumption of a residential house in Canada from 2012 to 2014. *Sci. Data* **2016**, *3*, 160037. [CrossRef]
146. Makonin, S. AMPds2: The Almanac of Minutely Power Dataset (Version 2). 2016. Available online: https://dataverse.harvard.edu/dataset.xhtml?persistentId=doi:10.7910/DVN/FIE0S4 (accessed on 10 December 2020).
147. DeOreo, W.B.; Mayer, P.; Dziegielewski, B.; Hazen, J.K.; Sawyer, P.C. *Residential End Uses of Water, Version 2*; Technical Report; Water Research Foundation: Denver, Co, USA, 2016. Available online: https://www.waterrf.org/research/projects/residential-end-uses-water-version-2 (accessed on 10 December 2020).
148. Carranza, J.C.I.; Morales, R.D.; Sánchez, J.A. Pattern recognition in residential end uses of water using artificial neural networks and other machine-learning techniques. In Proceedings of the CCWI 2017—15th International Conference on Computing and Control for the Water Industry, Sheffield, UK, 5–7 September 2017.
149. Vitter, J.S., Jr.; Leibowicz, B.; Webber, M.; Deetjen, T.; Berhanu, B. Data for: Optimal Sizing and Dispatch for a Community-Scale Potable Water Recycling Facility, Mendeley Data, V1, 2018. Available online: https://data.mendeley.com/datasets/5bvmy4kj25/1 (accessed on 10 December 2020).[CrossRef]

150. Kofinas, D.T.; Spyropoulou, A.; Laspidou, C.S. A methodology for synthetic household water consumption data generation. *Environ. Model. Softw.* **2018**, *100*, 48–66. [CrossRef]
151. EC FP7 project ISS-EWATUS. Dataset. Integrated Support System for Efficient Water Usage and Resources Management. Website under Maintenance. 2016. Available online: validation.issewatus.eu/data-re-use/ (accessed on 10 December 2020).
152. Clifford, E.; Mulligan, S.; Comer, J.; Hannon, L. Flow-Signature Analysis of Water Consumption in Nonresidential Building Water Networks Using High-Resolution and Medium-Resolution Smart Meter Data: Two Case Studies. *Water Resour. Res.* **2018**, *54*, 88–106. [CrossRef]
153. Omaghomi, T.; Buchberger, S.; Cole, D.; Hewitt, J.; Wolfe, T. Probability of Water Fixture Use during Peak Hour in Residential Buildings. *J. Water Resour. Plan. Manag.* **2020**, *146*, 04020027. [CrossRef]
154. Pacheco, C.J.; Horsburgh, J.S.; Tracy, J.R. A low-cost, open source monitoring system for collecting high temporal resolution water use data on magnetically driven residential water meters. *Sensors* **2020**, *20*, 3655. [CrossRef] [PubMed]
155. Di Mauro, A.; Di Nardo, A.; Santonastaso, G.F.; Venticinque, S. Development of an IoT System for the Generation of a Database of Residential Water End-Use Consumption Time Series. *Environ. Sci. Proc.* **2020**, *2*, 20. [CrossRef]
156. Nguyen, K.A.; Stewart, R.A.; Zhang, H. An autonomous and intelligent expert system for residential water end-use classification. *Expert Syst. Appl.* **2014**, *41*, 342–356. [CrossRef]
157. Makki, A.A.; Stewart, R.A.; Beal, C.D.; Panuwatwanich, K. Novel bottom-up urban water demand forecasting model: Revealing the determinants, drivers and predictors of residential indoor end-use consumption. *Resour. Conserv. Recycl.* **2015**, *95*, 15–37. [CrossRef]
158. Pastor-Jabaloyes, L.; Arregui, F.J.; Cobacho, R. Water end use disaggregation based on soft computing techniques. *Water* **2018**, *10*, 46. [CrossRef]
159. Godoy, D.; Amandi, A. A conceptual clustering approach for user profiling in personal information agents. *AI Commun.* **2006**, *19*, 207–227.
160. Cominola, A.; Nguyen, K.; Giuliani, M.; Stewart, R.A.; Maier, H.R.; Castelletti, A. Data Mining to Uncover Heterogeneous Water Use Behaviors From Smart Meter Data. *Water Resour. Res.* **2019**, *55*, 9315–9333. [CrossRef]
161. Cufoglu, A. User Profiling—A Short Review. *Int. J. Comput. Appl.* **2014**, *108*, 1–9. [CrossRef]
162. Kanoje, S.; Girase, S.; Mukhopadhyay, D. User Profiling Trends, Techniques and Applications. *Int. J. Adv. Found. Res. Comput.* **2014**, *1*, 2348–4853.
163. Beckel, C.; Sadamori, L.; Staake, T.; Santini, S. Revealing household characteristics from smart meter data. *Energy* **2014**, *78*, 397–410. [CrossRef]
164. Stagge, J.H.; Rosenberg, D.E.; Abdallah, A.M.; Akbar, H.; Attallah, N.A.; James, R. Assessing data availability and research reproducibility in hydrology and water resources. *Sci. Data* **2019**, *6*, 190030. [CrossRef] [PubMed]
165. Trabucchi, D.; Buganza, T. Data-driven innovation: Switching the perspective on Big Data. *Eur. J. Innov. Manag.* **2019**, *22*, 23–40. [CrossRef]
166. Adamala, S. An Overview of Big Data Applications in Water Resources Engineering. *Mach. Learn. Res.* **2017**, *2*, 10–18.
167. Van Tuyl, S.; Whitmire, A.L. Water, water, everywhere: Defining and assessing data sharing in Academia. *PLoS ONE* **2016**, *11*, e0147942. [CrossRef] [PubMed]
168. Zipper, S.C.; Stack Whitney, K.; Deines, J.M.; Befus, K.M.; Bhatia, U.; Albers, S.J.; Beecher, J.; Brelsford, C.; Garcia, M.; Gleeson, T.; et al. Balancing Open Science and Data Privacy in the Water Sciences. *Water Resour. Res.* **2019**, *55*, 5202–5211. [CrossRef]
169. Fernandes, F.; Morais, H.; Faria, P.; Vale, Z.; Ramos, C. SCADA house intelligent management for energy efficiency analysis in domestic consumers. In Proceedings of the 2013 IEEE PES Conference on Innovative Smart Grid Technologies, ISGT LA 2013, Washington, DC, USA, 24–27 February 2013.
170. Cominola, A.; Ghetti, A.; Castelletti, A. Building an open high-resolution residential water end-use dataset with non-intrusive metering, intrusive metering, and water use diaries. In Proceedings of the Geophysical Research Abstracts, EGU General Assembly, Vienna, Austria, 8–13 April 2018.
171. Giurco, D.P.; White, S.B.; Stewart, R.A. Smart metering and water end-use data: Conservation benefits and privacy risks. *Water* **2010**, *2*, 461–467. [CrossRef]
172. Cominola, A.; Giuliani, M.; Castelletti, A.; Abdallah, A.M.; Rosenberg, D.E. Developing a stochastic simulation model for the generation of residential water end-use demand time series. In Proceedings of the 8th International Congress on Environmental Modelling and Software, Toulouse, France, 10–14 July 2016.
173. Mostafavi, N.; Gándara, F.; Hoque, S. Predicting water consumption from energy data: Modeling the residential energy and water nexus in the integrated urban metabolism analysis tool (IUMAT). *Energy Build.* **2018**, *158*, 1683–1693. [CrossRef]
174. Yang, A.; Zhang, H.; Stewart, R.A.; Nguyen, K.A. Water end use clustering using hybrid pattern recognition techniques—Artificial Bee Colony, Dynamic Time Warping and K-Medoids clustering. *Int. J. Mach. Learn. Comput.* **2018**, *8*, 483–487.
175. Nguyen, K.; Zhang, H.; Stewart, R.A. Development of an intelligent model to categorise residential water end use events. *J. Hydro-Environ. Res.* **2013**, *7*. doi:10.1016/j.jher.2013.02.004. [CrossRef]
176. Chini, C.M.; Stillwell, A.S. The metabolism of U.S. cities 2.0. *J. Ind. Ecol.* **2019**, *23*, 1353–1362. [CrossRef]
177. Hamiche, A.M.; Stambouli, A.B.; Flazi, S. A review of the water-energy nexus. *Renew. Sustain. Energy Rev.* **2016**, *65*, 319–331. [CrossRef]

178. Klemenjak, C.; Makonin, S.; Elmenreich, W. Towards comparability in non-intrusive load monitoring: On data and performance evaluation. In Proceedings of the 2020 IEEE Power and Energy Society Innovative Smart Grid Technologies Conference, ISGT 2020, Washington, DC, USA, 17–20 February 2020.
179. McCartney, D. *Water Power: Measuring Household Water and Energy Intensity*. Technical Report; 2016. Available online: http://blogs.edf.org/texascleanairmatters/files/2016/10/Water-Power-Measuring-Household-Water-and-Energy-Intensity.pdf (accessed on 10 December 2020).
180. Barai, G.R.; Krishnan, S.; Venkatesh, B. Smart metering and functionalities of smart meters in smart grid—A review. In Proceedings of the 2015 IEEE Electrical Power and Energy Conference: Smarter Resilient Power Systems, EPEC 2015, London, ON, Canada, 26–28 October 2015; Institute of Electrical and Electronics Engineers Inc.: Piscataway, NJ, USA, 2016; pp. 138–145.
181. Cominola, A.; Giuliani, M.; Piga, D.; Castelletti, A.; Rizzoli, A.E. A Hybrid Signature-based Iterative Disaggregation algorithm for Non-Intrusive Load Monitoring. *Appl. Energy* **2017**, *185*, 331–344. [CrossRef]
182. Voulis, N.; Warnier, M.; Brazier, F.M. Understanding spatio-temporal electricity demand at different urban scales: A data-driven approach. *Appl. Energy* **2018**, *230*. [CrossRef]
183. Wang, Y.; Chen, Q.; Hong, T.; Kang, C. Review of Smart Meter Data Analytics: Applications, Methodologies, and Challenges. *IEEE Trans. Smart Grid* **2019**, *10*. [CrossRef]
184. Alahakoon, D.; Yu, X. Smart Electricity Meter Data Intelligence for Future Energy Systems: A Survey. *IEEE Trans. Ind. Inf.* **2016**, *12*, 425–436. [CrossRef]
185. Han, W.; Ang, L. Analysis of the Use of Smart Meter Data by Energy Companies in Various Countries. In *IOP Conference Series: Materials Science and Engineering*; Institute of Physics Publishing: Bristol, UK, 2018; Volume 439.
186. Yun, J.; Lee, S.S.; Ahn, I.Y.; Song, M.H.; Ryu, M.W. Monitoring and control of energy consumption using smart sockets and smartphones. In *Communications in Computer and Information Science*; Springer: Berlin/Heidelberg, Germany, 2012; Volume 339.
187. Abubakar, I.; Khalid, S.N.; Mustafa, M.W.; Shareef, H.; Mustapha, M. Application of load monitoring in appliances' energy management—A review. *Renew. Sustain. Energy Rev.* **2017**, *67*, 235–245. [CrossRef]
188. Najafi, B.; Moaveninejad, S.; Rinaldi, F. Data Analytics for Energy Disaggregation: Methods and Applications. In *Big Data Application in Power Systems*; Elsevier: Amsterdam, The Netherlands, 2018; pp. 377–408.
189. Zhuang, M.; Shahidehpour, M.; Li, Z. An Overview of Non-Intrusive Load Monitoring: Approaches, Business Applications, and Challenges. In Proceedings of the 2018 International Conference on Power System Technology, POWERCON 2018—Proceedings, Guangzhou, China, 6–8 November 2018; Institute of Electrical and Electronics Engineers Inc.: Piscataway, NJ, USA, 2019; pp. 4291–4299.
190. Froehlich, J.; Larson, E.; Gupta, S.; Cohn, G.; Reynolds, M.; Patel, S. Disaggregated end-use energy sensing for the smart grid. *IEEE Pervasive Comput.* **2011**, *10*, 28–39. [CrossRef]
191. Ellert, B.; Makonin, S.; Popowich, F. Appliance water disaggregation via non-intrusive load monitoring (NILM). In *Lecture Notes of the Institute for Computer Sciences, Social-Informatics and Telecommunications Engineering, LNICST*; Springer: Berlin/Heidelberg, Germany, 2016.
192. Rustad, M.L.; Koenig, T.H. Towards a Global Data Privacy Standard. *Fla. Law Rev.* **2019**, *71*, 365–454.
193. Nieswiadomy, M.L.; Molina, D.J. Comparing Residential Water Demand Estimates under Decreasing and Increasing Block Rates Using Household Data. *Land Econ.* **1989**, *65*, 280. [CrossRef]
194. Glenn, D.T.; Endter-Wada, J.; Kjelgren, R.; Neale, C.M. Tools for evaluating and monitoring effectiveness of urban landscape water conservation interventions and programs. *Landsc. Urban Plan.* **2015**, *139*, 82–93. [CrossRef]
195. Mini, C.; Hogue, T.S.; Pincetl, S. The effectiveness of water conservation measures on summer residential water use in Los Angeles, California. *Resour. Conserv. Recycl.* **2015**, *94*, 136–145. [CrossRef]
196. Porse, E.; Mika, K.B.; Litvak, E.; Manago, K.F.; Naik, K.; Glickfeld, M.; Hogue, T.S.; Gold, M.; Pataki, D.E.; Pincetl, S. Systems Analysis and Optimization of Local Water Supplies in Los Angeles. *J. Water Resour. Plan. Manag.* **2017**, *143*, 04017049. [CrossRef]
197. Gal, M.S.; Rubinfeld, D.L. Data standardization. *N. Y. Univ. Law Rev.* **2019**, *94*, 737–770. [CrossRef]
198. Rahm, E.; Do, H. Data cleaning: Problems and current approaches. *IEEE Data Eng. Bull.* **2000**, *23*, 3–13.

 *water*

Review

Water Network Partitioning into District Metered Areas: A State-Of-The-Art Review

Xuan Khoa Bui, Malvin S. Marlim and Doosun Kang *

Department of Civil Engineering, College of Engineering, Kyung Hee University, Gyeonggi-do 17104, Korea; khoabx@khu.ac.kr (X.K.B.); malvinmarlim@hotmail.com (M.S.M.)
* Correspondence: doosunkang@khu.ac.kr; Tel.: +82-31-201-2513

Received: 28 February 2020; Accepted: 27 March 2020; Published: 1 April 2020

Abstract: A water distribution network (WDN) is an indispensable element of civil infrastructure that provides fresh water for domestic use, industrial development, and fire-fighting. However, in a large and complex network, operation and management (O&M) can be challenging. As a technical initiative to improve O&M efficiency, the paradigm of "divide and conquer" can divide an original WDN into multiple subnetworks. Each subnetwork is controlled by boundary pipes installed with gate valves or flow meters that control the water volume entering and leaving what are known as district metered areas (DMAs). Many approaches to creating DMAs are formulated as two-phase procedures, clustering and sectorizing, and are called water network partitioning (WNP) in general. To assess the benefits and drawbacks of DMAs in a WDN, we provide a comprehensive review of various state-of-the-art approaches, which can be broadly classified as: (1) Clustering algorithms, which focus on defining the optimal configuration of DMAs; and (2) sectorization procedures, which physically decompose the network by selecting pipes for installing flow meters or gate valves. We also provide an overview of emerging problems that need to be studied.

Keywords: clustering; district metered area; network sectorization; water distribution network; water network partitioning

---

1. Introduction

A water distribution network (WDN) supplies drinking water by maintaining pressures and flow rates. As most of a WDN's components are buried and comprise thousands to tens of thousands of elements, operation and management (O&M) can be complex [1]. Increasing urbanization means WDNs are constantly being upgraded and expanded. In large cities with aging networks, O&M is becoming more challenging than ever before.

A critical O&M objective for utilities working on WDNs is improving the efficiency and efficacy of the supply for a specified water demand at the lowest cost. In particular, efficacy requires reducing water leakage and nonrevenue water, controlling uniform pressure, and ensuring sufficient pressure. Leakage control is the most effective way to reduce water prices. The quantity of leakage is related to system pressure, and reducing pressure reduces leakage. Utilities can apply a "divide and conquer" paradigm to this challenge by dividing the original complex network into independently controlled subnetworks called district metered areas (DMAs).

Most researchers agree that partitioning a network into DMAs offers many benefits [2–6]. These actions may include but are not limited to: (i) Substantially reduce nonrevenue water by active leakage control; (ii) simplify pressure management by setting off pressure reducing valves (PRVs); (iii) rapidly identify burst pipes; (iv) district isolation in order to protect the rest of network from accidental or malicious contamination events; and (v) potential creation of independent DMAs which exclusively supplied from its own water sources for better control of water quality (e.g., there is no mixing of water from different sources). Moreover, for intermittent WDNs, where water is only supplied during a certain time of a day, DMAs are transition processes that allow evolving intermittent WDNs to continuous systems by enabling equitable water supply in each DMA [7]. Ciaponi et al. [8] recently revealed the benefits of WNP not only for WDNs monitoring from contamination events but also for the effectiveness of optimal sensor placement.

Despite these advantages, they come with trade-offs, such as the reduced redundancy in network connectivity and the demotion of system pressure, which results in lower network preparedness for emergencies such as fire-fighting, water suspension due to burst pipes. An additional concern is water quality deterioration (i.e., water age growth) due to the reduction of available flow paths [9,10].

Because of the benefits brought about by DMAs, many utilities consider them an effective way to achieve O&M objectives [2]. However, dividing an original network into suitable DMAs can be challenging because of the intrinsic complexity of the WDN. In the past, before mathematical methods were applied to DMA configuration, utilities designed DMAs according to administrative boundaries (districts), main roads, the number of inhabitants, the economic level of leakage, or reservoir (tank) locations [4,5], which did not account for global perspectives. However, with the advent of mathematical models, hydraulic solvers can simplify the process and provide various approaches to optimizing the creation of DMAs while considering operational constraints and objectives.

Today, water network partitioning (WNP) is used to divide networks into DMAs. WNP is a heuristic process controlled by two phases: clustering and sectorization. The clustering phase is the preformation of DMAs based on network connectivity and topology. It is implemented through various algorithms such as graph theory, community structure, modularity-based algorithms, and spectral algorithms [2,3] to form feasible DMAs and minimize the number of connections to each other. Sectorization is an optimization process to locate flow meters and gate valves to maintain as high as possible network performances while minimizing the economic costs [6].

In recent years, WNP has been explored in various studies. In the 18th Water Distribution System Analysis Conference held in Colombia in July 2016, the "Battle of Water Networks" competition focused on creating DMAs. The main objective was to optimize the design and operation of a system's main components by determining new DMAs for an existing WDN in Colombia, the E-Town network, by taking into account costs, pressure uniformity, and water quality [3]. It was an opportunity for researchers and practitioners around the world to solve a challenging problem in a full-scale WDN. In addition, several methods have been proposed over the last decade for dividing a network into DMAs. Various benchmark networks have also been developed to test the state-of-the-art methods.

A literature review identified more than 100 published studies that focused applying various methodologies to WNP. After reviewing the main discussions and approaches in each paper, we selected 95 papers to study in-depth and 27 related articles; they are all cited here. We found that the methodologies proposed to date still have certain limitations in real-life applications. This paper provides a comprehensive review of WDN management using DMAs to help water utilities improve efficiency in O&M as well as support decision- and strategy-making processes. With this goal in mind, we first reviewed the rules of DMA and analyzed the merits and demerits of the O&M of DMAs. Second, we classified the methodologies proposed in recent studies of WNP processes into two major categories, clustering algorithms and sectorization processes, and analyzed the advantages and disadvantages of each method. Then, we highlighted the main indicators developed to quantify the segmentation performances. Finally, we discussed some limitations of the approaches.

This paper is organized as follows. Section 2 presents the principle of DMA and its function in the O&M of WDNs. Section 3 mainly reviews the clustering phase for the methods developed so far, and Section 4 describes the sectorization procedure and discusses the currently applied algorithms. Various indicators of DMA performance are described in Section 5. Finally, Section 6 draws conclusions and discusses possible improvements in WNP approaches.

2. Principles of District Metered Areas

The concept of DMA management was introduced by the United Kingdom water industry in the early 1980s [2,5,11] (Figure 1). At that time, DMA was an area of a distribution system that was specifically defined by the closure of valves and measurement of the quantities of water entering and leaving the area. The first goal of DMA is early detection and management of water leakage in a WDN [11]. Specifically, the measurement of night flow is analyzed to determine the level of leakage within each DMA and locate the most beneficial places for leakage probes [4,10].

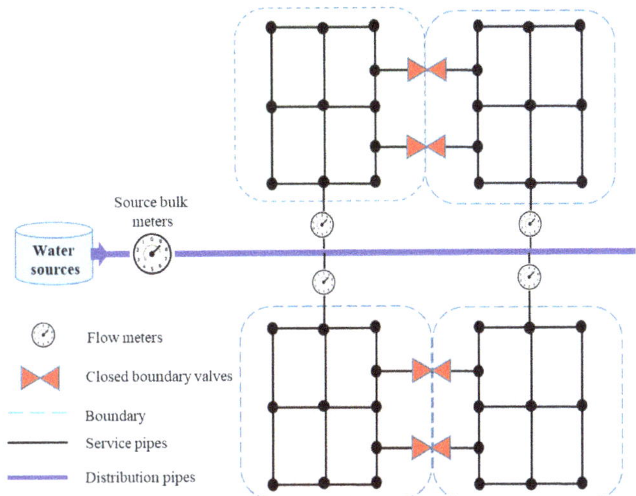

Figure 1. A schematic diagram of district metered areas.

Water leakage is a major concern for water utilities [12]. Leakage rate varies widely depending on the country, region, and age of the system. It is reported to be as low as 3 to 7% in a well-maintained system in the Netherlands, ranging from 10 to 30% in the United States and the United Kingdom, however as high as 70% in some undeveloped countries [13]. Water loss in WDNs can be classified as real loss and apparent loss in nature [14]. The real loss occurs from burst pipes or background leaks due to continual seepage of water from network properties, such as pipe and valve fittings, or to corrosion-induced perforation of pipes. The apparent loss includes the unauthorized consumption, a product of meter-reading errors, water theft, and accounting errors. To estimate the leakage in a DMA, the operator must monitor net minimum night flows in the system (when most consumers are inactive) and compare it with legitimate night flows to assess the rate of real losses.

One of the major factors influencing leakage is the pressure in the water network [12]. To reduce water losses, many utilities have changed from a passive approach (i.e., detection and repair) to proactive approaches (i.e., heuristic processes and pressure-leakage relationships as developed by Allan Lambert [15]) that indicate that the leakage rate of flow may increase or decrease with changing pressure levels. The DMA concept was introduced to help proactively manage the number of invisible water losses and detect the locations of failures based on the hydraulic characteristics of the WDN.

Researchers agree that dividing a network into DMAs is useful [4,16]. Most research assumes that the benefits of DMAs are greater than the drawbacks [6,17–21]. WDN management through DMAs has proven highly successful for leakage reduction, reportedly controlling up to 85% of national water leakage in the UK [11,22]. Gomes et al. [23] showed that dividing a network into DMAs allows for stable pressure management, which increases asset lifespans. Reduced pressure lowers the frequency of potential pipe breaks, which consequently reduces real water losses. Gomes et al. also proposed a method based on the minimum night-flow relationship with pressure to evaluate the benefits of pressure management using DMAs by predicting water losses before and after pressure drops, estimating the reduction in energy consumption through billed water as well as the estimated direct benefit of the pressure reduction process with DMAs [23]. Huang et al. [24] reported that DMAs allow for rapid detection of burst pipes by studying the uniformity of daily water demand. They applied a supervised learning algorithm to improve the positive effect of burst-event detection in real-time operations. Savic and Ferrari [20,25] and Lifshitz and Ostfeld [26] have also illustrated the effectiveness of implementing DMAs in WDNs with respect to reducing the frequency of pipe breaks. To quantify the benefits of reducing burst frequency, Lambert et al. [15] proved that the percentage of burst-frequency reduction relies on the proportion of pressure reduction obtained after setting up the DMAs. Their study also revealed that controlling pressure not only reduces burst frequencies, but also reduces leakage flow rates, extending the life of residual devices and reducing costs for both the water utility and the customer.

Ferrari and Savic [25] showed that, depending on the specific alternative DMA layout used, burst frequency can be reduced by approximately 53% to 60%. They also found that leakage reduction ranged between approximately 26% and 59% after DMA set-up. Furthermore, as the closure of valves completely isolates DMAs, it is possible to reduce the risk of chemical attacks or accidental events throughout a network [6]. Isolating DMAs is also useful in component maintenance, replacement, and repairs because closing boundary valves disconnects districts from other areas. Lifshitz and Ostfeld [26] demonstrated that combining DMAs with PRVs creates a "knowledge and action" approach to detecting and managing water leaks. PRVs reduce excess pressure and consequently reduce potential water leaks without prior information on the positions of the leaks. Meanwhile, DMAs enable the identification of possible locations of leaks and their potential amounts. A combination of the DMAs and PRVs will complement each other to provide a better solution for leak management.

The main drawbacks of DMAs are deterioration of water quality compared with that of the original network and the loss of system resilience against abnormal events. Water age is regarded as a surrogate for simulations for evaluating the reduction of water quality [27]. Grayman et al. [10], Diao et al. [19], and Di Nardo et al. [28] found that there was no significant change in the overall water age metric before and after dividing a large-scale, looped WDN into DMAs. This is consistent with previous studies by UKWIP [29] and WRc [30], which investigated the impact of WNP on water quality management. Armand et al. [31] utilized surrogate hydraulic variables to evaluate the impact of WNP on water quality and the likelihood of discoloration incidents. They reported that DMAs can compromise overall water quality by increasing the average water age for a set of nodes with dead-end-like hydraulic behavior. This also increased the likelihood of sedimentation in pipes due to low flow velocity. However, water quality is reportedly not a critical criterion when designing DMAs and water age is not a binding constraint [3,28]. Javier et al. [32] and Salomons et al. [33], who conducted water balance analyses in a WDN, pointed out that the water volume stored in the network was nearly half of the daily water consumption. It was therefore reasonable to assume that water would be replaced twice a day in the network, which is a good indicator of water quality. By running a hydraulic model to compare the network before and after DMA installation, no significant variations of water age were seen throughout the whole network.

One of the other weaknesses when dividing networks into DMAs is the reduction in a system's redundancy [19,28] due to reduced availability of flow paths to connect supply sources and demand nodes. The insertion of multiple gate valves and flow meters to isolate a DMA leads to increased head loss due to increased friction [34]. This change can reduce system redundancy in terms of available pressure throughout the network. Typically, several emergent cases, such as fire-flow supply and water suspension due to a burst pipe would be issued in system operations. Table 1 summarizes the main advantages and disadvantages of installing DMAs in WDNs.

Table 1. Main advantages and disadvantages of district metered areas (DMAs).

Advantages	Disadvantages
Improved burst detection and leakage identification	Reduced resilience to failures
Advantaged subarea management and reduced NRW	Reduced operational flexibility
Improved subarea pressure control	Potential negative impact on water quality
Improved protection against contamination	Security issues in peripheral areas and emergency cases
Reduced maintenance and repair costs	High initial investment cost
Characterized demand curve, especially at night	Reduced hydraulic redundancy

Several criteria should be considered when designing DMAs [11], such as

1. Maximum percentage of leakage allowed by the water utility;
2. Topography and number of properties per DMA;
3. Characteristics and topological taxonomy of WDNs;
4. Variations in nodal elevation, water demand, and pressure;
5. The number of flow meters and gate valves; and
6. Water quality considerations.

Depending on the existing network situation and leakage rates, each utility will have its own criteria to set up economically efficient levels of leakage for each DMA. Once the level of leakage has been determined, the utility can select the type of policy best suited for controlling leakage in the future, the size and number of DMAs, and the staff required for the required policy. Dividing a network into small DMAs will identify bursts quickly, maintain total leakage at a lower level, and reduce the time required to identify device failures. However, this also leads to increased investment and operational costs in terms of new flow meters and valves [11]. The international water association (IWA), as corroborated by previous studies, reports that DMA size is expressed by the number of properties (user flow meters) and varies between 500 and 5000 properties in urban areas [24]. Individual DMA size can vary depending on local factors and system characteristics. While a DMA with fewer than 500 properties requires much more initial investment and incurs a higher maintenance cost, a large DMA will face difficulty in discriminating small bursts and will suffer increased leakage location times [4,5,11].

From a topological connectivity point of view, a set of complex network metrics was proposed by Giudicianni et al. [35] to analyze the relationship between the metrics values and the topological structures of WDNs. To optimize the number of DMAs in the network, the eigengap heuristic was used to maximize the jump in spectrum of the Laplacian matrix. The study revealed that correlation between the number of DMAs and network size approximatively follows a power law. Hence, the optimal number of DMAs does not grow significantly with the network size. Such a relationship hints that, from a connectivity point of view, the increase of WDN size has more effects on the size of the DMA rather than the number of DMAs.

The number of water sources supplying each DMA also needs to be considered in the design process, as each source must be fitted with a flow meter. Depending on the network type (branched or looped), a DMA may be supplied by single or multiple sources and delivered consecutively or in parallel to adjacent DMAs. As suggested by Di Nardo et al. [2], a technical and economic rule is to minimize the number of installed flow meters and have a single flow meter for each DMA to simplify the calculation of the synchronous water balance. To isolate a DMA from adjacent DMAs, gate valves are installed in boundary pipes. However, installing gate valves may create more dead-ends and reduce the pathways of water to the nodes, which may lead to deteriorating water quality [11]. Therefore, optimizing the number and location of flow meters and valves while decomposing the original network into DMAs is necessary to minimize costs and optimize operational benefits.

Determining and optimizing the number of DMAs is essential. However, defining the configuration of DMAs is a demanding task because many different aspects of WDN performance must be considered [20]. This is usually approached as a multi-objective optimization problem. Traditionally, DMA design has been based on empirical data combined with trial-and-error methods. Recently, the concept and approach for WNP have been explored in the literature. Several smarter and more efficient approaches have been proposed to create optimal DMA layouts. Although the algorithms applied in each study are different, the WNP process commonly consists of two phases, clustering and sectorization [36,37].

3. Clustering to Create Feasible DMAs

Figure 2 summarizes the general procedures for WNP. The clustering phase is the initial process that designs the shape and dimensions of DMAs based on the network topology. The goal is to determine the optimal number of DMAs to balance the number of nodes in each cluster and to minimize the number of boundary pipes (i.e., pipe cuts where gate valves or flow meters will be installed). The algorithms applied include graph theory such as depth-first search (DFS) and breadth-first search (BFS) [6,9,38,39], community structure [19,34,37,40], modularity-based procedures [41–44], multilevel partitioning [17,37,45,46], spectral approaches [47–49], and multi-agent approaches [50–52]. This paper focuses on explaining six major algorithms and how they are handled in clustering WDNs to automatically create DMA configurations.

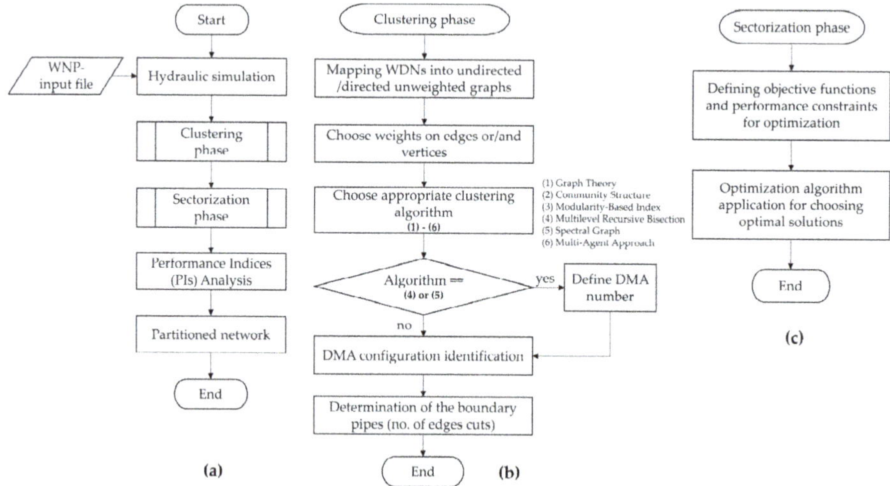

Figure 2. Steps of water network partitioning: (**a**) Overall main procedures, (**b**) steps for clustering, and (**c**) steps for sectorization.

3.1. Graph Theory

Most of the existing clustering algorithms developed for WNP relate to graph theory. To gain a deeper understanding of clustering algorithms, it is necessary to generalize some of the topological characteristics and properties of a WDN. Readers unfamiliar with graph theory should refer to previous studies [53,54].

WDN is a social infrastructure that allows water to flow along the pipes and communicate between nodes in the network. The topology structure of WDNs is mapped onto an undirected or directed graph and characterized by a pair of sets $G = (V, E)$, where V is the vertex set representing junctions, reservoirs, and tanks and $n = |V|$ is the total number of vertices. E is the set of edges in response to pipes, valves, pumps and $m = |E|$ is the total number of edges. An undirected graph with edges is an unordered pair $\{v_1, v_2\}$, while a directed graph with edges is an ordered pair and the vertices v_1, v_2 are called the endpoints of the edge.

A given network graph, and a WDN in particular, can be converted by an adjacent matrix $\mathbf{A}$, which is an $n \times n$ matrix, where A_{ij} is the (i, j) element equal to 1 if v_i is adjacent with v_j, otherwise, $A_{ij} = 0$. A weighted graph can be represented mathematically by an adjacency matrix that has a certain weight W_{ij} assigned for each pair of vertices (i, j). The weights are usually non-negative, real numbers, and they must satisfy $W_{ij} = W_{ji} \geq 0$, if i and j are connected. Otherwise, $W_{ij} = W_{ji} = 0$. The nodal degree, k_i, is the number of edges attached to a vertex i. The degree of node i is defined as $k_i = \sum_{j=1}^{n} A_{ij}$ for the adjacency matrix $\mathbf{A}$, and $k_i = \sum_{j=1}^{n} W_{ij}$ for the weighted adjacency matrix $\mathbf{W}$. From a topological point of view and complex network theory, Giudicianni et al. [35] treated the WDN as a graph by using several complex network metrics to characterize the topology of typical WDNs. It was a preliminary process for better understanding the network itself, and provided the classical approach for partitioning or/and designing the WDNs.

One of the graph theory algorithms applied to network clustering is DFS, which, as proposed by Tarjan [55], allows for the exploration of the connectivity of a graph by traversing a node in the network. The DFS algorithm is a recursive approach based on backtracking. It starts by picking a root node in the network and then searches for nodes as far as possible along each path (in-depth dimension) until there are no more adjacent nodes in the current path to traverse after backtracking to the next path. In contrast, the BFS algorithm proposed by Pohl [56] starts at a root node and traverses the graph broad-wise by first moving horizontally and exploring all the nodes of the current path and then moving to the next path. Figure 3 shows how a DFS and a BFS work.

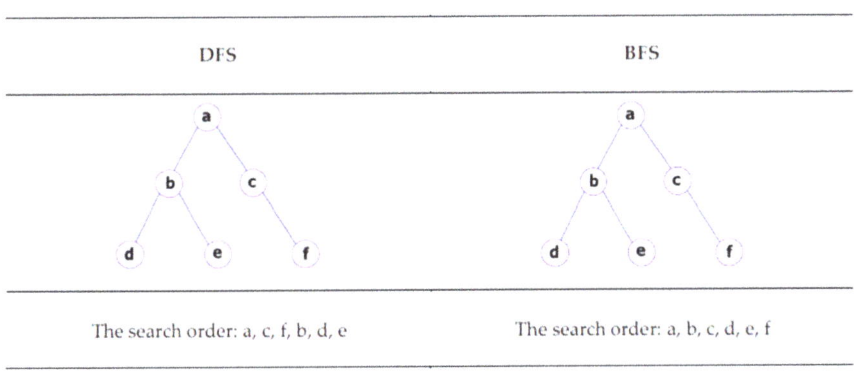

Figure 3. Diagram of depth-first search (DFS) and breadth-first search (BFS).

Tzatchkov et al. [38] applied the DFS and BFS to a WNP project in Mexico. DFS was used to segment a whole network into independent sectors by identifying nodes belonging to each sector (i.e., each sector is supplied exclusively from its own water sources, and it is not connected to other sectors in the network), and BFS was used to exam the set of disconnected nodes from any water sources, thus obtaining the size and configuration of independent sectors in the WDN. More specifically, Perelman and Ostfeld [39] and Lifshitz and Ostfeld [1] proposed a procedure for topology clustering based on the DFS algorithm to identify strongly connected clusters that had at least one path in both directions between them, while the opposing BFS algorithm was used to classify weakly connected clusters that had only one directed path between a set of nodes (i.e., from node u to node v, but not from node v to node u). The results were utilized for various purposes, such as contaminant prediction from a source and spread of infection in a WDN [1]. Di Nardo et al. [6] proposed a method for optimizing water network sectorization based on graph theory. DFS was used to find the independent sector combined with a hierarchical approach developed by Di Battista et al. [57] to draw hierarchical levels of a tree graph corresponding to each source, creating isolated DMAs, each of which was supplied by its own source and was disconnected from the rest of a network through gate valves.

Campbell et al. [58,59] proposed a more advanced orderly combination of a series of graphs to generate a flexible method for defining feasible DMA layouts. They proposed dividing a network into two components, a trunk network and a distribution network. To determine the scope of the trunk network, the shortest path and the BFS concept were implemented. Once the trunk network was determined, it was detached from the network, while the community detection algorithm was adopted for the rest of network (distribution network) to define the best structural communities in the distribution network, which is the configuration of sectors. The innovation of this study was that the trunk pipes acted as entrances to each DMA and were not considered candidates for sectorization, ensuring the reliability of the WDN. In a similar methodology proposed by Alvisi and Franchini [60], BFS defined the location of possible nodes to form an assigned number of DMAs and then the shortest path distance from each source to the nodes was simultaneously estimated to determine the set of boundary pipes for each DMA. In the case of a WDN with numerous water sources, Scarpa et al. [9] successfully applied a BFS algorithm to identify elementary DMAs in which each one was supplied only by its own source.

Gomes et al. [61] proposed a systematic way to divide a WDN into suitable DMAs based on the Floyd–Warshall algorithm [62] and user-defined criteria (e.g., pipe length and number of users in each DMA). This method facilitated the creation of appropriate DMAs by finding the shortest distance from source to nodes depending on the network flow direction at peak flow conditions. Compared with BFS, this algorithm provided superior results, as it considered the shortest path of sources to every other node in the network and identified the best path. The algorithm was repeated until the target number of DMAs and user-defined constraints were met. Further adjustment could be carried out by combining adjacent DMAs to minimize the number of boundary pipes as long as the user-defined criteria are fulfilled.

3.2. Community Structure Algorithm

The community structure detection algorithm is a bottom-up hierarchical approach based on graph theory and proposed by Newman and Girvan [63] and Clauset et al. [64]. It uses greedy optimization of a quantity known as modularity (Q), which is defined in Equation (1). They used the quality measure of network density to define the clusters, assuming that the density of a network division was effective if there were many edges within communities (intraclusters) and only a few between them (interclusters). Modularity index is a network property used as an indicator to quantify the quality of graph division in the community. The clustering method is based on maximizing the modularity index. Higher values of that metric are related to a community structure of the network, which is significant if $Q \geq 0.3$ [63,64]:

$$Q = \frac{1}{2m} \sum_{ij} \left[A_{ij} - \frac{k_i k_j}{2m} \right] \delta(C_i, C_j) \tag{1}$$

where $\delta(C_i, C_j)$ is the Kronecker delta coefficient, and $\delta(C_i, C_j) = 1$ if vertices i and j are the same community; otherwise $\delta(C_i, C_j) = 0$.

If we assume that the fraction of pipes that have both start and end nodes belonging to the same community is e_{ii}, and a_i is the portion of pipes with at least one end node in the community i, then the modularity can be formulated as:

$$Q = \sum_c e_{ii} - a_i^2. \tag{2}$$

The change in the two communities i and j to increase modularity can be computed by [63]:

$$\Delta Q = 2(e_{ij} - a_i a_j). \tag{3}$$

The community structure algorithm is implemented following the steps listed in Figure 4.

Step 1: Concept of WDN as a graph $G = (V, E)$.
Step 2: Define adjacency matrix **A** or weighted undirected adjacency matrix **W**.
Step 3: Start with each node as a community.
Step 4: Compute the change in modularity ΔQ resulting from merging each pair of communities.
Step 5: Merge the pair of communities with the highest value in modularity change ΔQ.
Step 6: Repeat steps 4 and 5 until one community remains.

Figure 4. Main steps for community structure algorithm clustering.

Diao et al. [19] first applied a community structure algorithm to detect clusters in a WDN. Their study used a community structure to automatically create boundaries for DMAs. WDN was mapped onto an undirected graph and community detection was implemented to maximize the modularity matrix and find the hierarchical community structure that represented the DMAs of the WDN. In the study, the authors determined the size constraint to be 300–5000 properties [10] for each community by applying a heuristic approach, known as oriented dendrogram cutting.

Instead of identifying network communities by maximizing the modularity index, Campbell et al. [34] proposed a procedure based on the idea that feedlines (i.e., a trunk network) should not be included in sectorization schemes. This was identified by means of determining the "betweenness" of edges, the flow, and the diameter analysis. The betweenness algorithm is a branch of graph theory that defines the edge (i.e., pipe) that connects to many pairs of vertices (i.e., nodes) [63]. A random-walk betweenness [19] can detect community segmentation with the highest modularity and a dendrogram can set the size constraint for each community.

Similarly, Ciaponi et al. [40] offered a different approach that combined convincing practical criteria when designing DMA as proposed by Morrison et al. [4]. Accordingly, automated identification of DMAs was performed by identifying the prevalent transport service (main transmission pipes) in WDNs and then each DMA, which was determined by the remaining distribution service pipes, was directly connected with the main transmission pipes. The procedure decomposed subsystems exceeding the threshold DMA size constraint owing to a modularity-based optimization algorithm. The two approaches brought the boundaries of identification of DMAs closer to reality and supported feasible alternative solutions to make more convincing decisions.

3.3. Modularity-Based Algorithm

The community structure algorithm uses a modularity index as a metric for the optimal design of DMAs. However, the modularity index may not be representative for the WDN because it is strongly affected by hydraulic properties (e.g., elevation, node demand, pipe diameter). Adopting the classic formulation of a modularity index without considering the physical and hydraulic constraints would therefore be artificial and misleading. Inspired by this approach, Giustolisi and Ridolfi [41] proposed a modularity-based method for WDN segmentation that accounts for hydraulic network properties to define WDN-oriented modularity. First, to formulate the modularity index for WDNs, the proposed method focused on conceptual segmenting of the network close to the ending nodes by using a topological incidence matrix and the number of pipes separating communities. This was done to minimize the number of required pipe cuts. Despite being tailored for a WDN, WDN-oriented modularity had an inherent limitation left over from the classic community detection algorithm. Fortunato and Barthelemy [65] stated that the modularity index proposed by Giustolisi and Ridolfi may fail to detect small communities if the community's total edge number is smaller than $\sqrt{2m}$, where m is the total number of edges in the network. To overcome such failures, Giustolisi and Ridolfi [66] proposed that an infrastructure modularity index can improve the negative effect of the inconsistency of modularity optimization. A new index is released through maximization of the classic modularity index in the framework of the two-objective optimization, modifying the framework to overcome the resolution limit. Laucelli et al. [67] took a further step by developing a flexible procedure for DMA planning based on Giustolisi and Ridolfi's achievement with a conceptual cut for segmentation. A two-step strategy was adopted for optimal sectorization design by maximizing the WDN-oriented modularity index versus minimizing the number of conceptual cuts, where the location of pipe cutting minimizes the number of devices to be installed. To determine the location of flow meters and gate valves, DMA design was optimized based on each conceptual cut and returned an optimal solution for each one, accounting for hydraulic behavior change in the network with respect to maximizing the reduction of background leakage in each DMA. Using the WDN-oriented modularity index, Simone et al. [68] developed a sampling-oriented modularity index to perform optimal spatial distribution and assess the optimal number of pressure meters needed in a network (i.e., sampling design) using a multi-objective optimization method to minimize pressure-meter cost versus sampling-oriented modularity.

As mentioned in Section 2, DMAs are designed to detect and actively manage leaks. To that end, pressure management is a fundamental and important factor affecting leak management. Zhang et al. [42] developed a hybrid procedure by combining node pressure with modularity-based community detection to segment a network into similar DMAs from a pressure aspect. However, to improve the resolution limits of classical modularity, they used a random-walk theory similar to that of Campbell et al. [34]. The random-walk theory allows for precise identification of communities with greater or smaller differences in size and the automatic creation of a multiscale community [42]. To illustrate the superiority of this method over previous methods, the results proposed by Diao et al. [19] were compared. They demonstrated that different partition schemes result at a variety of random-walk time periods because the variances of node pressure are integrated into the community. In Diao et al. [19], variance was made immutable using a top-down search. Additionally, in the aspect of boundary pipes proposed by the two respective methods, Zhang et al. [42] showed that the traditional modularity-based community detection introduced by Diao et al. had more boundary pipes.

Most recently, Perelman et al. [37] combined three branches of graph theory to evaluate the performance of each method. Global clustering, community structure, and graph partition were applied to two WDNs in Singapore. Global clustering is a bottom-up algorithm for grouping points concerning a measure of similarity defined for each pair of points. Community algorithms detect the community structure in the network focused on the concept of edge betweenness. Graph partitioning divides the graph into a predefined number of groups such that the number of edges crossing between the groups is minimal [69]. The authors showed that the methods were compatible and applicable to large networks, but the performance of each method was completely different and depended on the number of clusters and the parameters selected for evaluation. They proposed multi-criteria metrics based on visual and quantitative performance measures. Accordingly, a better approach would be to minimize four metrics, such as (a) worst cut size, (b) total cut size, (c) cluster size, and (d) running time, and maximize the metric in regard to (e) recurrence of inter-cluster edges [37]. The results demonstrate that graph-partitioning generally outperforms clustering and the community structure methods in terms of (a), (b), and (d), which implies that the number of flow meters needed to monitor the flow will be minimized. On the contrary, the global clustering method indicated a good expectation in terms of (e), while in terms of (c), the three methods showed similar results. Therefore, community structure and the graph partitioning methods were more flexible and outperformed global clustering under particular budget constraints.

Similarly, Di Nardo et al. [70] conducted a comprehensive analysis of two popular clustering algorithms, such as the graph partitioning based on multilevel recursive bisection (MLRB) and the spectral clustering based on the normalized cut algorithm. Applications to a real-life WDN in South Italy revealed that the graph partitioning outperformed the spectral clustering in balancing the number of nodes in each DMA. On the contrary, the spectral algorithm showed better performance than the graph partitioning to minimize the number of edges cuts, thus it was more efficient in both hydraulic and economic aspects. A similar study conducted by Liu et al. [71] explored the performance of three partitioning methods, including fast greedy [64], random walk [63], and multilevel recursive bisection (MLRB) [72] using a spectrum of topology-based indicators.

As mentioned earlier, WDNs exhibit dynamic hydraulic behavior changes in the spatial and temporal mode that are completely different compared to others. Most of the partitioning algorithms lack exhaustive analyses of the similarity of the hydraulic and physical aspects in DMAs, such as the number of nodes and balance in terms of water demand and pressure. It is therefore not sufficient to offer a universal O&M solution to a utility. Awareness will make DMA segmentation more reliable when physical properties and hydraulic behavior are considered in network partitioning. Realizing the limitations of Diao et al. [19] and Ciaponi et al. [40], Creaco et al. [73] incorporated engineering aspects (i.e., demand supplied along the pipe and pipe length) into WNP processes. However, unlike Giustolisi and Ridolfi [41,66], they focused on applying heuristic procedures to improve the original fast greedy partitioning algorithm to maximize the modularity index developed by Clauset et al. [64]. Two heuristic optimization techniques were developed and applied to the formulation of modularity to perform different merging combinations. In the first technique, randomness was added to the DMA merging process, which allows for the acquisition of numerous WDN-partitioning probabilistic solutions while generating a higher modularity increment during the merging steps and a lower number of boundary pipes compared with the traditional deterministic approach. The second technique illustrated the trade-offs between various engineering aspects by embedding the former technique inside a multi-objective genetic algorithm optimization [74].

Evaluation of DMAs scenarios after sectorization must also guarantee that hydraulic indicators are at an acceptable or higher threshold compared with the original network. Because different criteria lead to various DMA layouts, Brentan et al. [43,44] proposed a method that considers the relationship between many technical criteria, such as demand and pipe length, to create different DMA scenarios. The social community detection algorithm was used to define DMAs. To assess the performance of DMA generation, a comprehensive analysis was proposed that considered performance indicators such as resilience index, demand similarity, pressure uniformity, water age, cost, and energy consumption, hopefully provides decision-makers with an optimal DMA configuration.

3.4. Multilevel Graph Partitioning

Multilevel partitioning [72] is a fundamental approach based on an analogy of graph theory and graph-partitioning principles that uses parallel computing to allocate workloads among processors to minimize communication and equally distribute the computational burden among them. Based on that approach, the objective is to create subzones by equally distributing loads, such as DMA size, pipe length, water demand, and flow [45]. Recently, much effort has been devoted to developing techniques and heuristic procedures for optimal segmentation of a water network into isolated DMAs by balancing pipe length, nodal demand, and flow within each DMA [16,37,45,46].

Sempewo et al. [45] presented an automated prototype tool for the analysis of network spatiality to create analogy subzones based on balancing pipe length and demand at each zone using distributed computing called multilevel recursive bisection (MLRB) for monitoring and controlling leakage in the WDN. The core purpose of the MLRB algorithm was to design a highly effective method to deal with parallel *k*-way partitioning of a graph in computer science. The successor application of MLRB was mentioned by Di Nardo et al. [17,18,75]. They proposed a procedure adapting the traditional phase of the MLRB to create an automated tool for smart water network partitioning (SWANP). In the MLRB algorithm, three phases illustrating the computation of *k*-way partitioning in the graph are: coarsening, partitioning, and uncoarsening with refinement [72]. The coarsening phase simplifies the original graph by collapsing adjacent vertices in terms of maximally matching a graph with different techniques. The next phase is partitioning. First, the network is subdivided into a two-way partition. Each subgraph is then divided into bisections to obtain *k*-way partitioning. The boundary pipes that have the start and end nodes in different subgraphs must be minimized for associated weights as well. Finally, the uncoarsening phase, also known as the recovering and refining process, is completed by returning to the graphs in the first phase to reconsider constituent nodes. During each recovery level, a local refinement optimization of the partition is applied to obtain more equal districts. Figure 5 visualizes the processes in the MLRB algorithm.

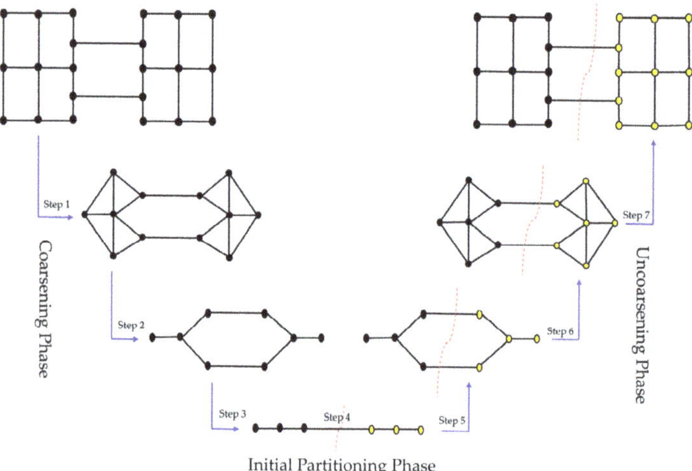

Figure 5. Different phases of a multi-level recursive bisection algorithm [72].

Alvisi [46] proposed a procedure for automated network sectorization using a combination of the MLRB graph partitioning algorithm and hydraulic simulation. However, unlike the traditional approach, it can simultaneously allocate the nodes to perform the best network partitioning into a given number of DMAs and identify the best locations for flow meters and valves in the network. Perelman et al. referred to a similar application [37] in which they used MLRB to compare performance in a WNP with other approaches, such as community structure and global clustering. This approach can assign weights to nodes and pipes that would otherwise not have been considered for the graph-clustering algorithm where the network clustering phase was considered a topology characteristic. Moreover, in terms of computational efficiency, MLRB showed advances in the uniform allocation of computation processes, considering more than one object simultaneously while minimizing the volume of information exchanged between them.

3.5. Spectral Graph Algorithms

Spectral graph theory is a mathematical approach to study the relationship of graph properties by associating both linear algebra and graph theory to determine the eigenvalue and eigenvector properties. Spectral clustering uses the spectrum in eigenvectors of the adjacent matrix to cluster groups of points into communities [49,54]. In spectral clustering, the row of eigenvectors of the Laplacian matrix for a pair of the nodes is similar if the nodes belong to the same cluster. Spectral-based graph clustering has been implemented in many fields over the last decade, especially in computer sciences, bioinformatics, and data analysis. Recently, in the field of WDN management, spectral graph theory has been used to define an optimal cluster configuration, a preliminary analysis of network vulnerability, and robustness through graph matrices eigenvalues [76,77] and a toolset for WDN management has been proposed [49]. Several types of research have applied spectral graph theory for WDN management, but in this subsection we focus on effective approaches that have been proposed for water network clustering. The core idea of spectral clustering is the Laplacian matrix as defined by three equations. The first is the non-normalized Laplacian relationship, which solves a relaxed version of the Ratiocut problem proposed by Von Luxburg [78]:

$$L = D - A, \tag{4}$$

where $\mathbf{D}$ is a diagonal matrix of nodal degrees k_i, $\mathbf{D} = diag(\mathbf{d})$, in which $\mathbf{d} = [k_1,\ k_2, \ldots, k_n]^{\mathrm{T}}$. $\mathbf{A}$ is the adjacency matrix.

The other two matrices are normalized graph Laplacians, which are closely related and can be defined as

$$L_{sym} = D^{-1/2}LD^{-1/2} \text{ and} \tag{5}$$

$$L_{rw} = D^{-1}L, \tag{6}$$

where L_{sym} is a symmetric matrix proposed to solve the NCut problem [78] and L_{rw} is closely related to a random walk, which can be used to solve the same problem.

The Laplacian matrix of an undirected graph has the following properties [78,79]

- $\mathbf{L}$ is symmetric and positive-semidefinite with eigenvalues $\lambda_i \geq 0$ for all i.
- Every row sum and column sum of $\mathbf{L}$ equals zero.
- The smallest eigenvalue λ_1 of $\mathbf{L}$ equals zero.
- $\mathbf{L}$ has n non-negative and the number of connected components in the graph equals the algebraic multiplicity of $\lambda_1 = 0 \leq \lambda_2 \leq \cdots \leq \lambda_n$.

Aim of spectral graph partitioning is to divide graph G into $p \leq n$ subgraphs $G_1, G_2, \ldots, G_p$. Then,

$$V = V_1 \cup V_2 \cup \cdots \cup V_p \text{ where } V_i \cap V_j = \varnothing, \ i \neq j, \tag{7}$$

Let $G_k = (V_k, E_k)$ represents for subgraph k, in which $k = 1, \ldots, p$ and V_k is the set of vertices of subgraph G_k. From Equation (7), an edge that has its endpoints in different vertex subsets is not contained in any of the formed subgraphs G_k and is called an intercluster edge. Let a set of the intercluster edges with one endpoint in V_k be denoted as Equation (8).

$$\partial(V_k) := \{ij : i \in V_k \text{ and } j \notin V_k\}, \tag{8}$$

Two different sets of edges can thus be distinguished as follows.

- intracluster edges: $E_1 \cup E_1 \cup \cdots \cup E_p$, and
- intercluster edges: $\partial(V_1) \cup \partial(V_2) \cup \cdots \cup \left(V_p\right)$.

From the optimal bipartitioning of a graph point of view, minimizing the *cut* values are objective functions. Von Luxburg [78] and Shi and Malik [80] proposed these functions to optimize *cut* value, called ratiocut method and normalized cut method, in Equations (9) and (10), respectively.

$$\min_{V_1, V_2, \ldots, V_p} \sum_{k=1}^{p} \frac{vol(\partial(V_k))}{|V_k|}, \tag{9}$$

$$\min_{V_1, V_2, \ldots, V_p} \sum_{k=1}^{p} \frac{vol(\partial(V_k))}{vol(V_k)}, \tag{10}$$

where $vol(\partial(V_k))$ is the sum of the weights on the all intercluster edges in $\partial(V_k)$; $|V_k|$ is the number of vertices in V_k; and $vol(V_k)$ is the sum of the weights on the vertices in V_k.

Equations (9) and (10) are NP-complete problems, however, they can be relaxed to find approximate solutions proved by Von Luxburg [78] and Shi and Malik [80] and reformed as Equations (11) and (12) following

$$\mathbf{LU} = \mathbf{U\Phi} \quad \text{for ratio cut} \tag{11}$$

$$\mathbf{LU} = \mathbf{DU\Phi} \qquad \text{for normalized cut} \qquad (12)$$

where $\mathbf{\Phi} := \mathrm{diag}\left(\lambda_1, \lambda_2, \ldots, \lambda_p\right) \in \mathbb{R}^{p \times p}$ and $\mathbf{U} := \left[u_1, u_2, \ldots, u_p\right] \in \mathbb{R}^{n \times p}$.

Equations (11) and (12) are eigenvalue problems for p smallest eigenvalues $\lambda_1 = 0 \leq \lambda_2 \leq \cdots \leq \lambda_p$ of the Laplacian matrix $\mathbf{L}$ and their corresponding eigenvectors $u_1, u_2, \ldots, u_p$.

The spectral clustering algorithm for a non-normalized Laplacian matrix can be described as shown in Figure 6. For other normalized spectral clustering, refer to Reference [78].

Step 1: Construction of WDN as a graph $G = (V, E)$.

Step 2: Define the adjacency matrix $\mathbf{A}$ or weighted undirected adjacency matrix $\mathbf{W}$.

Step 3: Compute the Laplacian L as Eq. (4).

Step 4: Compute the first k eigenvectors $u_1, \ldots, u_k$ of L.

Step 5: Delineate matrix $\mathbf{U}$ containing clustering information and the first k eigenvectors as columns.

Step 6: From the $\mathbf{U}$ matrix, the $k - mean$ algorithm is used to cluster nodes of the network into clusters $C_1, \ldots, C_k$ for each row.

Step 7: Continuity checking of the obtained clusters C_k. The links that have start nodes and end nodes belonging to different clusters are defined as the set of edge cuts, N_{ec} (i.e., boundary pipes).

Figure 6. Flowchart of a non-normalized spectral clustering algorithm.

Using the spectral clustering methods mentioned above, many studies have adopted the spectral graph theory [47–49,81]. Di Nardo et al. [47] defined the optimal layout of DMAs in a real WDN. The authors took into account both geometric features (i.e., connectivity) and hydraulic pipe features (i.e., diameter, length, conductance, flow) through weight-adjacency matrices, which led to significantly different layouts of the DMAs. In particular, they compared different weighted spectral clustering (i.e., normalized versus non-normalized Laplacian) to determine the effectiveness of those approaches and the optimal choice of weights.

One of the most useful approaches for handling WDN complexity is a graph spectral technique (GST). Di Nardo et al. [49] pointed out that GST can analyze network topology by taking advantage of the properties of some graph matrices, providing a complete toolset to evaluate the performance and the evolution of networks. Based on two graph matrices (i.e., adjacency and Laplacian), the authors highlighted that GST metrics and the algorithms accomplish some crucial tasks of WDN management using topological and geometric information. In addition to the inherent ability to define the optimal clustering layout proposed in the literature, GST assisted in the calculation of a surrogate index for assessing topological WDN robustness using two indices, such as spectral gap and the algebraic connectivity. Additionally, the spectral technique also provides a framework that ranks important nodes in WDN to provide a useful approach to identify the location of valves or sensors or even determine the most influential nodes in a network [49].

Similarly, Liu and Han [48] also proposed a strategy for automatic DMA design based on spectral clustering and graph theory. The spectral algorithm was used to determine the best node clusters, which correspond to the DMAs' configuration based on steady-state simulations using the peak-hour demand. The study proposed a method for DMA design that combines spectral clustering, graph theory, and network centrality analysis. First, a combination of graph spectral theory and k-mean clustering was implemented to generate the initial DMAs. Then, to improve the cluster quality, a genetic algorithm (GA) was added to converge on a global optimum. To measure network centrality, the eigenvector centrality [82] was used to identify the critical nodes, and edges betweenness centrality [83] were adopted to measure the important pipes, creating a high-quality cluster. Most recently, Zevnik et al. [81,84] proposed spectral graph partitioning based on a generalized and normalized cut method and compared it with two known spectral methods (ratio cut and normalized cut).

In the field of machine learning, graph Laplacians are used not only for clustering, but also for many other tasks, such as semi-supervised learning. Herrera et al. [52,85] demonstrated that graph-based semi-supervised learning methods [86] can take into account various criteria for segmentation of WDNs into DMAs. In this method, the kernel matrix [87] was first defined and then the adjacency matrix was enriched by adding hydraulic data such as weight factors to transform the results into a kernel matrix. The spectral clustering algorithm was adapted to this new matrix. Finally, graph-based semi-supervised learning methods were conducted. A similar method was found in a study by Giudicianni et al. [36,88] in which semi-supervised multiscale clustering was used to create dynamic DMAs. Compared with methods that use only topological connectivity or vector information, semi-supervised clustering showed improvement by integrating both forms, leading to the efficient development of robust DMAs.

Spectral clustering can take into account topological, geometrical, or hydraulic aspects as weight factors, which allows for a careful consideration of alternative factors that can affect the goals of the DMA design and provide a multidimensional view to help managers make better decisions. However, for large-scale networks that have thousands of nodes and links, spectral clustering has limited applicability and becomes infeasible due to the computational complexity of $Q(n^3)$ [52], where n is the number of nodes.

3.6. Multi-Agent Approach

A multi-agent system (MAS) [89] is a loosely coupled network of autonomous problem solvers composed of multiple interacting intelligent agents. Each agent works independently but can also interact with others to solve potential conflicts through negotiation. The properties of MAS can be described as follows.

1. Each agent has an imperfect standard or may lack the capacity for problem solving, and therefore has a somewhat limited and unbalanced perspective;
2. There is no global information;
3. Data is decentralized; and
4. Computation is asynchronous.

MAS networks are suitable for handling multiple-problem approaches or multiple-agent solving entities. Known as a complex system due to the joining of many physical devices, a WDN comprises multiple parties with different goals, actions, and information and is a dynamic system. A small change in behavior of the parties may result in unpredictable patterns in the entire system. WDNs and multi-agents exhibit a strong similarity, and MASs can therefore provide solutions to distributed applications, such as the problem of network partitioning, which is known to be complex and has multiple constraints. MASs have been successfully applied to heterogeneity problems in the water field. They have proven to be highly efficient at optimizing water networks, control systems for municipal water, water pollution diagnosis, water quality enhancement, and water demand management [90].

In terms of WDN clustering, many elements must be simultaneously considered. A network can be divided into elements, which are considered as agents that communicate with each other. Izquierdo et al. [90] were the first to develop a suitable software environment to formulate DMA segmentation in a WDN using a multi-agent approach. They proposed a likelihood method by running a simulation as verification to divide networks into subsectors based on sources, nodes, and pipe properties, which consider nodes and pipes as agents of a separate breed. This can be seen as a premise to improve as well as implement the multi-agent methods in different studies. Herrera et al. [52] assumed an a priori set of DMAs based on the homogeneity of the districts, which was related to the source tanks in the network, where each reservoir was seeded for the corresponding DMAs. These agents adopted a method of clustering by elicitation, linking their adjacent nodes to the source points, and scanning the likelihood of each being assimilated into the corresponding DMA.

On the other side, Hajebi et al. [51] combined a *k*-means clustering method and multi-agent approach to WNP. In particular, *k*-means graph clustering was used to divide the network topology into a predetermined number of clusters and then a MAS was implemented to negotiate the configuration of the network by adjusting nodes on the boundary pipes of the corresponding clusters while considering the hydraulic constraints. Compared with previous studies by Herrera et al. [50], differences in the approaches are evident. In the former study, DMA layouts were determined based on the source points of the network and expanded by negotiation, while the latter started from the geographical clustering of the network and boundary pipes were modified to obtain the best hydraulic performance.

4. Sectorization to Locate Flow Meters and Valves

Immediately after forming DMAs from a clustering phase, it is important to optimize the state of the boundary pipes, namely the position and number of gate valves and flow meters required to achieve reliable DMA operation. This is also known as a decision support tool to help utilities solve optimization problems while investigating the best trade-offs between the sectorization cost of investment versus indicators of the benefit of DMA installation. The position of these devices is important because closing a valve impedes hydraulic behavior and reduces network reliability. After segmentation of a network into districts, the standard requirement of a network is still to ensure adequate quantity and quality as well as appropriate pressure. Many algorithms and heuristic procedures have been proposed to find the optimal solution for this phase, which is concerned primarily with optimizing hydraulic performance and leak-reduction efficiency.

Many heuristic procedures are available to maximize the benefits of physical demarcation of a water network into DMAs, but this work has been implemented largely with evolutionary algorithms, which include single or multi-objective functions and are constrained by hydraulic or economic conditions. In this section, we focused on the optimization approaches to discuss the sectorization phase in WNP.

4.1. Single-Objective Optimization Approach

After a set of boundary pipes N_{bp} are defined in the clustering phase, the first objective is to determine how many flow meters N_{fm} and gate valves N_{gv} to insert along the boundary pipes. Most researchers agree that fewer flow meters will reduce reconstruction and operating costs, as well as the initial cost of installing the flow meters, which are often more expensive than gate valves [6,17,47,49].

In addition, the positions of the gate valves and flow meters have a significant effect on network properties such as hydraulic performance, resilience index, leakage rate, and water quality. WDN sectorization should therefore be considered as a multi-objective optimization problem to maximize the benefits of implementing DMAs. However, to simplify computational demands, some hypotheses or heuristic processes have been proposed to convert a multi-objective problem into a single objective and apply evolutionary algorithms to achieve feasible or optimal solutions.

Because the number of feasible solutions is large, various heuristic optimization techniques have been studied [6,91–94]. Although the objective functions and constraints are different among the various approaches, they all aimed to achieve as high as possible a network performance after sectorization. The total power of a WDN is classified into the dissipated power at pipes (i.e., internal power loss) and the supplied power at node (i.e., external power supplied). Di Nardo et al. [6,17,18,49,92,95,96] suggested the objective function to maintain the hydraulic performance of the network at the lowest dissipated power that consequently maximizes the nodal supplied power by maintaining the nodal head as high as possible after sectorization. The objective was defined in the following equation.

$$max\left(\gamma \sum_{i}^{n} (z_i + h_i)Q_i \right),$$
(13)

where γ is the specific weight of water and z_i, h_i, and Q_i are the elevation, pressure, and water demand at node i, respectively. For a large and complex network, it is not easy to decide how many flow meters should be positioned among boundary pipes due to the trade-offs between hydraulic performance and investment cost. To deal with this problem, Shao et al. [97] proposed a function that converted a dual-objective problem (i.e., hydraulic performance and cost) to a single-objective problem by considering the master-subordinate relationship of the two objective functions, which improved the computational efficiency.

In addition, a changing flow due to pipe failure can cause changes in velocity, energy losses in pipes, and pressure at nodes, especially in a looped network. This will cause changes in the pathway of water particles to the nodes. Moreover, if a node is being supplied at the minimum required pressure, it will not be able to provide the necessary flow and pressure. In the worst-case scenario, the network must ensure a capacity to provide a surplus power to overcome system failures. This is an approach proposed by Todini [98] to measure system resilience when redesigning a system or when system malfunction occurs. Based on that criterion, when reconstructing the system by creating isolated DMAs, several studies [34,36,46,60] have used a resilience index (Equation (14)) as an objective function for sectorization optimization. The objective function can be maximized to indicate that a greater surplus of available power leads to a higher network resilience such that:

$$I_r = \frac{\sum_{i=1}^{n_n} Q_i(h_i - h_{min})}{\sum_{r=1}^{n_r} Q_r H_r - \sum_{i=1}^{n_n} Q_i h_{min}}, \tag{14}$$

where n_n and n_r are the numbers of demand nodes and reservoirs, respectively; Q_i and h_i are water demand and pressure at node i; Q_r and H_r are the water discharge and total head of the source or tank r; and h_{min} is the minimum required pressure for adequate service.

For cost analysis, Gomes et al. [99] proposed an optimization model to design DMAs based on different decision-makers' options to reduce the total cost. Referring to different future scenarios for water demand (that will increase) and the infrastructure degradation forecasts, the cost of WNP was assessed. An objective function aims to minimize the cost of DMA redesign and first considers the cost of pipe reinforcement or replacement with flow meters and gate valves. Second, to ensure that the model approximates reality, the cost function is multiplied by the weight or probability of occurrence for each of the scenarios. A similar study that considers economic and energy criteria for DMA design can be found in Di Nardo et al. [75].

For reducing the leakage in WDNs, Creaco and Haidar [100] proposed a linear programming framework to optimize control valve settings. Accordingly, isolation valve closures, control valve installations, and DMAs creation are simultaneously optimized to search optimal solutions in the trade-off between installation costs, leakage, and demand uniformity across DMAs.

To solve the optimization problems mentioned above, evolutionary search algorithms have been applied. The GA [101] has been widely implemented by Di Nardo et al. [6,47,75,91,95]. Meanwhile, Shao et al. [97] improved the GA for faster and superior layout of flow meters and valves by modifying crossover and mutation mechanisms. In addition, the simulated annealing algorithm was presented by Gomes et al. [61,99,102].

4.2. Multiple-Objective Optimization Approach

WNP is a complicated task and must achieve many goals. Zhang et al. [42] proposed a multi-objective optimizing approach for sectorization, in which three objective functions were used: the number of boundary pipes, network pressure uniformity, and water age uniformity. Zhang et al. [103] proposed a multi-objective optimization to obtain more reasonable schemes for sectorization of a WDN by simultaneously considering pressure stability, water quality safety, and system reconstruction costs. For pressure stability, the average pressure was minimized, but was still above the minimum pressure. Water age is the time spent by a water parcel as it travels from a source to nodes in the network, which represents the water quality in a WDN. Additionally, the costs of installing flow meters and valves should be minimized depending on the size and number of DMAs. However, if only the initial investment cost of these devices is considered, it is impossible to comprehensively evaluate the expense of the WNP. De Paola et al. [104] presented an objective function to deal with the total cost of sectorization, which also involves water leakage costs and energy consumption by pump operations.

Even if we consider all the criteria in the process of network sectorization, it is impossible to provide an optimal result due to trade-offs. The current partitioning techniques prioritize only a few representative sets of criteria, and do not fully address the best practical problems of DMA design. In an attempt to provide a comprehensive review of the criteria when dividing the network as close as possible to reality, Hajebi et al. [105] considered two sets of objectives in the sectorization task, the structural objective and hydraulic objective. For the structural objective, they considered the minimum cut size and minimum boundary pipe diameter. For the hydraulic objective of the network after segmentation, they considered four objectives, including minimization of the average excessive pressure at nodes, minimization of dissipated power, minimization of elevation differences in each DMA, and maximization of network resilience.

In another combination, a series of energy, operative, and economic criteria were optimized in the sectorization process [58,59]. Five objectives were addressed concerning the minimum deviation of the resilience index [95], which measures the capacity of the network to conquer system failures, the ability of the system to ensure an appropriate service pressure in the whole network, and minimization of the variation of the operational power, which assessed the reliability of a sectorization layout based on a pressure target. Operative criteria were also formulated as objective functions of pressure at nodes. When pressure dropped, a reduction in the leakage was expected. Variation of nodal pressure should therefore be minimized. However, pressure at nodes after sectorization needs to be higher than the minimum threshold required for service. To accommodate this constraint, a penalty cost for a nodal pressure deficit was added. Finally, the cost criteria when performing sectorization are also important. The cost of positioning and operation of flow meters are expected to be higher than that for boundary valves. Therefore, an objective function to minimize the cost for installing and operating the devices needs to be considered. Similarly, Brentan et al. [43] adopted a multilevel optimization concept to reduce the complexity of sectorization. In their approach, two groups of the objectives were minimized. The first one corresponded to structural costs, which were related to valve and flow meter installation, while the second group reflected hydraulic performance, such as minimum pressure and maximum resilience index.

Giudicianni et al. [88] recently developed a heuristic framework for dynamic partitioning of WDNs using multi-objective functions to address different goals for saving energy, water, and costs. Specifically, they proposed a method for zero-net energy management of a WDN using microhydropower stations [106] along the boundary pipes during the day and a reduction of water leakage at night.

To provide a comprehensive method for optimal DMA design, Galdiero et al. [107] proposed a decision-support tool that focused on water network segmentation by considering two objective functions. A total cost function including the initial cost for device investment and a daily cost due to water leakages were considered to minimize and compared as trade-offs with changes in hydraulic performance in terms of the maximum resilience index. To integrate different algorithms and multi-objective functions to the development of a decision support tool, Di Nardo et al. [18] developed advanced software called SWANP. A clustering model was implemented based on MLRBs, which are multiagent approaches to water network clustering. In the sectorization phase, an optimization algorithm was proposed using multi-objective functions to find optimal DMA configurations that complied with the level of customer service and considered the minimum pressure and maximum resilience index, and balanced the cost of investment and operation by minimum devices inserted to achieve isolated DMAs. SWANP was written on a Python environment with a user interface and was evaluated as an effective decision support system providing the manager with different optimal layout solutions.

Many optimization algorithms have been applied to deal with discrete nonlinear combinations and solve the multi-objective optimization of water network sectorization. NSGA-II [74] has been widely applied to multi-objective optimization problems. In terms of water network sectorization, NSGA-II has been used in many studies [103–105,107] to obtain the Pareto front, which contains a set of Pareto optimal solutions, thus providing support for managers charged with making more accurate and reasonable decisions based on their priorities and objectives. Zhang et al. [42] implemented an auto-adaptive many-objective algorithm [108] to solve the sectorization problem that shows some new features compared with NSGA-II. Giustolisi and Ridolfi [41] used a multi-objective GA to support network segmentation. Campbell et al. [58], and Gilbert et al. [94] applied an agent-swarm optimization algorithm [109]. In addition, the combination of three optimization algorithms of GA, particle-swarm optimization [110], and soccer-league competition [111] was suggested in Brentan et al. [44].

4.3. Iterative Approach

In addition to the described optimization methods, iterative methods were applied to the placement of flow meters and valves [19,41,48,96]. An iterative method is a mathematical procedure that can generate a feasible solution using an initial guess to generate a sequence of solutions. The result is considered convergent when the initial set of criteria is met. Diao et al. [19] considered DMA size and minimum pressure as criteria, and used them as constraints in the heuristic-based iterative method to define the feedlines for each DMA. The approach determines the location of flow meters among boundary pipes between DMAs. In addition, Liu and Han [48] proposed an iterative method based on a heuristic procedure to determine the best location of flow meters subject to constrain head pressure at nodes. The iterative method permits the selection of one flow meter based on the shortest path from the source that can improve the pressure in each iteration. Di Nardo and Di Natale [96] inserted a certain number of flow meters on boundary pipes and then designed a procedure to alternately change the quantity and position of flow meters to achieve an optimal solution based on hydraulic performance constraints testing.

4.4. Adaptive Sectorization for Dynamic DMAs

In normal working conditions, a DMA layout is permanent and optimized by WNP processes to satisfy the hydraulic constraints and network performance indices. In abnormal cases, such as pipe breaks, fire-fighting, and unexpected increases in water demand, permanent DMAs may produce failures in preserving or maintaining regular water supplies. To adapt to such conditions and overcome the drawbacks that a permanent DMA can cause, Giudicianni et al. [36,88] proposed creating dynamic DMAs that allow for expansion of existing DMAs. That is, the small DMAs are dynamically aggregated into larger ones using a semi-supervised clustering algorithm. This approach allows for a new configuration that always includes former DMAs and maintains the set of boundary pipes at each subzone. In some cases, by controlling the dynamic gate valves, the operator restores connectivity to its original configuration and consequently helps the utility periodically desegregate.

In addition, Wright et al. [112] proposed a method of integrating the advantages of DMAs in reducing leakage while improving network resilience and water quality by dynamically reconfiguring network topology and pressure control through optimizing valve settings and boundary pipes status using a sequential convex programming approach. The proposed approach leaned on the self-powered multifunction network controllers that allowed adjustments of the network topology and continuously monitored the dynamic hydraulics based on consumers' actions (i.e., the varieties of the system's water demands). In low demand periods, original DMAs were preserved to capture the minimum night flow within small isolated areas and maximize the ability to detect leaks. In peak demand periods, DMAs were then aggregated into larger pressure-controlled zones to maximize the resilience index and improve energy efficiency due to reduced internal losses that come with using larger DMAs. The core idea here was inserting the network controller associated with dynamically reconfigurable DMAs that allows a utility to monitor high-resolution, time-synchronized, dynamic pressure conditions of the network. Similarly, Perelman et al. [113] used a linear programming approach to automate reconfiguration of an existing WDN into DMAs. The network was reorganized into a star-like topology by identifying and decomposing the existing network into a main network and subnetworks based on graph theory. Center nodes were located in main pipes and played an important role as key connections between transmission main pipes with water sources and other nodes in the rest of the network. The proposed method provided a flexible tool for water utilities by allowing only existing valves to be closed, saving investment and operation costs for additional valve installations.

Ideas derived from DMAs' limitations in emergencies, especially in the case of fire-fighting, overcome this drawback. Di Nardo et al. [18,114] recently proposed a method that allows for the redesigning of static DMAs to dynamic layouts. A heuristic procedure based on a GA was developed to determine the number of gate valves that have to be motorized and remotely controlled to satisfy hydraulic performance in a fire-fighting event. This practical technique provided system operators with a quick decision-making tool to respond to unexpected incidents in the network and eventually leads to a smart water management paradigm. Unlike the approaches proposed above, Santonastaso et al. [115] developed a dynamic scheme for adjusting a WNP by accounting for the real positions of isolation valves present in the WDN. To do this, the adjacency matrix of the WDN was changed and replaced with a dual topology based on WDN sectorization and isolation valves. DMAs obtained in this approach allowed topology matrix segments to merge while inter-DMA boundary pipes were forced to be selected among the valve-fitted pipes that separated segments. Feasible DMAs were generated that did not require additional isolation valves.

To visualize the procedure of WNP technique in the abovementioned different phases, Figure 7 illustrates the procedure of WNP for a real-life WDN in Parete town, South Italy [47,49]. In this case, 4 DMAs were generated in the clustering phase based on the normalized spectral algorithm, and a heuristic procedure based on GA is applied in sectorization phase to locate the control devices while maximizing the total nodal power of the network.

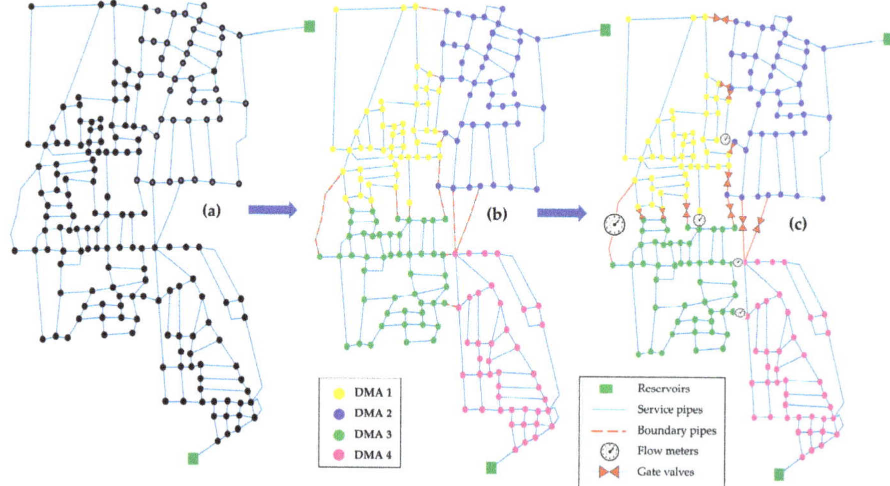

Figure 7. Illustration of the water network partitioning (WNP) procedure implemented to a water distribution network (WDN) in Parete, South Italy (adapted from Di Nardo et al. [47,49]); (**a**) original water network, (**b**) clustering phase, and (**c**) sectorization phase.

5. Performance Assessment of Water Network Partitioning

As mentioned in Section 2, the water network segmentation to DMAs is expected to bring many benefits along with effective reduction measures for invisible water losses, manage pressure uniformity, and prevent network contamination. However, in some cases, it may also decrease the hydraulic performance and reliability of the network. To measure how this change affects network hydraulic behavior, performance indices (PIs) can quantify the benefits and drawbacks that DMAs bring. A PI test allows for the evaluations of the performance of the original networks compared with those of the divided networks. Most were estimated using a hydraulic simulation solver based on the demand-driven analysis. Most studies applied multiple PIs to evaluate the effectiveness of DMAs, such as resilience indices, pressure indices, uniformity indices, water quality indices, and fire protection indices.

The first to be mentioned is the resilience metric, which monitors the power balance of a water network, as proposed by Todini [98] in the form of Equation (14). According to this metric, WDN resilience is defined as the capacity to overcome sudden system (hydraulic or mechanical) failure. The resilience index is often used to evaluate the performance of a WNP as a comparison of network power before and after the sectorization. Most studies have stated that the resilience index is not significantly affected by network sectorization compared to its benefits [3]. Herrera et al. [116] proposed a graph-theoretic approach by adopting the K-shortest paths algorithm [117] to assess the resilience of larger-scale partitioned WDNs. To do this, all nodes in every DMA are aggregated into a sector-node, where a new DMA-graph is represented by sector-nodes and edges that are abstracted by sector-to-sector connectivity. A mapping function was used to transform the resilience of nodes to a sector-scale resilience. They showed that the resilience of individual nodes in the DMAs closely links to the corresponding sector-nodes resilience. This establishes a different way to identify DMA configurations that have a major impact on the resilience index.

If a resilience index evaluates the overall performance of the network, hydraulic statistical indices allow for the evaluation of the level of service that a water system supplies to its customers, providing managers information on pressure change in terms of mean, minimum, maximum, and spatio-temporal deviations. More specifically, Di Nardo et al. [118] proposed several indices, such as a mean pressure surplus and mean pressure deficit compared with the design pressure.

On the other hand, water quality is measured by water age in a network and is influenced by network topology, flow velocities, and pipe lengths. The age of the water affects residual chlorine levels. Lower chlorine induces bacterial growth, and higher values indicate worse performance. Many studies used the water age index as an indicator to assess the impact of DMA design on water quality. Grayman et al. [10] and a series of studies by Di Nardo et al. [118] illustrate that after incorporating DMAs into a WDN, there was no systematic difference in the computed average water age between alternative scenarios. Although there can be significant variations in water age by node due to valve closures, when considered as a whole, no homogeneous difference was found.

Meanwhile, partitioning a WDN into subnetworks with gate valves can prevent the spread of contamination in the case of malicious attacks. Di Nardo et al. [28,119] proposed a method that uses a simple backflow attack with cyanide to investigate the effects of network partitioning. Grayman et al. [10] proposed an index to quantify the potential health impacts from contamination incidents in the WDN.

Several hydraulic uniformity indices [44,48,120,121] have been developed to evaluate the performance of DMAs. The size uniformity index reflects the cumulative demand deviation of all DMAs compared with a hypothetical DMA with average demand, for which a smaller value indicates a better performance. Pressure uniformity was suggested to guarantee that all nodes belonging to a certain district would have similar pressure patterns. A lower index value indicates better performance. Total head uniformity is also used to measure the variance of total heads along the nodes, which has the same meaning as pressure uniformity. Liu and Han [48] proposed a decision-making framework to determine the optimal DMA design by quantifying various indices, such as DMA uniformity, modularity index, and resilience index. Similarly, evaluating the benefits brought by DMAs in terms of cost-benefit analysis allows managers to make sensible decisions and create functional and efficient DMAs. Ferrari and Savic [25] proposed a comprehensive method that considers alternative DMA configurations to show the savings that utilities can obtain by considering three indices related leakage reduction, burst-frequency reduction, and pressure-sensitive demand reduction compared with the original network. Pressure reduction across the network was the main factor leading to reducing leakage and burst frequency. The study provided a decision-support tool for economic performance analysis of various DMA layouts.

In the case of a fire, while water demand is high for fire-fighting at a few nodes, the network must still have the capacity to supply enough water to users, especially during peak demand hours. This superposition of demand creates energy-loss leaps in pipes, leading to lower pressures at that time. Moreover, when creating isolated DMAs, some pipes feeding a district are closed and this could have negative impacts on the amount of flow entering a DMA. To test this situation, Grayman et al. [10] and Di Nardo et al. [118] developed a fire protection index based on the number of nodes with a pressure lower than the required pressure designed for the fire-fighting event. Those results indicated that some negative pressure values were occurring while most of the nodes had acceptable pressure. However, a significant difference was found between looped and branched networks.

WDN is a dynamic system in which pressure can vary significantly due to variations in water demand at nodes. Addressing spatial-temporal variability of water demand in the network, Di Nardo et al. [122] proposed a procedure for WNP under stochastic water demand and quantified its effects on hydraulic performance. The study revealed that by applying random variability of water demand, the magnitude of pressure distribution within the network was affected significantly. This led to a decrement of surplus pressure and network resilience compared with the constant-demand condition. To create feasible DMAs, especially for a WDN characterized by a small deviation between the surplus pressure and required standard pressure, spatial-temporal variability of water demand should be considered in WNP.

6. Discussion and Future Work

This paper provided a comprehensive review of the relevant studies on WNP over the last decade. The WNP procedure consists of two basic phases. First, the clustering phase involves the formation of the sizes and dimensions of DMAs as well as the definition of the boundary pipes that feed or interconnect DMAs. This phase is commonly associated with use of a clustering algorithm. In this study, six commonly applied algorithms such as (i) Graph theory, (ii) community structure algorithm, (iii) modularity-based algorithm, (iv) multilevel graph partitioning, (v) spectral graph algorithm, and (vi) multi-agent approach, are presented and discussed in-depth to understand how they work and handle in formation of the feasible DMA configurations. These algorithms are commonly based on the graph theorem that relies primarily on the network's topology. Since WDN is an infrastructure system with particular properties, the water network clustering algorithm allows for tailoring by appending weights to pipes or/and nodes to mimic distributing loads across the WDN. Many criteria for DMA design, such as hydraulic performance, network topology, system reliability, water quality, and cost-benefit ratio, are considered in this phase to minimize the number of boundary pipes and its goal is to define the reasonable size and configuration of DMAs.

Second, the sectorization phase is a physical segmentation process that identifies the position of gate valves and flow meters among the set of boundary pipes to satisfy operational constraints. This task requires the designer to apply an optimization algorithm or heuristic procedures to ensure that the locations of devices will have the least negative impact on the hydraulic performance of the network, minimize the energy use and leakage and be cost-effective.

The improvements and innovations in WNP developed to date often come from innovative approaches, either with the clustering algorithm or sectorization optimization. Those innovative features have emerged from combinations of clustering algorithms and alternating with flexibility or/and developing various objective functions based on the different criteria designs to propose a heuristic procedure for creating the most reliable DMAs. Many different approaches have been proposed for the automated creation of DMAs. However, several shortcomings remain to pursue in the future.

- Clustering is the crucial phase for WNP. Several algorithms and software tools were developed to deal with the large-scale networks that are burdensome to tackle manually. Various engineering aspects were embedded as weights to modulate WDN characteristics. More extensions of the existing graph clustering algorithms to weighted networks would be of great interest, as well as novel methods for clustering directed graphs.

- While there were many different approaches for the identification of DMAs in water networks, few studies tackled to determine the optimal number of DMAs for a given network. It is an open question and requires a decision-making procedure utilizing various network performance quantification metrics.

- In the sectorization phase, it still lacks how to assess the pump and tank operations in the partitioned network. Moreover, an approach to consider the consequences of device placements to the leakage, energy use, and post-damage restoration should be studied quantitatively in this phase.

- A demand-driven analysis (DDA) is generally used for WNP under the normal working condition at peak hour demand. In DDA, the supplied demand is assumed to be independent of pressure and this approach is valid when the pressure is above the minimum pressure requirement. In reality, a WDN works more likely as a pressure-driven analysis (PDA), in which the nodal consumption depends on the nodal pressure. Therefore, in pressure-deficient conditions (e.g., pipe failures, fire-fighting, unexpected water demand increase), a PDA should be applied for the novel dynamic WNP that adapts flexibly under the abnormal operating conditions.

- Last but not least, a WDN is supplied by single or multiple sources, with different elevations, divergent intended pressure in each zone. It also can be expanded or replaced according to urban planning needs. Further research should address the change of network's topology, controlling hydraulic uniformity in each zone as well as improving system resilience. Future research needs to be conducted to improve the abovementioned limits and eventually to provide optimal DMA layouts for efficient network operation and management.

Author Contributions: Conceptualization, X.K.B.; writing—original draft preparation, X.K.B.; writing—review and editing, D.K. and M.S.M.; supervision, D.K. All authors have read and agreed to the published version of the manuscript.

Funding: This research was supported by the EDISON* Program through the National Research Foundation of Korea (NRF) funded by the Ministry of Science, ICT & Future Planning (Grant No. 2017M3C 1A6075016). * Education-research Integration through Simulation On the Net.

Conflicts of Interest: The authors declare no conflict of interest.

References

1. Lifshitz, R.; Ostfeld, A. Clustering for Analysis of Water Distribution Systems. *J. Water Resour. Plan. Manag.* **2018**, *144*, 04018016. [CrossRef]

2. Di Nardo, A.; Di Natale, M.; Di Mauro, A. *Water Supply Network District Metering: Theory and Case Study*; CISM Courses and Lectures; Springer: Wien, Austria; New York, NY, USA, 2013.

3. Saldarriaga, J.; Bohorquez, J.; Celeita, D.; Vega, L.; Paez, D.; Savic, D.; Dandy, G.; Filion, Y.; Grayman, W.; Kapelan, Z. Battle of the Water Networks District Metered Areas. *J. Water Resour. Plan. Manag.* **2019**, *145*, 04019002. [CrossRef]

4. Morrison, J.; Tooms, S.; Rogers, D. *DMA Management Guidance Notes*; IWA Publishing: London, UK, 2007.

5. UK Water Industry Research Limited. *A Manual of DMA Practice*; UK Water Industry Research Limited: London, UK, 1999.

6. Di Nardo, A.; Di Natale, M.; Santonastaso, G.F.; Tzatchkov, V.G.; Alcocer-Yamanaka, V.H. Water Network Sectorization Based on Graph Theory and Energy Performance Indices. *J. Water Resour. Plan. Manag.* **2014**, *140*, 620–629. [CrossRef]

7. Ilaya-Ayza, A.; Martins, C.; Campbell, E.; Izquierdo, J. Implementation of DMAs in Intermittent Water Supply Networks Based on Equity Criteria. *Water* **2017**, *9*, 851. [CrossRef]

8. Ciaponi, C.; Creaco, E.; Di Nardo, A.; Di Natale, M.; Giudicianni, C.; Musmarra, D.; Santonastaso, G. Reducing Impacts of Contamination in Water Distribution Networks: A Combined Strategy Based on Network Partitioning and Installation of Water Quality Sensors. *Water* **2019**, *11*, 1315. [CrossRef]

9. Scarpa, F.; Lobba, A.; Becciu, G. Elementary DMA Design of Looped Water Distribution Networks with Multiple Sources. *J. Water Resour. Plan. Manag.* **2016**, *142*, 04016011. [CrossRef]

10. Grayman, W.M.; Murray, R.; Savic, D.A. Effects of Redesign of Water Systems for Security and Water Quality Factors. In *World Environmental and Water Resources Congress 2009*; American Society of Civil Engineers: Kansas City, MO, USA, 2009; pp. 1–11. [CrossRef]

11. Farley, M. *Leakage Management and Control: A Best Practice Training Manual*; No. WHO/SDE/WSH/01.1; World Health Organization: Geneva, Switzerland, 2001.

12. Puust, R.; Kapelan, Z.; Savic, D.A.; Koppel, T. A Review of Methods for Leakage Management in Pipe Networks. *Urban Water J.* **2010**, *7*, 25–45. [CrossRef]

13. Beuken, R.H.S.; Lavooij, C.S.W.; Bosch, A.; Schaap, P.G. Low Leakage in the Netherlands Confirmed. In *Water Distribution Systems Analysis Symposium 2006*; American Society of Civil Engineers: Cincinnati, OH, USA, 2008; pp. 1–8. [CrossRef]

14. Lambert, A.O. International Report: Water Losses Management and Techniques. *Water Sci. Technol. Water Supply* **2002**, *2*, 1–20. [CrossRef]

15. Lambert, A. Relationships between pressure, bursts and infrastructure life-an international perspective. In Proceedings of the Water UK Annual Leakage Conference, Coventry, UK, 18 October 2012.

16. Giugni, M.; Fontana, N.; Portolano, D.; Romanelli, D. A DMA design for "Napoli Est" water distribution system. In Proceedings of the 13th IWRA World Water Congress, Montpellier, France, 1–4 September 2008.

17. Di Nardo, A.; Di Natale, M.; Santonastaso, G.F.; Venticinque, S. An Automated Tool for Smart Water Network Partitioning. *Water Resour. Manag.* **2013**, *27*, 4493–4508. [CrossRef]

18. Di Nardo, A.; Di Natale, M.; Musmarra, D.; Santonastaso, G.F.; Tuccinardi, F.P.; Zaccone, G. Software for Partitioning and Protecting a Water Supply Network. *Civ. Eng. Environ. Syst.* **2016**, *33*, 55–69. [CrossRef]

19. Diao, K.; Zhou, Y.; Rauch, W. Automated Creation of District Metered Area Boundaries in Water Distribution Systems. *J. Water Resour. Plan. Manag.* **2013**, *139*, 184–190. [CrossRef]

20. Savić, D.; Ferrari, G. Design and Performance of District Metering Areas in Water Distribution Systems. *Procedia Eng.* **2014**, *89*, 1136–1143. [CrossRef]

21. Rajeswaran, A.; Narasimhan, S.; Narasimhan, S. A Graph Partitioning Algorithm for Leak Detection in Water Distribution Networks. *Comput. Chem. Eng.* **2018**, *108*, 11–23. [CrossRef]

22. Kunkel, G. Committee Report: Applying worldwide BMPs in water loss control. *J. Am. Water Work. Assoc.* **2003**, *95*, 65–79.

23. Gomes, R.; Sá Marques, A.; Sousa, J. Estimation of the Benefits Yielded by Pressure Management in Water Distribution Systems. *Urban Water J.* **2011**, *8*, 65–77. [CrossRef]

24. Huang, P.; Zhu, N.; Hou, D.; Chen, J.; Xiao, Y.; Yu, J.; Zhang, G.; Zhang, H. Real-Time Burst Detection in District Metering Areas in Water Distribution System Based on Patterns of Water Demand with Supervised Learning. *Water* **2018**, *10*, 1765. [CrossRef]

25. Ferrari, G.; Savic, D. Economic Performance of DMAs in Water Distribution Systems. *Procedia Eng.* **2015**, *119*, 189–195. [CrossRef]

26. Lifshitz, R.; Ostfeld, A. District Metering Areas and Pressure Reducing Valves Trade-Off in Water Distribution System Leakage Management. In Proceedings of the WDSA/CCWI Joint Conference Proceedings, Kingston, ON, Canada, 23–25 July 2018; Volume 1.

27. Marchi, A.; Salomons, E.; Ostfeld, A.; Kapelan, Z.; Simpson, A.R.; Zecchin, A.C.; Maier, H.R.; Wu, Z.Y.; Elsayed, S.M.; Song, Y.; et al. Battle of the Water Networks II. *J. Water Resour. Plan. Manag.* **2014**, *140*, 04014009. [CrossRef]

28. Di Nardo, A.; Di Natale, M.; Musmarra, D.; Santonastaso, G.F.; Tzatchkov, V.; Alcocer-Yamanaka, V.H. Dual-Use Value of Network Partitioning for Water System Management and Protection from Malicious Contamination. *J. Hydroinform.* **2015**, *17*, 361–376. [CrossRef]

29. UKWIR. *Effect of District Meter Areas on Water Quality*; UK Water Industry Research Limited: London, UK, 2000.

30. WRc. *The Effects of System Operation on Water Quality in Distribution*; WRc: Swindon, UK, 2000.

31. Armand, H.; Stoianov, I.; Graham, N. Impact of Network Sectorisation on Water Quality Management. *J. Hydroinform.* **2018**, *20*, 424–439. [CrossRef]

32. Javier Martínez-Solano, F.; Iglesias-Rey, P.L.; Mora Meliá, D.; Ribelles-Aguilar, J.V. Combining Skeletonization, Setpoint Curves, and Heuristic Algorithms to Define District Metering Areas in the Battle of Water Networks District Metering Areas. *J. Water Resour. Plan. Manag.* **2018**, *144*, 04018023. [CrossRef]

33. Salomons, E.; Skulovich, O.; Ostfeld, A. Battle of Water Networks DMAs: Multistage Design Approach. *J. Water Resour. Plan. Manag.* **2017**, *143*, 04017059. [CrossRef]

34. Campbell, E.; Ayala-Cabrera, D.; Izquierdo, J.; Pérez-García, R.; Tavera, M. Water Supply Network Sectorization Based on Social Networks Community Detection Algorithms. *Procedia Eng.* **2014**, *89*, 1208–1215. [CrossRef]

35. Giudicianni, C.; Di Nardo, A.; Di Natale, M.; Greco, R.; Santonastaso, G.; Scala, A. Topological Taxonomy of Water Distribution Networks. *Water* **2018**, *10*, 444. [CrossRef]

36. Giudicianni, C.; Herrera, M.; di Nardo, A.; Adeyeye, K. Automatic Multiscale Approach for Water Networks Partitioning into Dynamic District Metered Areas. *Water Resour. Manag.* **2020**. [CrossRef]

37. Sela Perelman, L.; Allen, M.; Preis, A.; Iqbal, M.; Whittle, A.J. Automated Sub-Zoning of Water Distribution Systems. *Environ. Model. Softw.* **2015**, *65*, 1–14. [CrossRef]

38. Tzatchkov, V.G.; Alcocer-Yamanaka, V.H.; Bourguett Ortíz, V. Graph Theory Based Algorithms for Water Distribution Network Sectorization Projects. In *Water Distribution Systems Analysis Symposium 2006*; American Society of Civil Engineers: Cincinnati, OH, USA, 2008; pp. 1–15. [CrossRef]

39. Perelman, L.; Ostfeld, A. Topological Clustering for Water Distribution Systems Analysis. *Environ. Model. Softw.* **2011**, *26*, 969–972. [CrossRef]

40. Ciaponi, C.; Murari, E.; Todeschini, S. Modularity-Based Procedure for Partitioning Water Distribution Systems into Independent Districts. *Water Resour. Manag.* **2016**, *30*, 2021–2036. [CrossRef]

41. Giustolisi, O.; Ridolfi, L. New Modularity-Based Approach to Segmentation of Water Distribution Networks. *J. Hydraul. Eng.* **2014**, *140*, 04014049. [CrossRef]

42. Zhang, Q.; Wu, Z.Y.; Zhao, M.; Qi, J.; Huang, Y.; Zhao, H. Automatic Partitioning of Water Distribution Networks Using Multiscale Community Detection and Multiobjective Optimization. *J. Water Resour. Plan. Manag.* **2017**, *143*, 04017057. [CrossRef]

43. Brentan, B.M.; Campbell, E.; Meirelles, G.L.; Luvizotto, E.; Izquierdo, J. Social Network Community Detection for DMA Creation: Criteria Analysis through Multilevel Optimization. *Math. Probl. Eng.* **2017**, *2017*, 1–12. [CrossRef]

44. Brentan, B.; Campbell, E.; Goulart, T.; Manzi, D.; Meirelles, G.; Herrera, M.; Izquierdo, J.; Luvizotto, E. Social Network Community Detection and Hybrid Optimization for Dividing Water Supply into District Metered Areas. *J. Water Resour. Plan. Manag.* **2018**, *144*, 04018020. [CrossRef]

45. Sempewo, J.; Pathirana, A.; Vairavamoorthy, K. Spatial Analysis Tool for Development of Leakage Control Zones from the Analogy of Distributed Computing. In *Water Distribution Systems Analysis 2008*; American Society of Civil Engineers: Kruger National Park, South Africa, 2009; pp. 1–15. [CrossRef]

46. Alvisi, S. A New Procedure for Optimal Design of District Metered Areas Based on the Multilevel Balancing and Refinement Algorithm. *Water Resour. Manag.* **2015**, *29*, 4397–4409. [CrossRef]

47. Di Nardo, A.; Di Natale, M.; Giudicianni, C.; Greco, R.; Santonastaso, G.F. Weighted Spectral Clustering for Water Distribution Network Partitioning. *Appl. Netw. Sci.* **2017**, *2*, 19. [CrossRef]

48. Liu, J.; Han, R. Spectral Clustering and Multicriteria Decision for Design of District Metered Areas. *J. Water Resour. Plan. Manag.* **2018**, *144*, 04018013. [CrossRef]

49. Di Nardo, A.; Giudicianni, C.; Greco, R.; Herrera, M.; Santonastaso, G. Applications of Graph Spectral Techniques to Water Distribution Network Management. *Water* **2018**, *10*, 45. [CrossRef]

50. Herrera, M.; Izquierdo, J.; Pérez-García, R.; Ayala-Cabrera, D. Water Supply Clusters by Multi-Agent Based Approach. In *Water Distribution Systems Analysis 2010*; American Society of Civil Engineers: Tucson, AZ, USA, 2011; pp. 861–869. [CrossRef]

51. Hajebi, S.; Barrett, S.; Clarke, A.; Clarke, S. Multi-agent simulation to support water distribution network partitioning. In Proceedings of the Modelling and Simulation 2013-European Simulation and Modelling Conference, ESM 2013, (Fernández), Lancaster, UK, 23–25 October 2013; pp. 163–168.

52. Herrera, M.; Izquierdo, J.; Pérez-García, R.; Montalvo, I. Multi-Agent Adaptive Boosting on Semi-Supervised Water Supply Clusters. *Adv. Eng. Softw.* **2012**, *50*, 131–136. [CrossRef]
53. Schaeffer, S.E. Graph Clustering. *Comput. Sci. Rev.* **2007**, *1*, 27–64. [CrossRef]
54. Song, S.; Zhao, J. Survey of Graph Clustering Algorithms Using Amazon Reviews. In Proceedings of the 17th International Conference on World Wide Web, Beijing, China, 21–25 April 2008.
55. Tarjan, R. Depth-first search and linear graph algorithms. *SIAM J. Comput.* **1972**, *1*, 146–160. [CrossRef]
56. Pohl, I.S. Bi-Directional and Heuristic Search in Path Problems. Ph.D. Thesis, Stanford Linear Accelerator Center, Stanford University, Stanford, CA, USA, 1969.
57. Di Battista, G.; Eades, P.; Tamassia, R.; Tollis, I.G. *Graph Drawing: Algorithms for the Visualization of Graphs*; Prentice Hall PTR: Upper Saddle River, NJ, USA, 1998.
58. Campbell, E.; Izquierdo, J.; Montalvo, I.; Ilaya-Ayza, A.; Pérez-García, R.; Tavera, M. A Flexible Methodology to Sectorize Water Supply Networks Based on Social Network Theory Concepts and Multi-Objective Optimization. *J. Hydroinform.* **2016**, *18*, 62–76. [CrossRef]
59. Campbell, E.; Izquierdo, J.; Montalvo, I.; Pérez-García, R. A Novel Water Supply Network Sectorization Methodology Based on a Complete Economic Analysis, Including Uncertainties. *Water* **2016**, *8*, 179. [CrossRef]
60. Alvisi, S.; Franchini, M. A Heuristic Procedure for the Automatic Creation of District Metered Areas in Water Distribution Systems. *Urban Water J.* **2014**, *11*, 137–159. [CrossRef]
61. Gomes, R.; Marques, A.S.; Sousa, J. Decision Support System to Divide a Large Network into Suitable District Metered Areas. *Water Sci. Technol.* **2012**, *65*, 1667–1675. [CrossRef]
62. Borgwardt, K.M.; Kriegel, H. Shortest-Path Kernels on Graphs. In Proceedings of the Fifth IEEE International Conference on Data Mining (ICDM'05), Houston, TX, USA, 27–30 November 2005; pp. 74–81. [CrossRef]
63. Newman, M.E.J.; Girvan, M. Finding and evaluating community structure in networks. *Phys. Rev. E* **2004**, *69*, 026113. [CrossRef]
64. Clauset, A.; Newman, M.E.J.; Moore, C. Finding Community Structure in Very Large Networks. *Phys. Rev. E* **2004**, *70*, 066111. [CrossRef]
65. Fortunato, S.; Barthelemy, M. Resolution Limit in Community Detection. *Proc. Natl. Acad. Sci. USA* **2007**, *104*, 36–41. [CrossRef]
66. Giustolisi, O.; Ridolfi, L. A Novel Infrastructure Modularity Index for the Segmentation of Water Distribution Networks. *Water Resour. Res.* **2014**, *50*, 7648–7661. [CrossRef]
67. Laucelli, D.B.; Simone, A.; Berardi, L.; Giustolisi, O. Optimal Design of District Metering Areas for the Reduction of Leakages. *J. Water Resour. Plan. Manag.* **2017**, *143*, 04017017. [CrossRef]
68. Simone, A.; Giustolisi, O.; Laucelli, D.B. A Proposal of Optimal Sampling Design Using a Modularity Strategy: Optimal Sampling Design. *Water Resour. Res.* **2016**, *52*, 6171–6185. [CrossRef]
69. Fortunato, S. Community Detection in Graphs. *Phys. Rep.* **2010**, *486*, 75–174. [CrossRef]
70. Di Nardo, A.; Di Natale, M.; Giudicianni, C.; Greco, R.; Santonastaso, G.F. Water Distribution Network Clustering: Graph Partitioning or Spectral Algorithms? In *Complex Networks & Their Applications VI*; Cherifi, C., Cherifi, H., Karsai, M., Musolesi, M., Eds.; Springer: Cham, Switzerland, 2017.
71. Liu, H.; Zhao, M.; Zhang, C.; Fu, G. Comparing Topological Partitioning Methods for District Metered Areas in the Water Distribution Network. *Water* **2018**, *10*, 368. [CrossRef]
72. Karypis, G.; Kumar, V. Multilevelk-Way Partitioning Scheme for Irregular Graphs. *J. Parallel Distrib. Comput.* **1998**, *48*, 96–129. [CrossRef]
73. Creaco, E.; Cunha, M.; Franchini, M. Using Heuristic Techniques to Account for Engineering Aspects in Modularity-Based Water Distribution Network Partitioning Algorithm. *J. Water Resour. Plan. Manag.* **2019**, *145*, 04019062. [CrossRef]
74. Deb, K.; Agrawal, S.; Pratap, A.; Meyarivan, T. A Fast Elitist Non-Dominated Sorting Genetic Algorithm for Multi-Objective Optimization: NSGA-II. In *Parallel Problem Solving from Nature PPSN VI*; Schoenauer, M., Deb, K., Rudolph, G., Yao, X., Lutton, E., Merelo, J.J., Schwefel, H.-P., Eds.; Springer: Berlin/Heidelberg, Germany, 2000; Volume 1917, pp. 849–858. [CrossRef]
75. Di Nardo, A.; Di Natale, M.; Giudicianni, C.; Santonastaso, G.; Tzatchkov, V.; Varela, J. Economic and Energy Criteria for District Meter Areas Design of Water Distribution Networks. *Water* **2017**, *9*, 463. [CrossRef]

76. Yazdani, A.; Jeffrey, P. Robustness and vulnerability analysis of water distribution networks using graph theoretic and complex network principles. In *Water Distribution Systems Analysis 2010, Proceedings of the 12th International Conference, WDSA 2010, Tucson, AZ, USA, 12–15 September 2010*; American Society of Civil Engineers: Reston, VA, USA, 2010; pp. 933–945. [CrossRef]

77. Yazdani, A.; Jeffrey, P. Complex network analysis of water distribution systems. *Chaos Interdiscip. J. Nonlinear Sci.* **2011**, *21*. [CrossRef]

78. Von Luxburg, U. A tutorial on spectral clustering. *Stat. Comput.* **2007**, *17*, 395–416. [CrossRef]

79. Mohar, B.; Alavi, Y.; Chartrand, G.; Oellermann, O.R. The Laplacian spectrum of graphs. In *Graph Theory, Combinatorics, and Applications*; Wiley: Hoboken, NJ, USA, 1991; pp. 871–898.

80. Shi, J.; Malik, J. Normalized Cuts and Image Segmentation. *IEEE Trans. Pattern Anal. Mach. Intell.* **2000**, *22*, 888–905. [CrossRef]

81. Zevnik, J.; Kramar Fijavž, M.; Kozelj, D. Generalized Normalized Cut and Spanning Trees for Water Distribution Network Partitioning. *J. Water Resour. Plan. Manag.* **2019**, *145*, 04019041. [CrossRef]

82. Bonacich, P. Power and centrality: A family of measures. *Am. J. Sociol.* **1987**, *92*, 1170–1182. [CrossRef]

83. Freeman, L.C. A set of measures of centrality based on betweenness. *Sociometry* **1977**, *40*, 35–41. [CrossRef]

84. Zevnik, J.; Kozelj, D. Partition of Water Distribution Networks into District Metered Areas Using a Graph Theoretical Approach. In Proceedings of the 13th International Conference on Hydroinformatics, Palermo, Italy, 1–6 July 2018; pp. 2408–2417. [CrossRef]

85. Herrera, M.; Canu, S.; Karatzoglou, A.; Perez-García, R.; Izquierdo, J. An approach to water supply clusters by semi-supervised learning. In *Modelling for Environment's Sake: Proceedings of the 5th Biennial Conference of the International Environmental Modelling and Software Society, IEMSs 2010, Ottawa, ON, Canada, 5–8 July 2010*; International Environmental Modelling and Software Society: Ottawa, ON, Canada, 2010; Volume 3, pp. 1925–1932.

86. Zhu, X.; Kandola, J.; Lafferty, J.; Ghahramani, Z. Graph Kernels by Spectral Transforms. In *Semi-Supervised Learning*; MIT Press: Cambridge, MA, USA, 2006; pp. 277–291.

87. Scholkopf, B.; Smola, A.J. *Learning with Kernels: Support Vector Machines, Regularization, Optimization, and Beyond*; MIT Press: Cambridge, MA, USA, 2001.

88. Giudicianni, C.; Herrera, M.; di Nardo, A.; Carravetta, A.; Ramos, H.M.; Adeyeye, K. Zero-Net Energy Management for the Monitoring and Control of Dynamically-Partitioned Smart Water Systems. *J. Clean. Prod.* **2020**, *252*, 119745. [CrossRef]

89. Sycara, K.P. Multiagent Systems. *AI Mag.* **1998**, *19*, 79. [CrossRef]

90. Izquierdo, J.; Herrera, M.; Montalvo, I.; Pérez-García, R. Agent-based division of water distribution systems into district metered areas. In Proceedings of the 4th International Conference on Software and Data Technologies, Sofia, Bulgaria, 26–29 July 2009; SciTePress-Science and and Technology Publications: Sofia, Bulgaria, 2009; pp. 83–90. [CrossRef]

91. Di Nardo, A.; Di Natale, M.; Santonastaso, G.F.; Tzatchkov, V.G.; Alcocer-Yamanaka, V.H. Water Network Sectorization Based on a Genetic Algorithm and Minimum Dissipated Power Paths. *Water Sci. Technol. Water Supply* **2013**, *13*, 951–957. [CrossRef]

92. Di Nardo, A.; Di Natale, M.; Greco, R.; Santonastaso, G.F. Ant Algorithm for Smart Water Network Partitioning. *Procedia Eng.* **2014**, *70*, 525–534. [CrossRef]

93. Ferrari, G.; Savic, D.; Becciu, G. Graph-Theoretic Approach and Sound Engineering Principles for Design of District Metered Areas. *J. Water Resour. Plan. Manag.* **2014**, *140*, 04014036. [CrossRef]

94. Gilbert, D.; Abraham, E.; Montalvo, I.; Piller, O. Iterative multistage method for a large water network sectorization into DMAs under multiple design objectives. *J. Water Resour. Plan. Manag.* **2017**, *143*, 04017067. [CrossRef]

95. Di Nardo, A.; Di Natale, M.; Giudicianni, C.; Musmarra, D.; Santonastaso, G.F.; Simone, A. Water Distribution System Clustering and Partitioning Based on Social Network Algorithms. *Procedia Eng.* **2015**, *119*, 196–205. [CrossRef]

96. Di Nardo, A.; Di Natale, M. A Heuristic Design Support Methodology Based on Graph Theory for District Metering of Water Supply Networks. *Eng. Optim.* **2011**, *43*, 193–211. [CrossRef]

97. Shao, Y.; Yao, H.; Zhang, T.; Chu, S.; Optimal, X. An Improved Genetic Algorithm for Optimal Layout of Flow Meters and Valves in Water Network Partitioning. *Water* **2019**, *11*, 1087. [CrossRef]

98. Todini, E. Looped Water Distribution Networks Design Using a Resilience Index Based Heuristic Approach. *Urban Water* **2000**, *2*, 115–122. [CrossRef]

99. Gomes, R.; Marques, A.S.A.; Sousa, J. District Metered Areas Design Under Different Decision Makers' Options: Cost Analysis. *Water Resour. Manag.* **2013**, *27*, 4527–4543. [CrossRef]

100. Creaco, E.; Haidar, H. Multiobjective Optimization of Control Valve Installation and DMA Creation for Reducing Leakage in Water Distribution Networks. *J. Water Resour. Plan. Manag.* **2019**, *145*, 04019046. [CrossRef]

101. Goldberg, D.E.; Holland, J.H. Genetic algorithms and machine learning. *Mach. Learn.* **1988**, *3*, 95–99. [CrossRef]

102. Gomes, R.; Sá Marques, A.; Sousa, J. Identification of the Optimal Entry Points at District Metered Areas and Implementation of Pressure Management. *Urban Water J.* **2012**, *9*, 365–384. [CrossRef]

103. Zhang, K.; Yan, H.; Zeng, H.; Xin, K.; Tao, T. A Practical Multi-Objective Optimization Sectorization Method for Water Distribution Network. *Sci. Total Environ.* **2019**, *656*, 1401–1412. [CrossRef]

104. De Paola, F.; Fontana, N.; Galdiero, E.; Giugni, M.; Savic, D.; Uberti, G.S. degli. Automatic Multi-Objective Sectorization of a Water Distribution Network. *Procedia Eng.* **2014**, *89*, 1200–1207. [CrossRef]

105. Hajebi, S.; Temate, S.; Barrett, S.; Clarke, A.; Clarke, S. Water Distribution Network Sectorisation Using Structural Graph Partitioning and Multi-Objective Optimization. *Procedia Eng.* **2014**, *89*, 1144–1151. [CrossRef]

106. Gallagher, J.; Styles, D.; McNabola, A.; Williams, A.P. Life Cycle Environmental Balance and Greenhouse Gas Mitigation Potential of Micro-Hydropower Energy Recovery in the Water Industry. *J. Clean. Prod.* **2015**, *99*, 152–159. [CrossRef]

107. Galdiero, E.; De Paola, F.; Fontana, N.; Giugni, M.; Savic, D. Decision Support System for the Optimal Design of District Metered Areas. *J. Hydroinform.* **2016**, *18*, 49–61. [CrossRef]

108. Hadka, D.; Reed, P. Borg: An Auto-Adaptive Many-Objective Evolutionary Computing Framework. *Evol. Comput.* **2013**, *21*, 231–259. [CrossRef] [PubMed]

109. Montalvo, I.; Izquierdo, J.; Pérez-García, R.; Herrera, M. Water Distribution System Computer-Aided Design by Agent Swarm Optimization: Water Distribution System Computer-Aided Design by Agent Swarm Optimization. *Comput. Aided Civ. Infrastruct. Eng.* **2014**, *29*, 433–448. [CrossRef]

110. Kennedy, J.; Eberhart, R. Particle Swarm Optimization. In Proceedings of the ICNN'95-International Conference on Neural Networks, Perth, WA, Australia, 27 November–1 December 1995; Volume 4, pp. 1942–1948. [CrossRef]

111. Moosavian, N.; Kasaee Roodsari, B. Soccer League Competition Algorithm: A Novel Meta-Heuristic Algorithm for Optimal Design of Water Distribution Networks. *Swarm Evol. Comput.* **2014**, *17*, 14–24. [CrossRef]

112. Wright, R.; Stoianov, I.; Parpas, P.; Henderson, K.; King, J. Adaptive Water Distribution Networks with Dynamically Reconfigurable Topology. *J. Hydroinform.* **2014**, *16*, 1280–1301. [CrossRef]

113. Perelman, L.S.; Allen, M.; Preis, A.; Iqbal, M.; Whittle, A.J. Flexible Reconfiguration of Existing Urban Water Infrastructure Systems. *Environ. Sci. Technol.* **2015**, *49*, 13378–13384. [CrossRef]

114. Di Nardo, A.; Cavallo, A.; Di Natale, M.; Greco, R.; Santonastaso, G.F. Dynamic Control of Water Distribution System Based on Network Partitioning. *Procedia Eng.* **2016**, *154*, 1275–1282. [CrossRef]

115. Santonastaso, G.F.; Di Nardo, A.; Creaco, E. Dual Topology for Partitioning of Water Distribution Networks Considering Actual Valve Locations. *Urban Water J.* **2019**, *16*, 469–479. [CrossRef]

116. Herrera, M.; Abraham, E.; Stoianov, I. A Graph-Theoretic Framework for Assessing the Resilience of Sectorised Water Distribution Networks. *Water Resour. Manag.* **2016**, *30*, 1685–1699. [CrossRef]

117. Eppstein, D. Finding the k shortest paths. *SIAM J. Comput.* **1998**, *28*, 652–673. [CrossRef]

118. Di Nardo, A.; Di Natale, M.; Santonastaso, G.F.; Tzatchkov, V.G.; Alcocer-Yamanaka, V.H. Performance Indices for Water Network Partitioning and Sectorization. *Water Sci. Technol. Water Supply* **2015**, *15*, 499–509. [CrossRef]

119. Di Nardo, A.; Di Natale, M.; Guida, M.; Musmarra, D. Water Network Protection from Intentional Contamination by Sectorization. *Water Resour. Manag.* **2013**, *27*, 1837–1850. [CrossRef]

120. Alvisi, S.; Franchini, M. A Procedure for the Design of District Metered Areas in Water Distribution Systems. *Procedia Eng.* **2014**, *70*, 41–50. [CrossRef]

121. Araque, D.; Saldarriaga, J.G. Water Distribution Network Operational Optimization by Maximizing the Pressure Uniformity at Service Nodes. In *Impacts of Global Climate Change*; American Society of Civil Engineers: Anchorage, AK, USA, 2005; pp. 1–10. [CrossRef]
122. Di Nardo, A.; Di Natale, M.; Gargano, R.; Giudicianni, C.; Greco, R.; Santonastaso, G.F. Performance of Partitioned Water Distribution Networks under Spatial-Temporal Variability of Water Demand. *Environ. Model. Softw.* **2018**, *101*, 128–136. [CrossRef]

Review

Rethinking the Framework of Smart Water System: A Review

Jiada Li [1],*, Xiafei Yang [2] and Robert Sitzenfrei [3]

[1] Department of Civil and Environmental Engineering, University of Utah, 201 Presidents' Cir, Salt Lake City, UT 84112, USA
[2] Department of Engineering Systems and Environment, University of Virginia, Charlottesville, VA 22903, USA; xy2wf@virginia.edu
[3] Unit of Environmental Engineering, University of Innsbruck, Innrain 52, 6020 Innsbruck, Austria; Robert.Sitzenfrei@uibk.ac.at
* Correspondence: jiada.li@utah.edu

Received: 19 December 2019; Accepted: 1 February 2020; Published: 4 February 2020

Abstract: Throughout the past years, governments, industries, and researchers have shown increasing interest in incorporating smart techniques, including sensor monitoring, real-time data transmitting, and real-time controlling into water systems. However, the design and construction of such a smart water system are still not quite standardized for massive applications due to the lack of consensus on the framework. The major challenge impeding wide application of the smart water network is the unavailability of a systematic framework to guide real-world design and deployment. To address this challenge, this review study aims to facilitate more extensive adoption of the smart water system, to increase effectiveness and efficiency in real-world water system contexts. A total of 32 literature pieces including 1 international forum, 17 peer-reviewed papers, 10 reports, and 4 presentations that are directly related to frameworks of smart water system have been reviewed. A new and comprehensive smart water framework, including definition and architecture, was proposed in this review paper. Two conceptual metrics (smartness and cyber wellness) were defined to evaluate the performance of smart water systems. Additionally, three pieces of future research suggestions were discussed, calling for broader collaboration in the community of researchers, engineers, and industrial and governmental sectors to promote smart water system applications.

Keywords: smart water system; framework; smartness; cyber wellness

---

1. Introduction

The world's urban population has grown rapidly from 1.019 billion in 1960 to 4.117 billion in 2017 [1]. It is estimated that the population will reach 9.7 billion by 2050 [2]. The excessive population growth will cause urgent water problems like water shortage and water quality degradation in urban areas. In the 21st century, the global water sector faces quality and quantity challenges, which are highly related to climate change and population growth [3]. The 2018 Global Risk Report shows that most of the high risks (high-likelihood and high-impact) issues are water-related either directly or indirectly and are currently being exacerbated by climate change [4]. Water crises have become one of the five most significant risks in terms of their societal impacts. Additionally, the breaking of economic growth and unbalanced urbanization can also contribute to water shortage [5,6]. It is predicted that due to population and industrial growth the percentage of water scarcity will increase by 50% in developing countries and decrease by 18% in developed countries by 2025 [7]. One upcoming water scarcity event will occur in Cape Town, where it is supposed to be the first city to experience day zero but will not be the last if these threats still hold [8,9].

However, due to the growing complications of water-related issues such as water shortage, water deterioration, and aging infrastructure, traditional techniques, and management for drinking water supply have gradually shown their drawbacks and incapability to address these water issues [10]. Climate change and anthropic activities exacerbate the water issues by reducing water quantity and deteriorating water quality. Especially, the limited effort on ecological maintenance results in the pollution being increasingly found in the water distribution system for the public around the world. There is an urgent need for modernized water supply technologies to alleviate current water concerns by improving water supply efficiency and approaching sustainable water management globally [11]. Traditionally, engineers and researchers get used to re-sizing water supply systems. However, upgrading the existing water distribution network is time-consuming and costly. Instead, retrofitting the water system with smart components such as sensors, controllers, and a data center can achieve real-time monitoring, transmitting, and controlling in water systems for decision-makers, which is a more cost-effective and sustainable approach to address the water challenges [12].

To date, automated control technology (ACT) and information communication technology (ICT) are applied to tackle existing problems in water distribution networks, where both technologies play critical roles in large-scale ACT and ICT applications. A number of study cases around the world consider using smart water metering to monitor the water consumption and further track leakage and pipe burst issues in water distribution networks [7]. The real-time measurements can be utilized to improve the accuracy of hydraulic model calibration and forecasting. Real-time control is commonly applied in pumping, valve operation, and scheduling. The water supply efficiency significantly benefits from automatic control technology but the electricity energy efficiency needs optimization in practical applicability. If matched with appropriate and effective ICT or ACT solutions, in the form of a smart water system (SWS), city-wise water issues can be appropriately addressed and managed [13]. In SWS, progress can be made via smart metering (real-time monitoring that transmits data to the utility) and intelligent controlling (real-time feedback and action). For example, the Western Municipal Water District (WMWD) of California utilities have used the SCADA system to manage real-time alarms and automatically operate plants and networks [14]. The implementation of SCADA has been associated with 30% savings on energy use, a 20% decrease in water loss, and a 20% decline in disruption [15]. In Brisbane City, Australia, the web-based communication and information system tools are used by governments and municipalities to deliver relevant water information to the public, as well as to provide early warnings [16]. Another SWS case is in Singapore, where a real-time monitoring system called WaterWiSe was built, utilizing wireless sensor networks and data acquisition platforms to improve the operational efficiency of the water supply system [17]. Moreover, in San Francisco, the automated real-time water meters are installed among those communities for more than 98% of their 178,000 customers to transmit hourly water consumption data to the billing system via wireless sensing networks [18]. This access to frequently updated water consumption information allows engineers to detect water quality events and localize pipe leaks faster than traditional water systems that are still using existing manually-read meters [19]. Given these ACT and ICT applications in water sectors, smart water concepts therefore emerge and inspire SWS to be widely accepted by large amounts of stakeholders.

The terms "smart water grid," "smart water supply system," "smart water system" or "smart water network" have been widely spread. The concept of SWS in the urban water field is gaining great impetus among academia, government, and industry, drawing attention from international communities (SWAN, EWRI, HIC, and CCWI) to top-level organizations (IWA, AWWA, AWC-Asian Water Council). Other international collaboration projects (e.g., i-WIDGT from EU [20], CANARY from US [21], SEQ from the Australian water resources department [22], and Smart City reports [23]) are providing professional support to smart urban water infrastructure all over the world [24–26]. Although researches on SWS are speeding up to meet the demand of industry and government, the conceptual, technical and practical gaps between providers and clients are still not well bridged. The influences of SWS could be more significant and essential if priorities are precisely defined and implemented into sensing technology domains in water contexts [27]. As a consequence of the lack of a systematic consensus from conceptual, technical, and practical perspectives, investigating the current architectures, initiatives, and applications of SWS around the world must be required to help generate a better understanding of the definitions, characteristics, and future trends of SWS. While review studies related to SWS and smart grid areas have already been undertaken [28–31], a review of SWS in conceptual and comprehensive prospects, with a further notification on the definition, architecture, and metrics, is missing. To address this research gap, this paper is going to analyze the existing SWS concepts, identify the possible metrics for SWS, and establish a more systematic SWS architecture, to enlighten future research on its implementations.

This review paper is structured as follows: firstly, the current definitions and historical development of SWS are presented and discussed in the introduction section, reviewing the evolution of SWS from the past to the date and analyzing the weakness of the current water system. Secondly, given the previous literature of SWS, a new architecture of SWS with five layers is put forward and demonstrated explicitly. Then, by reconciling definitions and architectures, two metrics of SWS are proposed to characterize its properties. Finally, recommendations on future research directions are given for smart water system development.

2. Literature Review of SWS

SWS is a multidisciplinary term. A ScienceDirect search for "Smart" or "Intelligent" in the title, abstract, and keyword gave a total of 31,527 article results. However, most of them belong to smart transportation, smartphone, and smart grid fields. If "water" is included in the search, the number goes down to 9847. Further searching "Smart Water System" made the results decline to 9517. By adding "framework" to the "Smart water system" only 4026 articles remained. Given that searching results, we conducted the literature review by considering relevant references from the selected papers. The step-by-step literature searching rules were summarized below: (1) "Smart" or "Intelligent" in title, abstract, and keyword with 31,527 article results; (2) "Smart Water" in title, abstract, and keyword with 9517 article results; (3) "Framework" and "Smart Water" in title, abstract, keyword, and body with 4026 article results; (4) "Structure", "Layer", "Framework" and "Smart Water" by manual filtering with 32 article results. These 32 final pieces of literature, including four article forms such as forums, papers, reports, and presentations gain high popularity in multiple sources like Google Scholar, SCOPUS, and ScienceDirect. All the papers come from the recent 10–15 years; they provide a wide presentation of the smart water system for the readers, which include the typical arguments of the framework of SWS. With insight from these literatures in Table 1, there are 17 papers, 10 technical reports from different well-known organizations including the International Telecommunication Union [32], U.S Environment Protection Agency [33], UK Department for International Development [34], UN Global Opportunity Committee [35–38], and Colorado State University [39]. Four key presentations and one International forum [40,41] are also taken into consideration in this review. As the number of studies that we reviewed is limited, this paper does not cover all aspects of SWS.

Table 1. Classification of the literature considering the framework in SWS.

Reference Number	Smart Water Definition	Smart Water Structure					Smart Water Metrics	Future Research
		Instrument Layer	Property Layer	Function Layer	Benefit Layer	Application Layer		
[42]	Smart water is also called Smart Water Grid (SWG4), Internet of Water5, Smart Water Management, etc.	•	•	•		•		•
[43]	Smart water grid is based on the Internet of Things and the structure of smart water system.	•	•	•				•
[44]	A smart water system (network) in water context comprises smart meters, smart valves, smart pumps, data communication, data management, data fusion, and analysis tools.	•		•		•		•
[45]	A smart water network should comprise smart meters, smart valves, and smart pumps.	•					•	•
[46]	An advanced and intelligent water supply system and includes ICT integration of the water supply network.	•			•	•		•
[17]	Water management technologies converging with ICT have been called Smart Water Management (SWM) distinguished from traditional water management technologies.	•	•	•	•	•		•
[47]	A Smart Water Grid system integrates information and communications technologies into the management of the water distribution system.	•	•	•		•	•	•
[48]	Smart water system illustrates many of the ways technology, middleware, and software help maximize the value of Smart Metering data to all stakeholders.	•			•	•		
[33]	Not specified			•		•		

Table 1. *Cont.*

Reference Number	Smart Water Definition	Smart Water Structure					Smart Water Metrics	Future Research
		Instrument Layer	Property Layer	Function Layer	Benefit Layer	Application Layer		
[34]	A smart water system offers a mechanism to capture and communicate data on water resources through hydro informatics systems.	•			•	•		
[36]	Not specified	•		•		•		•
[37]	Not specified	•		•		•		•
[38]	Not specified	•		•		•		•
[39]	Smart water networks is a system composed of automation, sensing and communication tools.	•	•		•	•		
[41]	A smart water system is an integrated set of products, solutions, and systems.	•			•			•
[49]	A smart water system is based on the smart metering system to collect real-time water data.	•		•	•		•	•
[40]	Smart water networks are layered, as any data ecosystem is, starting from sensors, remote control, and enterprise data sources, through data collection and communications, data management and display, and up to data fusion and analysis.	•	•	•	•			
[50]	Not specified				•		•	•
[51]	Not specified							
[23]	Not specified					•		
[52]	A smart water system is designed to gather meaningful and actionable data about the flow, pressure, and distribution of a city's water.				•	•		•
[53]	A smart water system uses data-driven components to help manage and operate the physical network of pipe.	•	•	•			•	

Table 1. *Cont.*

Reference Number	Smart Water Definition	Smart Water Structure					Smart Water Metrics	Future Research
		Instrument Layer	Property Layer	Function Layer	Benefit Layer	Application Layer		
[54]	Smart water system aims to deploy the Internet of Things (IoT) technology throughout the water supply infrastructure and consumers' usage.	•		•	•	•		
[43]	Smart water grid is based on the Internet of Things, mainly including hierarchy framework, technical system, and function framework.	•	•	•	•	•		•
[20]	An assortment of components and procedures for the continuous monitoring and evaluation of water use.				•	•		•
[55]	A smart water architecture can be characterized by five layers: physical layer, sensing, and control layer, communication layer, data management layer, and data fusion layer.			•		•		•
[56]	A smart water grid is an innovative way to monitor water distribution networks.		•		•		•	•
[57]	A water smart grid would direct the innovative technologies suite to create a data-driven system for intelligent water resources management.			•	•		•	•
[58]	Smart water networks need online water monitoring for the collection and analysis of data.			•	•	•		•
[59]	Not specified	•		•		•		•
[60]	Not specified			•		•		
[61]	Smart water system requires many digital devices (sensors and actuators) to be deployed across the water distribution network to enable near real-time monitoring and control of the water grid components.			•	•			•

A graphical statistical data overview of the 32 selected literature is presented in Figures 1–4. In Figure 1, these pieces of literature are classified into four types including the forum, presentation, report and paper based on their literature formats. In Figure 2, they are reclassified into three types—academia, industrial, and governmental based on their published organizations. As Figure 1 shows, paper and report take up the highest percentage while key presentations and forums occupy only a small portion, which shows that current SWS studies are mainly documented by papers and reports for efficient sharing. The 22 publications out of a total of 32 from academia in Figure 2 imply that most SWS researchers are from universities and academic research institutions. There is a lack of industrial and governmental inputs for such interdisciplinary work in the SWS field. In Figures 3 and 4, it is clear how the percentages of each type of publication weight in each classification group. For example, in Figure 3, among those SWS papers, the highest percentage of publications are from academia. In contrast to the data presented in Figure 3, Figure 4 shows the different picture that there are equal numbers of all four literature formats in industrial publications.

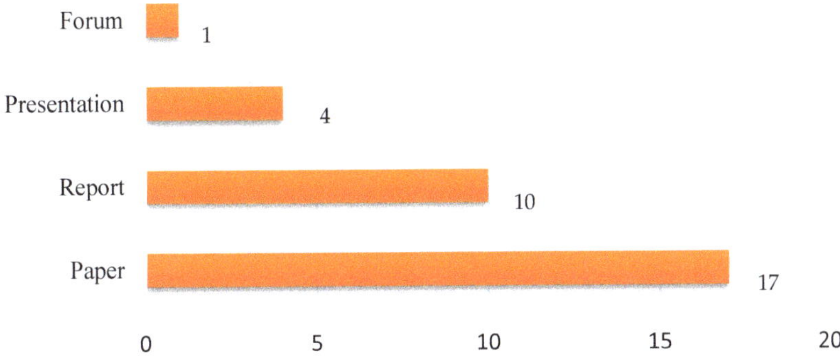

Figure 1. Literature overview: the number of publications for smart water systems (SWS's) definition.

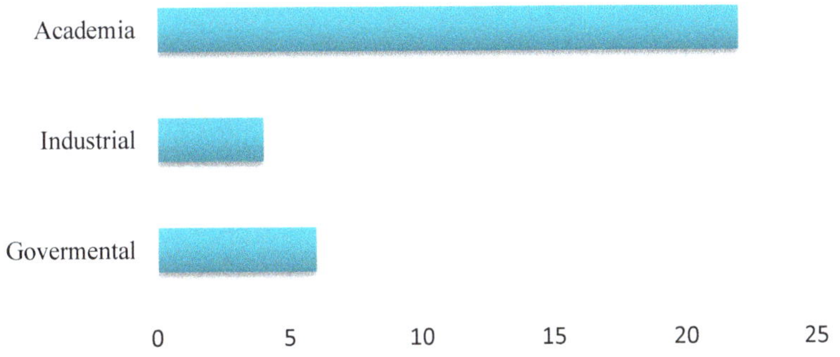

Figure 2. Literature overview: the number of the organization for SWS's definition.

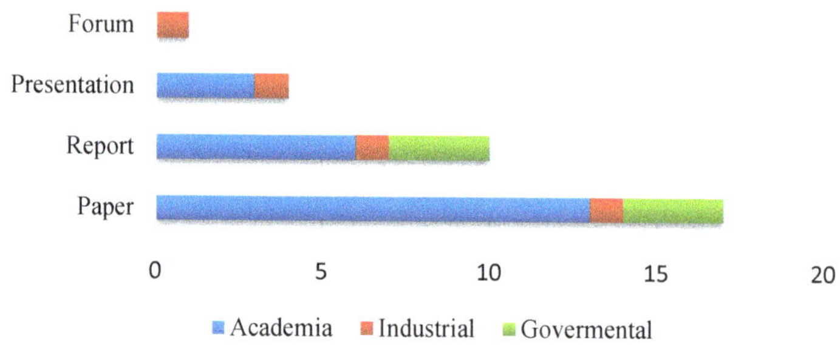

Figure 3. Literature overview: the number of literature types for different organizations.

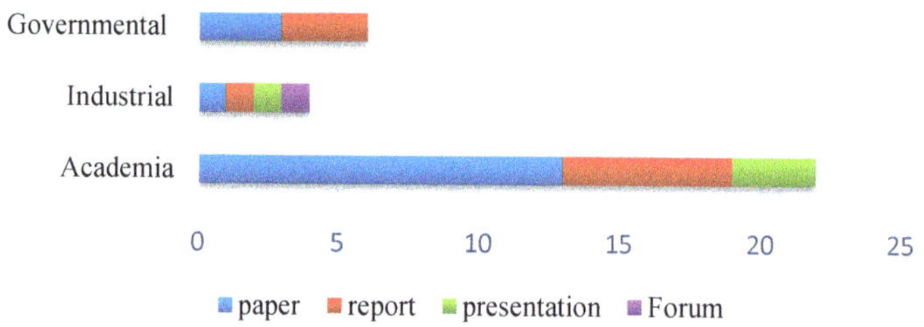

Figure 4. Literature overview: number of studies for literature type.

The following is a list of works of literature related to the framework of the smart water system in Table 1. We categorized these papers according to the smart water definition, structures (Instrument layer, Property layer, Function layer, Benefit layer, Application layer), and metrics, which correspond with the components of the proposed SWS framework below. The definition provides theoretical support for SWS structures, and the metrics can be used to evaluate the SWS performance. These five basic layers inside structures consist of a comprehensive SWS architecture. The relationship among definitions, layers, and metrics make SWS work, which is considered as the reason for the literature classification in this study. In Table 1, "Instrument" represents the instrument layer of SWS including the physical and ACT and ICT components. "Property" means the property layer of SWS containing the components like systems attributes. "Function" represents the function layer of SWS such as data fusion in the data center. "Benefit" represents the benefit layer including features like water quality security and energy saving. "Application" represents the application layer such as commercial and educational applications. "Metrics" relate to the methods applied to evaluate the smart water system. "Future research" stands for research direction recommendations regarding the smart water system. Each reference might include but not all components (definition, layers, and metrics) of the smart water system. A black solid circle was used to mark the elements that references have covered. It can be found that most of the pieces of literature have definitions of SWS. However, none of them have covered all key layers. Since SWS is built for different purposes, the structure of SWS may vary from case to case. Furthermore, among all those pieces of literature that we reviewed, only 7 of them discussed metrics for SWS. How to assess the performance of SWS was not fully explored based on currently available literature. It was evident that there is still great potential to improve the consensus and understanding of SWS. Table 1 also shows that over 90% of the relevant literature has recommended future research directions for SWS. This paper will summarize these suggestions later in Section 5.

3. A Systematic Framework of SWS

A systematic architecture of SWS is comprised of various layers working synergistically to perform useful functions and applications [62]. Such a system can be represented as a set of components, with specific properties and benefits. In past years, previous studies proposed various versions of SWS to meet their particular demands. The combinations of SWS that are water management technologies and ICT distinguished from traditional water management technologies were put forward [54]. However, the scopes and characteristics of such SWS were not identified. Further, the term "SWG" refers to an advanced smart water grid that includes real-time information sharing through smart measurement and networking and a sustainable water distribution infrastructure [46]. The smart components in SWG imply that a smart water network should comprise smart meters, smart valves, and smart pumps by definition [28]. These smart elements including physical electronic parts, like sensors and microcontrollers, communication protocols, and embedded systems are all folded in the concept of the Internet of Things (IoT), which is the foundation of SWS [46]. The structure of SWS, therefore, should contain three frameworks: the hierarchy framework, technical system, and function framework [61].

In the hierarchy framework and technical system, there are also numerous pieces required. An easy-to-understand architecture of SWS would be preferred. The principals of the smart water network were then explained [44]. This research can be segmented in various layers: (1) physical layer (like pipes); (2) sensing and control layer (like flow) sensors and remote control; (3) data collection and communication layer (like data transfer); (4) data management and display; (5) data fusion and analysis (like analysis tool and even detection, leakage detection, and decision making). Nonetheless, these layers still only contain physical and cyber components and a lack of improvement to the service level. It was proposed that SWS contains 5 layers: physical layer, sensing layer, and control layer, collection and communication layer, data management and display layer, and data fusion and analysis layer [53]. They also put forward a bottom-up framework of SWS with 5 layers: sensing layer, transport layer, processing layer, application layer, and unified portal layer, which are based on IoT and cloud computing [53]. Another SWS composed of 4 stages was established to secure the vast amounts of high-resolution assumption data and customized information [20].

The most widely accepted smart water architecture is characterized by five layers: the physical layer, sensing, and control layer, collection and communication layer, data management and display layer, and data fusion and analysis layer. Each segment covers a distinct function in the network [62]. However, all SWS introduced above are under debate since most of them are defined for one particular purpose without complete demonstration. Some of them are for smart water targets, some stress the innovation of mechanism, while others emphasize the application of ICT. Very few of in situ frameworks for understanding SWS are comprehensive and directly applicable for education, research, and public. They lack some critical elements like properties, metrics and case studies, and the ability to guide future research directions. Hence, it is necessary to build a systematic framework of SWS to further the understanding of SWS and accelerate the implementation of SWS. In this study, we adopt and integrate some of the existing architectures to propose systematic architecture. Figure 5 illustrates the authors' conceptual representation of an orderly architecture of SWS within a systematic smart water framework. There are five layers (from bottom to top: instruments layer, function layer, property layer, benefits layer, and application layer) that are proposed in order to understand how systematic architecture is implemented in the SWS framework. Although such a conceptual framework has not been tested in the field, this provides the guidance for engineers to replicate the SWS according to their purposes and application. For instance, a smart water test-bed for educational purposes can be built on the lab by following the SWS framework, while the application layer might be unnecessary in this case [63].

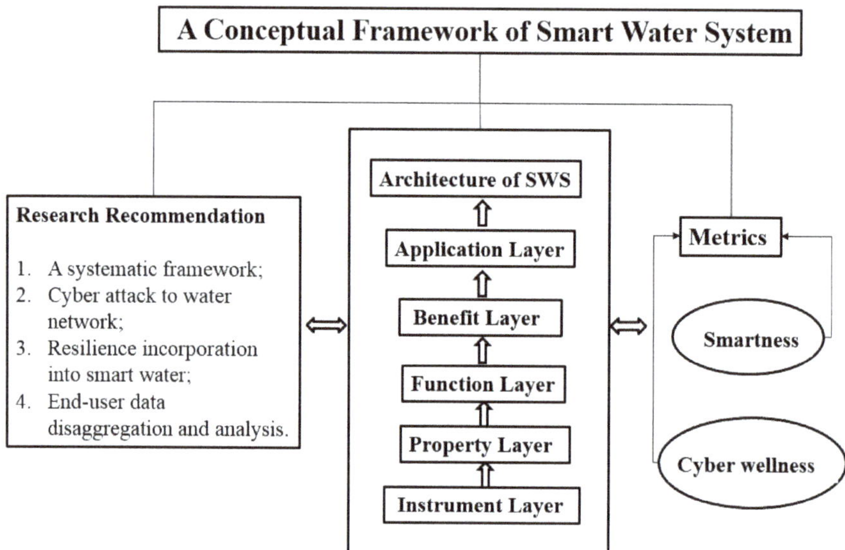

Figure 5. A New Framework of Smart Water System.

3.1. Instruments Layer

Typically, the instruments layer of SWS should be composed of physical infrastructure (network-level components) and cyberinfrastructure (internet-related hardware, software, and services). Although the physical instruments are the basic structure of the water system, they cannot make it smart or even data-enables only by itself. The cyberinfrastructure includes multiple intelligent devices like smart sensors, smart pumps, and smart valves, etc. Their primary roles and application goals are summarized in Table 2. The smart components of cyberinfrastructure are the elements in which SWS differs from the traditional water distribution system. For example, traditional water distribution systems with only physical instruments carry on pressure or flow data. Conversely, the SWS with cyber instruments not only sends a flow or pressure signal, but their data steam including diagnostic information also makes the SWS detect leaks more efficiently and automatically. Additionally, for the integrated SWS, the interaction and relationship between physical and cyberinfrastructure should not be ignored. Showing in Figure 5 below, physical instruments, including pipes, valves, and pumps provide the structural basement for the placement and installation of cyber instruments like smart meter and intelligent sensors (e.g., electromagnetic or ultrasonic). Meanwhile, physical infrastructures are elements that produce the required data and information, which would be collected, transferred, processed, and fused by internet-related hardware, software, and services. In return, the cyber instruments can instruct the operation and maintenance of physical components by analyzing the newly produced data and forecasting the system condition. For example, the automated meters are bi-direction communication devices that can execute actions on devices (e.g., valve turn off and on) [23]. Furthermore, the different smart sensors might be designed to solve various problems (shown in Figure 6) by operating systems discriminatively. Therefore, the components layer of SWS should achieve both of the roles of physical infrastructure and cyber-infrastructure.

Table 2. Components of Cyber-Infrastructure.

Components	Roles	Problems Solved
Smart Flow Sensors	Monitor flow	Water leakage, Pipe burst
Smart Pressure Sensor	Monitor pressure	Pressure instability, Water Loss, Energy Loss
Smart Valves	"Bi-direction" operation	Water leakage, Pipe burst
Smart Pumps	"Bi-direction" operation	Pressure unbalance, Energy loss
Smart Irrigation Controllers	"Bi-direction" operation	Water loss, Energy loss, Water Overuse
Smart Contaminant Sensor	Monitor water quality	Pipe deterioration, Water aging, Contaminant intrusion
Smart Flood Sensor	Monitor flood volume	Flood disaster, Water quality issues

Note: Bidrirection communication denotes the ability of the meter operator to "at a minimum, obtain meter reads on-demand, to ascertain whether water has recently been flowing through the meter and onto the premises, and to issue commands to the meter to perform specific tasks such as disconnecting or restricting water flow" [58].

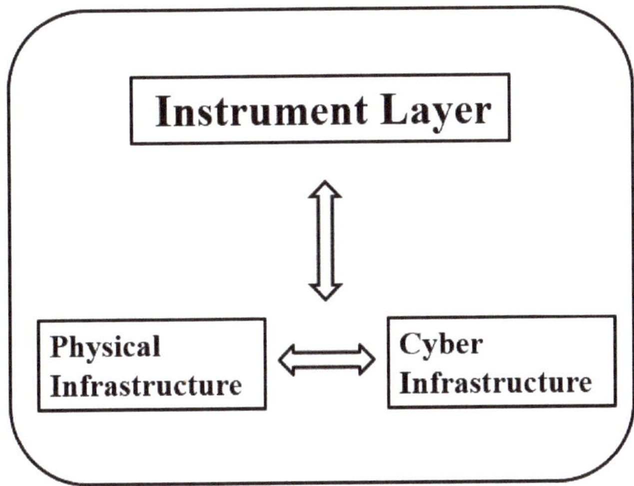

Figure 6. Designing of Instruments layer.

3.2. Property Layer

The property of SWS can be understood as the ability to respond to threat, withstand attack, and adapt more readily to failure risks, like connectivity, real-time, security, resourcefulness, and robustness. These properties can be considered indicators of how smart the SWS is, and have to be quantified either qualitatively or quantitatively through metrics. The metrics applied in the SWS assessment, such as smartness, how efficient the SWS is towards real-time, and cyber wellness, how defensible the SWS is against cyberattack, would be discussed in the next section. In this work, we proposed that SWS based on components layer should have 4 properties: (1) Automation; (2) Resourcefulness; (3) Real-time; (4) Connectivity. We designed the interaction between each other within the property loop in Figure 7. For instance, automation is the foundation to achieve real-time and also real-time is facilitated by connectivity. Resourcefulness is ensured by the automation and connectivity among various IoT. One common sense is that these four properties might not be applied to each case of SWS, and also, some researchers may consider SWS with other extra features, but these four demonstrate what most SWS look like.

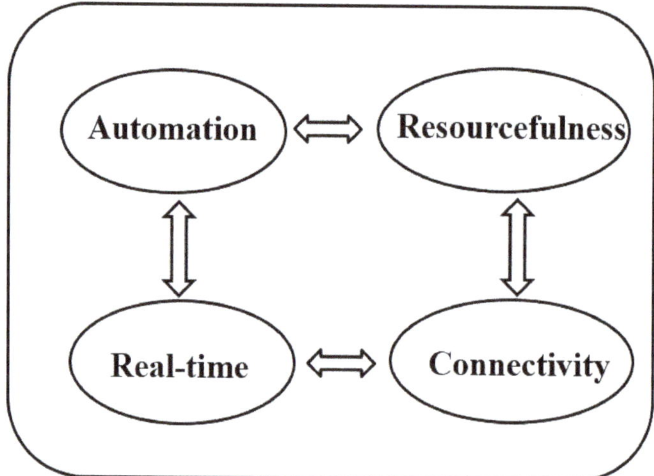

Figure 7. Property layer designing.

Automation is the fundamental property, which means that SWS can perform the physical and cyber components' operation automatically once the relevant parameters are set manually. This is a near-automatic process in the SWS since the application of ACT and ICT enables the SWS to execute one-way or two-way orders without too many operators' involvement. For the one-way control case (e.g., by setting the sampling interval in the Arduino sketch), the smart sensor can modify the data collecting frequency. As to the two-way control example, the real-time metering data fusion can help provide feedback for the control center, where the data analytic tools, in turn, give instructions to the opening status of smart meters, smart valves, or intelligent pumps. Further, automation can be implemented not only for components providing functionality but also for operation mechanisms. For example, when water problems happen in the operation process or the element itself, the smart components notify the system center and then take action to avoid a crash. Additionally, with automatic self-verification, the water utility can know when the sensor needs maintenance or re-installation [64].

Connectivity is also a fundamental property, which means the degree of interconnectedness or duplication [56]. In a cyber-physical system like SWS, connectivity can be implemented by deploying multiple sensors and software for monitoring the same physical processes [65]. SWS should be qualified to connect with different software and hardware to make the system collect data, analyze data, and share information publicly. For example, the SWS can be connected with the hydraulic model, GIS platform, billing system, and Database Model.

Real-time is the core property of SWS, which can also be called system efficiency. The real-time performance of SWS is characterized as online steps like online data monitoring, online data assimilation, online modeling, online plotting, and online results output, despite offline performance included in SWS. Real-time is the property that SWS obtains to achieve the required smart features [66]. However, current research mainly focuses on real-time modeling. The real-time modeling of SWS is structured as 6 steps: (1) Communicate SCADA datasets; (2) Updating the network model boundary conditions and operational statuses; (3) Pausing execution; (4) Generating the corresponding networks analysis; (5) Waiting for the new SCADA measurements to reload the network model; (6) Return the network simulation. Mentioned in the connectivity property, real-time modeling can be performed only by connecting with data sources systems like SCADA with modeling tools like EPANET-RTX [60]. This newly produced data from real-time hydraulic modeling can forecast results and calibrate the model by comparing measurements and predicted values.

Resourcefulness is the final property, which means the SWS not only owns massive data storage but also aims to timely exchange data for further analysis. This property of data exchange can also be interpreted as interoperability, which refers to the capability of units of an SWS to exchange and use information and services with one another and interfaced external units [67]. SWS provides massive information to an automation and security system, compared with the traditional instruments [58]. Typically, there are three types of source data open to the processor: spatial data, attribute data, and multimedia data, which determine the database model designing. However, these data can only be shared with the public and business after being analyzed by experts. Those processed data would be input into the hydraulic model to produce forecasting data, and those visualized data would be interpreted as valuable information such as early warning system, assessment of pipe leakage/breaks, or identification of cyberattacks or for decision support [68]. Moreover, the processed information or data would be transmitted back to the SCADA system and stored as instrument status and diagnostic information.

3.3. Function Layer

Functions of SWS can be determined by the instrument layer and property layer since different components and properties lead to different functions. For example, one SWS installed with flow and pressure sensor would consider being featured with resourcefulness and can verify the pressure-driven modeling analysis with enormous data collection resources [69]. In contrast, SWS is equipped with temperature sensors functions predicting the infiltration rate in the water systems with available temperature data [70]. Within the architecture of SWS, the function layer is localized in the connection point between system's property and metrics and plays a role in linking these two (shown in the framework Figure 5). Thus, the function layer can be interpreted as the backbone of SWS that includes functionalities of intelligent sensing, simulation, diagnosis, warning, dispatching, disposal, and control [43]. However, this study does not demonstrate the status of a function layer on the whole framework. Based on this function layer, this paper re-designs and specifies the function layer shown in Figure 8, which includes data producing, data sensing, data processing, simulation operating, and application supporting.

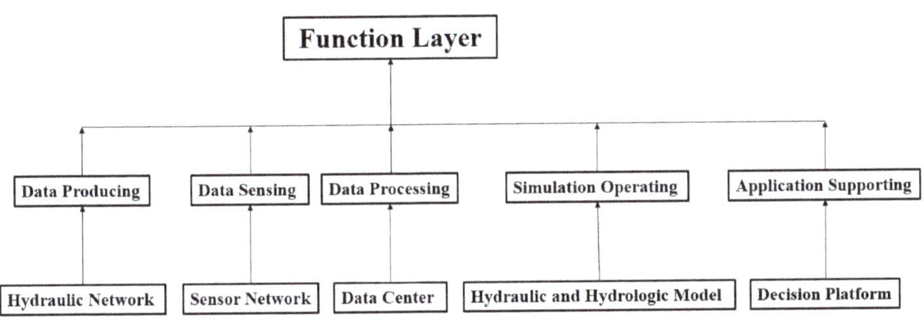

Figure 8. Function layer designing.

3.4. Benefit Layer

Retrofitting the traditional water supply system with smart devices brings many benefits such as bill reduction for consumers, operation cost decreases for utility, and water loss declines [10]. In this study, the benefit layer mainly contains four aspects including prolonging the asset life cycle, increasing energy sustainability, optimizing pressure and water quality, obtaining real-time water consumption shown in Figure 9 below. According to the benefits taxonomy by [71–73], the first two can be considered as business benefits while the third one belongs to shared benefits. The last benefit is classified as customer benefits.

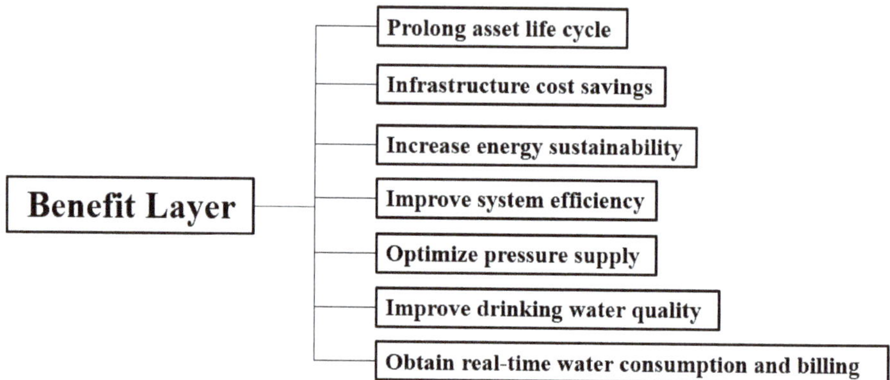

Figure 9. Benefits layer designing.

3.4.1. Prolong Asset Life Cycle and Cost Saving

As real-time property states, an SWS can integrate and analyze real-time monitoring data from various instruments for decision-making support. Especially, monitoring data regarding pipelines, valves, pumps, and tank conditions can be used to develop a risk-evaluated model for instrument replacement or maintenance. This allows SWS to plan and schedule the replacement and rehabilitation of the asset program efficiently as well as effectively so that the right assets can be replaced, repaired, and rebuilt in a timely manner. Thus, the real-time monitoring of asset status could be regarded as indicators of preventive maintenance and predictive replacement to prolong the assets' life cycle and cost savings [74–76].

3.4.2. Reduce Energy Loss and Improve System Efficiency

Energy is needed to extract, deliver, treat, and heat water for municipal, industrial, and agricultural uses [77,78]. In the United States, the energy required to move and handle the water is estimated to comprise 4% of total electricity consumption nationwide [79]. Knowing the quantity of water needed, and accounting for any losses, utilities can produce less water. This will be reflected in lower energy consumption as well as in water conservation. The energy loss and costs can be reduced by reducing the amount of water needed to be pumped, transported, heated, or treated [80]. Leakage in the water distribution system leads to additional energy required to pump and carry water to consumers as a result. Smart water sensors can monitor online pressure and warn pressure changes or significant pressure losses along with the water network, where utilities can remotely optimize the network pressure to help save energy [57].

3.4.3. Optimize Pressure Supply and Water Quality

Real-time data allows SWS to enhance system planning and operations by monitoring the hydraulic and water quality situation throughout the system [81]. On the one hand, the deficit pressure in the water distribution system can be detected so that actions can be taken promptly to alleviate the risk of pipe bursts. On the other hand, automated valve or smart gate operations can adjust the operating status to prevent pipe burst and water leakage, which might compensate for instruments changing process conditions or water treatment plant requirements [47]. Those data from SCADA and hydraulic integration can also be used for foreseeing the negative operating consequences, predicting future boundary conditions, and knowing the current system status in real-time, which empowers operators and engineers to control the water systems more effectively.

The implementation of cyber instruments into the water system can be utilized for warning pollutants intrusion. Continuous online monitoring for water quality indicators, such as free chlorine, total organic carbon, pH, conductivity, and turbidity, assist the effective response to a water contamination incident and to mitigate further consequences. For example, the real-time water quality issue detection system called CANARY can use advanced monitoring technologies and enhanced surveillance instruments to collect, integrate, analyze, and communicate information that provides a timely warning of potential contamination events [82].

3.4.4. Obtain Real-Time Water Consumption and Consumer Billing

SWS connected with the billing system and consumption system can display the metering information on the end-user's platform, like laptops, smartphones, or tablets via internet connection. This real-time water consumption information helps customers to save water and costs [83,84]. As the crux elements of SWS, the automated meter readings (AMR) and automatic meter infrastructure (AMI) provide real-time feedback on water usage for customers. This enables customers to make informed choices towards water-consuming habits and join in the water management activity. In reverse, the change of demand pattern would help the engineers and operators to calibrate their model, optimize pump and valves schedules [85], and modify the boundary conditions timely as efficiently as accurately [47].

3.5. Application Layer

In the application layer, there are three aspects generally classified as the Public application, Government application, and Business application shown in Figure 10. The application layer is mainly user-oriented for decision-making support finally. These three kinds of applications can receive feedback from the corresponding users like consumers, utilities and markets, or other terminals shown in Figure 10 below.

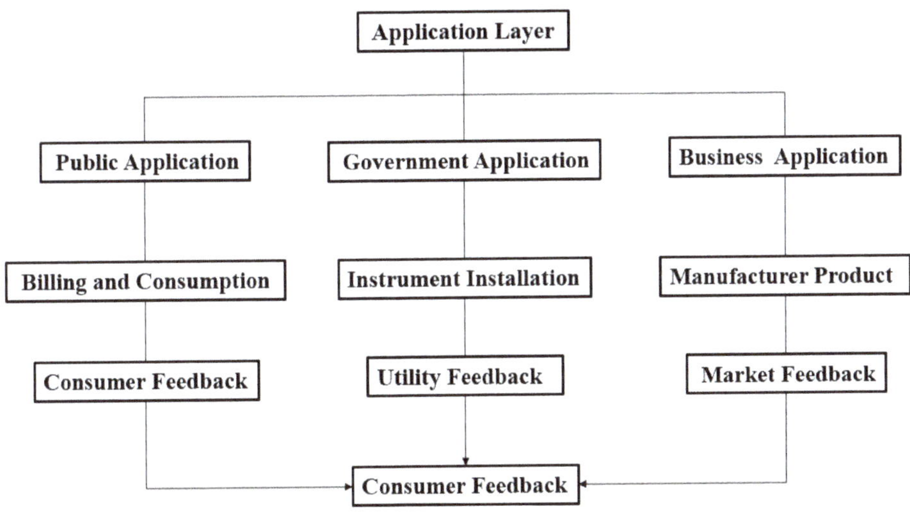

Figure 10. Application layer designing.

The public application aims to share real-time data and information with consumers for their feedback on billing and consumption, which is beneficial for water conservation and costs saving. Besides, public awareness about this SWS and water conservation should be stressed. More community involvement might be eliminated by providing citizens with more information regarding water conservation and SWS. Manual flow meters reading and installation may cause problems like the embarrassing access of individuals and invasion of government on private property [86]. SWS's technology development, combined with public awareness improvement, can reduce those risks.

The government application targets water utilities, different users from public applications. As the construction work of the water system, such as pipe replacement, instruments installation, and data center establishment, is the responsibility of the water utility, the sensors deployed by the water utility are helpful to take an insight of what is happening in SWS, and the collected sensor data would be adopted to support government decision.

The last application is the business application. The current smart water market mainly focuses on developing sensors and real-time software. But there are only a few standards or guidelines to evaluate these intelligent products. The application and development of SWS can accelerate the speed of evaluating the system on market feedback and build the rules of the smart water market. Additionally, the application layer is dependent on the purpose of data acquisition. However, there is no single protocol that can adequately suit all the applications and communication processes, and an overall application protocol would be too complex to support efficient business processes. Therefore, another business application is the on-going multi-protocol handing devices for specific purposes since one proprietary single-manufacturer protocol would not be the best for interoperability in SWS. These devices can stimulate communications by using the existing or coming protocols and also allow open programmable interfaces that can be customized for market tests and feedback [55].

4. Metrics for SWS

The technical structure of SWS has a pyramid structure with core information on the top to ensure system efficiency and security [87]. Figure 11 illustrates the features of such a technical structure. In this general structure of SWS, the configuration of components and connections can be interpreted as a network of cyber information (e.g., leak detection, discharge control, and noise recognition), data compiling (e.g., real-time modeling, real-time controlling, and real-time sampling), and physical instruments (e.g., sensors and loggers) domain. In Figure 11, nodes represent system components while the links stand for the functional relationship between nodes. For instance, the bottom nodes are connected with the intermedia nodes, which optionally means that the data from the sensor is transmitted to SCADA via links. To better assess the SWS's efficiency and security within these domains, it is necessary to develop the metrics [45].

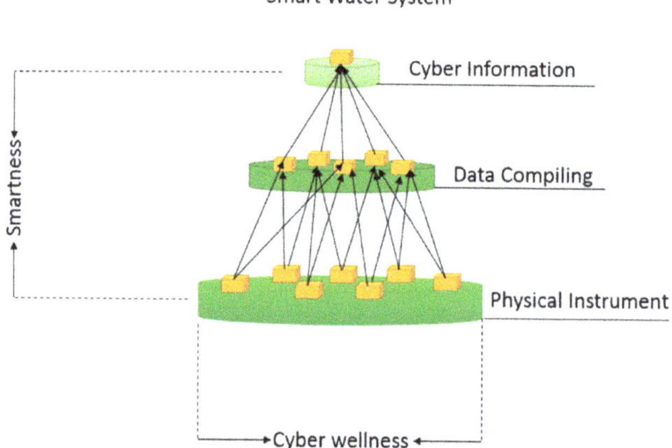

Figure 11. Illustration of a smart water system technical structure.

Before moving to the metrics discussion, the relationship between property and metrics should be clarified. While metrics are refined from properties, and both metrics and properties might be connected by functions, the application of SWS ultimately aims to assess the performance of SWS. Therefore, properties can be seen as the inherent components of SWS whilst metrics are the manual product. Additionally, properties might determine the assessment indexes on a given SWS, while metrics are those elements to achieve the terminal performance. For example, real-time modeling is a crucial property of SWS, which makes measuring the efficiency of SWS one indicator for smartness.

Furthermore, the performance of data processing in the context of resourcefulness is related to informational security. However, the effects of property layers on metrics are not certain without specific analysis of a given system. In this section, the paper proposes two new conceptual metrics (Smartness and Cyber wellness) for assessing two essential properties of SWS, efficiency and security, and discusses how to define these two metrics and how they can be objectively built to deal with threats of SWS.

A brief investigation of 27 reviewed academic studies was conducted to analyze the SWS metrics shown in Figure 12, showing the number of studies (report and paper) for smartness scope and cyber wellness scope. Smartness and cyber wellness are seldom discussed directly in previous articles and reports. Most of them mention relationships with metrics or present the features of these two metrics. Thus, we consider that these papers and reports listed should be included in the scope of the metric. In the cyber wellness scope, cyber wellness only comes from the electrical and telecommunication fields [30], which makes it necessary to translate the cyber wellness into water systems. Although smartness has been described in the previous environmental studies, the vision is a little broader as water systems are only a small part of the environment [87]. More efforts would be required to narrow down the scope of smartness if it is applied in the water system sector.

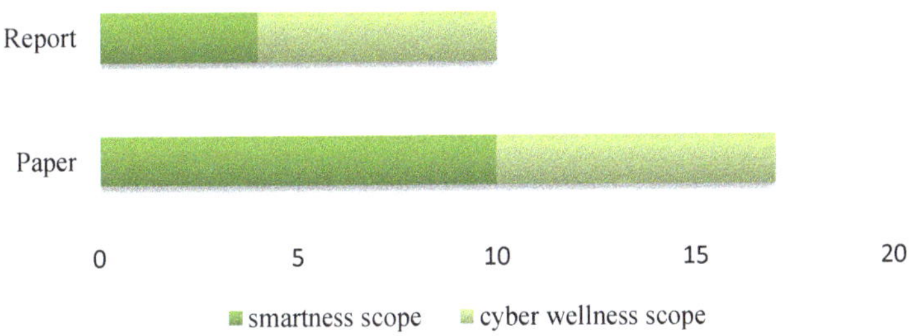

Figure 12. The number of studies for different metrics scope.

4.1. Smartness

In previous studies, researchers made many efforts in achieving smart performance in water systems. The smart performance of SWS ought to expand to real-time modeling, real-time sampling, and real-time controlling, etc., which aim to minimize the time delay between system input and system output. It was revealed that the efficiency of data transmission would be promoted significantly by using SCADA [88,89]. Nevertheless, SCADA system enables multiple connections with various database and real-time modeling tools; the connection between the SCADA and offline or time-delay modeling tools still makes the water system not so smart as to reduce the data acquisition time. Some offline modeling tools like Hydraulic CAD and WATSYS can only process historical data, even though they are commonly used for the hydraulic model. Even if the SCADA system and so-called real-time simulation tools have been integrated into the water distribution system, it is still necessary to further understand what makes the water system smart. Real-time modeling tool applications like EPANET-RTX [90], LVVWD [91], and EPANET-CPS [92] actually need two steps to finish work: pausing execution and waiting for the new SCADA measurements to reload and to update the boundary condition [93]. The Pausing and Waiting takes typically 10 seconds and 14 minutes, respectively, which makes smart modeling is close to being near real-time process [94].

Thus, the time interval between system input and output, which is characterized as a time lag equation, can be used to assess how much smart it is. Overall, to define the SWS is to establish the mechanism for reducing time lag among those processes like real-time monitoring, real-time transporting, real-time processing, real-time sampling, and real-time simulation in SWS. In this study, the term "Smartness" is introduced to quantify the time lag reduction. The optimal levels of smartness require trade-offs from the source to the end [87]. This way, the smartness used in Equation (1) aims to minimize the system time costs from start input and terminal output. As smartness ensures performance, efficiency, and expediency, the maximum of smartness will revolutionize the interaction relationship between systems and engineers.

$$\text{Smartness} = \min (\text{Time lag: system input, system output}) \tag{1}$$

The unit for smartness is minutes; a smartness below 15 min can be acceptable while a smartness over 15 min is unfavorable. Smartness can be used to assess the efficiency of the smart water system when dealing and interpreting measurements from real-world systems.

4.2. Cyber Wellness

The scale IoT devices are growing even faster than the world's population. There will be 20.4 billion connected devices by 2020, compared with 7.8 billion global population [95]. These numbers highlight IoT's significance in the fourth industrial revolution, also known as the digital economy, where IoT is expected to deliver the "smarts" needed to address common everyday challenges in areas such as education, healthcare, utilities, transportation, and public and residential buildings [96].

SWS has a number of logical sub-layers in the integration of physical components and IoT, where sensors and actuators are deployed across the water distribution network to enable significant data processes and real-time performance. However, the pursuit of smartness without considering data authenticity and reliability would lead to information security issues. Many information safety issues are from system integration [42]. Thus, security at both IoT and network-level is critical to the operation of SWS [61]. Compared with physical network level components, these data-related smart sensors and intelligent devices with larger importance are the heart of SWS, occupying the decision-making resources. In comparison with the physical network level, the related data are easier to be chosen as the hitting points and even more vulnerable to cyberattacks because the hackers or cyberattackers might tend to crush the critical components of SWS or to steal the most crucial data saved in the database.

In this paper, we propose cyber wellness as one of the metrics to evaluate the information security of SWS. Cyber wellness is first put forward in the International Telecommunication Union (ITU) to describe the health level of information and communication technologies [97]. Continuously, cyber wellness is also introduced into the education field to describe the health level of Internet users [88]. In terms of information security, the emphasis from all cases is on how many of the IoT works well to extract data and how long it will last to defend itself when a cyberattack happens. In this work, we define the well-being level of information as "cyber wellness" in SWS. The goal of cyber wellness is, therefore, to store data as much as possible before a cyberattack and to withstand cyberattacks as long as possible at the same time shown in Equation (2):

$$\text{Cyber wellness} = \max(\text{Information security: capacity, endurance}) \tag{2}$$

Cyber wellness is unitless, usually ranging from 0% to 100%. The higher cyber wellness reflects that the smart water system has more capacity to be against system failure and to prevent information loss.

5. Future Research Recommendations

A few challenges in the application of SWS are still waiting to be solved before the wide application of a smart water system [47]. More coordination is needed for collective work from the academia, industry, and government to enable smart techniques for public adoption. Based on the results of this review paper and the work from different water organizations, several research directions are recommended to help engineers, researchers, and the public to work on those challenges in a more efficient and focused way.

5.1. Cyberattack to Smart Water Micro-Components

As the latest information and data center systems are vulnerable to hackers switching off the power easily, the smart equipment being connected to the wider internet is revealed to be exposed to cyberattack risks [98]. This finding reflects the current security problems in the IoTs are a significant public concern. SWS, which is also on the basis of IoT, should not ignore this concern, and it is required to build a strong defensive wall to cyberattacks [52]. However, there are few studies discussing the protection of smart water micro-components. Different smart devices such as sensors, loggers, samplers, and controllers play roles of varying importance in water systems, meaning that cyberattackers might select the attacking intentionally in one specific case. In this way, we need to develop efficient methods and technologies to benchmark the smart water micro-components protection efficiency. This will budget the investment in the water system retrofitting where identical smart water micro-components are required for different defense costs.

5.2. Resilience Incorporation into Smart Water System

The resilience concept has been incorporated into the traditional water distribution system, rainwater harvesting system, storm drainage system, and wastewater system in some papers and projects [99–102]. However, there are few papers discussing how to implement resilience in the smart water system. The smart water system needs to be resilient since it involves more recovery not only in physical components but also in the intelligent ones. In contrast to traditional water systems, a smart water system has more complex connections with automatic and online operations. Thus, the performance of SWS is determined by water, energy, and electricity availability. The necessity of incorporating resilience into the smart water system to evaluate their recovery ability and function efficiency are therefore very important. Incorporating the sub-resilience into the smart water micro-components would also be a better way to quantify the recovery capacity of the whole water system.

5.3. Smart Water (End-User) Data Disaggregation and Analysis

Real-time meter readings generate massive data that require good organization and powerful analytics to extract useful information [103]. The high-resolution water data sampling can enable end-user disaggregation to efficiently recognize the peak demand, pipe leakage, and breakage [104,105]. However, the smart water system field is still lacking proper data disaggregation and analysis tools.

The development of data analysis tools that utilize the processed data to obtain water consumption information such as peak hourly consumption, end-uses, or comparative analysis is still at the formative stage [106]. Multi-Resolution data availability can assist utilities in obtaining benefits from data sampling and the cost of high-resolution metering, real-time data-model coupling, and maintaining metering infrastructures [67,72]. These are the most critical challenges requiring urgent research attention as the efficient end-user data analysis approach can benefit the customers' billing and costs. Such work will provide insight into future big water data management research and commercial applications [107–109].

6. Conclusions

This paper has conducted a critical review of studies that deal with smart water techniques applied in water systems, with a particular focus on the understanding of how to address the key components for the framework of the smart water system (SWS). Four critical components composed of the SWS (instrument layer, property layer, function layer, benefit layer, and application layer) and two metrics (smartness and cyber wellness) are proposed to characterize SWS. In this review, a total number of 32 literature including 1 international forum, 17 peer-reviewed papers, 10 reports, and 4 presentations, explicitly supporting the smart water system's definition, architecture, and metrics, are analyzed. The main conclusions drawn from this study can be summarized as follows:

1. The lack of consensus in the definition and architecture of a smart water system and metrics of intelligent water system assessment is hindering the process of smart techniques entering the water sector;
2. A systematic and comprehensive smart water system framework is put forward including critical elements like the definition, architecture, metrics, and research directions, which can be directly applicable for education, research, and public knowledge;
3. Two conceptual metrics (smartness and cyber wellness) to evaluate the performance of the smart water system are proposed to characterize system efficiency and information security;
4. Existing challenges drive concentration on future research directions, and these future tasks can be viewed as a synthetic work where the academia, industry, and government will join in together.

Overall, this review has defined what SWS is and established a systematic framework for SWS, including architecture and metrics of SWS, which also shows that SWS has great potential to maximize the benefits in water sectors over the coming decades. This study is useful for designing assessing, and rehabilitating SWS when different goals are required in practical applicability in the field or lab. Future research directions are also clarified for this cross-disciplinary work, to assist the water areas to move towards a smarter future. As smart water technologies are under development, more real-world tests will be needed to realize the full benefits of smart water system.

Author Contributions: Conceptualization, J.L.; methodology, J.L.; formal analysis, J.L. and X.Y.; investigation, J.L. and R.S.; resources, J.L.; data curation, J.L.; writing—original draft preparation, J.L.; writing—review and editing, J.L. X.Y. and R.S.; visualization, J.L.; financial support, R.S. All authors have read and agreed to the published version of the manuscript.

Funding: This work is mainly supported by the Teaching Assistance Scholarship of the Civil and Environmental Engineering Department at the University of Utah. The publication fee is financially covered by the Climate and Energy Fund of the University of Innsbruck.

Acknowledgments: We would like to show our great thanks to the financial support Climate and Energy Fund and also the assistance from the program "Smart Cities Demo-Living Urban Innovation 2018" (project 872123) of the University of Innsbruck.

Conflicts of Interest: The authors declare no conflict of interest.

References

1. UN DESA. *World Urbanization Prospects: The 2018 Revision*; United Nations: New York, NY, USA, 2019.
2. The World Bank Group. *The World Bank Population Growth (Annual %)*; The World Bank: Washington, DC, USA, 2018.
3. Butler, D.; Farmani, R.; Fu, G.; Ward, S.; Diao, K.; Astaraie-Imani, M. A New Approach to Urban Water Management: Safe and Sure. *Procedia Eng.* **2014**, *89*, 347–354. [CrossRef]
4. Collins, A. *The Global Risks Report 2018*, 13th ed.; World Economic Forum: Geneva, Switzerland, 2013; ISBN 978-1-944835-15-6.
5. Tao, T.; Xin, K. Public health: A sustainable plan for China's drinking water. *Nature* **2014**, *511*, 527. [CrossRef] [PubMed]
6. Tao, T.; Li, J.; Xin, K.; Liu, P.; Xiong, X. Division method for water distribution networks in hilly areas. *Water Sci. Technol. Water Supply* **2016**, *3*, 727–736. [CrossRef]
7. Lee, S.W.; Sarp, S.; Jeon, D.J.; Kim, J.H. Smart water grid: The future water management platform. *Desalin. Water Treat.* **2015**, *55*, 339–346. [CrossRef]
8. Maxmen, A. As Cape Town water crisis deepens, scientists prepare for 'Day Zero'. *Nature* **2018**, *554*, 7690. [CrossRef]
9. Booysen, M.J.; Visser, M.; Burger, R. Temporal case study of household behavioural response to Cape Town's "Day Zero" using smart meter data. *Water Res.* **2019**, *149*, 414–420. [CrossRef]

10. Nguyen, K.A.; Stewart, R.A.; Zhang, H.; Sahin, O.; Siriwardene, N. Re-engineering traditional urban water management practices with smart metering and informatics. *Environ. Model. Softw.* **2018**, *101*, 256–267. [CrossRef]

11. Choi, G.W.; Chong, K.Y.; Kim, S.J.; Ryu, T.S. SWMI: New paradigm of water resources management for SDGs. *Smart Water* **2016**, *1*, 3. [CrossRef]

12. Sonaje, N.P.; Joshi, M.G. A Review of Modeling an Application of Water Distribution Networks (WDN) Softwares. *Int. J. Tech. Res. Appl.* **2015**, *3*, 174–178.

13. Liu, X.; Tian, Y.; Lei, X.; Wang, H.; Liu, Z.; Wang, J. An improved self-adaptive grey wolf optimizer for the daily optimal operation of cascade pumping stations. *Appl. Soft Comput. J.* **2019**, *75*, 473–493. [CrossRef]

14. Leitão, P.; Colombo, A.W.; Karnouskos, S. Industrial automation based on cyber-physical systems technologies: Prototype implementations and challenges. *Comput. Ind.* **2016**, *81*, 11–25. [CrossRef]

15. NET Plateform. Smart cities applications and requirements. *City*, 20 May 2011; 1–39.

16. Hayes, J.; Goonetilleke, A. Building Community Resilience ± Learning from the 2011 Floods in Southeast Queensland, Australia. In Proceedings of the 8th Annual Conference of International Institute for Infrastructure, Renewal and Reconstruction: International Conference on Disaster Management (IIIRR 2012), Kumamoto, Japan, 24–26 August 2012.

17. Allen, M.; Preis, A.; Iqbal, M.; Whittle, A.J. Case study: A smart water grid in Singapore. *Water Pract. Technol.* **2012**, *7*, 4. [CrossRef]

18. Barsugli, J.; Anderson, C.; Smith, J.B.; Vogel, J.M. *Options for Improving Climate Modeling to Assist Water Utility Planning for Climate Change*; University of Colorado at Boulder: Denver, CO, USA, 2009.

19. Arregui, F.J.; Gavara, F.J.; Soriano, J.; Pastor-Jabaloyes, L. Performance analysis of aging single-jet water meters for measuring residential water consumption. *Water* **2018**, *10*, 612. [CrossRef]

20. Ribeiro, R.; Loureiro, D.; Barateiro, J.; Smith, J.R.; Rebelo, M.; Kossieris, P.; Gerakopoulou, P.; Makropoulos, C.; Vieira, P.; Mansfield, L. Framework for technical evaluation of decision support systems based on water smart metering: The iWIDGET case. *Procedia Eng.* **2015**, *119*, 1348–1355. [CrossRef]

21. Mounce, S.; Machell, J.; Boxall, J. Water quality event detection and customer complaint clustering analysis in distribution systems. *Water Sci. Technol. Water Supply* **2012**, *12*, 580–587. [CrossRef]

22. Anzecc, A. Australian and New Zealand Guidelines for Fresh and Marine Water Quality. *Natl. Water Qual. Manag. Strategy* **2000**, *1*, 1–103.

23. Albino, V.; Berardi, U.; Dangelico, R.M. Smart cities: Definitions, dimensions, performance, and initiatives. *J. Urban Technol.* **2015**, *22*, 3–21. [CrossRef]

24. Hall, R.E.; Bowerman, B.; Braverman, J.; Taylor, J.; Todosow, H. The Vision of a Smart City. In Proceedings of the 2nd International Life Extension Technology Workshop, Paris, France, 28 September 2000.

25. Höjer, M.; Wangel, J. *Smart Sustainable Cities*; Springer International Publishing: Cham, Switzerland, 2014; pp. 333–349.

26. Owen, D.L. The Singapore water story. *Int. J. Water Resour. Dev.* **2013**, *29*, 290–293. [CrossRef]

27. Gourbesville, P. Key Challenges for Smart Water. *Procedia Eng.* **2016**, *154*, 11–18. [CrossRef]

28. Boyle, T.; Giurco, D.; Mukheibir, P.; Liu, A.; Moy, C.; White, S.; Stewart, R. Intelligent metering for urban water: A review. *Water* **2013**, *5*, 1052–1081. [CrossRef]

29. Fang, X.; Misra, S.; Xue, G.; Yang, D. Smart grid—The new and improved power grid: A survey. *IEEE Commun. Surv. Tutor.* **2012**, *14*, 944–980. [CrossRef]

30. Balijepalli, V.S.K.M.; Pradhan, V.; Khaparde, S.A.; Shereef, R.M. Review of Demand Response under Smart Grid Paradigm. In Proceedings of the 2011 IEEE PES International Conference on Innovative Smart Grid Technologies-India, ISGT India 2011, Kollam, Kerala, India, 1–3 December 2011.

31. Gelazanskas, L.; Gamage, K.A.A. Demand side management in smart grid: A review and proposals for future direction. *Sustain. Cities Soc.* **2014**, *11*, 22–30. [CrossRef]

32. ITU. *Global Cybersecurity Index & Cyberwellness Profiles*; International Telecommunication Union: Geneva, Switzerland, 2015; ISBN 9789261250614.

33. Hagar, J.; Murray, R.; Haxton, T.; Hall, J.; McKenna, S. Using the CANARY Event Detection Software to Enhance Security and Improve Water Quality. In Proceedings of the World Environmental and Water Resources Congress 2013: Showcasing the Future—Proceedings of the 2013 Congress, Cincinnati, OH, USA, 19–23 May 2013.

34. University of Oxford Smart Water System. Available online: https://assets.publishing.service.gov.uk/media/57a08ab9e5274a31e000073c/SmartWaterSystems_FinalReport-Main_Reduced__April2011.pdf (accessed on 9 June 2018).
35. DNV GL. Global Opportunity Report 2015. Available online: https://issuu.com/dnvgl/docs/globalopportunityreport/14 (accessed on 10 December 2019).
36. DNV GL. Global Opportunity Report 2016. Available online: https://issuu.com/dnvgl/docs/the-2016-global-opportunity-report (accessed on 10 December 2019).
37. United Nations Global Compact Global Opportunity Report 2017. Available online: https://www.unglobalcompact.org/library/5081 (accessed on 10 December 2019).
38. DNV GL. Global Opportunity Report 2018. Available online: https://www.dnvgl.com/feature/gor2018.html (accessed on 12 October 2019).
39. Colorado State University. Smart Water Grid-Plan B Technical Final Report. Available online: https://www.engr.colostate.edu/~{}pierre/ce_old/Projects/RisingStarsWebsite/Martyusheva,Olga_PlanB_TechnicalReport.pdf (accessed on 12 October 2019).
40. The Smart Water Networks Forum What is a Smart Water Network? Available online: https://www.swan-forum.com/swan-tools/what-is-a-swn/ (accessed on 18 October 2019).
41. Dragan Savic Intelligent/Smart Water System. Available online: https://www.slideshare.net/gidrasavic/intelligent-smart-water-systems (accessed on 12 October 2019).
42. Li, J.; Lee, S.; Shin, S.; Burian, S. Using a Micro-Test-Bed Water Network to Investigate Smart Meter Data Connections to Hydraulic Models. In Proceedings of the World Environmental and Water Resources Congress 2018: Hydraulics and Waterways, Water Distribution Systems Analysis, and Smart Water—Selected Papers from the World Environmental and Water Resources Congress 2018, Minneapolis, MN, USA, 3–7 June 2018.
43. Ye, Y.; Liang, L.; Zhao, H.; Jiang, Y. The System Architecture of Smart Water Grid for Water Security. *Procedia Eng.* **2016**, *154*, 361–368. [CrossRef]
44. Günther, M.; Camhy, D.; Steffelbauer, D.; Neumayer, M.; Fuchs-Hanusch, D. Showcasing a smart water network based on an experimental water distribution system. *Procedia Eng.* **2015**, *119*, 450–457. [CrossRef]
45. Mutchek, M.; Williams, E. Moving Towards Sustainable and Resilient Smart Water Grids. *Challenges* **2014**, *5*, 123–137. [CrossRef]
46. Li, J.; Bao, S.; Burian, S. Real-time data assimilation potential to connect micro-smart water test bed and hydraulic model. *H2Open J.* **2019**, *2*, 71–82. [CrossRef]
47. McKenna, K.; Keane, A. Residential Load Modeling of Price-Based Demand Response for Network Impact Studies. *IEEE Trans. Smart Grid* **2016**, *7*, 2285–2294. [CrossRef]
48. Oracle. *Smart Metering for Water Utilities*; Oracle Corporation: Redwood Shores, CA, USA, 2009.
49. Beal, C.; Flynn, J. The 2014 Review of Smart Metering and Intelligent Water Networks in Australia & New Zealand. In Proceedings of the Report prepared for WSAA by the Smart Water Research Centre, Griffith University, Queensland, Australia, 1 November 2014.
50. Murray, R.; Haxton, T.; McKenna, S.; Hart, D.; Umberg, K.; Hall, J.; Lee, Y.; Tyree, M.; Hartman, D. Case Study Application of the CANARY Event Detection Software. In Proceedings of the Water Quality Technology Conference and Exposition 2010, Savannah, GA, USA, 14–18 November 2010.
51. U.S. Department of Energy. *2014 Smart Grid System Report*; The U.S. Department of Energy: Washington, DC, USA, 2014; ISBN 2025863098.
52. Batty, M.; Axhausen, K.W.; Giannotti, F.; Pozdnoukhov, A.; Bazzani, A.; Wachowicz, M.; Ouzounis, G.; Portugali, Y. Smart cities of the future. *Eur. Phys. J. Spec. Top.* **2012**, *214*, 481–518. [CrossRef]
53. Kartakis, S.; Abraham, E.; McCann, J.A. WaterBox: A Testbed for Monitoring And Controlling Smart Water Networks. In Proceedings of the 1st ACM International Workshop on Cyber-Physical Systems for Smart Water Networks, CySWater 2015, Seattle, WA, USA, 14–16 April 2015.
54. Koo, D.; Piratla, K.; Matthews, C.J. Towards Sustainable Water Supply: Schematic Development of Big Data Collection Using Internet of Things (IoT). *Procedia Eng.* **2015**, *118*, 489–497. [CrossRef]
55. Luciani, C.; Casellato, F.; Alvisi, S.; Franchini, M. From Water Consumption Smart Metering to Leakage Characterization at District and User Level: The GST4Water Project. *Proceedings* **2018**, *2*, 675. [CrossRef]
56. Gabrielli, L.; Pizzichini, M.; Spinsante, S.; Squartini, S.; Gavazzi, R. Smart Water Grids for Smart Cities: A Sustainable Prototype Demonstrator. In Proceedings of the EuCNC 2014—European Conference on Networks and Communications, Bologna, Italy, 23–26 June 2014.

57. Chen, Y.; Han, D. Water quality monitoring in smart city: A pilot project. *Autom. Constr.* **2018**, *89*, 307–316. [CrossRef]

58. Helmbrecht, J.; Pastor, J.; Moya, C. Smart Solution to Improve Water-energy Nexus for Water Supply Systems. *Procedia Eng.* **2017**, *186*, 101–109. [CrossRef]

59. Wu, Z.Y.; El-Maghraby, M.; Pathak, S. Applications of deep learning for smart water networks. *Procedia Eng.* **2015**, *119*, 479–485. [CrossRef]

60. Hatchett, S.; Uber, J.; Boccelli, D.; Haxton, T.; Janke, R.; Kramer, A.; Matracia, A.; Panguluri, S. Real-Time Distribution System Modeling: Development, Application, and Insights. In Proceedings of the Urban Water Management: Challenges and Oppurtunities—11th International Conference on Computing and Control for the Water Industry, CCWI 2011, Exeter, UK, 5–7 September 2011.

61. Ntuli, N.; Abu-Mahfouz, A. A Simple Security Architecture for Smart Water Management System. *Procedia Comput. Sci.* **2016**, *83*, 1164–1169. [CrossRef]

62. The Smart Water Networks Forum Swan Workgroups. Available online: https://www.swan-forum.com/workgroups/ (accessed on 10 December 2019).

63. Li, J.; Lee, S.; Shin, S.; Burian, S. Using a Micro-Test-Bed Water Network to Investigate Smart Meter Data Connections to Hydraulic Models. In Proceedings of the World Environmental and Water Resources Congress 2018: Hydraulics and Waterways, Water Distribution Systems Analysis, and Smart Water, Reston, VA, USA, 3–7 June 2018; 2018; pp. 342–350.

64. Kamienski, C.; Soininen, J.P.; Taumberger, M.; Dantas, R.; Toscano, A.; Cinotti, T.S.; Maia, R.F.; Neto, A.T. Smart water management platform: IoT-based precision irrigation for agriculture. *Sensors* **2019**, *19*, 276. [CrossRef] [PubMed]

65. Bragalli, C.; Neri, M.; Toth, E. Effectiveness of smart meter-based urban water loss assessment in a real network with synchronous and incomplete readings. *Environ. Model. Softw.* **2019**, *112*, 128–142. [CrossRef]

66. Alvisi, S.; Casellato, F.; Franchini, M.; Govoni, M.; Luciani, C.; Poltronieri, F.; Riberto, G.; Stefanelli, C.; Tortonesi, M. Wireless middleware solutions for smart water metering. *Sensors* **2019**, *19*, 1853. [CrossRef]

67. Cominola, A.; Giuliani, M.; Castelletti, A.; Rosenberg, D.E.; Abdallah, A.M. Implications of data sampling resolution on water use simulation, end-use disaggregation, and demand management. *Environ. Model. Softw.* **2018**, *102*, 199–212. [CrossRef]

68. Jha, M.K.; Sah, R.K.; Rashmitha, M.S.; Sinha, R.; Sujatha, B.; Suma, K.V. Smart Water Monitoring System for Real-Time Water Quality and Usage Monitoring. In Proceedings of the International Conference on Inventive Research in Computing Applications, ICIRCA 2018, Coimbatore, Tamil Nadu, India, 11–12 July 2018.

69. Siew, C.; Tanyimboh, T.T. Pressure-Dependent EPANET Extension. *Water Resour. Manag.* **2012**, *26*, 1477–1498. [CrossRef]

70. Pai, T.Y.; Lin, K.L.; Shie, J.L.; Chang, T.C.; Chen, B.Y. Predicting the co-melting temperatures of municipal solid waste incinerator fly ash and sewage sludge ash using grey model and neural network. *Waste Manag. Res.* **2011**, *29*, 284–293. [CrossRef]

71. Monks, I.; Stewart, R.A.; Sahin, O.; Keller, R. Revealing unreported benefits of digital water metering: Literature review and expert opinions. *Water* **2019**, *29*, 838. [CrossRef]

72. Stewart, R.A.; Nguyen, K.; Beal, C.; Zhang, H.; Sahin, O.; Bertone, E.; Vieira, A.S.; Castelletti, A.; Cominola, A.; Giuliani, M.; et al. Integrated intelligent water-energy metering systems and informatics: Visioning a digital multi-utility service provider. *Environ. Model. Softw.* **2018**, *105*, 94–117. [CrossRef]

73. Beal, C.D.; Stewart, R.A. Identifying residential water end uses underpinning peak day and peak hour demand. *J. Water Resour. Plan. Manag.* **2014**, *140*, 04014008. [CrossRef]

74. Romano, M.; Kapelan, Z. Adaptive water demand forecasting for near real-time management of smart water distribution systems. *Environ. Model. Softw.* **2014**, *60*, 265–276. [CrossRef]

75. Public Utilities Board Singapore. Managing the water distribution network with a Smart Water Grid. *Smart Water.* **2016**, *1*, 1–13.

76. Aşchilean, I.; Giurca, I. Choosing a water distribution pipe rehabilitation solution using the analytical network process method. *Water* **2018**, *10*, 484. [CrossRef]

77. Fontana, N.; Giugni, M.; Glielmo, L.; Marini, G.; Zollo, R. Hydraulic and electric regulation of a prototype for real-time control of pressure and hydropower generation in a water distribution network. *J. Water Resour. Plan. Manag.* **2018**, *144*, 04018072. [CrossRef]

78. Sitzenfrei, R.; Rauch, W. Optimizing small hydropower systems in water distribution systems based on long-time-series simulation and future scenarios. *J. Water Resour. Plan. Manag.* **2015**, *141*, 04015021. [CrossRef]

79. Torcellini, P.A.; Long, N.; Judkoff, R.D. Consumptive Water Use for U.S. Power Production. In Proceedings of the 2004 Winter Meeting—Technical and Symposium Papers, American Society of Heating, Refrigerating and Air-Conditioning Engineers, Anaheim, CA, USA, 24–28 January 2004.

80. Ton, D.T.; Smith, M.A. The U.S. Department of Energy's Microgrid Initiative. *Electr. J.* **2012**, *25*, 84–94. [CrossRef]

81. Di Nardo, A.; Di Natale, M.; Santonastaso, G.F.; Venticinque, S. An Automated Tool for Smart Water Network Partitioning. *Water Resour. Manag.* **2013**, *27*, 4493–4508. [CrossRef]

82. U.S. EPA. *CANARY User's Manual Version 4.3.2*; Sandia National Laboratories (SNL-NM): Albuquerque, NM, USA, 2012.

83. Sønderlund, A.L.; Smith, J.R.; Hutton, C.J.; Kapelan, Z.; Savic, D. Effectiveness of smart meter-based consumption feedback in curbing household water use: Knowns and unknowns. *J. Water Resour. Plan. Manag.* **2016**, *142*, 04016060. [CrossRef]

84. Cardell-Oliver, R.; Wang, J.; Gigney, H. Smart meter analytics to pinpoint opportunities for reducing household water use. *J. Water Resour. Plan. Manag.* **2016**, *142*, 04016007. [CrossRef]

85. Brentan, B.; Meirelles, G.; Luvizotto, E.; Izquierdo, J. Joint operation of pressure-reducing valves and pumps for improving the efficiency of water distribution systems. *J. Water Resour. Plan. Manag.* **2018**, *144*, 04018055. [CrossRef]

86. Murphy, J.J.; Dinar, A.; Howitt, R.E.; Rassenti, S.J.; Smith, V.L. The design of "Smart" water market institutions using laboratory experiments. *Environ. Resour. Econ.* **2000**, *17*, 375–394. [CrossRef]

87. Marchese, D.; Linkov, I. Can You Be Smart and Resilient at the Same Time? *Environ. Sci. Technol.* **2017**, *51*, 5867–5868. [CrossRef]

88. Jazri, H.; Jat, D.S. A Quick Cybersecurity Wellness Evaluation Framework for Critical Organizations. In Proceedings of the 2016 International Conference on ICT in Business, Industry, and Government, ICTBIG 2016, Indore, India, 18 November 2016.

89. Dong, X.; Lin, H.; Tan, R.; Iyer, R.K.; Kalbarczyk, Z. Software-Defined Networking for Smart Grid Resilience: Opportunities and Challenges. In Proceedings of the CPSS 2015—1st ACM Workshop on Cyber-Physical System Security, Part of ASIACCS 2015, Denver, CO, USA, 16 October 2015; pp. 61–68.

90. Hatchett, S.; Boccelli, D.; Uber, J.; Haxton, T.; Janke, R.; Kramer, A.; Matracia, A.; Panguluri, S. How Accurate Is a Hydraulic Model? In Proceedings of the Water Distribution Systems Analysis 2010—12th International Conference, WDSA 2010, Tucson, AZ, USA, 12–15 September 2010; pp. 1379–1389.

91. Boulos, P.F.; Jacobsen, L.B.; Heath, J.E.; Kamojjala, S. Real-time modeling of water distribution systems: A case study. *J. Am. Water Works Assoc.* **2014**, *106*, E391–E401. [CrossRef]

92. Taormina, R.; Galelli, S.; Tippenhauer, N.O.; Salomons, E.; Ostfeld, A. Characterizing cyber-physical attacks on water distribution systems. *J. Water Resour. Plan. Manag.* **2017**, *143*, 04017009. [CrossRef]

93. Okeya, I.; Kapelan, Z.; Hutton, C.; Naga, D. Online modelling of water distribution system using Data Assimilation. *Procedia Eng.* **2014**, *70*, 1261–1270. [CrossRef]

94. Sadler, J.M.; Goodall, J.L.; Behl, M.; Morsy, M.M.; Culver, T.B.; Bowes, B.D. Leveraging open source software and parallel computing for model predictive control of urban drainage systems using EPA-SWMM5. *Environ. Model. Softw.* **2019**, *120*, 104484. [CrossRef]

95. International Telecommunication Union. ITU Guide to Developing a National Cybersecurity Strategy—Strategic Engagement in Cybersecurity. 2018. Available online: https://www.itu.int/dms_pub/itu-d/opb/str/D-STR-CYB_GUIDE.01-2018-PDF-E.pdf (accessed on 20 November 2019).

96. Muhammad, A.; Haider, B.; Ahmad, Z. IoT Enabled Analysis of Irrigation Rosters in the Indus Basin Irrigation System. *Procedia Eng.* **2016**, *154*, 229–235. [CrossRef]

97. Tonge, A.M. Cyber security: Challenges for society- literature review. *IOSR J. Comput. Eng.* **2013**, *2*, 67–75. [CrossRef]

98. Sharpe, R.G.; Goodall, P.A.; Neal, A.D.; Conway, P.P.; West, A.A. Cyber-Physical Systems in the re-use, refurbishment and recycling of used Electrical and Electronic Equipment. *J. Clean. Prod.* **2018**, *170*, 351–361. [CrossRef]

99. Lansey, K. Sustainable, Robust, Resilient, Water Distribution Systems. In Proceedings of the 14th Water Distribution Systems Analysis Conference 2012, WDSA 2012, Adelaide, Australia, 24–27 September 2012.

100. Juan-García, P.; Butler, D.; Comas, J.; Darch, G.; Sweetapple, C.; Thornton, A.; Corominas, L. Resilience theory incorporated into urban wastewater systems management. State of the art. *Water Res.* **2017**, *115*, 149–161. [CrossRef]

101. Todini, E. Looped water distribution networks design using a resilience index based heuristic approach. *Urban Water* **2000**, *2*, 115–122. [CrossRef]

102. Youn, S.G.; Chung, E.S.; Kang, W.G.; Sung, J.H. Probabilistic estimation of the storage capacity of a rainwater harvesting system considering climate change. *Resour. Conserv. Recycl.* **2012**, *65*, 136–144. [CrossRef]

103. Abdallah, A.M.; Rosenberg, D.E. Heterogeneous residential water and energy linkages and implications for conservation and management. *J. Water Resour. Plan. Manag.* **2014**, *140*, 288–297. [CrossRef]

104. Horsburgh, J.S.; Leonardo, M.E.; Abdallah, A.M.; Rosenberg, D.E. Measuring water use, conservation, and differences by gender using an inexpensive, high frequency metering system. *Environ. Model. Softw.* **2017**, *96*, 83–94. [CrossRef]

105. Nguyen, K.A.; Stewart, R.A.; Zhang, H. An autonomous and intelligent expert system for residential water end-use classification. *Expert Syst. Appl.* **2014**, *41*, 342–356. [CrossRef]

106. Cominola, A.; Nguyen, K.; Giuliani, M.; Stewart, R.A.; Maier, H.R.; Castelletti, A. Data mining to uncover heterogeneous water use behaviors from smart meter data. *Water Resour. Res.* **2019**, *55*, 9315–9333. [CrossRef]

107. Stewart, R.A.; Willis, R.; Giurco, D.; Panuwatwanich, K.; Capati, G. Web-based knowledge management system: Linking smart metering to the future of urban water planning. *Aust. Plan.* **2010**, *47*, 66–74. [CrossRef]

108. Yuan, Z.; Olsson, G.; Cardell-Oliver, R.; van Schagen, K.; Marchi, A.; Deletic, A.; Urich, C.; Rauch, W.; Liu, Y.; Jiang, G. Sweating the assets—The role of instrumentation, control and automation in urban water systems. *Water Res.* **2019**, *155*, 381–402. [CrossRef] [PubMed]

109. Cominola, A.; Giuliani, M.; Piga, D.; Castelletti, A.; Rizzoli, A.E. Benefits and challenges of using smart meters for advancing residential water demand modeling and management: A review. *Environ. Model. Softw.* **2015**, *72*, 198–214. [CrossRef]

MDPI

St. Alban-Anlage 66

4052 Basel

Switzerland

Tel. +41 61 683 77 34

Fax +41 61 302 89 18

www.mdpi.com

Water Editorial Office

E-mail: water@mdpi.com

www.mdpi.com/journal/water